Archimedes

NEW STUDIES IN THE HISTORY AND PHILOSOPHY
OF SCIENCE AND TECHNOLOGY

VOLUME 59

SERIES EDITOR

Archimedes has three fundamental goals: to further the integration of the histories of science and technology with one another; to investigate the technical, social and practical histories of specific developments in science and technology; and finally, where possible and desirable, to bring the histories of science and technology into closer contact with the philosophy of science.

The series is interested in receiving book proposals that treat the history of any of the sciences, ranging from biology through physics, all aspects of the history of technology, broadly construed, as well as historically-engaged philosophy of science or technology. Taken as a whole, Archimedes will be of interest to historians, philosophers, and scientists, as well as to those in business and industry who seek to understand how science and industry have come to be so strongly linked.

Submission / Instructions for Authors and Editors: The series editors aim to make a first decision within one month of submission. In case of a positive first decision the work will be provisionally contracted: the final decision about publication will depend upon the result of the anonymous peer-review of the complete manuscript. The series editors aim to have the work peer-reviewed within 3 months after submission of the complete manuscript.

The series editors discourage the submission of manuscripts that contain reprints of previously published material and of manuscripts that are below 150 printed pages (75,000 words). For inquiries and submission of proposals prospective authors can contact one of the editors:

Please find on the top right side of this webpage a link to our *Book Proposal Form.*

More information about this series at http://www.springer.com/series/5644

Robert B. Waide • Sharon E. Kingsland
Editors

The Challenges of Long Term Ecological Research: A Historical Analysis

Editors
Robert B. Waide
Department of Biology
University of New Mexico
Albuquerque, NM, USA

Sharon E. Kingsland
Department of History of Science and Technology
Johns Hopkins University
Baltimore, MD, USA

ISSN 1385-0180   ISSN 2215-0064 (electronic)
Archimedes
ISBN 978-3-030-66932-4   ISBN 978-3-030-66933-1 (eBook)
https://doi.org/10.1007/978-3-030-66933-1

This Springer imprint is published by the registered company Springer Nature Switzerland AG
The registered company address is: Gewerbestrasse 11, 6330 Cham, Switzerland

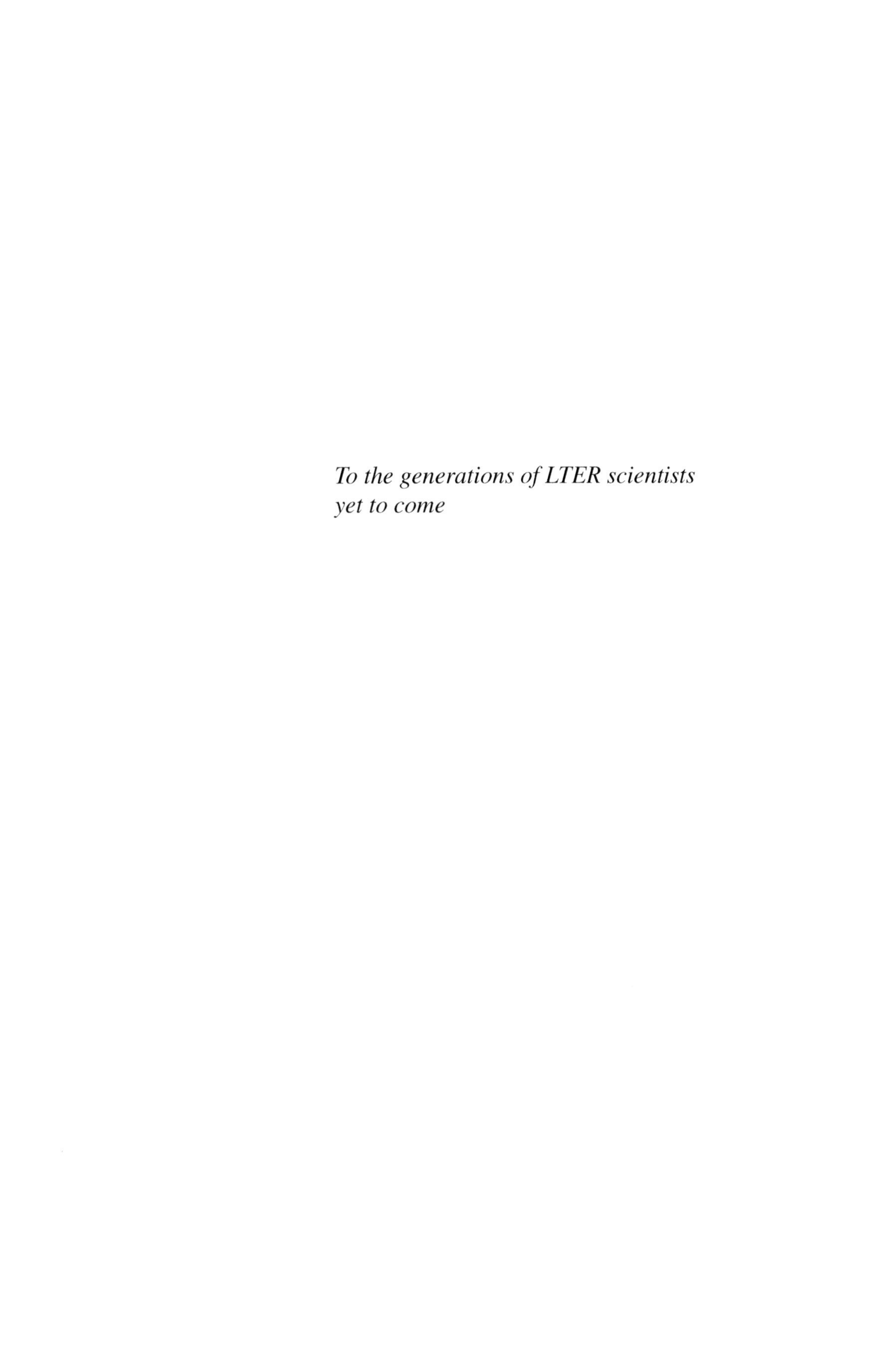

To the generations of LTER scientists yet to come

Contents

Contributors

Merryl Alber has been a participating principal investigator on the Georgia Coastal Ecosystems LTER since it started in 2000, and has been lead principal investigator since 2006. She is a professor in the Department of Marine Sciences and the director of the University of Georgia Marine Institute at the University of Georgia.

John Blair is Edwin G. Brychta professor of biology at Kansas State University. He is a terrestrial ecosystem ecologist who joined Kansas State University in 1992 and was recruited with the intention that he would eventually assume the role of lead principal investigator of the Konza Prairie LTER site. He took over following the site's mid-term review in 1999, and led his first Konza Prairie LTER renewal proposal in 2002. He continued as lead principal investigator through two additional KNZ renewals until the 2017 mid-term review. He remains an investigator at the Konza Prairie LTER site and is director of the Konza Prairie Biological Station.

Scott L. Collins moved in 1992 from the University of Oklahoma to the National Science Foundation, where he served as a program director in various programs, including the Long Term Ecological Research (LTER) Program. In 2003, he moved to the University of New Mexico where he is now a distinguished professor of biology. He served as principal investigator on the Sevilleta Long-Term Ecological Research Program (LTER) from March 2003 through March 2014. He has also worked extensively in tallgrass prairie as part of the Konza Prairie Long-Term Ecological Research Program since 1988.

Charles T. Driscoll is university professor of environmental systems and distinguished professor of civil and environmental engineering at Syracuse University, Syracuse, New York. He has worked at the Hubbard Brook Experimental Forest since 1976 and was a principal investigator on the Hubbard Brook LTER from 1988 to 2018. His research interests include the chemistry of soils and drainage waters, environmental modeling, and biogeochemistry. He investigates long-term biogeochemical patterns in forest and aquatic ecosystems and their response to disturbance, including climate change, air pollution, and land disturbance. He has primary responsibility for long-term data sets of longitudinal stream chemistry and soil water chemistry.

Hugh Ducklow is professor in the Department of Earth and Environmental Sciences at Columbia University in New York City and a senior scientist at Lamont-Doherty Earth Observatory, Palisades, New York. He is a biological oceanographer and was lead principal investigator of the Palmer Station Antarctica LTER project for three funding cycles from 2002 to 2020. He was beset in heavy sea ice with PAL colleagues aboard the research icebreaker Nathaniel B. Palmer for over 20 days in September 2001.

Timothy Fahey is professor in the Department of Natural Resources at Cornell University, Ithaca, New York. He is a forest ecologist who has long been an investigator at the Hubbard Brook Ecosystem Study in New Hampshire.

David R. Foster has been a faculty member in Organismic and Evolutionary Biology at Harvard University since 1983 and director of the Harvard Forest, the University's 4000-acre ecological laboratory and classroom in central Massachusetts, since 1990. He participated as a co-investigator on the original proposals for the Harvard Forest and Luquillo LTER sites in 1988. David has had the great pleasure to have developed much of his appreciation for collaborative science and many of his favorite colleagues through the LTER Network. He served as principal investigator of the Harvard Forest LTER program for 30 years (1990–2020), joined the executive board for three stints, and has had the great pleasure of working with Alan Knapp and Dan Reed on the publication committee since its formation.

Jerry F. Franklin is professor emeritus of Ecosystem Analysis at the College of Forest Resources, University of Washington, Seattle, Washington. He directed the H.J. Andrews Ecosystem Project, which joined the LTER program in 1980. Professor Franklin chaired the LTER Program's Coordinating Committee from 1982 to 1995, and established and directed LTER's coordinating office from 1982 to 1996. He is a leading authority on sustainable forest management and the maintenance of healthy forest ecosystems, and was responsible for integrating ecological and economic values into harvest strategies. Professor Franklin has been president of the Ecological Society of America and has received numerous awards, including an Award for the Environment from the Heinz Foundation, and a LaRoe Award for lifetime contributions to conservation biology from the Society for Conservation Biology. In 2016, he received the Eminent Ecologist Award from the Ecological Society of America and was awarded the Pinchot Medallion by the Pinchot Institute for Conservation.

William R. Fraser has worked in the Western Antarctic Peninsula region since 1974, and has been a co-principal investigator of the Palmer Station Antarctica LTER project since its start in 1990. He is president and lead investigator of the Polar Oceans Research Group, a non-profit institution in Sheridan, Montana, that he founded in 2000, and does research on the ecology of Southern Ocean seabirds and penguins in particular.

Lissy Goralnik is assistant professor in the Department of Community Sustainability at Michigan State University. Previously she was a postdoctoral scholar in the Department of Forest, Ecosystems, and Society at Oregon State University, where she primarily worked in the H.J. Andrews Experimental Forest on projects related to arts, humanities, and environmental science interactions. Her work lies in the areas of environmental ethics and experiential and environmental learning. She uses the tools of qualitative social science—interviews, questionnaires, focus groups, participatory methods, ethnography, and text analysis—to explore relationships, learning, and engagement with self, communities, and place.

Peter M. Groffman is a senior research scientist at the Cary Institute of Ecosystem Studies in Millbrook, New York, and professor at the City University of New York, Advanced Science Research Center, and Brooklyn College Department of Earth and Environmental Sciences. He has a long history of benefiting from the LTER network, beginning with trips to the Coweeta LTER site while a graduate student at the University of Georgia in the early 1980s, participation in early discussions of the Kellogg site in the mid-1980s, and more than 20 years of research at the Hubbard Brook LTER and Baltimore Ecosystem Study sites beginning in the 1990s. Most relevant to the chapter on governance in the LTER program co-authored with Ann Zimmerman, he was chair of the LTER Science Council from 2014 to 2019.

J. Morgan Grove's exposure to the LTER network began during his graduate studies when he took courses at the Yale School of Forestry & Environmental Studies from Herb Bormann, who co-founded the Hubbard Brook LTER. Morgan was part of the initial writing team for the first BES proposal and has led the social science team for BES since its founding in 1997. He was the co-chair for the LTER Network Social Science Committee and organized numerous workshops on social science and interdisciplinary science in the LTER Network. Morgan has worked for the USDA Forest Service since 1996 and is the team leader for the Baltimore Field Station.

John E. Hobbie is distinguished scientist and senior scholar at The Ecosystems Center, Marine Biological Laboratory, Woods Hole, Massachusetts. Between 1984 and 2006, he served as acting director, director, and co-director of the Center. He is an aquatic ecologist whose primary interest is the role that natural assemblages of microbes play in ecosystems. John has long been involved in ecological research in the Arctic tundra, including the Tundra Biome study of the International Biological Program during the 1970s, and later the Arctic LTER project.

Julia Jones began her engagement with LTER in 1989 through work with long-term Andrews Forest data and has been a co-principal investigator of the Andrews Forest LTER since 2002. She has collaborated with many other sites, including leading cross-site synthesis efforts and publications, being a visiting fellow at Harvard Forest LTER, and serving on NSF panels and site reviews for LTER sites.

David M. Karl is the Victor and Peggy Brandstrom Pavel professor of microbial oceanography and director of the Daniel K. Inouye Center for Microbial Oceanography: Research and Education (C-MORE) in the School of Ocean and Earth Science and Technology, University of Hawai'i, Manoa, Hawai'i. In 1988, he co-founded the Hawai'i Ocean Time-series (HOT) program that has conducted sustained physical, biogeochemical, and microbial measurements and experiments at Station ALOHA for the past 25 years. He has participated in over 70 major oceanographic cruises, including numerous trips to Antarctica in connection with the Research on Antarctic Coastal Ecosystem Rates and Palmer Antarctica LTER programs.

Sharon E. Kingsland is a historian of science at the Johns Hopkins University in Baltimore, Maryland. She has been writing about the history of ecology and related sciences for over 40 years and has published two books on the history of ecology: *Modeling Nature: Episodes in the History of Population Ecology* (2nd ed., 1995) and *The Evolution of American Ecology, 1890–2000* (2005).

Alan Knapp is a university distinguished professor in the Department of Biology and senior ecologist with the Graduate Degree Program in Ecology at Colorado State University, Fort Collins, Colorado. Trained as a plant physiological ecologist, he was lead principal investigator at the Konza Prairie LTER site from 1991 to 1999. Professor Knapp has had a three-decade long association with large-scale ecosystem research through NSF's LTER Program. His research seeks to understand how grass-dominated ecosystems are responding to a changing climate as well as other environmental changes.

Mary Beth Leigh serves as the coordinator of arts and humanities activities for the LTER Network and is the founding director of Bonanza Creek LTER's "In a Time of Change" collaborative arts-humanities-science program. She is a professor of microbiology and coordinator of arts-humanities-STEMM integration at the University of Alaska, Fairbanks.

Ariel E. Lugo is director of the International Institute of Tropical Forestry, within the U.S. Forest Service, U.S. Department of Agriculture, in Puerto Rico. His connection to the Luquillo LTER site in Puerto Rico began with the proposal-writing effort in the early 1980s. He was co-principal investigator for four cycles of the program, from 1988 to 2012.

John J. Magnuson participated in the three NSF workshops in the 1970s to establish an LTER Program. He led the North Temperate Lakes LTER site from 1980 to 2000 and was a member of LTER Network's steering committee from 1980 to 1999 and its executive committee from 1988 to 1991. His numerous other roles include: co-chair of the NSF workshop on marine sites in 1996, NSF panel member on urban sites review in 1997, and chair of the LTER Network from December 2005 to May 2007. He participates in North Temperate Lakes LTER by publishing scholarly papers on inter-site research within and outside the network of LTER sites.

Michael Paul Nelson has served as the lead principal investigator for the H.J. Andrews Experimental Forest LTER Program since 2012, the first environmental philosopher to serve in that role. He has also been a close collaborator with a non-LTER long-term study since 2004, namely the Isle Royale Wolf-Moose Project, which is the longest continuous study of a predator/prey system in the world, beginning in 1958. Prior to arrival at Oregon State University and Andrews LTER, he was a writer in residence at the Andrews in 2006 and attended a number of workshops prior to his arrival as well. The idea for his book, *Moral Ground: Ethical Action for a Planet in Peril* (co-edited with Kathleen Dean Moore) was "born" at the Andrews Forest (the book will come out in a 10th anniversary edition in 2020).

Debra P. C. Peters is a research scientist and acting chief science information officer with the U.S. Department of Agriculture, Agricultural Research Service. She was a graduate student and scientist with the Shortgrass Steppe LTER from 1984 to 1997, a scientist with the Sevilleta LTER from 1995 to 2010, and has been the lead PI on the Jornada Basin LTER in Las Cruces, New Mexico, since 2003. Debra is a landscape ecologist with interests in the drivers of alternative states and pattern–process relationships interacting across spatial and temporal scales to create surprising system dynamics. She is a co-lead on an inter-disciplinary Grand Challenge project within the USDA to develop and apply big data-model integration strategies for predictive disease ecology using an infectious disease as a model system. Her interests in catastrophes and cross-scale interactions are reflected by her recent research on the historic Dust Bowl in the 1930s from an ecological perspective.

Steward T. A. Pickett was the founding lead principal investigator of the Baltimore Ecosystem Study LTER, and served as its director from 1997 to 2016. He was for a time a member of the Scientific Initiatives Committee of the LTER Network, and he participated in numerous workshops on LTER synthesis and on integration of social sciences into the network. He has also served on the diversity committee of the network.

Edward B. Rastetter is senior scientist at The Ecosystems Center, Marine Biological Laboratory, Woods Hole, Massachusetts. He is lead principal investigator of the Arctic LTER project. His research focuses on how ecosystems are regulated through the interactions among carbon, nutrient, energy, and water cycles and how this regulation maintains the life-support system of the Earth.

Lindsey E. Rustad is a research ecologist with the U.S. Forest Service. She has worked at long-term ecological research sites her entire career. Since the past decade, she has been the Forest Service team leader at the Hubbard Brook Experimental Forest in New Hampshire. Her research addresses forest ecosystem response to natural and human-caused disturbance, such as acid rain and global climate change.

Timothy Seastedt is emeritus professor in the Ecology and Evolutionary Biology Department at the University of Colorado, Boulder, Colorado, and a fellow at the Institute of Arctic and Alpine Research, University of Colorado. His graduate training was at the Coweeta Hydrologic Laboratory just before it joined the LTER program, followed by a postdoctoral fellowship at the Konza Prairie LTER site. He was lead principal investigator at Konza Prairie from 1988 to 1991. In 1992, he became principal investigator of the Niwot Ridge LTER site in Colorado, where his research has involved many aspects of terrestrial ecosystem studies, including factors influencing biodiversity, productivity, soil carbon dynamics, decomposition and mineralization processes, and how these processes affect short- and long-term ecosystem–atmosphere interactions. Most recently, his research has involved studies of invasive plant species and invasibility of ecosystems within the context of other components of global change.

Gaius R. Shaver is senior scientist at The Ecosystems Center, Marine Biological Laboratory, Woods Hole, Massachusetts. For many years, his research has focused on Alaskan tundra ecosystems. His current focus is on a series of long-term, whole-ecosystem experiments at the Arctic LTER at Toolik Lake, Alaska. A second major component of his research is the regulation of terrestrial carbon accumulation and exchanges of carbon with the atmosphere. Much of this work involves the use of computer simulation models, working at a range of scales from an experimental plot in Alaska to whole continents to the entire globe. With Ed Rastetter, he has developed theoretical models of the interaction of multiple limiting factors in control of terrestrial stocks of soil and plant organic matter.

Susan G. Stafford is professor and dean emerita of the Department of Forest Resources at the University of Minnesota. During her career, she has been affiliated with two LTER sites, the Andrews Forest LTER at Oregon State University and the Shortgrass Steppe LTER at Colorado State University, and co-located with the Cedar Creek Reserve LTER at the University of Minnesota. The information management protocols that she and her colleagues developed were instrumental in helping the LTER sites to form a network. Their innovations helped LTER to become a leader in developing best practices in information management and archiving.

Frederick J. Swanson has been part of the HJ Andrews Experimental Forest LTER program since its inception in 1980, including serving as the second principal investigator beginning in the late 1980s. He contributed to several early proposals for Luquillo LTER and was a visiting fellow at Harvard Forest LTER for a year. He co-authored multi-site synthesis papers, including several in the decadal sets of LTER articles in *BioScience*.

Jonathan R. Thompson is a senior ecologist at the Harvard Forest at Harvard University and is the current principal investigator of the Harvard Forest LTER Program. He also leads the New England Landscape Futures project, which collaborates with diverse stakeholders from throughout the region to build and evaluate scenarios that show how land-use choices and climate change could shape the landscape over the next 50 years.

Kristin Vanderbilt is an information manager who has worked at three LTER sites during her career: the H.J. Andrews Forest as a Ph.D. student, the Sevilleta from 2000 to 2016, and the Florida Coastal Everglades since 2016. She has engaged in many International LTER (ILTER) activities during the past 22 years, first as a graduate student, then as a post-doc at Síkfökút LTER site in Hungary (2000–2002), and later as the first chair of the ILTER's Information Management Committee (2006–2010) and co-chair of the U.S. LTER's International LTER Committee (2010–2014). She collaborates with US LTER and ILTER informatics specialists on scholarly papers and co-organizes information management training workshops for ILTER audiences.

Robert B. Waide is professor emeritus in the Department of Biology at the University of New Mexico. As senior scientist at the Center for Energy and Environment Research (University of Puerto Rico), he was one of the founding lead principal Investigators (with Ariel Lugo) of the Luquillo LTER project from 1988 to 1997. He was executive director of the LTER Network Office in Albuquerque, New Mexico, from 1997 to 2016, and in this role, he was a member of the LTER Executive Board, Science Council, and most of the LTER standing committees. Professor Waide continues his association with the LTER Network as a scientist with the Luquillo LTER and as a member of the Environmental Data Initiative. He claims to have written or co-written more proposals (successful and unsuccessful) to NSF's LTER Program than anyone else.

Ann Zimmerman is a social scientist who chaired the LTER Governance Working Group, which was convened as part of the LTER planning process and was active in 2005–2006. Between 2013–14, she was a member of the task force charged by NSF to evaluate approaches for implementing the network-level activities of LTER.

List of Abbreviations

ACERE	Advisory Committee for Environmental Research and Education
AEC	Atomic Energy Commission (U.S.)
AINA	Arctic Institute of North America
ALTER-Net	A Long-Term Biodiversity, Ecosystem and Awareness Network
AMERIEZ	Antarctic Marine Ecosystem Research at the Ice Edge Zone
AND	H. J. Andrews Forest LTER Project
ARC	Arctic Tundra LTER Project
ARL	Arctic Research Laboratory
ARRA	American Recovery and Reinvestment Act
ARS	Agricultural Research Service
BES	Baltimore Ecosystem Study
BIO	Biological Sciences Directorate (NSF)
BIOAC	Biological Sciences Directorate Advisory Committee
BNZ	Bonanza Creek LTER Project
CC	Coordinating Committee (of LTER Program)
CDRRC	Chihuahuan Desert Rangeland Research Center
CERN	Chinese Ecological Research Network
CPER	Central Plains Experimental Range
DataONE	Data Observation Network for Earth
DEB	Division of Environmental Biology (NSF)
DEIMS	Drupal Ecological Information Management System
DEIMS-SDR	Dynamic Ecological Information Management System – Site and Dataset Registry
DIRT	Detritus Inputs and Removals Treatments
DPP	Division of Polar Programs (NSF)
EAP ILTER	East Asia-Pacific International Long Term Ecological Research
EC	Executive Committee (of LTER Program)
ECN	Environmental Change Network
EDI	Environmental Data Initiative

ELTOSA	Environmental Long Term Observatories of Africa
EML	Ecological Metadata Language
ENSO	El Niño Southern Oscillation
ESA	Ecological Society of America
eSAH	Environmental Sciences, Arts, and Humanities
GEO	Geosciences Directorate (NSF)
GIS	Geographic Information System
GLEON	Global Lake Ecological Observatory Network
HBES	Hubbard Brook Ecosystem Study
HBRF	Hubbard Brook Research Foundation
HPS	Hantavirus Pulmonary Syndrome
IBP	International Biological Program
IGY	International Geophysical Year
ILTER	International Long Term Ecological Research
IPCC	Intergovernmental Panel on Climate Change
ISSE	Integrative Science for Society and Environment
ITEX	International Tundra Experiment
LIAG	Lake Ice Analysis Group
LIDET	Long-Term Inter-site Decomposition Experiment Team
LNO	LTER Network Office
LTER	Long-Term Ecological Research
LTREB	Long-Term Research in Environmental Biology
LTSER	Long-Term Socio-Ecological Research
LUQ	Luquillo LTER project
MSI	Minimum Standard Installation
NARL	Navy Arctic Research Laboratory
NCO	LTER Network Communications Office
NEON	National Ecological Observatory Network
NEPA	National Environmental Policy Act
NERP	National Environmental Research Park
NRC	National Research Council
NSF	National Science Foundation (USA)
NSWG	Network Science Working Group
OCE	Division of Ocean Sciences (NSF)
ONR	Office of Naval Research
OPP	Office of Polar Programs (NSF)
PASTA	Provenance Aware Synthesis Tracking Architecture
PBI	Partnership for Biodiversity Informatics
PNG	Pawnee National Grasslands
RATE	Research on Arctic Tundra Environment
RNA	Research Natural Area
SBE	Directorate of Social, Behavioral and Economic Sciences (NSF)
SIP	Strategic Implementation Plan

SPE	Science Policy Exchange (New England)
TERN	Taiwan Ecological Research Network
TIE	The Institute of Ecology
TRACE	Tropical Responses to Altered Climate Experiments
ULTRA	Urban Long-Term Research Area program
URGE	Urban-Rural Gradient Ecology
USDA	United States Department of Agriculture

Chapter 1
Introduction

Sharon E. Kingsland and Robert B. Waide

Abstract This volume explores the history of the Long-Term Ecological Research (LTER) Program that began in the U.S. in 1980, with funding from the National Science Foundation. We examine challenges involved in sustaining long-term research and contributions made to the advancement of ecological science. Chapters discuss the early history of the LTER program and provide an overview of its maturation over four decades and the challenges it faced. The history of selected sites (representing tropical forest, arid and semi-arid grasslands, and arctic tundra and boreal forest ecosystems) links LTER to earlier scientific traditions. Themes include how individual sites expanded and broadened the vision of ecological science by influencing environmental policy, promoting global perspectives, expanding into urban environments and forging links to social sciences, and developing programs in the arts and humanities. The question of what it meant to constitute a "network" as opposed to a collection of individual sites is a central theme, and is addressed by examining the histories of cross-site collaborative research, information management, efforts to create a new social-ecological framework, the evolution of network governance structures, and the globalization of long-term ecological research. Published and archived resources for further research are reviewed.

Keywords LTER program · LTER archive · Long-term ecological research · History of ecology · National Science Foundation · Ecological networks · International ecology · Big science

---

S. E. Kingsland (✉)
Department of History of Science and Technology, Johns Hopkins University, Baltimore, MD, USA
e-mail: sharon@jhu.edu

R. B. Waide
Department of Biology, University of New Mexico, Albuquerque, NM, USA
e-mail: rbwaide@unm.edu

R. B. Waide, S. E. Kingsland (eds.), *The Challenges of Long Term Ecological Research: A Historical Analysis*, Archimedes 59,
https://doi.org/10.1007/978-3-030-66933-1_1

1.1 Introduction

In the spring of 1993, a new and frightening illness emerged in the Four Corners region of the United States (where New Mexico, Arizona, Utah, and Colorado meet). The illness began with influenza-like symptoms, then progressed rapidly to an acute stage leading to severe pulmonary edema. Medical staff were alerted after the sudden unexplained deaths of a small cluster of young adults, all previously healthy, from the Navajo tribe. Over 2 months, ten people died from the illness. There was no cure, and the mortality rate of the initial outbreak was 75% (Van Hook 2018). News about an illness that seemed to be hitting young tribal members created panic and led to discrimination against the Navajo and Hopi tribes. Speculations abounded about its cause. Was it a new form of influenza, or caused by a new pathogen entirely? *Scientific American* reported on a rumor that it was caused by a bio-warfare agent, which people believed had been accidentally released when a nearby army base was decommissioned (Horgan 1993).

Experts from the Centers for Disease Control and Prevention (CDC) quickly got involved and a multi-disciplinary investigative team was assembled. Within 3 weeks the team found the cause: a previously unknown hantavirus and its rodent vector, the deer mouse, *Peromyscus maniculatus*. People became ill when they inhaled aerosolized particles from the urine and feces of infected animals that had invaded human dwellings. Over the next few months the study teams succeeded in culturing the virus and worked out the clinical course of this disease, named Hantavirus Pulmonary Syndrome (HPS). Initially the CDC recommended naming the virus itself Muerto Canyon (Death Canyon) virus, thinking this was the location on the New Mexico reservation where it was discovered, but Navajo elders protested that there was no Muerto Canyon in New Mexico, although there was a Canyon del Muerto on the reservation in Arizona. It was instead named Sin Nombre (no-name) virus, or SNV.

Although prompt action by alert medical staff had led to quick identification of the unknown virus and syndrome, the virus itself was not new, nor was the illness, as further medical and ecological research would reveal (Morse 1994). While the medical and epidemiological researchers made rapid progress in understanding the disease in the year following the outbreak, accounting for the timing of the outbreak was an ecological problem. There were no data sets on rodent populations within the infected area, but there were two locations that had long-term data sets both on climate and on rodent populations in the Four Corners region, including in habitat types similar to those of the infected areas. One of these was in Canyonlands National Park near Moab, Utah. The other was on the Sevilleta National Wildlife Refuge near Socorro, New Mexico, where data were being collected as part of the Long Term Ecological Research (LTER) Program that the U.S. National Science Foundation had started in 1980.

The Sevilleta project, whose academic base was at the University of New Mexico in Albuquerque, had become part of the LTER Program in 1988, and in June 1993 its researchers joined the hantavirus study. Their data on rodent populations at the wildlife refuge from 1989 to 1993 revealed that certain rodent species, including

Peromyscus, had increased significantly in the spring of 1993, compared to previous years. Those data also suggested a reason: the fairly wet "El Niño" winter and spring in 1992 had stimulated the growth of vegetation, producing an abundance of seeds, berries, acorns and nuts on which rodents fed. Green vegetation also supported high populations of insects such as grasshoppers and beetles, also consumed by rodents. The result of this cornucopia was high reproduction and survival in the rodent population, creating exceptionally large populations in the spring of 1993 (Parmenter et al. 1993).

By drawing on cryogenically preserved rodent blood samples in two museum collections, as well as samples from the Sevilleta research sites, scientists were able to show that SNV was present in the region before 1993. They found that the virus existed in just a small area of Sevilleta in 1989, but by 1991 it had spread across the wildlife refuge to rodents in all adjacent habitats. Archived human medical records, as well as preserved tissues of patients with symptoms similar to HPS, revealed that HPS had also appeared before 1993. In 1993 the tenfold increase in the rodent population stimulated by the wet weather a year earlier had brought humans into closer contact with the animals, producing the outbreak. As rodent populations dropped during the summer of 1993, the number of cases of HPS also declined.

The ecologists' work was not yet done. It was also important to try to predict what might occur in the future, so that the public could be prepared for future outbreaks and take appropriate action. The ecological analysis therefore continued with longitudinal studies of rodent populations and a closer look both at general trends and local variations. These longer-term studies supported the initial hypothesis, but also introduced refinements to the hypothesis that took into account other variables, such as changes in the rate of infection in the rodent populations (Yates et al. 2002). This long-term research underscored the need to understand the complexity of natural and human-dominated ecosystems, as a basis for forecasting zoonotic disease outbreaks and safe-guarding the public. In the early 2000s, satellite imagery of northern New Mexico and northeastern Arizona, combined with algorithms developed in the 1990s, helped scientists identify risk areas for illness in the spring and summer of 2006 (Michener et al. 2009). The accumulated knowledge needed to predict outbreaks was based on long-term data on four phenomena: patterns of El Niño/La Niña cycles, the effects of these cycles on rainfall, the cascading effects of increased rainfall on vegetation and rodent populations, and the timing of the spread of SNV among rodents.

The experience of scientists involved in the hantavirus investigation helped also when West Nile virus (WNV), which affects humans, birds, and wildlife, emerged in 1999 and started spreading westward from New York. Ecologists at the Sevilleta Research Field Station, along with the Albuquerque Environmental Health Department, established a surveillance network along the Rio Grande from Texas to Colorado. They identified the mosquito species that carried the virus, tracked the pattern and rate of spread of the virus, and identified which habitats would support chronic infections. Ecologists concluded that "the fortuitous combination of the availability of local field-station infrastructure in a central location and collaborating, experienced personnel who had mobilized 8 years earlier for the hantavirus research was paramount to the success of the WNV study" (Michener et al. 2009). Local research infrastructure, experienced scientists, and long-term data records: these were essential to understanding and predicting disease outbreaks.

The purpose of this volume is to explore in depth and from multiple perspectives the Long Term Ecological Research program of which this study was a part, in order to examine the challenges involved in sustaining long-term research and the contributions that this research has made to the advancement of ecological science on many fronts. NSF's LTER Program, begun in 1980, was at the time considered an experimental and pioneering program, and there was skepticism about it both within NSF and within the ecological community. Initially modest in scale, its underlying ambitions were to transform ecology by making it into a more rigorous and predictive science. As the LTER program reaches its fortieth anniversary in 2020, it has indeed helped to transform ecological science, including in ways not anticipated at the outset.

The fortieth anniversary of the LTER program is a landmark that will be accompanied by scientific stock-taking and evaluation of its strengths, current challenges, and future prospects. Already, ecologists and environmental philosophers involved in long-term research are suggesting that the old arguments explaining the need for such research may not be adequate to secure future funding, and that new ways of advocating for long-term research may be needed (Vucetich et al. 2020). At this moment of evaluation and planning, a historical examination can provide some perspective.

This book is intended to contribute a historically grounded analysis of aspects of this program, and to suggest some fruitful research problems that might interest historians, philosophers, and anthropologists and sociologists of science. While not a complete history of the LTER program, we identify some of the central questions that the program raises, many of which are common to other multi-disciplinary scientific projects. We analyze the development of the LTER program from different perspectives: most of the chapter authors have long been involved in the program and their areas of expertise cover a wide range of disciplines, including several fields within ecology, earth sciences and oceanography, environmental engineering, social sciences, environmental philosophy, and the performing arts. Adding to these diverse areas of expertise we also bring in the perspective of the history of science. Thus this volume is also an experiment in collaboration across disciplines, and we hope it can be a stimulus for other collaborative research involving the humanities, social sciences, and ecology.

1.2 What Is the LTER Program and Why Should We Study It?

By the late 1970s, the National Science Foundation (NSF), the main government funding agency for basic sciences in the United States, recognized that generating meaningful results in ecological science often required many years of research, and that the normal 2–3-year funding cycles common to NSF impeded long-range planning of research (Callahan 1984). The creation of the Long Term Ecological Research Program in 1980 was meant to remedy this problem by offering multi-year funding that was better adjusted to the needs of ecologists.[1] Although NSF limited its initial funding commitment to 5 years, the projects themselves were expected to continue past 5 years, and funding could be renewed if sites passed muster following a review process. Moreover, the individual projects were expected to constitute a "network", broadly understood as engaging in cooperative inter-site research. However, just how this "networking" was to be achieved was not well articulated at the outset, but rather evolved over time.

The program also had higher ambitions to transform the science of ecology in a fundamental way. When Tom Callahan, the first program officer who oversaw the creation and expansion of the program, issued guidelines for evaluating the progress of LTER projects in 1980, he wrote that "The LTER network of research projects … and the Foundation are entering into an experiment. The results of this experiment can promote an advance in ecosystem science which will cause the field to change from a largely descriptive discipline to a predictive science" (quoted in Magnuson et al. 2006). But once again, how these transformations were to be achieved by a "network" of sites, as opposed to a collection of individual projects, was not spelled out.

Given initial uncertainty about the success of what was viewed as a pilot project, the start of the experiment was modest, but it grew fairly quickly in its first decade. In 1979 NSF held its first competition to identify sites for the program, selecting six sites in 1980. A second competition added five more sites, starting in January 1982.[2] In 1987 five sites were added, and three more in 1988, while two sites were withdrawn from the program, leaving a total of 17 projects within the continental United States (including two in Alaska), and in Puerto Rico. Sites in Antarctica were added in 1991 and 1993. An International LTER network was founded in 1993, also initiated by NSF, and it now includes 700 sites across all continents (Mirtl et al. 2018). In the late 1990s, the program veered off in a new direction, moving into two urban environments (Phoenix, Arizona, and Baltimore, Maryland), and bringing in a whole new set of problems concerning the relationship between biophysical and socio-economic-cultural systems, problems that were absent in the first set of sites

[1] National Science Foundation, "A new emphasis in long-term research," Request for Proposals, announcement, 1979.

[2] The first sites were: North Temperate Lakes (Wisconsin), H. J. Andrews Experimental Forest (Oregon), Coweeta Hydrologic Laboratory (Southern Appalachia), Konza Prairie (Kansas), North Inlet Marsh (South Carolina), and Niwot Ridge (Colorado). The second group of sites was: Central Plains Experimental Range (later Shortgrass Steppe) (Colorado), Okefenokee (bordering Georgia and Florida), Illinois Rivers (Illinois), Cedar Creek Natural History Area (Minnesota), and Jornada Basin (New Mexico). Five sites from that original list were subsequently withdrawn (Illinois Rivers and Okefenokee in 1988, North Inlet Marsh in 1993, Shortgrass Steppe in 2014, and Coweeta in 2019).

selected in 1980. Since 1998, addition of six coastal and three marine sites have brought the network to its current 26 sites.[3]

The relatively smooth establishment of the LTER program and its robust growth during the first decade owed a lot to the fact that the individual projects also encouraged partnerships, not just with university departments, field stations, and other kinds of research and educational institutes, but with other government agencies that had been involved in these regions well before the start of the LTER projects. Of those government agencies, the U.S. Forest Service, a division of the U.S. Department of Agriculture, was one of the most important partners. Many of the lead investigators in LTER projects were Forest Service employees, and several chapters in this book highlight the role of the Forest Service in strengthening and broadening the impact of the LTER program.

By 2013 the LTER program, which had grown to 26 sites, plus a central coordinating office at the University of New Mexico, was hailed as "the largest and longest-lived ecological network in the United States" (Waide and Thomas 2013). Over 1500 scientists worked at the different sites, and the annual budget for the program approached $30 million. Currently, the 26 sites in the network include forests, tundra, desert, grassland and agricultural systems, as well as freshwater, marine, coastal, and urban systems. Some of the projects selected in 1980 continue to receive NSF support today, as the LTER program approaches its fortieth anniversary. The program is now recognized for being "one of the first governmental programs to catalyze long-term, site-based, multidisciplinary, and collaborative research" (Willig and Walker 2016).

In this book, we explore several dimensions of this evolving experiment. We first examine its deeper historical roots and general trends, and then analyze in some depth the way the scientific community met the challenges not just of doing and sustaining long-term research, but the challenges of doing comparative work across different ecosystems, of expanding the vision of ecology to include human society, and of conveying the value of ecological thinking to the broader public. The LTER program is worthy of our attention, first of all, because of its breadth and geographical reach. The scientific research conducted under LTER embraces almost every question that modern ecologists must grapple with, across a range of ecological systems.

Second, the program is worthy of attention because it is an unusual type of hybrid enterprise. It is, on the one hand, a collection of diverse individual projects that have certain research themes in common, but that also pursue quite different research questions depending on the ecosystem being studied. On the other hand, it also strives to be a large-scale network. To the extent that it operates as a network, it resembles a "big science" approach to ecology, although not at the level of big science projects in the physical sciences.

---

[3] NSF has designated 33 LTER sites and seven of these have been terminated (Illinois Rivers, Okefenokee, North Inlet Marsh, Shortgrass Steppe, Sevilleta, Coweeta, and Baltimore). The Sevilleta site rejoined the Network through a subsequent competition.

What became increasingly evident as the LTER program approached its tenth anniversary, was that a key criterion for evaluating its success, and its continuation as a program, was the degree to which a collection of hitherto independent sites could function as a network, implying a higher level of collaboration between projects than was the norm for the original sites. The evolution from a collection of individual projects into a network involved many adaptations on scientific, technological, institutional, and cultural levels. The result today is a different kind of ecological science from what was common 40 years ago, one that is far more collaborative and multidisciplinary, and where the emphasis is on cooperation among projects as opposed to competition between projects. How the sites evolved into a network is one of the central themes of this book.

The idea that ecological science needed to be transformed was not confined to the LTER program, but was increasingly felt by ecologists to be important as the 1980s wore on. Parallel to the formation of the LTER program, and eventually intersecting with it, there were intensifying discussions within the scientific community about how to achieve global sustainability. Sustainability means meeting future human needs while also conserving the planet's life-support systems. For many years, and concurrently with the development of the LTER program, American ecologists have discussed ways to bolster support for ecological sciences, to face the complex environmental challenges of the late twentieth and twenty-first centuries.

In 1988 the Ecological Society of America (ESA) began a strategic planning process, culminating in a report published in 1991 called "The Sustainable Biosphere Initiative," which set out an ambitious agenda for the support of basic research (Lubchenco et al. 1991). Two ecologists involved in that initiative, Jane Lubchenco and Paul Risser, also co-chaired the committee that conducted the first 10-year review of the LTER program in the early 1990s. The committee's report, completed in 1993, envisioned a greatly expanded LTER program, arguing that the challenge of "designing and operating a sustainable biosphere could be most effectively and economically confronted with a newly defined LTER program."[4] The existing LTER program, in their view, was the nucleus of a "vitally important national effort." The report recommended significant expansion of the program to encompass more sites and a broader suite of scientific disciplines.

In 2002 ESA again created a committee tasked with producing an action plan that would advance the science of ecology and project a vision for the future of the science as well as the Society itself. The committee's report, presented in 2004, aimed to refocus the discipline "on questions that explicitly address how to sustain nature's services in the midst of burgeoning human populations" (Palmer et al. 2005). To achieve sustainability, the report emphasized not only the need to promote innovative science and new technologies, but the need to foster the kinds of cultural changes within science that would make it easier to conduct multi-disciplinary team science and international collaborations. More active engagement with

[4]The 10-year review is available from the LTER document archive, under the category "LTER History," at: https://lternet.edu/documents/

policy-makers and the public was also considered a priority: producing the science was important, but communicating the science was equally in need of attention.

More than a decade later, this conversation is continuing within the international scientific community. Removing barriers to multidisciplinary science and fostering communication between scientists and the public are seen as even more urgent goals today. An international forum on "Re-thinking interdisciplinary research for global sustainability," held in Beijing in January 2018, prompted reflections on how to improve the ability of academic institutions to "support society's effort to achieve global sustainability" (Irwin et al. 2018). These recent discussions have called for transcending disciplinary boundaries, creating new frameworks for interdisciplinary and goal-oriented work, linking academics with practitioners outside of academia, linking natural science disciplines with the medical and public health disciplines, and linking science with the humanities. The LTER program created four decades ago also had to address these kinds of challenges. In the recent call to action cited above, the LTER program and the International LTER efforts that grew from it were cited as mechanisms for breaking down disciplinary barriers and moving toward the achievement of a sustainable future.

Since the LTER program is now perceived as a step in the right direction for ecological science because it seeks the integration of knowledge across disciplines, this is a good time to investigate how this program developed, how this form of long-term research was sustained, and how it fostered a culture that was open to collaborative research as well as community outreach. These are among the several themes that our chapters explore.

The initiators of the LTER program did not know how it would develop or how long it would last, did not at first have the conceptual framework of "sustainability science" in mind, and did not explicitly link these new directions to the solution of any particular environmental problem. The underlying motive appears to have been more general, simply the notion that long-term studies were needed to advance the science of ecology. The criteria for selecting appropriate candidates for the network, deciding what the optimal size for such a network was and what it meant to function as a network, were not spelled out. This was an experiment in which the experimental subjects – the project managers, support staff, scientific researchers, information managers, and others associated with the projects – would have to figure out how this experiment should unfold as time went on, and what exactly it meant to try to transform ecology. What emerged was a vision of ecological science as a collaborative and integrated enterprise, a vision that today's advocates of sustainability science believe will be crucial for solving the immense environmental problems that we face in the twenty-first century.

Thus while the LTER program can be seen as a form of "big biology," it differs in important ways from other kinds of "big science" projects which are focused on achieving very specific goals, such as building a space telescope or mapping the human genome. This "experiment" did not have a specifically defined goal, nor did it have an endpoint. Long-term ecological projects were not meant to come to an end, they were meant to continue indefinitely. A portion of the research at each site was expected to address five designated core areas that NSF specified in advance, but that did not mean all sites were asking identical questions or aiming at the same goal. The fact that the program was vague in its definition meant it could be envisioned in different ways over time, and could move in different directions. As noted by Waide and Thomas (2013), scientific research at LTER sites is "dynamic and evolves continuously in response to increasing knowledge and new opportunities":

> The commitment of LTER sites to long-term observations and experiments does not imply rigidity in focus and approach. The ability of LTER researchers to respond nimbly to new opportunities results from a flexible network structure with a minimum of requirements and uniform site activities. This flexibility would not be possible in a more monolithic network design.

Our goal is to investigate this evolutionary process in greater depth and explore the cultural changes that it fostered. What new way of doing science has emerged from nearly four decades of long-term ecological science? We investigate how those involved in the LTER program met the challenge of creating a network, and what impact their solutions had on ecology and its relationship to other disciplines. We present here different perspectives on the challenges and accomplishments of this novel research enterprise, while also situating it in its longer historical context.

1.3 Themes and Problems

Early History, Growth, and Expansion of the LTER Program The next three chapters provide a general overview of the creation, development, and expansion of the LTER concept. Chapter 2 by Sharon Kingsland, Jerry Franklin, and Robert Waide provides background on the development of the LTER program in the U.S. Starting in the 1960s, it explores discussions occurring within ecology about the need to strengthen ecological science in light of new environmental problems, and also in response to fears that many sites important for ecological research were in danger of being lost to development. This chapter provides a personal account of the origins and development of the LTER program by Jerry F. Franklin, who was involved in early discussions about the need for long-term research and had a leading role in shaping the LTER program during its first 15 years. The chapter discusses also the consequences of the shift of the LTER Network Office to the University of New Mexico in 1997, and the development of the LTER program after 2000. Chapter 3 by Julia Jones and Michael Paul Nelson provides an overview of the entire 40 years of the LTER program, analyzing trends in the requests for proposals and changes in the selection of sites as the program evolved, in order to draw some broad conclusions about shifts in direction and emphasis as the program matured. Chapter 4 includes contributions by several past and present Principal Investigators representing six sites that became part of the LTER Network from 1980 to 2000. This chapter examines the unique challenges associated with long-term research, including initial development, long-term sustainability, mission creep, budgetary planning, and organizational continuity.

Each Site Reveals the Broader Story of Ecology's Development Many of the LTER sites built on existing foundations that had been laid decades earlier. The stories of individual sites do not begin in 1980 or in the year that the sites were added to the LTER network. Most sites were selected because they already had a strong history of place-based research. Four of the early sites had connections to the biome projects initiated as part of the International Biological Program (1968–1974), and many sites benefited from ecological research programs going back much further. Once we place these sites within a longer time frame, we can appreciate how the ecological work done under the LTER program was building on or responding to earlier ecological advances, and we can recognize and appreciate the contributions of earlier generations to this later program.

One of the goals of this book is to capture that longer-term history of LTER, so that we can situate the work of the late-20th and early twenty-first century in its broader context. In particular we will highlight how momentum was building toward the LTER program for decades, and how earlier decisions that helped to broaden and strengthen ecological science were crucial foundations for the later program. The earlier efforts that helped to make ecology a stronger discipline are therefore part of the stories we will tell.

Our individual case studies in Chaps. 5, 6, and 7 focus on regions with long histories of research. In Chap. 5, Ariel Lugo takes a broad historical perspective on the Luquillo LTER in Puerto Rico, linking the LTER to the very long history of tropical forestry in Puerto Rico, highlighting the important influence of both the New York Botanical Garden and the U.S. Forest Service in establishing a research tradition in Puerto Rico, and examining the legacy of Howard T. Odum, who developed a major ecosystem research program under the auspices of the Atomic Energy Commission. Debra Peters in Chap. 6 offers a comparative study of two arid and semi-arid sites that were early contributors to LTER: the Jornada Basin LTER in the desert of southern New Mexico, and the Central Plains Experimental Range LTER (later Shortgrass Steppe LTER) in northern Colorado. She shows how research at these sites led to completely new ways of thinking about problems of drought, over-grazing, and desertification, yielding new concepts that could be applied more broadly to problems of desertification elsewhere. Sharon Kingsland's study in Chap. 7 of the two Alaskan projects, the Arctic Tundra LTER and Bonanza Creek LTER, links their successes to the Cold War imperatives that prompted the U.S. Navy to create and maintain a new Arctic Research Laboratory in northern Alaska shortly after the Second World War. This laboratory fostered a strong research tradition, not only in the ecology of plants and animals, but also in human ecology, and its presence provided an important training ground for the scientists involved in the LTER program. These three case studies collectively show how LTER projects drew on very rich legacies of research, and together they underscore the idea that ecological science is fundamentally a long-term enterprise.

An Expanding Vision of Ecology Given that the LTER Program was an experiment without a fixed or specific goal, it also had the capacity to expand and evolve in unanticipated ways. This aspect of the Program is the subject of several of the chapters in this book. One result was to open up new connections between science and society and extend the impact of scientific research on environmental policy. In Chap. 8, Frederick Swanson, David Foster, Charles Driscoll, Jonathan Thompson, and Lindsey Rustad examine three LTER projects involving forested ecosystems – Andrews Forest LTER in Oregon, Hubbard Brook LTER in New Hampshire, and Harvard Forest LTER in Massachusetts – to examine the broader policy outcomes and social significance of the LTER program. Work at these sites had major impact on land use, conservation, and environmental policies, although these outcomes were not anticipated when the LTER program began, and went well beyond the directives and review criteria of site proposals and programs. The chapter analyzes the attributes of these three LTER projects to explain this outcome, including strong interdisciplinary research communities with cultures of openness, dispersed leadership within those communities, a commitment to carry scientific perspectives to society through multiple governance processes, strong public-private partnerships, and communications programs that facilitated the exchange of information and perspectives among scientific communities, policy-makers, land managers, and the public.

In Chap. 9, John Magnuson explains how site scientists (especially Magnuson and Swanson) introduced the metaphors of the "invisible present" and "invisible place," which they used to develop arguments and concepts to explain to NSF, scientists, and the public how important it is to set ecological results in a broader context of both time and place. Magnuson uses the study of lake ice cover to show how scientists tackled the joint problems of time and place by greatly expanding time scales to over a century, and expanding spatial scales to include not just the North Temperate Lakes LTER in Wisconsin, but also North America and the Northern Hemisphere. Research conducted by an international Lake Ice Analysis Group was instrumental in providing strong evidence for warming winters in the northern hemisphere and the impact of the second industrial revolution on climate.

The broadening of LTER beyond its original focus on ecosystems relatively removed from human settlements was nowhere more evident than in the decision in the late 1990s to expand the network to two urban sites in Phoenix, Arizona, and Baltimore, Maryland. While many of the LTER sites moved toward the incorporation of human-dominated landscapes into their projects in an incremental way, for the urban sites, the relationship between natural and socio-ecological systems had to take center stage. This expanding vision of ecology did not fit well within a funding agency where natural sciences and social sciences were in different directorates. Morgan Grove and Steward Pickett describe in Chap. 10 how difficult it was to shift the culture at NSF so that social science could fit more easily within the LTER framework. They explain the conceptual evolution that occurred within the Baltimore Ecosystem Study, as it went from being a more traditional ecological project to one that more fully integrated natural and social sciences. Also of interest in this chapter is the way their ideas about cross-disciplinary, interdisciplinary, and transdisciplinary science draw on social theory, especially the concept of "boundary objects" developed by sociologist Susan Leigh Star and philosopher James Griesemer, and on approaches to complex problems developed in military contexts, especially General Stanley McChrystal's concept of a "team of teams".

Another significant aspect of the LTER program is the way that these long-term place-based research projects generated interactions between scientists and the public and fostered creative exchanges with the arts and humanities. These interactions took different forms. An extensive and innovative educational program touched all levels of education from kindergarten through to graduate education. Other interactions, as Mary Beth Leigh, Michael Paul Nelson, Lissy Goralnik, and Frederick Swanson discuss in Chap. 11, fostered links to the arts and humanities, creating an impressively wide range of projects and programs that helped to express and promote ecological understanding. Environmental programs in the arts and humanities are not only ways of engaging with and educating the public. More than that, such creative endeavors are seen as ways to help society address the grand ecological-social challenges of the twenty-first century, on the understanding that these are problems that science alone cannot solve.

The expansion outward of long-term ecology, moving from its center in biological science toward a wide range of professions and audiences, illustrates a type of creative evolutionary growth that was not initially planned, but was and is nonetheless important in its contributions to society. At the workshop that led to this volume, this idea of outward expansion was expressed in the question "Has LTER outgrown NSF?" Participants of the workshop realized that this idea needed to be rephrased, that it was not so much a question of LTER "outgrowing" its funding agency as it was a question of how, in any long-term place-based research, new kinds of relationships can form that broaden the impact of science on society. The unexpected outcomes of the LTER program amount to an expanded vision of ecology and underscore the relevance and importance of ecology to society.

What Does It Mean to Be a Network? One of the central themes that is addressed by several of the chapters concerns the problem of what it means for a group of sites to operate as a "network", and how this network came to be. The announcement of the new program implied that the transformative aspects of this proposed experiment were tied to the idea that ecological studies were going to be conducted in a new way, one that explicitly fostered collaborative as well as multi-disciplinary and possibly also interdisciplinary science. These projects were envisioned as involving "groups of investigators" who would "coordinate their studies across sites, utilize documented and comparable methods, and be committed to continuation of work for the required time."[5] The hope of the National Science Foundation and the ecological science community was that the Long Term Ecological Research Program would move ecosystem science to a new level by developing approaches to and

[5] National Science Foundation, "A new emphasis in long-term research, Request for Proposals, announcement, 1979.

conducting comparative analyses over longer time and broader space scales than had previously been possible.

Despite these lofty aspirations, NSF had only a vague idea of what it meant to operate as a network when the program started. Selection of the initial sites in the program did little to facilitate comparison. Proposals for sites were judged on the strength of site science, and little weight was given to approaches that might foster collaboration. The LTER program did not have the idea of comparative research as one of its founding principles. The transformation from a group of independent research sites into something resembling a coordinated network would largely fall on the community of ecologists involved in long-term ecological research.

Only after the program was formed and sites selected did the push for comparative research and networking begin. The first two cohorts of sites were taken by surprise by NSF's dictate about networking, and some investigators pushed back fiercely. The dilemma was straightforward: How could research programs designed to accomplish individual goals adopt a common set of goals after the fact? What would have to be sacrificed to make such a dramatic shift in direction? The LTER Network was caught between conflicting priorities, with few sites willing to risk abandoning their site-based investigations to shift towards comparative research.

The LTER program contained inherent obstacles to the formation of a research network. Some of these obstacles arose from the way NSF responded to, dealt with, or related to the sites, and from the fact that funding resources were limited and at times static. The initial selection of six sites from a diverse array of biomes (e.g., forests, prairie, lakes, coastal lagoon, alpine tundra) did not provide much common ground for comparison. Subsequent selection of additional sites increased the diversity of biomes, and rapid loss of three of the first 11 sites selected hindered the formation of strong collaborations.[6] The budget of the LTER program would not always allow for the level of action needed to operate as a network, so there were significant financial constraints. The LTER Network developed strategic plans to address some of these obstacles, but NSF's response to these plans was not always enthusiastic.

NSF officers and directors tried to encourage the formation of a network with increasing urgency as the 1980s wore on. John Langdon Brooks, who in 1981 became Director of the Division of Environmental Biology (where the LTER Program was located), was unequivocal in his message to ecologists. In 1986, at a meeting of the LTER Network Coordinating Committee, he made it clear that the program had to demonstrate that it was obtaining new results from "network science," results that could not otherwise be obtained. His message was that the future of the program depended on showing that the network concept was really producing something novel. But as Tom Callahan noted in an assessment published in 1991, "Persons at all levels of the US/LTER activity sense that collaboration among projects and scientists has not yet been fully developed. The missing level of development, which has come to be called 'networking,' has become a major

[6] Illinois Rivers and Okefenokee were withdrawn in 1988, and North Inlet was withdrawn in 1993.

objective on all planes of LTER operations" (Callahan 1991). William Schlesinger (2016) pointed out that even the 30-year review of the program stressed the "need to develop cross-site experiments more broadly."

Although NSF exerted pressure, the sites themselves had to figure out how to accomplish this task. Solutions to this problem involved three general approaches. *Intellectual* (i.e., scientific) approaches involved creative thinking to identify questions or problems that could be answered using comparative methods. Some of the more successful examples of this approach used new technologies (e.g., remote sensing) to make common measurements across sites. *Organizational or institutional* approaches brought people with similar interests together through committees, All Scientists Meetings, Science Council activities, or distance communication. These interactions aimed to build the trust that underlies successful collaboration. *Cultural* approaches attempted to create an interdependent culture that was receptive to data sharing and group problem solving.

The community's response was wide-ranging, highly productive, but at times frustrating for scientists who were trying to take the LTER program in new directions. The *scientific* challenge involved figuring out what kinds of cross-site comparisons made sense, and using these comparisons to generate conclusions about ecosystem properties and how environmental sustainability can be achieved. Chapter 12 by John Magnuson and Robert Waide discusses several efforts to create meaningful cross-site comparisons at different LTER sites, some in direct response to John Brooks's exhortation in 1986 to function like a network. The chapter also examines the transition of the LTER Network from a collection of independent sites to a community to a research network during five periods over 35 years of the program.

Another challenge involved innovations in *information management*, which were generated in the LTER program and flowed into other areas of ecology. Devising novel ways to manage, store, and communicate information responded to the demand for "network science" and had broader impact on ecological science. In Chap. 13, Susan Stafford recounts this history in detail, starting from the early stages of design, testing and implementation of a data and information management system that helped to create standards and protocols across the Long Term Ecological Research Program. These innovations, she points out, influenced the entire field of interdisciplinary ecological research. Schlesinger (2016) also stresses the importance of these initiatives: "The LTER network has emerged as a leader in the best practices to manage and archive large data sets." Stafford's chapter highlights the role of a particular group within the LTER program, originally known as "data managers" and later "information managers", showing how this group took the initiative to solve these problems of standardization and archiving, helping to bring about a new culture of data sharing and collaborative science.

One of the most challenging problems was to figure out an acceptable way of *integrating the natural and social sciences*, in other words to create a new conceptual model that placed humans firmly within the picture. This was a difficult problem because it entailed not only altering the culture within NSF, but also finding approaches that all LTER sites were comfortable adopting. Scott Collins in Chap.

14 discusses some of the efforts – and frustrations – of trying to create an integrated social-ecological framework in the early 2000s. Innovations in *governance structures* were also central for a well-functioning network, and these innovations were generated by the members of the ecological community. Chapter 15 by Ann Zimmerman and Peter Groffman explores the history of LTER network governance, showing how governance has influenced the creation, sharing, and dissemination of knowledge in the LTER network and how it has adapted as the network has grown, while responding to changed priorities at NSF.

The emergence of global environmental problems like climate change and biodiversity loss reinforces the importance of comparative, multi-national studies. Development of predictive models that anticipate changes in the Earth's ecosystems and the services they provide to humans requires long-term data and understanding of fundamental ecological principles. The scope of these current environmental issues demands cooperation at a global scale. In Chap. 16, Robert Waide and Kristin Vanderbilt examine the international reach of the LTER program with the creation of the International LTER program in the 1990s. This chapter is important for understanding the global importance of long-term ecological studies and the value of collaborative international networks of sites that now characterize change in almost every biome on the planet.

1.4 Opportunities for Future Studies of the LTER Program

Opportunities exist for additional studies of the development and evolution of the LTER program. Our work here describes historical activities from the beginning of LTER in 1980 through 2015, when the Cooperative Agreement for the LTER Network Office (LNO) at the University of New Mexico ended. Although we cover a broad range of topics, we did not attempt an exhaustive assessment of LTER. We hope that this volume will encourage studies of additional topics that we were not able to cover.

One subject that requires further examination is the management of the LTER program by NSF over the last 40 years. Many of NSF's goals for LTER focus on the short term. We document changes in the NSF's attitudes and approach to LTER in Chap. 3, but the logic behind some of these changes is poorly understood. Although NSF encouraged strategic planning by the LTER Network and the Network Office, there is no evidence that NSF had its own strategic plan with regard to LTER. In particular, the growth of the Network after 1988 seems to have been driven by opportunities to expand the base of funding rather than any overarching framework. Fundamental questions of optimum network size and composition have never been resolved. As far as we have been able to determine, NSF has never projected the long-term cost of the LTER program, information which would be critical for planning. The lack of clear, consistent guidelines for the LTER program has left the responsibility for program direction in the hands of the program officers (Gholz et al. 2016), who most recently have served relatively short rotations in the LTER program. A history of NSF's perception of the development of LTER, parallel to the material we provide in this volume, would be a valuable complement to our understanding of the LTER program.

The integration of LTER research and education efforts (Gosz et al. 2010) is another subject that is ripe for exploration. The long-term nature of the LTER program provides the opportunity for students to mature into research roles in LTER, and there have been examples of researchers whose first association with LTER was as undergraduate students. The LTER Network sponsors a Schoolyard LTER program and book series, remote learning resources, research experiences for undergraduates, and graduate education, allowing association with LTER sites and scientists from K-12 through professional training. The benefits to students and educators as well as to the LTER program itself are themes that might be explored profitably by future historical studies.

1.5 Resources for the History of Long-Term Ecological Research

Two collections produced by scientists interested in long-term ecological research provide insights into how ecologists, both within and outside of the LTER program, perceived the value of long-term research and its challenges in the 1980s. One is a volume edited by Gene Likens, *Long-Term Studies in Ecology: Approaches and Alternatives* (1989), and the other is a volume edited by Paul Risser, *Long-term Ecological Research: An International Perspective* (1991). Both provide useful context for understanding scientific support for the idea of long-term research at the time the LTER program was getting started.

Systematic historical examination of the LTER program has only just begun. Robert Waide and McOwiti Thomas published a succinct overview in 2013, which provides a good starting place for anyone interested in the LTER program (Waide and Thomas 2013). David Coleman's *Big Ecology: The Emergence of Ecosystem Science* (2010) links the LTER program to the earlier biome studies that were part of the International Biological program, and provides an overview of the LTER program. A scholarly study by Elena Aronova, Karen Baker, and Naomi Oreskes examines the connection between the International Geophysical Year's programs and the later International Biological Program, with particular attention to data collection, storage, and retrieval, and ends by considering the LTER program's links to the IBP (Aronova et al. 2010). A volume edited by Michael R. Willig and Lawrence R. Walker, *Long-term Ecological Research: Changing the Nature of Scientists* (2016), provides an overview of the LTER program, discusses the challenge of sustaining long-term research, and offers a view of LTER from the perspective of NSF's program directors, while also giving voice to the personal stories of people who have been involved in the LTER program. Several of those authors are contributing to this volume, building on their earlier reflections to provide a deeper analysis of what this program has achieved, and how challenges to building and sustaining an ecological network were met.

The LTER program has also caught the attention of scholars interested in multidisciplinary collaborations in biology. An edited volume by John N. Parker, Niki Vermeulen, and Bart Penders, *Collaboration in the New Life Sciences* (2010), has several chapters on collaborative work in ecology, two of which deal with the LTER program directly, and one indirectly. Stephen Bocking's chapter on collaboration in ecology includes discussion of the Hubbard Brook study led by F. H. Bormann and Gene Likens, but focuses on the decades before Hubbard Brook joined the LTER network. A chapter by Karen Baker and Florence Millerand considers the problem of data management in the LTER network, which is relevant to Susan Stafford's chapter on information and data management in this volume. A chapter by Ann Zimmerman and Bonnie Nardi contrasts LTER's approach to big science with the National Ecological Observatory Network (then still in the planning stage) and is relevant to Zimmerman's chapter co-authored with Peter Groffman in this volume.

The LTER program has created a useful website which includes both a timeline of key events that provides a snapshot of the program's development over 40 years (https://lternet.edu/network-organization/lter-a-history/), as well as an archive that contains a wide range of documents available digitally, including decadal reports, reports from committees, and mini-symposia, although the coverage of the first decade is relatively sparse compared to later decades (https://lternet.edu/intranet/). Individual sites all have websites, with varying amounts of historical information available. Some (such as H. J. Andrews Forest, Harvard Forest, and North Temperate Lakes) have document archives, but there is great variation in historical record-keeping among the sites. Andrews Forest also has an oral history project for its LTER site, held at Oregon State University and available digitally at this website: http://scarc.library.oregonstate.edu/omeka/exhibits/show/forestryvoices/collections/. Records from the LTER Network Office in Albuquerque, including some documents from the previous office in Seattle, are archived in the University of New Mexico libraries under collection number, UNMA 082. The finding aid to the paper files is available online: https://rmoa.unm.edu/docviewer.php?docId=nmuunma082.xml.

Digital copies of much of this material will be available online in the near future.

1.6 Concluding Thoughts

As we point out above, the LTER program was an experiment designed by NSF and the ecological community and aimed at making ecology a more rigorous and predictive science. However, LTER also represented a new approach for many of the individuals engaged in the program. Academic scientists, whose research often addressed short-term, narrowly focused questions examined by a laboratory populated by undergraduate and graduate students, were asked to work closely with colleagues from other disciplines to address common questions of broad scope.

They were expected to share data, not hoard it, and cooperate in a network of unfamiliar sites and scientists. The funding agency, NSF, took a greater interest in their research directions than it did for individual research grants and often urged particular approaches. Individual investigators were challenged to fit their ideas into an overarching conceptual framework, and the success of that joint framework determined individual success. LTER projects succeeded in part because they were able to create and become part of communities that shared common goals. Instead of competing for grant money, they cooperated for research results. The results of this sociological experiment (Willig and Walker 2016) are fully as interesting as the results of the research experiment that fostered it.

One metaphor that has frequently been used to describe LTER sites is that of a research platform, similar to a research vessel on the open ocean. The metaphor is particularly apt for those LTER sites that are isolated patches representing previously widespread habitats. In this sense, the LTER program has achieved one of the primary goals set out for the program by NSF. LTER sites serve as protected sites whose existence allows LTER and non-LTER researchers to conduct measurements and experiments on intact ecosystems. The importance of these research platforms can be gauged by the fact that on average LTER sites support research projects valued at 2.9 times the investment by NSF in each LTER site. The broad range of research conducted at each site and the commitment to openly accessible data combine to achieve one of the most important goals of LTER: to create a legacy of well-designed and documented long-term observations, experiments, and archives of samples and specimens for future generations.

We have already described the value of such research legacies in recounting the importance that LTER data played in understanding the hantavirus outbreak in New Mexico. Investigators who first set up long-term rodent population studies and decided to collect blood samples from captured animals could not have anticipated the value that these data would have just a few years later. The same is true of many long-term studies. Data collected with one question in mind have great potential to address new, unanticipated questions. Such serendipitous outcomes may not be frequent, but they are certain to be important. Thus, the contribution made by long-term ecological data to understanding the spread of hantavirus has been identified as one of the 50 discoveries made with NSF funding that have had the most influence on the lives of Americans (Gosz et al. 2010). This story represents a classic example of the value of long-term ecological data in resolving societal problems.

Acknowledgments We are grateful to Maria Portuondo, who as Chair of the History of Science and Technology Department at Johns Hopkins University provided financial support for the workshop held in June 2018 in Baltimore, Maryland, which laid the groundwork for this book. Gina Rumore, who attended the workshop, also kindly allowed us access to interviews pertaining to the history of the LTER Program that she had conducted in 2010. Stephen Bocking's participation in our workshop provided valuable historical perspective as we identified and debated the central themes and goals of the book. Frank Harris, who was at NSF in the early days of LTER and who co-chaired the 20-Year Review of the program, shared his recollections of the development of long-term research at NSF. Marty Downs, Director of the LTER Network Office at the National Center for Ecological Analysis and Synthesis, provided information on recent developments in the

LTER Network. The National Science Foundation provided support to organize and digitize the archives of the LTER Network Office through a supplement to award DEB-1440478 to the University of New Mexico.

References

Aronova, E., K.S. Baker, and N. Oreskes. 2010. Big science and big data in biology: From the international geophysical year to the international biological program to the long term ecological research network, 1957-present. *Historical Studies in the Natural Sciences* 40: 183–224.

Callahan, J.T. 1984. Long-term ecological research. *Bioscience* 34 (6): 363–367.

———. 1991. Long-term ecological research in the United States: A federal perspective. In *Long-term ecological research: An international perspective*, ed. Paul G. Risser, 9–21. Chichester: Wiley.

Gholz, H.L., R. Marinelli, and P.R. Taylor. 2016. Reflections on long-term ecological research from the National Science Foundation program directors' perspectives. In *Long-term ecological research: Changing the nature of scientists*, ed. M.R. Willig and L.R. Walker, 43–51. New York: Oxford University Press.

Gosz, J.R., R.B. Waide, and J.J. Magnuson. 2010. Twenty-eight years of the US-LTER program: Experience, results, and research questions. In *Long-term ecological research – Between theory and application*, ed. F. Müller, C. Baessler, H. Schubert, and S. Klotz, 59–74. Dordrecht: Springer.

Horgan, John. 1993. Were four corners victims biowar casualties? *Scientific American* 269 (November): 206.

Irwin, Elena G., P.J. Culligan, M. Fischer-Kowaldki, K.L. Law, and R. Murtugudde. 2018. Bridging barriers to advance global sustainability. *Nature Sustainability* 1: 324–326.

Likens, Gene, ed. 1989. *Long-term studies in ecology: Approaches and alternatives*. New York: Springer.

Lubchenco, J., A.M. Olson, L.B. Brubaker, S.R. Carpenter, M.M. Holland, S.P. Hubbell, S.A. Levin, J.A. MacMahon, P.A. Matson, J.M. Melillo, H.A. Mooney, C.H. Peterson, H.R. Pulliam, L.A. Real, P.J. Regal, and P.G. Risser. 1991. The sustainable biosphere initiative: An ecological research agenda. *Ecology* 72 (2): 371–412.

Michener, W.K., K.L. Bildstein, A. McKee, R.R. Parmenter, W.W. Hargrove, D. McClearn, and M. Stromberg. 2009. Biological field stations: Research legacies and sites for serendipity. *Bioscience* 99: 300–310.

Mirtl, M., E. T. Borer, I. Djukic, M. Forsius, H. Haubold, W. Hugo, J. Jourdan, D. Lindenmayer, W.H. McDowell, H. Muraoka, D.E. Orenstein, J.C. Pauw, J. Peterseil, H. Shibata, C. Wohner, X. Yu, and P. Haase. 2018. Genesis, goals and achievements of long-term ecological research at the global scale: A critical review of ILTER and future directions. *Science of the Total Environment* 626: 1439–1462.

Morse, S.S. 1994. Hantaviruses and the hantavirus outbreak in the United States: A case study of disease emergence. *Annals of New York Academy of Sciences* 740: 199–207.

Palmer, Margaret, E. Bernhardt, E. Chornesky, S.L. Collins, A. Dobson, C. Duke, B. Gold, R. Jacobson, S. Kingsland, R. Kranz, M. Mappin, F. Micheli, J. Morse, M. Pace, M. Pascual, S. Palumbi, J. Reichman, W.H. Schlesinger, A. Townsend, M. Turner, and M. Vasquez. 2005. Ecological science and sustainability for the 21st century. *Frontiers in Ecology and the Environment* 3 (1): 4–11.

Parker, John N., Niki Vermeulen, and Bart Penders, eds. 2010. *Collaboration in the new life sciences*. Farnham: Ashgate.

Parmenter, Robert R., J.W. Brunt, D.I. Moore, and S. Ernest, 1993. The Hantavirus epidemic in the Southwest: Rodent population dynamics and the implications for transmission of Hantavirus-associated Adult Respiratory Distress Syndrome (HARDS) in the Four Corners region. Report submitted to the Federal Centers for Disease Control and Prevention and the New Mexico State Department of Health. Albuquerque, New Mexico: Sevilleta LTER Publication No. 41. https://doi.org/10.13140/RG.2.1.2623.1841.

Risser, Paul G., ed. 1991. *Long-term ecological research: An international perspective*. Chichester: John Wiley & Sons.

Schlesinger, W.H. 2016. Coda: Some reflections on the long-term ecological research program. In *Long-term ecological research: Changing the nature of scientists*, ed. M.R. Willig and L.R. Walker, 391–396. New York: Oxford University Press.

Van Hook, Charles J. 2018. Hantavirus pulmonary syndrome – The 25th anniversary of the four corners outbreak. *Emerging Infectious Diseases* 24: 2056–2060.

Vucetich, John A., Michael P. Nelson, and Jeremy Bruskotter. 2020. What drives declining support for long-term ecological research? *Bioscience* 70: 168–173.

Waide, Robert B., and McOwiti O. Thomas. 2013. Long-term ecological research network. In *Earth system monitoring: Selected entries from the Encyclopedia of sustainability science and technology*, ed. J. Orcutt, 233–268. New York: Springer.

Willig, Michael R., and Lawrence R. Walker, eds. 2016. *Long-term ecological research: Changing the nature of scientists*. New York: Oxford University Press.

Yates, T.L., J.N. Mills, C.A. Parmenter, T.G. Ksiazek, R.R. Parmenter, J.R. Vande Castle, S.H. Calisher, S.T. Nichol, K.D. Abbott, J.C. Young, M.L. Morrison, B.J. Beaty, J.L. Dunnum, R.J. Baker, J. Salazar-Bravo, and C.J. Peters. 2002. The ecology and evolutionary history of an emergent disease: Hantavirus pulmonary syndrome. *Bioscience* 52: 989–998.

Part I
Background and General Overview of the LTER Program

Chapter 2
The Origins, Early Aspects, and Development of the Long Term Ecological Research Program

Sharon E. Kingsland, Jerry F. Franklin, and Robert B. Waide

Abstract The chapter explores a series of overlapping discussions from the 1960s to the mid-1970s concerning the need to preserve natural areas for ecological research and observation, as well as the need for ecologists to be advocates for ecological science in the light of modern environmental problems. The U.S. Forest Service was also actively promoting the creation of Research Natural Areas and making them available for ecological research. A report prepared under the auspices of The Institute of Ecology in the mid-1970s drew attention to the desirability of creating a large network of experimental ecological reserves across the U.S. and its territories. This report led to three workshops during which ecologists debated what such a network of research sites might look like. The third workshop's proposals led directly to the creation of the Long Term Ecological Research (LTER) Program, which began in 1980 with funding from the National Science Foundation (NSF). The chapter follows these conversations with particular emphasis on the participation of Jerry F. Franklin, who took part in early discussions about expanding ecological research infrastructure while serving as Ecosystem Studies Program Officer at NSF during 1973–1975. He subsequently was director of the H.J. Andrews Ecosystem Research Project, which joined the LTER Program in 1980, chaired the LTER Program's Coordinating Committee from 1982 to 1995, and established and directed LTER's coordinating office from 1982 to 1996. In 1997 the Network Office shifted

S. E. Kingsland (✉)
History of Science and Technology Department, Johns Hopkins University, Baltimore, MD, USA
e-mail: sharon@jhu.edu

J. F. Franklin
School of Environmental and Forest Sciences, University of Washington, Seattle, WA, USA
e-mail: jff@uw.edu

R. B. Waide
Department of Biology, University of New Mexico, Albuquerque, NM, USA
e-mail: rbwaide@unm.edu

R. B. Waide, S. E. Kingsland (eds.), *The Challenges of Long Term Ecological Research: A Historical Analysis*, Archimedes 59,
https://doi.org/10.1007/978-3-030-66933-1_2

to the University of New Mexico, beginning a new phase in the coordination of the developing network.

Keywords LTER program · LTER network · LTER network office · Long-term ecological research · History of ecology · Conservation movement · Ecosystem ecology · International biological program

2.1 Introduction

This chapter explores the evolving discussions leading to the creation of the Long Term Ecological Research Program (LTER), which began in 1980 with the selection of the first group of six sites. Discussions among ecologists during the preceding two decades provide insight into how a portion of the ecological community was thinking about ways to strengthen the discipline of ecology. The intensive planning process that led directly to the Program took place over a 4-year period from 1976 to 1979. It began with a study commissioned by the National Science Foundation (NSF) in 1976, prepared under the auspices of an advisory body known as "The Institute of Ecology" (TIE), whose origins will be discussed in more detail below. TIE's report, titled "Experimental Ecological Reserves: A Proposed National Network," was completed in December 1976 and was published in June 1977. That report envisioned an extensive network of 71 sites for ecological research in 67 locations, a highly ambitious plan that was designed to expand on the existing place-based research already being conducted in the United States and its territories. Although that plan was not implemented, the report led to three NSF-sponsored workshops held in 1977, 1978, and 1979, which explored ways to support long-term ecological research and observation. The third workshop, held in June 1979 in Indianapolis, Indiana, was convened by The Institute of Ecology.

The reports from these meetings reveal concerns in the ecological community about strengthening the science in the 1970s, including programs that might be useful. The reports also show how the community worked to translate the ambitious vision outlined in TIE's report into a specific set of proposals that would be acceptable to NSF. Many in the ecological community had been critical of the International Biological Program and were not in favor of such a long-term ecological program, fearing it would draw off funds from more meritorious science. However, by 1979 long-term research had many strong and influential supporters. NSF responded to these ideas and largely adopted the recommendations of the third workshop, although the LTER program that emerged from this process was much smaller than the network concept originally envisioned in TIE's report.

However, the history of the LTER program does not begin with this planning process. We also need to understand how the ecological community arrived at this point in the mid-1970s. For this purpose, we must go back to about 1960 and seek answers in the discussions that ultimately generated the strategic planning process.

Because the decision to fund the program was based on a belief that many in the ecological community supported it, we need to probe how those members of the community were articulating the needs of their science in the 1970s. Since the vision presented in TIE's report was much more ambitious than the LTER program, it will be instructive also to explore how that vision arose from a decade and a half of intense discussion about the future of ecology. What were the imperatives to which ecologists were responding? We aim in this chapter to answer this question, not as it applies to individual sites – each of which has its own story related to its traditions of place-based research -- but to the program as a whole.

This chapter emphasizes three inter-related developments that helped create the rationale for the LTER program. One was a resurgent *preservation movement* that emerged in the postwar period and focused on the need to preserve natural areas for scientific research. The U.S. Forest Service took on a leadership role as this movement expanded in the 1970s. The second, closely related development, was the emerging *environmental movement* of the 1960s, which challenged ecologists to articulate more clearly the value of ecology to American society, and federal land management agencies to adopt an ecological perspective in their work. Ecologists' response to the environmental movement helps to explain the timing of LTER, that is, why ecologists believed that the time was ripe in the late 1970s for this kind of program, and for a systematic reorganization of ecosystem ecology. The third development was the emergence and gradual dominance of *ecosystem ecology*, which provided ecologists with a reason to seek resources for long-term ecological research at the ecosystem level (while including research at community and population levels), justified in part by its relevance to resource management.

The chapter also provides a personal account of the origins and development of the LTER program by Jerry F. Franklin, who was involved in discussions about long-term research as a program officer at the National Science Foundation during the 1970s, and went on to take important leadership roles in the program from its inception to 1995. From 1975 to 1986 he served as director of the H. J. Andrews Ecosystem Research Project. This project, at the H. J. Andrews Forest in Oregon, became part of the first group of LTER sites in 1980. From 1982 until 1995, Franklin also served as Chair of what was initially called the Steering Committee of the LTER program, and later the Coordinating Committee. While based at Oregon State University in the early 1980s, Franklin obtained a coordination grant to establish an LTER Network Office, which in 1986 moved with him to the University of Washington, Seattle. As David Coleman recalls in his history of the LTER program, "Much of the growth of the LTER Network occurred during Franklin's leadership" (Coleman 2010, p. 98). Franklin retired as Chair after 12 years, when James Gosz became the Coordinating Committee Chair. In 1997 the Network Office moved to the University of New Mexico in Albuquerque, and Robert Waide became its executive director. The chapter draws on the experiences of Franklin and Waide, providing an insider's account of the early years of the program, as seen from the perspective of those in the Network Office.

2.2 Recognizing a Need

Before we delve into the details of The Institute of Ecology's report on "Experimental Ecological Reserves," a bit of background is needed to set this report in historical context. This context will help us to understand how the creation of the LTER program was the culmination of ecology's growth and maturation as a discipline, as well as an attempt to respond to the impacts that human population growth and rapid development were having in the United States. In the 1960s ecologists were becoming aware that a sea-change was underway in postwar American society, and that they needed to recognize it, respond to it, and come up with a plan. These changes were transforming the land and were threatening to eliminate the last vestiges of what was viewed as a pristine natural environment. That sense of the finiteness of "Spaceship Earth," and the impending loss of species, habitat, and entire natural worlds, engendered profound concern, which was transformed into a call for action.

TIE's report on "Experimental Ecological Reserves," which articulated the scientific community's desire to preserve and maintain areas for ecological research, was the culmination of a preservation movement within ecology that had been building support for decades.

As Gina Rumore (2012) has noted in her study of the preservation of Glacier Bay National Monument in Alaska, this scientific preservation movement has a long history. The Ecological Society of America created a Committee for the Preservation of Natural Conditions for Ecological Study as early as 1917, an idea promoted by Victor Shelford, the Society's first president. Rumore argues that the protection of Glacier Bay in the 1920s truly fulfilled the Committee's goal of preserving land for ecological research, unlike some of its other campaigns, which were aimed more at protecting lands from development. She suggests that preserving land for research reflected ecologists' desires to do long-term, place-based research, even when funding for such long-term research was hard to obtain.

Shelford argued that research in natural areas was essential also to the work of wildlife managers. The job of management was to control populations, which meant that managers worked in highly unnatural environments. As he pointed out in 1933, their work was hampered by an absence of guiding scientific principles and accurate information (Rumore 2012). Thus, scientific research in protected natural areas could have direct practical value. The Committee was disbanded in 1945, but Rumore sees a continuous historical thread linking the earlier interests of ecologists in long-term study and the eventual creation of the LTER program. This thread can be discerned more clearly if we track how discussion about preservation of land for ecological research continued in the postwar years. That discussion led directly to the planning stages for the LTER program.

The link between scientific research on protected areas and practical problems of resource management and conservation was nowhere more evident than in forestry, a subject we will return to later in this chapter. By the early twentieth century, Americans were becoming alarmed at the devastating impact of the logging industry on eastern and Midwestern forests, where destruction of forests increased the risks of wildfires, erosion, and flooding. The early conservation movement was largely driven by concern to preserve forests.

Creation of reserved forest areas began in 1891 with a Congressional provision that allowed Presidents of the United States to establish forest reserves out of the public domain. During the period of 1892 to 1905, 63 million acres of reserved had been set aside by presidential proclamation. Theodore Roosevelt aggressively added to that total and together with his successor, William Howard Taft, left the protected forest lands at 187 million acres.

Initially administered as the Forest Reserves in the US Department of Interior, Roosevelt transferred them to the Department of Agriculture in 1905, to the jurisdiction of a newly-created Forest Service. A dynamic conservationist and close friend of Roosevelt, Gifford Pinchot, led this agency and renamed the reserves the National Forests. Passage of the Weeks Act in 1911, after a long and heated campaign, granted the federal government the right to create forest reserves through the purchase of private forest lands containing headwaters of navigable streams (Johnson and Govatski 2013). Pinchot and other conservationists argued vehemently that deforestation contributed to catastrophic floods, making the case that protection of forests had many important ecological and economic consequences beyond ensuring a supply of lumber.

The Forest Service included active research almost from its establishment, creating a research branch that reported directly to the Chief of the Forest Service. As a result, the National Forests became home to experiment stations focused on understanding all aspects of these forests, including their management. The role of forests in regulating stream flow was a part of the research program right from the start, because of the need for much more scientific information about the relationship between forest cover and streamflow and water quality – what would now be broadly categorized as "ecosystem services". Many experimental forests and ranges were established to provide sites for long-term studies, which included experimental manipulations of watersheds.

The creation of this research infrastructure within the Forest Service ultimately had a direct bearing on the selection of sites for the Long Term Ecological Research Program. For example, in 1926 the Appalachian Forest Research Station in North Carolina started a long-term program of research on forest cover, erosion, and streamflow under the direction of forest ecologist Charles R. Hursh (Douglass and Hoover 1988). It became the Coweeta Experimental Forest in 1934, and during the 1930s benefited from Franklin Roosevelt's New Deal programs, especially the creation of the Civilian Conservation Corps, which provided the manpower needed to expand its facilities.

With a strong research tradition that continued into the 1960s, Coweeta was selected in 1969 as one of five study areas for the Eastern Deciduous Forest Biome Project of the International Biological Program. In 1980, it became one of the first sites of the new Long Term Ecological Research Program (Swank et al. 2001). This example supports Rumore's argument that there is a strong historical thread linking early interest in long-term research (partly in relation to specific resource management problems) and the LTER program of the 1980s. Thus, when the Coweeta LTER site was only 4 years old, it celebrated the fiftieth anniversary of research in forest hydrology and ecology at that location, and as Eugene Odum remarked, by 1984 it was already the "longest continuous environmental study on any landscape in North America" (Odum 1988).

The H. J. Andrews Experimental Forest in Oregon and Hubbard Brook Experimental Forest in New Hampshire (both discussed in Chap. 8 in this volume) provide two additional examples of important long-term research sites established by the Forest Service that ultimately became locations of LTER projects. Both were established after World War II and benefited from the pioneering watershed research at Coweeta. The extraordinary productive collaboration between Forest Service and academic scientists at Hubbard Brook demonstrated the value of such joint efforts in using long-term research, including experiments, to elucidate fundamental ecosystem principles. Research at H. J. Andrews extended such research into the massive coniferous forests of western North America. The three properties – Coweeta, Hubbard Brook, and Andrews – have sometimes been described as the "Crown Jewels" of the Forest Service's Experimental Forests and Ranges.

Following the Second World War, the scientific preservation movement that can be traced to the early years of ecology took on new life, broadening and intensifying. The goal was to make ecology into a more predictive science, so that it could better serve the needs of resource managers, much as Shelford had argued decades earlier. This movement also involved traditional conservation groups like the Sierra Club, who saw the advantage of linking their interests in preserving wilderness to scientific needs. For example, the sixth Wilderness Conference sponsored by the Sierra Club in 1959 adopted as its theme "the meaning of wilderness to science" (Brower 1959). As was customary for these conferences, it ended with recommendations. One supported the passage of the Wilderness Bill then under discussion (but not passed into law until 1964), on the grounds that natural areas were needed for benchmark scientific studies to assess how humans had affected the planet. The second recommendation urged agencies administering public lands to undertake long-term research programs involving every branch of science and focused on natural or unaltered reserves.

By the 1960s, ecologists were also thinking harder about their responsibilities in the wake of the environmental movement, which was putting the word "ecology" into everyone's vocabulary. As environmental problems steadily mounted in the 1950s, the tipping point was the appearance of Rachel Carson's book, *Silent Spring*, in 1962, which galvanized the modern environmental movement with a call to curb our excessive use of pesticides. Carson's book got the attention of the Kennedy administration, and especially of Secretary of the Interior Stewart Udall, who was also an impassioned conservationist. The growth of the environmental movement would be a challenge to ecology, a relatively small scientific discipline in the early 1950s, to enlarge its social role. In the words of plant ecologist F. Herbert Bormann, "ecology" was becoming "more powerful than ecologists" (Bormann 1996). This time of crisis and dispute, as Bormann and others realized, demanded a response from ecologists. These responses to the environmental movement, as well as revitalization of the movement to preserve natural areas for scientific research, helped lay the groundwork for the creation of the LTER program. Bormann's career at this time illustrates particularly well how these two threads were intertwined.

In the late 1950s and early 1960s Bormann was on the faculty at Dartmouth College in New Hampshire. Later he would become known for the important results that emerged from the Hubbard Brook Ecosystem Study in New Hampshire, which he co-founded in 1963 with Gene Likens, Robert S. Pierce, and N. M. Johnson, and which led to the discovery that acid precipitation was affecting the United States (Likens and Bormann 1974; Likens and Bailey 2014). Hubbard Brook Experimental Forest would join the LTER program in 1987 and is still part of that program (see Chap. 8 in this volume for further discussion). But Bormann is also important for the history of LTER because he was among the earliest to recognize that a new age of environmentalism was dawning, and ecologists needed to take note and realize what this all meant for ecology.

His own research brought this home to him. In 1957 Bormann was studying mineral transfer between roots using radioactive tracers, and found that his greenhouse experiments were ruined because the control plants, which were not supposed to be radioactive, had been contaminated with radioactivity. A sample of leaves from the garden showed that all the plants outside were radioactive, and the most likely source was the nuclear bomb test series conducted in Nevada that year. Bormann thought the results were important enough for publication in *Science*, but the manuscript was rejected. His suspicion was that reviewers from the Atomic Energy Commission did not wish this information to come to public attention. Instead he published it in *Ecology*, a journal of more limited readership and impact, where it drew no attention (Bormann et al. 1958; Bormann 1996).

These kinds of experiences, he recalled, coupled with his awareness of public concern about the environmental impact of nuclear and chemical technologies, began to enlarge his view of ecology (Bormann 1996). At the time *Silent Spring* was published in 1962, robins on the Dartmouth campus were showing traumatized behaviors. Bormann and his students followed the scientific literature linking these behaviors to ingestion of DDT. He spent a year in 1962–1963 with George Woodwell at Brookhaven National Laboratory, where studies of the ecological cycling of radioactive nuclides were being conducted. Ecology conducted at the national laboratories focused on ecosystem science, which emphasized the connection of living and non-living components of the ecosystem. Making these connections was crucial to understanding the effects of such contaminants as radioactivity and pesticides within and between ecosystems.

Bormann and Woodwell discussed these environmental issues and what they meant for ecology. Together they embarked on a "missionary trip" around the country with the message that "ecology is more powerful than ecologists" and that "ecologists needed to emerge from their cocoon" (Bormann 1996, p. 22). Bormann remembered getting few converts, but making many enemies. But ecology was starting to change under the leadership of Bormann, Woodwell, and others who recognized the relevance of these environmental problems for ecology, and for understanding our relationship to the world. After moving to Yale University in

1966, Bormann continued to spread awareness of environmental problems to students and the broader public.

In 1966, the year that he moved to Yale's School of Forestry, Bormann also spoke out about the implications of human population growth and uncontrolled development for scientific research, publishing an editorial in *BioScience* appealing to biologists to support measures to protect natural areas for research (Bormann 1966). His experiences, including 8 years in the Piedmont region of the southeast and another 10 years in the hardwood forest ecosystem of northern New England, revealed that it was getting harder and harder to locate old-age, relatively undisturbed forests. That meant that it was getting harder to find natural areas to serve as biological baselines. Bormann's colleagues were having the same difficulty, and it was affecting both their research and teaching. As he observed, "Destruction or alteration of important biological research areas is unquestionably accelerating as our population grows and as the tools for environmental manipulation become more powerful" (Bormann 1966, p. 585). Without knowing how an undisturbed ecosystem functioned, it would be difficult for a land manager to know how to implement and judge the efficacy of practices relating to such things as timber production, disease control, wildlife management, recreation, water yield, or conservation of species. Baseline studies, he argued, were needed both to understand how ecosystems functioned, but also to help people evaluate the changes they were making, and would continue to make, in their environments.

He advocated setting up a federal system of natural biological reserves in order to preserve a range of biologically important ecosystems across North America, and to use these reserves both for scientific research and teaching. At that time, he noted, there was "no thoughtfully planned national system of representative ecosystems set aside for the purpose of descriptive and experimental research" (Bormann 1966, p. 585). Instead, there was a "potpourri of biologically interesting areas set aside by thoughtful citizens both within and without the government." A more systematic approach to the study of these ecosystems, especially those that were already protected, was clearly needed.

Bormann was not alone in making this plea for a system of reserves, and pointed out that in the previous 2 years five separate groups had "called for intensified study of our natural environment" (Bormann 1966, p. 585). These groups were: the President's Science Advisory Committee (which had issued a report on "Restoring the quality of our environment" that included the idea of studying naturally occurring ecosystems as baselines); the U.S. National Committee for the International Hydrologic Decade (which similarly advocated the study of natural environments); the planning groups for the International Biological Program (IBP), which also called for the study and conservation of natural ecosystems; biologists who were concerned that plans for weather modification were being made without adequate prior study of environments that would be affected by such modifications; and finally the Ecological Society of America, which was promoting the creation of biological stations in major biomes. The American Association for the Advancement of Science had also written a report in 1963 calling for a larger and better coordinated natural area program.

Bormann also drew attention to a Senate bill introduced by Senator Gaylord Nelson of Wisconsin in 1965, called the Ecological Reserves and Surveys Bill (S.2282), which advocated creating a federal system of natural areas. Nelson is perhaps best known for leading the organization of the first Earth Day on April 22, 1970. His deep commitment to environmentalism was shared by the Secretary of the Interior, Stewart Udall. The bill would have authorized the Secretary of the Interior to conduct a program of research, study and surveys of the natural environmental systems of the United States (U.S. Senate 1966). Nelson wanted to set aside representative natural environments on federal lands for scientific study, as well as assist in setting up similar reserves on state and private lands. The purpose was in part to provide information to natural resource managers. The plan was not to encroach on other federal agencies that were supporting ecological research (amounting to about $90 million per year), but those agencies were seen as being more mission-oriented rather than focused on basic research. Nelson hoped to zero in on the study of "basic processes," of the kind that would be needed to make policy decisions about resource use.

The bill was referred to the Senate's Committee on Interior and Insular Affairs, and Nelson presided over the hearings conducted in April 1966. Although the bill never got out of committee, the hearings, which involved testimonies of many leading ecologists, including Bormann, provide a snapshot of ecological and environmental thinking at this critical time of change (U.S. Congress 1966). Bormann's statement was the basis of the editorial he subsequently published in *BioScience*. Several leading ecologists from the Ecological Society of America (ESA) similarly supported the bill, including Bostwick Ketchum (President of ESA), Stanley Auerbach (Secretary of ESA), LaMont Cole (Chairman of ESA's Public Affairs Committee), and ESA's Committee on Applied Ecology. Roger Revelle, Chairman of the National Academy of Sciences Committee for the International Biological Program, argued that the bill's provisions would help to strengthen the activities being planned in connection with the IBP, due to start the following year. Secretary Udall, speaking on behalf of the Department of the Interior, supported the Bill strongly, arguing that such a program of ecological research would help to accomplish the goals that President Lyndon Johnson had outlined in a special message to Congress on "Conservation and Restoration of Natural Beauty," delivered on February 8, 1965 (Johnson 1965).

However, both the Department of Agriculture and the National Science Foundation opposed the bill, partly on the grounds that it would duplicate work already being done. In fact, a federal regulation had just been issued in March 1966 directing the Forest Service (part of the Department of Agriculture) to designate a series of "Research Natural Areas" (RNAs) especially in areas that had unique characteristics of scientific importance. These Research Natural Areas were to be maintained as much as possible in their pristine or unmodified condition and used for scientific research. (The Forest Service had established its first research natural area as early as 1927, the Santa Catalina RNA in the Coronado National Forest in Arizona.)

There might have been legitimate cause for thinking that Nelson's bill would duplicate these and other activities within the USDA, but the idea that it would also duplicate work by NSF was less obvious. Leland Haworth, a particle physicist and director of NSF, presented the agency's brief report. He argued that the bill's wording needed clarification about which activities were to be authorized, and that a section of the bill meant to authorize grants to universities and colleges for the training of ecologists was not necessary, because there were already sufficient training grants and fellowships available in the sciences. John Cantlon, an ecologist who was serving for 1 year as program director in NSF's Division of Environmental Biology, offered a personal statement at the Senate hearings, arguing that the Department of the Interior was in fact the logical agency to conduct these kinds of preservation and research activities, because they were precisely aligned with the mission of the Department. He did, however, add that research supported by the National Science Foundation in the past and future would also be indispensable over the long range.

Despite the bill's failure, the pressure to preserve ecological diversity for scientific purposes continued to intensify in the 1960s. Ecologists realized they needed to take a more active role in educating the public about the value of ecological science and bringing about the kinds of conservation policies that would be beneficial both to science and to American society. As Udall put it, scientists had new roles as "midwives of conservation" (Miller 1968). A symposium on "The Role of the Biologist in Preservation of the Biotic Environment", published in *BioScience* in May 1968, gave ecologists an opportunity to reflect more deeply on what had been accomplished and what remained to be done. Stanley Cain, an ecologist and Assistant Secretary of the Interior for Fish and Wildlife and Parks, viewed the need to preserve natural areas as a matter of national urgency (Cain 1968). He noted that the U.S. lagged behind many other countries in establishing natural areas for research. One of the writers was Orie Loucks, Wisconsin ecologist, who wrote about Wisconsin's innovative program of natural area preservation, which went back to the 1940s and which he hoped would be a model for other states (Loucks 1968). Wisconsin had been home to the renowned conservationist Aldo Leopold, who in his posthumous book *A Sand County Almanac*, published in 1949, articulated the need for a new "land ethic", or a statement of human responsibility to preserve the land's capacity for self-renewal (Leopold 1968). Loucks later became the Science Director at The Institute of Ecology and was involved in the planning process leading to the LTER program.

The mid-to-late-1960s, therefore, were years of increasing activism on the part of ecologists, as they realized they needed to emerge from the cocoon and guide the environmental movement and also take leadership in the resurgent preservation movement. One idea was to create a coordinating institution, a National Environmental Institute, to guide and inform the newly created Environmental Protection Agency (Bowers et al. 1971). The creation of The Institute of Ecology in 1971 was another expression of this activism, and TIE's study and recommendation in 1977 to create a network of experimental ecological reserves was the culmination of discussions that had been building for a decade. If we ask why many ecologists believed that the time was ripe in the late 1970s for a long-term ecological research program, part of the answer lies in understanding the origins and *raison d'être* of The Institute of Ecology, which was so heavily involved in the planning process.

The Institute of Ecology was formed to draw attention to the importance of ecological research and of the discipline of ecology to American society (Doherty and Cooper 1990). It was a response to the realization that with the rise of the environmental movement in the 1960s, the federal government was ignoring or overlooking the expertise of ecologists. Ecologists began to see that having a concrete plan for expanding resources for ecological science was important in representing the interests of the ecological profession to Congress. The Institute of Ecology originated in a strategic plan developed within the Ecological Society of America, but it was incorporated as an independent body in 1971. TIE was actually a consortium that involved many institutions, and it operated much like a think tank or general advisory body. But instead of having a small group of thinkers who operated from a central institute, on the model of other think-tanks, it was dispersed across a wide range of academic institutions and included people outside of academia. Until it ended in 1984, it served as a vehicle to foster cooperative relations between ecologists and people having other skills and perspectives (The Institute of Ecology 1977). Each of its projects included several institutions and disciplines, and any given project might involve 200 or more people. Working through numerous advisory and study groups, TIE obtained grants from agencies such as the NSF, which it used to sponsor symposia and workshops, and from these generated reports on a broad range of topics. In selecting TIE to prepare an advisory report to NSF on the future directions of ecological science in 1976, it was a foregone conclusion that TIE would strongly back the expansion of resources for the support of ecology, and its report of 1977 did exactly that.

Yet another source of support for the idea of creating and preserving natural reserves was the Atomic Energy Commission, especially through the work of the national laboratories established under its control after the Second World War. The national laboratories were surrounded by large reservations, some created during the Manhattan Project to act as buffer zones for the secret work being done on the atomic bomb. When the national laboratories were created, these buffers became ecological reserves, and in 1971 were officially designated as "National Environmental Research Parks" (NERPs). By the late 1970s (by which time the Atomic Energy Commission had been dissolved and replaced by the Department of Energy) several NERPs had been designated at Savannah River (South Carolina), Hanford (Washington), Idaho, and Los Alamos Scientific Laboratory (New Mexico). Others were under discussion at Oak Ridge National Laboratory and the Nevada Test site. These locations were highly disturbed sites, and one of the goals of ecological research was to measure the effects of disturbance on ecosystems (Hinds 1979).

Among these many initiatives, discussions, and calls for the preservation of reserves that were important for ecological research, the Forest Service stood out in the 1970s as it began to greatly expand its network of Research Natural Areas. In the next section we will consider in more detail the role of ecologists within the Forest

Service, focusing especially on the career of Jerry F. Franklin, who was on the planning committee for TIE's 1977 study and was a strong voice in the discussions about the need for long-term ecological research in the 1970s.

2.3 Expanding Ecology's Role

In 1968 Jerry Forest Franklin, then Research Forester with the Forest Service's Northwest Research Station in Portland, Oregon, explained the logic behind the Research Natural Areas in an article co-authored with James Trappe. They argued that the forestry profession should continue its past leadership role in building a system of natural areas (Franklin and Trappe 1968). While many research problems that they named were closely tied to specific practical issues, such as evaluating pollution, they also recommended research on more basic ecosystem processes and pointed out that natural areas could serve as gene reservoirs for species. They urged foresters not to think of natural areas as "museum pieces," but as "outdoor laboratories for applied and basic research" (Franklin and Trappe 1968, p. 461).

Although research natural areas were an old concept within the Forest Service dating to 1927, the designation of such areas had remained fairly flat through to the 1960s. That changed after the passage of the National Environmental Policy Act (NEPA), which went into effect in January 1970. NEPA contained a provision for the preservation of important historic, cultural, and natural aspects of the American heritage. As Franklin explained in 1972, the Federal government was involved in an "intensive program to greatly expand the number of Research Natural Areas in the next several years, in order to have at least a minimal system incorporating all major ecosystem types" (Franklin et al. 1972, p. 135). From 1968 to 1972 the number of natural preserves increased from 336 to over 500. Franklin, along with Robert E. Jenkins (ecology advisor to the Nature Conservancy, Arlington, Virginia) and Robert M. Romancier (also from the Pacific Northwest Research Station in Portland) appealed to ecologists to make more use of these Research Natural Areas, which, being permanently protected, were ideal for obtaining baseline information of natural systems and for doing integrated ecological studies (Franklin et al. 1972). Yet they reported that there were few natural areas being used for these kinds of integrated studies.

This period of rapid expansion of Research Natural Areas coincided with the expansion of ecosystem ecology and with the biome studies conducted in the U.S. in connection with the International Biological Program. Eugene P. Odum, along with his brother Howard T. Odum, had been working to make the ecosystem concept into a central organizing concept for ecology since the 1950s, and Eugene Odum became the chair of the Ecological Society of America's Committee on IBP. Some American biologists had reservations and even aggressively opposed IBP as it was first conceived. However, as Golley (1993) has discussed, by 1968 many were more strongly in favor of the program and eager to start the biome studies, which ran from 1968 to 1974. The IBP biome projects analyzed five biomes: grasslands, tundra, coniferous and deciduous forests, and deserts (a tropical forest study was planned but did not materialize). These studies helped promote ecosystem ecology because of their relatively large scale, their focus on the study of biogeochemical cycles, and their emphasis on ecological modeling.

As Michigan ecologist Frederick E. Smith explained in 1968, the term "ecosystem" was understood to mean "the total system of populations together with all nonliving components that interact in a defined region of space and time" (Smith 1968). Despite uncertainty about whether it was possible to analyze whole ecosystems and compare them meaningfully, Smith saw potential in what the IBP might bring to this area of ecology. He had just switched from an interest in population ecology to ecosystem ecology around the time of the IBP, and served as Director of the IBP Analysis of Ecosystems Program. Until about 1968, Smith explained, ecosystems had been "regarded primarily as backgrounds against which studies of components are executed" (Smith 1968, p. 7). As the IBP got underway, he noted that ecologists were feeling optimistic that it would be possible to analyze something as complex as an ecosystem, and that there was a growing conviction "that ecosystem analysis should be at the forefront of ecology, rather than serving as the background" (Smith 1968, p. 7).

Smith's discussion sheds light on how a contemporary ecologist saw in the IBP some hope for the future of his profession, a future that meant change both in intellectual directions and in the culture of science. In Smith's appeal to his colleagues to recognize and embrace the potential of the IBP, we can discern both a sense of frustration at the slow pace of ecology's development until that time, and a hopeful sense that this snail's pace of development was about to speed up.

Smith admitted that "despite some courageous attempts, we have no way to interrelate different aspects of ecosystems, such as energy flow, species diversity, and vegetational structure," and that "we have only the haziest of ideas on why some particular set of results is found." Those continuing uncertainties meant that "ecosystem analysis is a discipline that is just being born" (Smith 1968, p. 7). But in 1968 he looked forward to the way the biome projects of the IBP would stimulate new concepts and principles in ecosystem ecology that would, he hoped, add up to "a whole new level of ecology" (Smith 1968, p. 10).

It is noteworthy that Smith's vision and his hopes for ecology included a strong element of human ecology as a counterweight to ecologists' tendency to study natural or undisturbed systems. Because the IBP included a program focused on human adaptability (mostly aimed at the disciplines of anthropology and physiology), he expected that human ecology would also be incorporated into ecosystem-level analysis, and that ecologists might be nudged to adopt a point of view that accepted the idea that humans were part of ecology. He observed that a shift toward human problems was already underway, "perhaps in response to congressional interest in ecology," and commented that there was strong interest among graduate students in the applications of ecology and in "man as a part of ecology" (Smith 1968, p. 11). He also imagined that the IBP would help the discipline of ecology become more unified or integrated, but that change would require cultural shifts: "There is even hope that we can be induced to accept such radical concepts as team research and data sharing. If this is accomplished, the eventual effect of the IBP will be establishment of an integrated profession based on regional centers of study, dedicated both to the development of basic science and to its applications to human welfare" (Smith 1968, p. 11). Although the IBP projects did not all live up to the grand expectations that were envisioned, we wish to highlight how the IBP stimulated ecologists like Smith to see it as a vehicle to strengthen and advance their discipline. In many respects the LTER program, which entered its planning stages a decade later, was part of the same conversation about expanding the role of ecology.

The potential of the IBP biome projects for advancing ecology scientifically was exemplified in the Coniferous Forest Biome project. This was the last of the biome studies to be funded and almost lost out because of the inability of interested parties at the University of Washington and Oregon State University to reach agreement on a collaborative program. Ultimately, under threat that there would be no funding if an accommodation was not forthcoming, the principals agreed to a program divided between the two institutions. Dr. Stanley P. Gessell became the Director of the Coniferous Forest Biome study, Jerry Franklin his Deputy Director, and Richard Waring and Dale Cole the respective leaders of the Oregon and the Washington efforts. The University of Washington utilized several research sites near Seattle, Washington and the group at Oregon State University, which included both university and Forest Service scientists, focused primarily on the old-growth coniferous forest in an experimental watershed in the H. J. Andrews Experimental Forest.

The primary goal at H. J. Andrews was to build carbon, nutrient, and hydrologic models of a pristine old-growth Douglas fir-western hemlock forest ecosystem. The project was a collaboration of Forest Service scientists at the Corvallis, Oregon Forest Sciences Laboratory and academic scientists, primarily at Oregon State University. Generic objectives at both Biome locations were to study the characteristics of ecosystems, the processes causing the transfers of matter and energy within the systems, how the systems responded to natural and human-induced stresses, and to understand the land-water interactions characteristic of each biome. The results of these and earlier studies would be synthesized into predictive models that would aid in resource management in each biome (U.S. National Committee for the International Biological Program 1971).

Franklin viewed this kind of integrated biome study as urgently needed to respond to what he called the "demand explosion," i.e., the tremendous need policy makers had for information as resource management problems became more complex (Franklin 1972). The relatively straightforward questions traditionally asked by forest managers were being replaced by more complicated questions involving the broad ecological impacts of such activities as clear-cutting forests or applying herbicides. Effects of management on all aspects of ecosystems were being considered, and information was needed on how the different parts of ecosystems were linked, how the natural and social sciences were linked, and how quickly ecosystems recovered from disturbances. All of this was driven by requirements of federal environmental legislation passed during the 1960s and 1970s, including the Endangered Species Act and the National Forest Management

Act of 1976 (see below). These laws made it imperative that policies would have to develop and adopt ecosystem-level approaches to management of federal forest lands (Skillen 2015).

The "whole systems" or ecosystem viewpoint was not traditional in either resource management or ecological science, and ecologists were challenged to develop interdisciplinary programs that would provide the relevant science. The biome projects were an initial attempt to confront these big problems directly, each struggling to learn to deal with a complicated new worldview and its unprecedented demand for information. The Coniferous Forest Biome project, Franklin argued, was critically important to provide the information needed to design and forecast the effects of ecosystem-based management practices as well as other stresses on the composition, stability, and short and long-term productivity of the forest. "At least as important as the increased predictive capabilities," he added, "are the insights into processes and the identification of additional research needs such as in the areas of below-ground processes, decomposition and extrapolation of predictive capabilities in time and space" (Franklin and Waring 1974, p. 232).

The environmental legislation of the 1960s and 1970s profoundly affected federal land agencies, as discussed by Skillen (2015) in his historical analysis of federal ecosystem management. The Forest Service had faced intense criticism and litigation over its management practices, which favored timber production as the most important forest value. The merit of its multiple-use management approach was debated throughout the early 1970s and its practices found to be illegal under existing law, which led to the passage of the National Forest Management Act of 1976. This Act required the Forest Service to prepare comprehensive forest plans for all resources on each national forest every 15 years. The planning process had to follow the procedures laid out in the National Environmental Policy Act, which included public review of and comment on proposed plans. All of these required more scientific information, more attention to environmental concerns; the agency had to hire "ologists" of many different specialties, including ecologists to meet these requirements. The Act also required the Forest Service to protect habitat in order to preserve biological diversity, and this mandate, as Skillen argued, "forced the agency to look at national forests through an ecological lens, because it had to look at forests through the perceived needs of various species" (Skillen 2015, p. 78). The research conducted as part of the IBP Biome projects was exactly the sort that the Forest Service needed to develop its forest plans to achieve ecological goals, not just traditional timber production goals. But this expanded vision of the Forest Service also carried over into the discipline of ecology more broadly.

By advocating an expanded role for ecology, especially in relation to forest management, Franklin and his associates were also arguing for an expanded research capability, so that ecological advances could be made simultaneously on different fronts. The availability of biological reserves, such as the Research Natural Areas (which by this time involved all federal land management agencies, not just the Forest Service), were providing new opportunities for ecological research, with an emphasis on collecting baseline data from relatively pristine, or undisturbed, systems. Similarly, several of the Forest Service Experimental Forests and Ranges were proving critical for manipulative types of research activities. Conservation efforts were underway in various branches of government, in the private sector, and internationally in connection with the International Biological Program. The Biome studies were a beginning in the effort to develop a more useful and predictive ecology, in an emerging era where managers were required to understand and address management of federal lands as ecosystems.

Franklin was also involved in the follow-up to the IBP, UNESCO's Man and the Biosphere program, where he chaired the U.S. Committee on Project 8, which was about Biosphere Reserves. This international program was meant to safeguard genetic diversity of species as well as provide areas for ecological research, education, and training. The U.S. program focused mainly on preserving natural areas representing major biomes or biotic divisions and pairing them up with areas that could be experimentally manipulated, such as the experimental forests and ranges. The list of eligible sites included many of the locations that would later become part of the LTER network, chosen because they were already experimental areas with histories of ecological research and monitoring. The U.S. Committee created working groups to encourage collaborative programs linking reserves of different types, with the goal of stimulating research and monitoring programs (Franklin 1977). However, resources for such expansion were still sparse; these discussions reflected what scientists aspired toward rather than what was achieved.

All these developments were converging in the mid-1970s, setting the stage for the creation of the Long Term Ecological Research program. Jerry Franklin was a "rotator" or temporary program officer at NSF's Ecosystem Studies Program from 1973 to June 1975. Tom Callahan, who would later be the program officer for the LTER program, was his assistant. Franklin recalls the climate at NSF at that time[1]:

> I arrived at NSF in September 1973 to take a rotator position as Program Officer for the Ecosystem Studies Program and continued in that capacity until June 1975. My boss was Dr. Eloise "Betsy" Clark (who was then a divisional director within the Biological and Medical Sciences Division, and from 1976–1983 was assistant director of the new Directorate for Biological, Behavioral, and Social Sciences created after a major reorganization of NSF). Permanent NSF staffer Tom Callahan was my assistant. My initial duties included reassuring folks in the Biome programs that their funding was not going to terminate abruptly with the end of the International Biological Program. My understanding was that this was because the funding for IBP that Congress had added to the NSF budget was going to be rolled over to support an expanded Ecosystem Studies Program in FY 1974. Other immediate activities included working with Hubbard Brook folks on continuation of their funding and dealing with complex policy issues related to NSF funding research activities at Oak Ridge National Laboratory. The job often got interesting around and during panel reviews as evaluations of ecosystem research proposals were being done by the regular ecology panel, which was composed predominantly of traditional scientists, many of whom were very unenthusiastic about ecosystem science!

Franklin remembers many discussions with other program officers in the division concerning long-term ecological research:

[1] All quotations of Franklin, except where otherwise noted, are from an essay written for inclusion in this chapter.

> I remember many sessions Tom [Callahan] and I had over lunch with Bill Sievers, the program director for Biological Research Resources (BRR Program) in the nearby General Services Administration cafeteria. There were also larger, more formal brain-storming sessions occurring within the staff of the Division of Environmental Biology. There were many converging concerns in NSF with developing mechanisms to provide for longer-term support for ecological research. This was because many questions in ecological science – particularly ecosystem science -- needed to be approached using long-term experiments and installations at field locations. This research also typically needed to be conducted by interdisciplinary teams. It was clear to NSF staff that research of this sort simply was not possible using the typical 2–3-year NSF grants that provided support for individual academic scientists with one or two graduate students. There also had long been a concern over the need to sustain field stations and other important field research facilities.

The Hubbard Brook long-term ecosystem studies in New Hampshire had provided a model of interdisciplinary research and had strong policy implications, especially on the problem of acid rain in the U.S., one of the key findings of this study (see also Chap. 8, this volume). As Franklin notes, it was "an early and outstanding example of problem-relevant science." The IBP had also "generated strong field-based interdisciplinary programs and there was a desire to provide continuing support for meritorious groups that had emerged from that and other initiatives." Finally, as Franklin recalled, "there was in NSF at this time an increased concern over support for more problem-oriented ('applied') research, as well as broader pressures within the federal government for a significant expansion in ecological research and education," which the Council on Environmental Quality, an advisory group to the President, had recommended in 1974 (Council on Environmental Quality 1974). These discussions occurring within NSF prompted the exploratory initiatives that constituted the immediate planning process leading to the formation of the LTER program. Franklin had a key role in all these discussions. We turn in the next section to these various reports and workshops.

2.4 Planning for LTER: The Institute of Ecology's Report and NSF Workshops

The Institute of Ecology's planning committee for its project on "Experimental Ecological Reserves" was the first stage directly leading to the LTER program. Franklin participated in this committee. The key recommendation of the report, which distilled over a decade of thought within the scientific community of ecologists, was to establish a "network" of field research facilities, or experimental ecological reserves, for long-term research. The report provides insight into how the "network" was conceived by the scientific community, and the level of consultation that went into its recommendations. Discussion of the study involved more than 40 groups before it was submitted to the National Science Foundation, and more than 300 scientists contributed to the project. The planning committee of 11 scientists represented a variety of government, academic, and private groups, including members from the Forest Service, the Nature Conservancy, the Organization of Biological Field Stations, and several universities (The Institute of Ecology 1977).

In this report, we see the fuller expression of the rationale behind creation of a network of ecological experimental reserves, as articulated by the community of ecologists. There are three outstanding differences between this report and the later LTER program. One was a difference of scale: the original network was imagined to be far larger and more comprehensive than the LTER program, a vision that was unrealistic given the budget constraints. The LTER Program in contrast was not originally designed as or expected to be a network (as will be discussed later in this chapter as well as in Chap. 12).The second was the synergistic relationship that was being promoted between basic and applied ecology, or between basic research and resource management, which was originally envisioned by the TIE planning group as central, but which was not emphasized in the later NSF program announcement. The third was the management strategy or administrative structure that would maintain the network. The report recommended creating a consortium of Federal agencies, universities, and private institutions, which would "implement and maintain" the reserve system "in the nation's best interest" (The Institute of Ecology 1977, p. 28). The eventual LTER program would abandon this concept of creating and managing reserves altogether, and also the idea of a consortium, which was modeled on TIE's structure, but could not have been adapted to the research-oriented LTER program.

The goal of the scientists writing this report was to raise the level of long-term place-based research across the discipline of ecology and across the United States and its territories. They envisioned a very large network of 71 existing field sites in 67 locations across the U. S. (including Alaska, Puerto Rico, and the marine environments of the Virgin Islands). These were selected to ensure that a wide variety of ecosystems would be included. About three-quarters of those sites had some research facilities and ongoing programs already, but not all had facilities to support long-term experimental research, so capital improvements were included in the plan.

Funding of the network was considered to be relatively modest given the scale of the enterprise, but was still substantial. Estimates (based on 1975 dollars) were that it would require over $73 million to make the capital improvements that would make all sites sophisticated, year-round experimental reserves, with annual operating costs over $15 million at that level. The lowest estimate, which envisioned all sites having minimal facilities for long-term research, was $270,000 for capital improvements and just over $390,000 for annual operating costs. The highest estimate would have made this program into a Big Science project for ecology, although at about $300 million over a 15 year period, it was less expensive than other Big Science projects being contemplated around that time, such as the Human Genome Project (an international project on which discussion began in the mid-1980s and which cost $2.7 billion in 1991 dollars over a 13 year period). In TIE's report we see a pattern of creating very ambitious plans that were too expensive to be executed, and that were significantly scaled back when implemented. (Much later, a larger and more expensive program did ultimately emerge, the National Ecological Observatory Network.)

The report advocated adopting a systems-level approach, one that aimed at understanding how ecosystems functioned and how they were affected by human activities. "As ecology moves from a descriptive toward a predictive discipline," it noted, "scientists need access to sites at which to test their hypotheses" (The Institute of Ecology 1977, p. 6). Being "predictive" therefore meant performing experiments and testing hypotheses. Those tests would involve experimental manipulations, perhaps perturbing a site to assess its response. Long-term studies were essential because experiments might involve changing the site environment, and ecological effects might take time to emerge.

The report strongly emphasized creation of a "network" for various reasons, but these fell short of fully articulating the meaning of the term "network," or providing a worked-out vision of a new kind of ecological enterprise that was functioning as a network. One justification was to help provide resources to sites, while eliminating redundancy. Another was to create a database that was long-term and that could coordinate or combine the results of independent environmental monitoring projects. The report also envisioned more collaborative and integrated research efforts, including large-scale field experiments made possible by having an enhanced database. However, mindful that ecologists would worry about loss of autonomy through greater centralization, it also recommended that proposals would be selected through a competitive peer-review process, and that funding would come from additional sources, not from existing research funds. These arguments, which rely on wishful thinking rather than realistic assessments of what was possible, suggest that the participants in the planning process were still struggling to develop a new vision of ecological science, did not want to alienate people in the process of planning an expanded scale of research, and could not yet foresee the kind of cultural shift that would be needed to develop a network of sites.

One of the most interesting aspects of this report was the way it advocated a network of ecological reserves specifically in order to link basic and applied ecology. As the report noted, "a new interface is emerging as the need for similar data at the ecosystem level brings basic and applied ecologists into cooperative interdisciplinary studies, which result in both the development of new scientific knowledge and improvement of resource management" (The Institute of Ecology 1977, p. 9). The value of this kind of extensive research network was related to the needs of policy makers, who were tasked with evaluating the impacts of new technologies, new products, and new management strategies on the nation's ecosystems.

The type of work envisioned for this network of sites included very specific practical questions. For example, experiments could be designed to assess the impact of toxic substances on the environment, as a supplement to laboratory tests of toxicity, and therefore guide regulatory agencies who had to set standards for the use of such substances. Another example was to assess the impact of coal and oil extraction on different ecosystems. The reason that a "network" of many sites was needed was because it was important to be able to assess environmental impacts across a range of ecosystem types. The report made a clear link between basic research and resource management or environmental assessment needs, and anticipated that one outcome would be better communication between scientists and resource managers and users.

The management or governance structure of this large-scale network was envisioned quite differently from the eventual LTER program. Since Eloise Clark, the Assistant Director of NSF's Biological, Behavioral, and Social Science Directorate, had not envisioned LTER to be a network or managed by NSF as a network, the LTER Program started with an informal governance structure, although a more formal one evolved later as part of the strategic planning process in the early 2000s. NSF always retained the power to select or to withdraw sites, and it determined the core research areas that all sites had to embrace. In contrast, TIE's report recommended the creation of a new consortium that included Federal agencies, universities, and private institutions, to coordinate research within the network. The rationale was that these institutions held the land containing the sites. But non-land-holding agencies such as the National Science Foundation and the Council on Environmental Quality (an advisory body to the U.S. President) were included, and it was thought that representatives of state and local governments, state and private universities, and private owners of ecological reserves would also be involved.

This vision of governance was extraordinarily inclusive, although it was potentially unwieldy for the same reason, and the report did not appear to recognize that a structure like this might militate against the pursuit of hypothesis-driven science. The consortium concept was similar to that adopted by TIE itself, and seemed intended to provide for a broad or democratic form of governance, one that allowed for input from any agency with a stake in ecological research at these sites, and a stake in how that ecological research would be applied to solve environmental problems. The details of how this consortium would operate were not spelled out, but the proposal included a core staff of people who would handle communications, travel, and coordination of the network.

Given that this was a Big Ecology proposal in terms of its budget, and that with so many sites identified it was also aiming to raise the level of ecosystem ecology across the board, it would have required substantial buy-in from the entire ecological community to pursue this agenda, and substantial lobbying efforts to ensure that funding for this kind of enterprise would flow from different agencies and institutions. But TIE's report was prepared for NSF's Division of Environmental Biology, which could not undertake such a broad and expensive project. Instead the Division responded by encouraging continued discussion, in the form of three workshops held in 1977, 1978, and 1979, which allowed ecologists to present the case for long-term research, and to identify some of the sites where such research would be feasible.

The first workshop did not try to duplicate the full ambitions of TIE's report, but instead adopted a more modest program based on measurement and monitoring. The workshop was chaired by Daniel B. Botkin, who was then at the Ecosystem Center of the Marine Biological Laboratory at Woods Hole, Massachusetts. It was entitled "Long-Term Ecological Measurements," and stressed the need for *measurement* instead of ecological *experiments*, which had been included as one facet of TIE's report (National Science Foundation 1977). NSF was to provide support through its Biological Research Resources Program, which had been established in 1973 and was mainly funding systematic biology, that is, the work of museums and botanical gardens (Appel 2000).

Experimental work was not excluded, but the report of the workshop placed the goal of long-term measurement front and center, and this emphasis made the proposal look like a monitoring program and little more. The repetition of the words "measurement" and "monitoring" in this workshop's report is striking. As Franklin noted, "A comprehensive list of measurements important for terrestrial, freshwater, and marine ecosystems was provided in the report, which was viewed by some critics as a typical ecologist's laundry list of data sets that they would find interesting."

The second workshop in 1978, also chaired by Botkin and convened at Woods Hole, was called "Pilot Program for Long-Term Observation and Study of Ecosystems in the United States," which again outlined "monitoring strategies" across a range of ecosystems (National Science Foundation 1978). One central goal was to distinguish between cyclical changes and unidirectional changes and to distinguish human-caused changes from natural ones. This report also presented a plan for organizing, developing and administering a long-term program, and recommended specific study sites. A minimum set of monitoring sites was proposed that would provide "a representative cross-section of the major ecosystems in the United States" with NSF funding through the Biological Research Resources Program.

NSF was not keen to support "monitoring", however, and the first two workshops did not gain traction because they did not appear to promise enough research. The third workshop, convened at the headquarters of The Institute of Ecology in Indianapolis, Indiana, in 1979, and chaired by Orie Loucks, Science Director for TIE, adopted a different approach. It was titled "Long-term Ecological Research: Concept Statement and Measurement Needs," and it opened with the statement that "the proposed program would have to answer significant ecological research questions if it were to be considered for support by the National Science Foundation" (National Science Foundation 1979).

The language of this report returned to the more forceful research-oriented language of the initial TIE report. It argued that "a strong program of long-term ecological research must encompass individuals, who provide creativity, and institutions, which maintain longevity and continuity." The database generated by such a program would be used to "generate new hypotheses in the future and to formulate and test hypotheses from long-term measurements in the present." The proposal advocated ensuring that these criteria would be met by proposing a "core research program" at a network of sites, as well as an "investigator-specific program" which could be done either at the primary sites or at other locations outside the core network. Data generated would be used to answer local research questions or to make broad comparisons across sites. In this report, scientific creativity, hypothesis testing, and the importance of placing ecological research in temporal and spatial context were emphasized. This vision found favor at NSF, and the LTER program that was subsequently approved adopted much of the same rhetoric and the same goals that the third workshop outlined.

In placing ecological research squarely at the center of the proposal, this workshop sought a balance between common or "network" activities and individual curiosity-driven research. There would be a common core of research areas pursued at all sites, an idea that appeared for the first time in this workshop report, but individuals were given freedom to pursue other questions related to their study sites. The report referred to the common core as "research questions," but they were really not specific questions, rather they were broad areas of investigation, and there was no requirement that LTER would provide comparable measurements of such phenomena of any kind. These dealt with five topics:

1. Dynamic patterns and control of primary production over time, and in relation to natural and induced stresses or disturbances;
2. Dynamics of selected populations of seed plants, saprophytic organisms, invertebrates, fish, birds, and mammals in relation to time as well as natural and induced stresses or disturbances;
3. Patterns and control of organic accumulation (biomass) in surface layers and substrate (or sediment), in relation to time or natural and induced stresses or disturbances;
4. Patterns of inorganic contributions (atmospheric or hydrological) and movement through soils, groundwater, streams and lakes in relation to time and natural or induced stresses or disturbances;
5. Patterns and frequency of apparent site interventions (disturbances) over space and time (drought, fire, windthrow, insects or other perturbations) that may be a product of, or induce, long-term trends.

In addition to these core areas, the report recognized that investigators would also wish to pursue long-term studies of other problems. Some would be site-specific and others might involve multiple sites. The report noted that the allowance for individual and exploratory research, over and above the five core areas, might in time prove to be significant nationally and therefore might even become part of the core research program in the future.

Recommendations for interagency cooperation, coordination and management of the program, and criteria for grant determination and funding periods also were part of the third workshop report, as was a second major section on the measurements recommended for core research at LTER sites. The report recommended that a network coordination office be created, and that overall policy and review would reside in an "LTER Council", which was envisioned as including members from land ownership and management agencies (such as the Forest Service or National Park Service), the Department of Energy, universities, foundations, and other agencies with research interests (such as the NSF). This idea harkened back to the inclusive approach of TIE's report, but was not adopted by the LTER program. The report also emphasized that the data generated by such a program could be used in many different ways, ranging from state and federal resource management and regulatory agencies, the academic research community, and private industry. In this way the report partly captured some of the breadth of vision and the inclusion of applied ecology in the TIE report, in which Loucks had also been involved.

These wide-ranging discussions form the backdrop to the creation of the LTER program. Franklin recalled that there was much internal debate at NSF before it decided to proceed, although he was not privy to these discussions. He did have a conversation in August 1979 with Frank Golley, ecosystem ecologist at the University of Georgia, who was about to become director of NSF's Division of Environmental Biology. Golley asked for his perspective on the LTER proposal because he was doubtful about its merit and unsure whether he could support it. Franklin was given to understand that the request for proposals that NSF finally issued for the program was approved by Eloise Clark (Assistant Director of NSF's Directorate of Biological, Behavioral, and Social Sciences), on the grounds that projects were to be selected upon their individual scientific merits and not as elements in a network. This concern about the intellectual merit of the proposals may have been in Golley's mind as well.

In 1979 Clark approved the release of a Request for Proposals for the initial set of sites, with the proviso that they would be selected on their individual scientific merit. This stipulation was followed in the selection of the sites, although not all staff in NSF's directorate agreed with it. The new program absorbed many of the ideas of the third workshop, including the requirement that sites address a set of five core areas. Unlike the vision set out in the workshop, however, the five core areas did not change in the entire history of the LTER program, despite later efforts from the community to introduce other themes. Nor did it require even a minimal set of comparable measurements that were to be made at all sites, as had been proposed in the TIE report. Otherwise, the vision Loucks and his colleagues presented in the third workshop report became the LTER program.

Unlike the large scope of the project originally imagined in TIE's report, LTER, when created in 1980, was not designed to raise the level of the whole of ecology, but was a smaller scale pilot program that began with six sites in 1980, to which five more were added in 1981 (Callahan 1984). The maximum funding for each site was set at $300,000 per year. Four of the original sites were connected to the biome projects of IBP, including to Franklin's Western Coniferous Forest Biome study.

A Coordinating Committee was created shortly after the program started. It was originally called a "steering" committee. The initial plan was that the chairmanship of this group would rotate yearly among the Principal Investigators (PIs) of the LTER sites. Dr. Richard Marzolf (PI of the Konza Prairie LTER in Kansas) was the chair of the committee for 1981 and was awarded a small coordination grant by NSF. Dr. Richard Waring (from the H. J. Andrews Experimental Forest LTER in Oregon) was scheduled to be the chair the next year, but he asked Franklin to do it in his place and Franklin agreed to do so. At the 1982 meeting at North Inlet, South Carolina, a decision was made on the need for a network office and that this could not very well move each year. Consequently, Franklin was asked to continue to chair the LTER Coordinating Committee for an indeterminate period, to which he agreed. He continued as chair through 1995, which was a period during which the network office and coordination activities made immense growth. At the time of his resignation, the LTER program had grown to 18 sites.

In the next section, Jerry Franklin offers his recollections of the early years of the LTER program, during the time that he chaired the Coordinating Committee. His comments also address the relationship between the individual sites and NSF, and the problem of fostering cross-site comparative research, or of behaving like a "network" as opposed to a collection of independent sites.

2.5 To Be or Not to Be a Network

Jerry Franklin recounts here his experiences and perceptions of the LTER program during his tenure as chair of the Coordinating Committee. His account starts with the Denver 1986 meeting of the Coordinating Committee. John Brooks, Director of the Division of Environmental Biology at the National Science Foundation, took the unusual step of attending the meeting and informing the participants that they had to pay more attention to cross-site comparative work.[2] He was concerned that LTER did not have specific products or results to showcase at a time when there was growing interest in long-term ecological research in Britain and Europe. Since the program was approaching a 10-year review, which was completed in 1993, Brooks hoped that the program would be able to demonstrate unique results that could not have been obtained in another way. In his view, the program needed to incorporate greater time spans, be geared up to include larger spatial dimensions, and develop an approach to comparative ecosystem analysis. As Franklin recollects, Brooks made explicit that if LTER did not begin to function as a network, it would not have a future! His remarks surprised and were taken seriously by all present.

2.5.1 Jerry Franklin on the Early Years of the LTER Program

The LTER sites viewed themselves as independent but with some common interests for the first several years of the program. The Coordinating Committee did agree, as a collective, to adopt the principle that the LTER sites would not compete against each other for advantage in funding. If NSF wanted to provide additional funding to supplement work at LTER sites, it would have to be at all sites. As new LTERs were added in the second and third Requests for Proposals they were appraised of this understanding. Efforts at integration or cross-site coordination were minimal, however, until John Brooks appeared at the Coordinating Committee

[2] Minutes of the LTER Coordinating Committee Meeting, November 8–9, 1986, Denver, Colorado. In the Center for Limnology Document Archive, Steenbock Library, University of Wisconsin Archives and Record Management, Madison, Wisconsin.

meeting in Denver in 1986 and directed, "that we would be a network or LTER would not continue!"

As a group the PIs of the LTER sites were not happy with the NSF direction that we were to become and function as a network. Gradually the sites began to reluctantly step up – on the basis that NSF was going to have to pay for any standardization and additional work, such as in cross-site analysis. However, a breakthrough occurred at the LTER Coordinating Committee Meeting at Kellogg Biological Station in November 1988. With encouragement from many data management personnel, representatives of all the LTERs endorsed adoption of common network hardware and software, which was called the minimum standard installation (MSI), and asked NSF to provide supplemental funding to each LTER for its acquisition. NSF readily agreed to do so.

Major "micromanagement" of the LTER network and supplemental funding, including of the LTER Network Office, really began at this point. Some NSF staff have asserted that they did not conduct such management direction. The majority of the initiatives undertaken by the LTER Network were, in fact, the result of direct requests by NSF. My job, as network office director and chair of the Coordinating Committee, was often functioning as the "rubber bumper" between NSF and the network. An NSF program manager would call me up and say, "Well, we think it is time for there to be an LTER newsletter" (or a book series or a new round of cross-site comparisons or a long-term plan for the network). In most cases I would communicate this to the members of the coordinating committee (a committee composed of one representative from each of the LTERs) to inform them and then develop a supplemental request for additional funding to pay for the proposed activity. In one case, the NSF Deputy Director Mary Clutter contacted me directly with her proposal that we establish an International LTER Committee, which we did.

Rarely was said agreement to such additional activities immediately forthcoming from the LTER sites, particularly when it might involve commitment of their time or funds. Chairing LTER Coordinating Committee meetings was always a personal challenge, particularly because the LTERs were generally led by well-known scientists who had their own strong opinions about things. Several of the LTER PIs clearly were not at all interested in being part of a network; they had their grant and they wanted to get on with their activities without the time and expense of having to be part of a network. There were several individuals with particularly strong egos and voices that generally could be expected to be naysayers regarding new initiatives and I had to learn how to contain them and insure that some of the LTERs represented by individuals with quieter voices also were heard from during discussions and the ultimate votes.

The idea for the All-Scientists meetings did emerge from the LTER group itself. Support for it was a roller-coaster with several ups-and-downs within the NSF hierarchy. I was one of the individuals that conceived and strongly supported it, believing that such meetings were critical if we were really going to have collaboration among the sites. It represented a major expense, however, and at least twice decisions were made in NSF that there wouldn't be any more All-Scientists meetings. These decisions were obviously later reversed.

My participation abruptly ended in 1995. We were preparing a grant proposal for a large new multi-year grant to the coordinating office. After extensive discussions with NSF staff, we had prepared and submitted a final proposal for continued funding of the LTER Network office in Seattle. Late on a Friday afternoon I received a call from Mike Allen, who was the responsible program officer at that time, saying that NSF was not going to fund this proposal. Dr. Mary Clutter, the Assistant Director in the Directorate for Biological Sciences, had decided that the next network office proposal needed to be competed. I burned all remaining bridges with Mary Clutter and NSF in challenging their decision, which would have resulted in an unacceptable gap in funding for network activities had it been implemented as NSF originally proposed. Sufficient interim funds were provided to bridge the time between ending of the current network office grant and awarding of the competed grant at the new location for the LTER office in Albuquerque. Dr. Clutter never forgave me for the Cain that the Vice Provost for Research at the University of Washington and I raised over their abrupt decision to decline the proposal which we had prepared – and following their direction!

2.5.2 *Changing of the Guard and the Next 20 Years*

James Gosz of the University of New Mexico succeeded Franklin as chair of the coordinating committee, while Franklin continued as director of the Network Office in Seattle. NSF announced an open competition for the LTER Network Office in 1995, and the University of Washington and the University of New Mexico submitted a joint proposal with Gosz and Franklin as lead investigators. The University of California-San Diego and the Santa Fe Institute were collaborators on the proposal. Although the open competition was designed to attract a broad range of applicants to host the Network Office, only two proposals were received. NSF decided to accept the proposal from New Mexico, but again declined to fund the Washington component. As a result, Jim Gosz led the transition of the Network Office from Seattle to Albuquerque in March 1997.

The second seismic shift in the LTER program occurred when Scott Collins replaced Tom Callahan as LTER program officer. The cause of this change is not known with certainty. Callahan himself, as David Coleman recalled, offered no explanation beyond the comment "I was a bad boy," but Coleman added that the shift might have reflected Mary Clutter's belief that permanent staff positions should rotate every decade or so (Coleman 2010, p. 103). Although Collins did an excellent job in the position, many people were disappointed that Callahan, who had shepherded the program through its formative period, was removed unceremoniously. Collins continued in the position until 2000, when he became program officer for the National Ecological Observatory Network (NEON). Henry Gholz then assumed responsibilities for the LTER program.

With the addition of two urban sites in 1997, the LTER Network comprised 20 sites when the Network Office moved to Albuquerque. Two of the original sites

(Coweeta and North Temperate Lakes) received augmentations to their budgets for regional-level and interdisciplinary research. The International LTER Network was growing, and Gosz was elected chair of its Network Committee. NSF initiated a competition for new land-margin LTER sites. LTER scientists increased their focus on cross-site research through two funding programs. NSF held a 1995 special competition that supported 13 awards for cross-site comparisons and synthesis at LTER and non-LTER sites. In the same year, seven LTER scientists received funding from the NSF/DOE/NASA/USDA Joint Program, Terrestrial Ecology and Global Change (TECO). The transition of the Network Office from Seattle thus took place in a very positive environment for the LTER program.

The award establishing the LTER Network Office (LNO) in Albuquerque had two important differences from previous awards for the Seattle Office. The new award established the position of Executive Director, whose function was to coordinate the Network Program, oversee the awards and supplements that supported LTER Network activities, and guide the Network Office activities and staff. This new position allowed the Chair of the Coordinating Committee to focus on long-term planning and development of site and cross-site research, education, and collaborations. Because the Coordinating Committee had instituted a policy of periodically rotating the Chair, assigning responsibility for the LNO to an Executive Director increased the stability of the Network Office and the continuity of its programs.

The LNO was supported by NSF through a Cooperative Agreement rather than a research grant, as had been the case in Seattle. Cooperative Agreements are designed to ensure coordination between the Foundation and the awardee and involve frequent communication with and oversight of the awardee. Theoretically, activities and funding under a Cooperative Agreement are re-negotiated annually. The use of this kind of award instrument provided NSF with the capacity to influence the direction of activities in the LTER Network. This hands-on approach resulted in confusion regarding the reporting structure for the LNO since up to then the Coordinating Committee had guided its activities. Under the new award structure, the LNO was viewed by NSF as being somewhat independent from the LTER Network. For example, NSF required a strategic plan for the LNO that was separate from the planning activities undertaken by the Network. As mentioned above, the LNO was set up as a kind of buffer between NSF and the LTER Network, which was a continuing source of tension on both sides.

Bob Waide was hired as Executive Director after a national search for the position. Waide was one of the founding principal investigators of the Luquillo LTER site which he led before moving to the LNO in October 1997. John Vande Castle, Associate Director for Technology, had moved from the Seattle office to Albuquerque, and provided continuity in the transition as well as critical institutional memory. James Brunt, Associate Director for Information Management, had been the data manager for the Sevilleta LTER site and was a trusted member of the LTER information management community. These three individuals continued in their positions until the LNO was moved to Santa Barbara in 2015. The staff of the LNO, which at one point reached 18 individuals, had a broad range of expertise directed at three major activities: development of the Network Information System, coordination and support of Network research programs, and communication and outreach. The staff was augmented by two NSF program officers who were temporarily assigned to the LNO. Christine French, from the Office of International Programs, was posted to the LNO from 1997–2000 and played an important role in the development of the International LTER Network. Sonia Ortega, who managed multiple education programs for NSF, joined the LNO from 2001–2004 and worked with the LTER Education Committee on strategic planning.

In 2000, the LTER Network defined the central, organizing intellectual aim of the LTER program as an effort "to understand long-term patterns and processes of ecological systems at multiple spatial scales."[3] Over the following 20 years, the LTER Network endeavored to achieve that aim through six interrelated themes. We summarize the results for each of these themes, noting chapters in this book that bear on each theme. For a more detailed description of the accomplishments of the LTER Network over the last decade, consult the Long Term Ecological Research Network Decadal Review Self Study.[4]

Understanding: Gaining Ecological Understanding The foundation of the LTER Network is place-based, long-term research conducted at each LTER site. Although sites share common research themes, each site creates a research program that addresses their particular environment. Overall, LTER sites have been extremely effective at developing in-depth understanding of a diverse range of ecosystems and extending that understanding to broader geographic scales. Rates of scientific publication have increased fourfold since the beginning of the LTER program, and LTER sites are considered as unique resources for scientists from many disciplines. Three examples of how LTER sites gain ecological understanding are presented in Chaps. 5, 6 and 7.

Synthesis: Create General Ecological Knowledge Comparative research took hold early in the development of the LTER program. The breadth and depth of these comparative studies increased over time until LTER became a closely integrated research network (Chap. 12). As a network, LTER developed a conceptual model integrating social and ecological sciences that has been adopted by many scientists outside the LTER community (Chap. 14). LTER science has contributed to the advancement of our understanding of many fundamental ecological principles.

Provide a Legacy for Future Scientists The LTER Network has provided two kinds of legacies for the benefit of future scientists. The first is the existence of well-designed and documented long-term observations, experiments, and samples that provide a baseline for detection of future change. Many of these observations and

[3] LTER 2000–2010: A Decade of Synthesis. https://lternet.edu/wp-content/uploads/2010/12/lter_2010.pdf. Accessed 31 May 2020.

[4] Long Term Ecological Research Network Decadal Review Self Study. https://lternet.edu/wp-content/uploads/2019/10/LTER_Self_Study_2019-10-04.pdf. Accessed 27 July 2020.

experiments are designed to continue under the leadership of generations of future scientists.

Information: Create Accessible Databases The second legacy is the archive of well-documented data from LTER sites representing many of the biomes of the United States (Chap. 13). In addition to the data itself, the LTER Network created a language for documenting ecological metadata that has been adopted by research networks around the world. The LTER Network Information System is the forerunner of the Environmental Data Initiative, which allows the broader scientific community to benefit from LTER's four decades of experience in managing ecological data. The influence of LTER information management has been felt in other ways. For example, William Michener of the LNO was the first Principal Investigator and Director of DataONE, a national facility that provides access to Earth and environmental data across multiple member repositories.

Education: Promote Training, Teaching and Learning Because most LTER sites are located at or managed by academic institutions, they provide excellent opportunities for undergraduate and graduate study and research. The LTER Network has also been very successful in attracting students from outside the host institutions to work at LTER sites. Many of these students have continued to work at LTER sites as technicians or researchers once they have received their degrees, and LTER leadership is sprinkled with scientists who began their academic careers at LTER sites. Equally impressive is the development of a Network-wide Schoolyard LTER program serving K-12 students. Despite minimal funding, Schoolyard LTER has become one of the best-known accomplishments of LTER.

Outreach: Provide Knowledge to Address Complex Environmental Challenges The dissemination and application of knowledge based on LTER research is a key goal of the program. Sites in the LTER Network, especially those sites associated with Federal agencies like the Forest Service, are well positioned to inform resource managers and policy makers and have had many notable successes with such collaborations (Chaps. 8 and 10). Informing the general public is a challenging task that LTER sites have approached in a variety of innovative ways, including framing the scientific message in ways that are more friendly to the user (Chap. 9) and utilizing non-traditional modes of communication (Chap. 11). Dissemination of information to the broader ecological community is done traditionally through publications (approaching 20,000 for the Network as a whole) and presentations at scientific meetings. The influence of the LTER community is felt by ecologists in other ways. For example, the valuable experience LTER has accumulated in managing ecological data is shared through training, publications, and access to the Environmental Data Initiative services. LTER scientists have often played important roles in developing other national (e.g., Critical Zone Observatories) or international networks (e.g., Global Lake Ecological Observatory Network, International LTER Network: Chaps. 15 and 16). The National Ecological Observatory Network (NEON) is a continental-scale observation facility that was designed to collect long-term ecological data. LTER scientists provided expertise in designing NEON, and two LTER scientists (Bruce Hayden of the Virginia Coast Reserve and William Michener of the LNO) led the first effort to formalize that design. These are just a few of the ways that the LTER Network has contributed to better understanding and management of key national ecosystems.

2.6 Conclusion

The LTER program both reflected and responded to ecologists' ideas, especially those articulated in the third NSF-funded workshop, but it also differed in crucial ways from the ecological community's original ambitions, as presented in TIE's report. That vision could not be realized in the funding climate of that time because of its scope and cost, but it was also in crucial ways different from LTER in conception. It was broader, more inclusive, and had a clear emphasis on applied problems. As an experiment and a pilot project, LTER was far less ambitious in its scope and budget. LTER did not adopt the somewhat unwieldy idea of a multi-institutional consortium to manage the network and did not place the goal of linking basic and applied science at its center in the way both the TIE report and the third workshop report sought to do. It did include the idea of partnering with other agencies (such as the Forest Service) and it adopted many of the recommendations of the third workshop, including the emphasis on research and the identification of five core research areas that would be common to all sites. LTER also did not adopt even a minimal set of common measurements, as the TIE report had proposed; had it done so, developing inter-site comparisons probably would have been much easier.

But even here there was an interesting difference between what NSF did, and what the ecological community thought would occur. The LTER program that was created was smaller and perhaps also less responsive to community input than the planning documents envisioned. But LTER did evolve over four decades, and in some respects it evolved more in the direction of the original vision proposed in TIE's report of 1977, but not without experiencing growing pains. This evolutionary process will be examined further in subsequent chapters. One of the central themes in the evolution of the LTER program, one that emerged especially strongly in the mid-1980s, was whether this group of independent sites could evolve into a network by promoting cross-site research and collaborations. It is clear from Franklin's comments in this chapter that this pressure to become a network largely came from NSF, and represented a shift toward greater control of the LTER program by NSF. But Johns Brooks' presentation at the 1986 meeting in Denver suggested that NSF was trying to protect the program by making demands for more cross-site research, as well as for research at expanded temporal and spatial scales; it probably also represented his original vision for LTER, which differed from his boss at the time, Eloise Clark.

The solution to this problem of how to function like a network had to be devised by the community of researchers at the sites, with help from the LTER Network

Office. The problem of how to behave as a "network", although implied in the early discussions of LTER, had not been thought through in a deliberate or systematic way from the start. Several of the chapters in this book – notably those by Magnuson and Waide (Chap. 12), Stafford (Chap. 13), Collins (Chap. 14), and Zimmerman and Groffman (Chap. 15) – explicitly address the question of what it meant to operate as a "network", and related to that, what new ideas, resources, technological developments, cross-disciplinary connections, governance structures, and cultural changes were needed to support this new notion of "network science."

References

Appel, Toby A. 2000. *Shaping biology: The National Science Foundation and American biological research, 1945–1975*. Baltimore: Johns Hopkins University Press.

Bormann, F.H. 1966. The need for a federal system of natural areas for scientific research. *Bioscience* 16 (9): 585–586.

———. 1996. Ecology: A personal history. *Annual Review of Energy and Environment* 21: 1–29.

Bormann, F.H., P.R. Shafer, and D. Mulcahy. 1958. Fallout on the vegetation of New England during the 1957 atomic bomb test series. *Ecology* 39 (2): 376–378.

Bowers, Raymond, Paul Hohenberg, Gene Likens, Walter Lynn, Dorothy Nelkin, and Mark Nelkin. 1971. A program to coordinate environmental research: What kind of organization can improve the level of research on the environment? *American Scientist* 59 (2): 183–187.

Brower, David, ed. 1959. *The meaning of wilderness to science: Proceedings of the 6th wilderness conference*. San Francisco: Sierra Club.

Cain, Stanley A. 1968. Natural area preservation: National urgency. *Bioscience* 18 (5): 399–401.

Callahan, J.T. 1984. Long-term ecological research. *Bioscience* 34 (6): 363–367.

Coleman, David C. 2010. *Big ecology: The emergence of ecosystem science*. Berkeley, California: University of California Press.

Congress, U.S. 1966. *Ecological research and surveys: Hearings before the United States senate committee on interior and insular affairs, 89th Congress, second session, April 27*. Washington, DC: U.S. Government Printing Office.

Council on Environmental Quality. 1974. *The role of ecology in the federal government. Report of the committee on ecological research*. Washington, DC: National Science Foundation.

Doherty, Josephine, and Arthur Cooper. 1990. The short life and early death of the Institute of Ecology: A case study in institution building. *Bulletin of the Ecological Society of America* 71 (1): 6–17.

Douglass, J.E., and M.D. Hoover. 1988. History of Coweeta. In *Forest hydrology and ecology at Coweeta*, ed. W.T. Swank and D.A. Crossley, 17–31. New York: Springer.

Franklin, J.F. 1972. Why a coniferous forest biome? In *Research on coniferous forest ecosystems: First year progress in the coniferous forest biome, US/IBP*, ed. J.F. Franklin, L.J. Dempster, and R.H. Waring, 3–5. Portland: Forest Service, U.S. Department of Agriculture.

———. 1977. The biosphere research program in the United States. *Science* 195 (4275): 262–267.

Franklin, J.F., and J.M. Trappe. 1968. Natural areas: Needs, concepts, and criteria. *Journal of Forestry* 66: 456–461.

Franklin, J.F., and R.H. Waring. 1974. Predicting short- and long-term changes in the function and structure of temperate forest ecosystems. In *Proceedings of first international Congress on ecological structure, functioning and management of ecosystems, the Hague, Netherlands*, 228–232. Wageningen: Centre for Agricultural Publishing and Documentation.

Franklin, J.F., Robert E. Jenkins, and Robert M. Romancier. 1972. Research natural areas: Contributors to environmental quality programs. *Journal of Environmental Quality* 1 (2): 133–139.

Golley, Frank B. 1993. *A history of the ecosystem concept in ecology: More than the sum of the parts*. New Haven/London: Yale University Press.

Hinds, W.T. 1979. The cesspool hypothesis versus natural areas for research in the United States. *Environmental Conservation* 6 (1): 13–20.

Johnson, Lyndon B. 1965. Special message to the Congress on conservation and restoration of natural beauty, February 8, 1965. Online in *The American Presidency project*: http://www.presidency.ucsb.edu/ws/?pid=27285. Accessed 28 Oct 2018.

Johnson, Christopher, and D. Govatski. 2013. *Forests for the people: The story of America's eastern national forests*. Washington, DC: Island Press.

Leopold, Aldo. 1968. *A Sand County almanac, and sketches here and there*. New York: Oxford University Press.

Likens, Gene E., and S.W. Bailey. 2014. The discovery of acid rain at the Hubbard brook experimental Forest: A story of collaboration and long-term research. In *USDA forest service experimental forests and ranges: Research for the long term*, ed. D.C. Hayes, S.L. Stout, R.H. Crawford, and A.P. Hoover, 463–482. New York: Springer.

Likens, Gene E., and F. Herbert Bormann. 1974. Acid rain: A serious regional environmental problem. *Science* 184 (4142): 1176–1179.

Loucks, Orie L. 1968. Scientific areas in Wisconsin: Fifteen years in review. *Bioscience* 18 (5): 396–398.

Miller, Richard S. 1968. Responsibility of environmental scientists and their professional societies. *Bioscience* 18 (5): 384–387.

National Science Foundation. 1977. *Long-term ecological measurements: Report of a conference*. Available from the National Science Foundation LTER Network website, LTER History Timeline at https://lternet.edu/network-organization/lter-a-history/. Accessed 28 Oct 2018.

———. 1978. *Pilot program for long-term observation and study of ecosystems in the United States*. Available from the National Science Foundation LTER Network website, LTER History timeline at https://lternet.edu/network-organization/lter-a-history/. Accessed 28 Oct 2018.

——— 1979. *Long-term ecological research: Concept statement and measurement needs*. Available from the National Science Foundation LTER Network website, LTER History timeline at https://lternet.edu/network-organization/lter-a-history/, Accessed 28 Oct 2018.

Odum, Eugene P. 1988. Preface. In *Forest hydrology and ecology at Coweeta*, ed. W.T. Swank and D.A. Crossley, vii–viii. New York: Springer.

Rumore, Gina. 2012. Preservation for science: The Ecological Society of America and the campaign for preservation of Glacier Bay National Monument. *Journal of the History of Biology* 45 (4): 615–650.

Skillen, James R. 2015. *Federal ecosystem management: Its rise, fall, and afterlife*. Lawrence: University Press of Kansas.

Smith, Frederick E. 1968. The International Biological Program and the science of ecology. *Proceedings National Academy Sciences USA* 60: 5–11.

Swank, W.T., J.L. Meyer, and D.A. Crossley Jr. 2001. Long-term ecological research: Coweeta history and perspectives. In *Holistic science: The evolution of the Georgia Institute of Ecology (1940–2000)*, ed. G.W. Barrett and T.L. Barrett, 143–163. New York: Taylor & Francis.

The Institute of Ecology. 1977. *Experimental ecological reserves: A proposed national network*. Washington, DC: U.S. Government Printing Office.

U.S. National Committee for the International Biological Program. 1971. *Research programs constituting U.S. participation in the International Biological Program*, Report no. 4. Washington, DC: National Academy of Sciences.

Chapter 3
Long-Term Dynamics of the LTER Program: Evolving Definitions and Composition

Julia Jones and Michael Paul Nelson

Abstract This chapter investigates how the National Science Foundation's (NSF) Long Term Ecological Research (LTER) Program has changed from 1980 to 2018. The LTER program is designed to balance persistence with response to change in science, society, and ecosystems through renewable 6-year grants subjected to peer review at the midterm and at renewal. The LTER program had an initial period of rapid growth with some terminations (1980s), a middle period of slower growth with no terminations (1990–2010), and a third period of no net growth, with added and terminated sites and an accelerated rate of site probations (2010s). Changes in the character and composition of the LTER program are associated with changes in leadership and research directions within individual LTER sites, as well as changes in the sources of funding for the LTER program within NSF, turnover in NSF program officers, and changes in review criteria used to renew LTER site funding. In the past decade, a focus on conceptual frameworks as a tool for integrating LTER research emerged from the LTER renewal review process. Given the accelerated pace of environmental change, the need for long-term ecological research is even more urgent today than when NSF established the pioneering LTER program. The LTER Program history reveals important lessons for how to structure and manage long-term ecological research.

Keywords LTER program · Long-term ecological research · National Science Foundation · NSF funding, Ecological networks · Network management · Site termination · Site probation · Conceptual frameworks

---

J. Jones (✉)
Geography, College of Earth, Ocean, and Atmospheric Sciences, Oregon State University, Corvallis, OR, USA
e-mail: Julia.Jones@oregonstate.edu

M. P. Nelson
Department of Forest Ecosystems and Society, Oregon State University, Corvallis, OR, USA
e-mail: mpnelson@oregonstate.edu

R. B. Waide, S. E. Kingsland (eds.), *The Challenges of Long Term Ecological Research: A Historical Analysis*, Archimedes 59,
https://doi.org/10.1007/978-3-030-66933-1_3

3.1 Introduction

Long-term ecological research is overwhelmingly valued by the scientific community for its capacity to provide ecological understanding (Kuebbing et al. 2018) and contribute to policy (Hughes et al. 2017). Yet large programs that fund long-term ecological research are difficult to establish and sustain. In the 1980s, Callahan (1984) captured the essence of the challenge involved in making long-term ecological research a reality when he wrote:

> There is a serious contradiction between the time scales of many ecological phenomena and the support to finance their study. The problem is a difficult one. Funding cannot be guaranteed to any research undertaking for even tens of years, let alone for centuries or more. How can this pattern be broken, a pattern that acts against the consistent and reliable accumulation of sets of long-term synoptic data?

To address the problems that Callahan noted, the US National Science Foundation (NSF), founded in 1950, has funded the US Long Term Ecological Research (LTER) program since 1980 (i.e., for more than half the lifespan of the agency).

For 40 years, the LTER program has supported a changing number of sites through a competitive process. The NSF LTER program has expanded over time, and new sites have been added. LTERs develop understanding of how ecosystem processes respond to long-term environmental change in the past and the future. LTER site-based research programs design and implement experiments and studies that will outlive any given grant period and even the researchers themselves. LTER sites operate as a network, with offices (the LTER Network Office and Network Communications Office) and regular meetings to promote network interactions and synthetic research.

The NSF LTER renewal process attempts to strike a balance between providing funding for periods long enough to embark on, and continue, "consistent and reliable" (sensu Callahan 1984) long-term ecological research, without guaranteeing funding to any particular LTER site. Each site is funded for periods of 6 years (5 years in the 1980s) and must submit a renewal proposal to NSF every 6 years. Some LTER sites have obtained renewal funding for as many as seven periods, while other sites have been terminated after as little as one to as many as six funding periods. Some LTER sites have been placed on "probation" at the time of renewal; these sites must submit a 4-year renewal proposal 2 years later. As of 2020, currently funded LTER sites vary from as little as 3 years to 40 years of continuous funding.

A number of publications have described research and other accomplishments of the LTER program (e.g., Kratz et al. 2003; Gosz et al. 2010; Foster 2012). However, no publications have addressed the funding and evaluation process of the LTER program. Here we examine the review criteria, the evolving composition of sites resulting from establishment and renewal decisions, and factors influencing these changes over the history of the NSF LTER program. The following questions are examined:

1. Over the history of the LTER program (1980–2018), how many sites have been funded, and what types of ecosystems have been studied?
2. What have been the review criteria for LTER sites? How have they changed?
3. What have been the rates of funding, renewal, probation, and termination of sites?

This analysis of program history and administration may be helpful for those seeking context for the development and management of long-term ecological research programs globally.

3.2 Methods

3.2.1 Data

Analyses were based on information provided by the LTER Network Office (LNO), LTER Network Communications Office (NCO), NSF, and LTER principal investigators (PIs). The LNO provided historical information on dates of site funding, renewal, probation, termination, and wind-down funding. NSF documents included requests for proposals (RFPs, now called solicitations at NSF, and hereafter collectively referred to as RFPs); invitations for renewal; proposal reviews and panel summaries; and program officer comments. RFPs and letters for renewal were obtained from archives compiled by the LNO, NCO, and individual LTER sites. Reviews, panel summaries, and program officer comments were provided with permission to use by current LTER principal investigators.

3.2.2 Analyses

Our analysis has three main components: (1) inventory of LTER sites and their funding status over the history of the program, (2) analysis of historical NSF requests for proposals (RFPs), and (3) analysis of the outcome of the renewal review process resulting in probation and termination decisions since 2010, a period in which 11 sites were placed on probation and four were terminated.

We compiled an inventory of LTER sites funded over the history of the program. A table of LTER site names, acronyms, funding periods, and ecosystem type was assembled from data provided by the LNO and NCO (Table 3.1). A master database was created including the dates of each event (funding, renewal, probation, termination, end of post-termination wind-down funding) throughout the funded period for each LTER site (Fig. 3.1). Data in the master database were grouped by type of event (i.e., funding, probation, termination) to show the growth and change in numbers of LTER sites (Fig. 3.2). These data were also grouped by decade and ecosystem type in order to show how the composition of the LTER network changed over

Table 3.1 Names, funded periods, and ecosystem types of LTER sites examined in this study

Name and abbreviation	Type of ecosystem	Period
Continued long-term funding		
Andrews Forest (AND)	Conifer forest and streams	1980–
Arctic (ARC)	Tundra, lakes, rivers	1986–
Beaufort Lagoon Ecosystem (BLE)	Coastal ocean	2017–
Bonanza Creek[a] (BNZ)	Boreal forest	1986–
California Current Ecosystem (CCE)	Coastal upwelling biome	2004–
Cedar Creek Ecosystem Science Reserve[a] (CDR)	Savanna/tallgrass prairie	1982–
Central Arizona – Phoenix[b] (CAP)	Urban	1997–
Florida Coastal Everglades[b] (FCE)	Freshwater marsh, mangroves, seagrass	2000–
Georgia Coastal Ecosystems (GCE)	Coastal rivers, marsh, barrier islands	2000–
Harvard Forest (HFR)	Temperate forest and wetlands	1988–
Hubbard Brook (HBR)	Temperate forest and streams	1986–
Jornada Basin[ab] (JRN)	Chihuahuan desert grassland, shrubland	1982–
Kellogg Biological Station[b] (KBS)	Row-crop agriculture	1986–
Konza Prairie (KNZ)	Native tallgrass prairie	1980–
Luquillo[ab] (LUQ)	Tropical forest and streams	1988–
McMurdo Dry Valleys (MCM)	Ice-covered lakes, streams, ice-free soil	1991–
Mo'orea Coral Reef[b] (MCR)	Coral reef and lagoon	2004–
Niwot Ridge[a] (NWT)	Alpine glacier and tundra	1980–
North Temperate Lakes (NTL)	Lakes in a forested landscape	1980–
Northeast US Shelf (NES)	Coastal ocean	2017–
Northern Gulf of Alaska (NGA)	Coastal ocean	2017–
Palmer Antarctica (PAL)	Pelagic marine	1990–
Plum Island Ecosystems[b] (PIE)	Coastal rivers, estuaries and marshes	1998–
Santa Barbara Coastal (SBC)	Coastal rivers, kelp forests	2000–
Virginia Coast Reserve[a] (VCR)	Coastal marsh, estuary, barrier islands	1986–
Terminatedd		
Baltimore Ecosystem Study[b] (BES)	Urban	1997–2018
Coweeta[b] (CWT)	Temperate forest	1980–2018
Illinois Rivers	River, riverine marsh, floodplain forest	1982–1988
North Inlet	Coastal rivers, estuaries and marshes	1980–1993
Okefenokee	Freshwater wetland	1982–1988
Shortgrass Steppe[a] (SGS)	Semi-arid grassland	1982–2012

(continued)

Table 3.1 (continued)

Name and abbreviation	Type of ecosystem	Period
Terminated and re-funded		
Sevilleta[abc] (SEV)	Desert grassland, shrubland, woodland	1988–

[a]Site experienced probation before 2010
[b]Site experienced probation between 2010 and 2018
[c]Terminated and then re-funded
[d]Terminated sites receive "wind-down" funding for several years after termination. This study uses the date when termination decision was made
Sources: https://lternet.edu/site/ and archives of the LTER Network Office, https://lternet.edu/?taxonomy=document-types&term=lter-network-office

time relative to type of ecosystem and source of funding (i.e., Directorates and Divisions) within NSF (Fig. 3.3).

Analysis of Historical NSF RFPs, 1980–2020 All RFPs from 1980 to 2020 were collected and placed in an electronic archive in the following locations.
New site solicitations:
https://lternet.edu/?taxonomy=document-types&term=new-site-solicitations
Renewal solicitations and guidance:
https://lternet.edu/?taxonomy=document-types&term=renewal-solicitations
Planning documents:
https://lternet.edu/?taxonomy=document-types&term=planning-documents
All three categories are posted on the LTER Network archive, under the heading LTER Organization:
https://lternet.edu/intranet/

The dates, titles, and names of program officers listed in each RFP from 1980 to 2020 were tabulated in order to determine the Directorates and Divisions of NSF and the numbers and turnover of program officers involved over the course of the program (Table 3.2). Each RFP was read, and six review criteria were identified from the original RFP in 1980, as well as six additional review criteria that appeared later and persisted in renewal RFPs. Each RFP was searched for each combination of these twelve sets of review criteria keywords/key phrases, and sentences containing these keywords or key phrases were excerpted, tabulated, and compared over time (Table 3.3). The dates of first appearance and revisions of review criteria were assembled into a timeline for the history of the LTER program (Fig. 3.4). The numbers of words in each RFP was counted (Fig. 3.5), and the turnover in the wording of review criteria in each successive RFP from 1980 to 2018 was calculated (Table 3.3).

An analysis was conducted of the outcome of the renewal review process in the fourth decade. A schematic diagram was developed to show the process for renewal of LTER sites (Fig. 3.6). The reviews, panel summaries, and program officer comments for 12 cases of probation or termination that occurred between 2010 and 2018 were read and searched for the key words/key phrases of review criteria identified

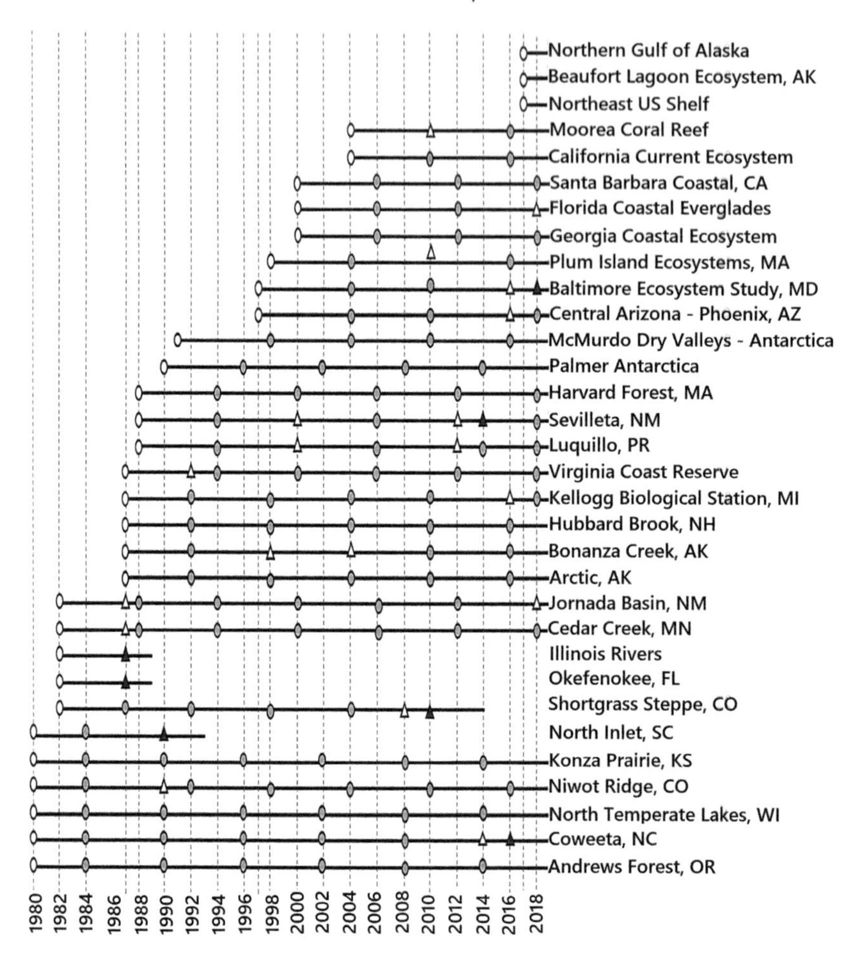

Fig. 3.1 Timelines of initial funding, renewal, probation, and termination of LTER sites, 1980–2018. Length of each line indicated the total funding period for that site, including wind-down funding after termination, and renewal funding to 2020, 2022, and 2024

from the analysis of historical RFPs. Sentences containing these keywords/key phrases were excerpted and grouped by review criterion and site. The frequency of use of these keywords in probation and termination decisions was tabulated by year from 2010 to 2018 (Table 3.4). The outcomes of LTER site renewals (shown in Fig. 3.1) were tabulated by decade and used to calculate the proportion of probations and terminations relative to numbers of proposals reviewed (Table 3.5). The outcomes of LTER site renewals (from Fig. 3.1) also were tabulated to determine the proportion of probations and terminations relative to the numbers of renewal proposals each site had submitted (Table 3.6).

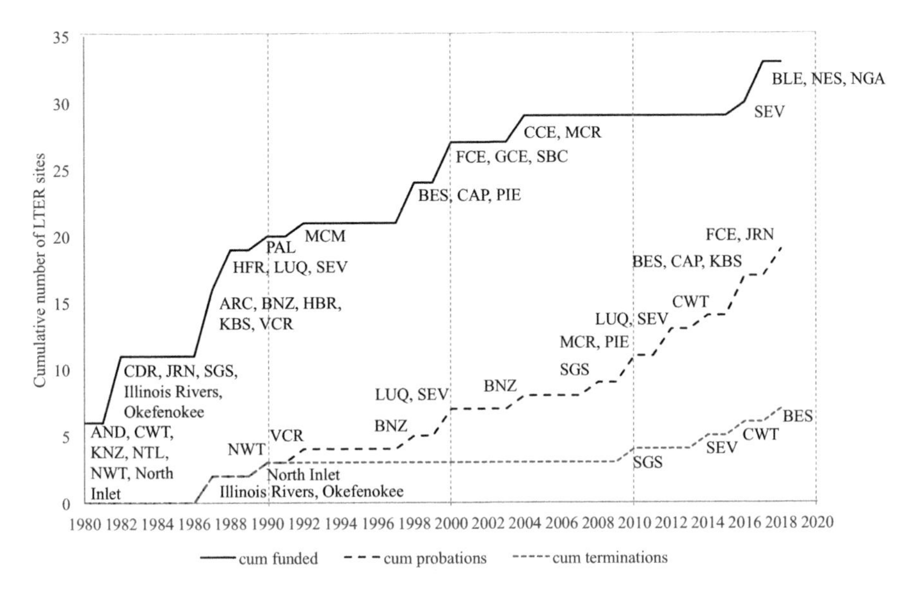

Fig. 3.2 Cumulative numbers of sites funded, on probation, and terminated in the NSF LTER Program. LTER site abbreviations are from Table 3.1

3.3 Results

3.3.1 Overview of LTER Program Sites by Ecosystem Type Over Time

Thirty-three awards were made to initiate funding to 32 LTER sites from 1980 to 2018 (one site was funded twice) (Table 3.1). Funding periods of individual LTER sites ranged from one to 38 years as of 2018 (Fig. 3.1). Overall, 133 renewal proposals have been reviewed. Nineteen of these proposals (14.3%) resulted in probation and seven proposals (5.3%) resulted in termination of an LTER site. Three sites were terminated in the first decade of the program after 5–10 years of funding. The first instance of probation was in 1990. Four sites were terminated in the most recent decade (2010–2018) after 20–38 years of funding.

The number of LTER sites increased rapidly in the first decade and more slowly in the second decade, and then leveled off (Fig. 3.2). In the first decade, twenty sites were funded, and three sites were terminated, leaving a net of 17 sites as of 1990 (Table 3.5). By 2000, including terminations and new sites, 24 LTER sites were funded. Since 2000 there has been little net change in numbers of funded LTER sites: four sites have been terminated, and six new sites were added, including one which had been terminated but was re-initialized. Four sites were placed on probation in each of the second and third decades (1990s and 2000s), and one of these eight sites was terminated in 2010. Seven sites were placed on probation in the

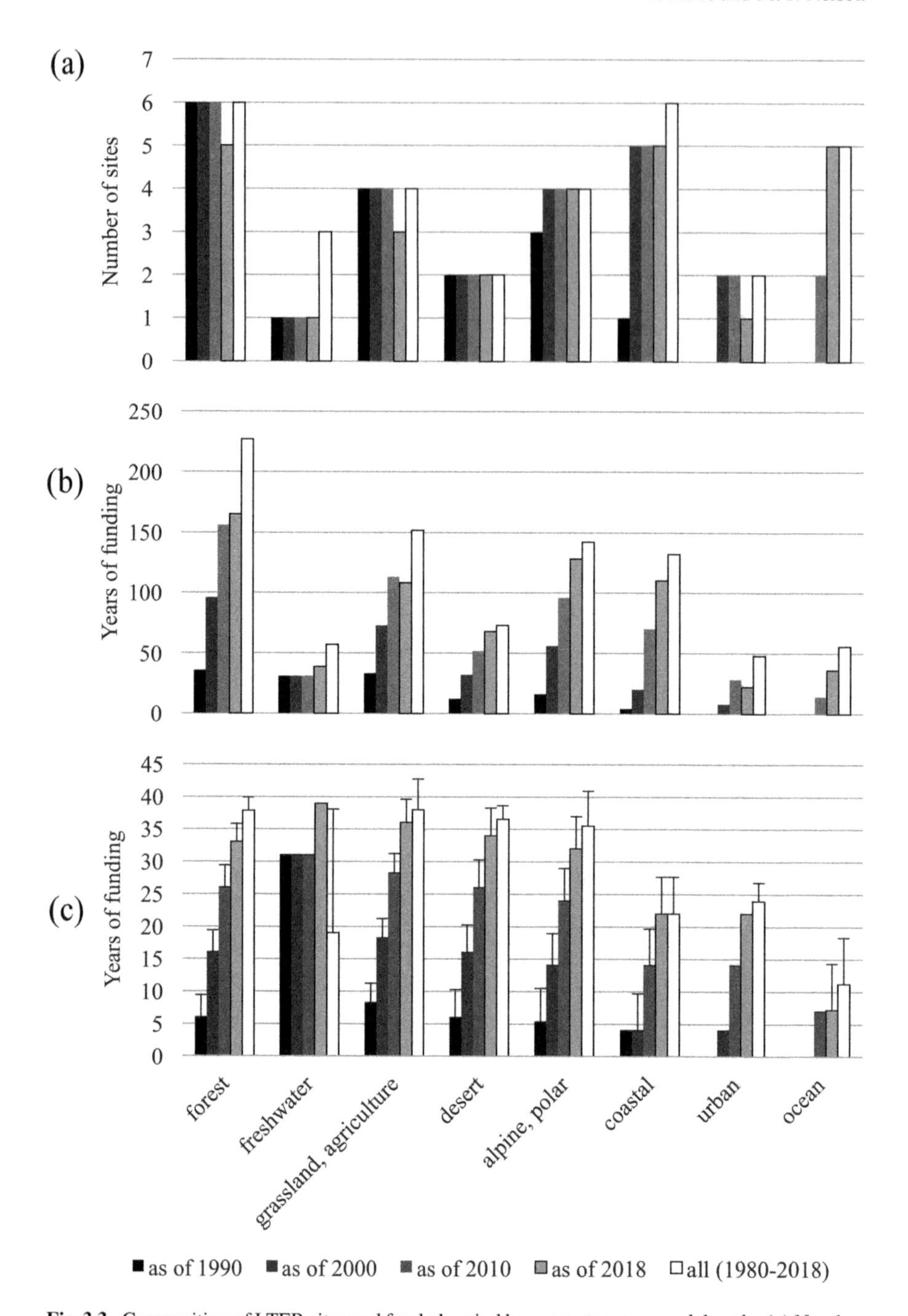

Fig. 3.3 Composition of LTER sites and funded period by ecosystem type and decade. (**a**) Number of sites funded, (**b**) total years of funding, and (**c**) average funded period. A total of 32 sites have been funded since 1980 (Table 3.1), broken into eight categories: Forest (n = 6): AND, BNZ, CWT, HBR, HFR, LUQ; freshwater (n = 3): Illinois Rivers, Okefenokee, NTL; Grassland, agriculture (n = 4): KNZ, SGS, CDR, KBS; desert (n = 2): JRN, SEV; alpine and polar (n = 4): NWT, ARC, MCM, PAL; coastal (n = 6): North Inlet, PIE, FCE, GCE, SBC, VCR; Urban (n = 2): BES, CAP; ocean (n = 5): CCE, MCR, BLE, NES, NGA

Table 3.2 Sources of information for analysis of NSF requests for proposals (RFPs) involving the LTER program. Site abbreviations are explained in Table 3.1

Date	Name of competition	Outcome	NSF Directorate/ Division	Program officers
1980	NSF 79–64 A new emphasis in long-term research (first competition)	AND, CWT, KNZ, NTL, NWT, North Inlet	BIO/DEB	J.T. Callahan
1981	NSF long-term ecological research (LTER)	CDR, JRN, SGS, Big Rivers, Okefenokee	BIO/DEB	J.T. Callahan
1986	n/a (renewal)		BIO/DEB	J.T. Callahan
1987	NSF 86–16 third competition for long-term ecological research (LTER)	ARC, BNZ, HBR, KBS, VCR	BIO/DEB	J.T. Callahan
1988	NSF 87–41 fourth competition for long-term ecological research (LTER)	HFR, LUQ, SEV	BIO/DEB	J.T. Callahan
1990,1991	n/a	MCM, PAL	GEO/OPP, BIO/DEB	n/a
1994	NSF 94–60 LTER project augmentation for regionalization, comprehensive site histories, and increased disciplinary scope	Augmented CWT, NTL	BIO/DEB	J.T. Callahan
1997	NSF 97–53 fifth competition for long-term ecological research (LTER): Urban LTER	BES, CAP	BIO/DEB, EHR, SBE	S.L. Collins, E. Hamilton, J.W. Harrington
1999	NSF 99–89 long-term ecological research (LTER) in land/ocean margin ecosystems	PIE, FCE, GCE, SBC	BIO/DEB, GEO/OCE	S.L. Collins, P. Taylor
2002	Guidelines for LTER 2002 renewal proposals	Renewal	BIO/DEB	H.L. Gholz
2004	NSF 03–599 long-term ecological research (LTER) in Coastal Ocean ecosystems	CCE, MCR	BIO/DEB, GEO/OCE	D.L. Garrison, G. Pugh
2006	Preparation Guidelines for LTER 2006 Renewal Proposals	Renewal	BIO/DEB, GEO/OCE	H.L. Gholz, P. Taylor

(continued)

Table 3.2 (continued)

Date	Name of competition	Outcome	NSF Directorate/ Division	Program officers
2008	Preparation Guidelines for LTER 2009 Renewal Proposals	Renewal	BIO/DEB, GEO/OCE, GEO/OPP, SBE	H.L. Gholz, M. Caldwell, P. Taylor, R. Marinelli, T. Baerwald
2010	Preparation Guidelines for LTER 2011 Renewal Proposals	Renewal, ~~SGS~~	BIO/DEB, GEO/OCE, GEO/OPP, SBE	T. Crowl, H.L. Gholz, D. Garrison, R. Marinelli, T. Baerwald
2012	NSF 12–524 long-term ecological research (LTER)	Renewal	BIO/DEB, GEO/OCE, GEO/OPP, SBE	S. Twombly, T. Baerwald, D.L. Garrison, P. Milne
2014	NSF 13–588 long-term ecological research (LTER)	Renewal, ~~SEV~~	BIO/DEB, GEO/OCE, GEO/OPP	S. Twombly, D.L. Garrison, L. Clough
2016a	NSF 16–509 long-term ecological research (LTER) new site competition	BLE, NES, NGA, SEV	BIO/DEB, GEO/OCE	S. Twombly, D.L. Garrison
2016b	NSF 15–596 long-term ecological research (LTER) renewal	Renewal	BIO/DEB, GEO/OCE, GEO/OPP	J. Schade, L. Kaplan, D.L. Garrison, L. Clough
2018	NSF 17–593 long-term ecological research (LTER) renewal	Renewal	BIO/DEB, GEO/OCE, GEO/OPP, SBE	D. Garrison, J. Schade, D. Levey, L. Kaplan
2020	NSF 19–593 long-term ecological research (LTER) renewal	Renewal	BIO/DEB, GEO/OCE, GEO/OPP, SBE	J. Burns, R. Delgado, D. Levey, C. St. Mary, J. Schade, D. Thornhill, J. Yellen

Program officers are those listed in the RFP. *BIO/DEB* Directorate for Biological Sciences/ Division of Environmental Biology, *SBE* Directorate of Social, Behavioral and Economic Sciences, *GEO/OCE* Directorate for Geological Sciences/Division of Ocean Sciences, *GEO/OPP* Directorate for Geological Sciences/Office of Polar Programs. Original funding to a site is shown in regular font and terminated sites are struck out, e.g. Data on funding outcomes is from records maintained by the LTER Network Office, 2015 onward

fourth decade (2010s) and three of these have been terminated (Figs. 3.1 and 3.2, Table 3.5).

The composition of LTER sites varies by ecosystem type and decade. Funding for LTER sites has been provided from several of the Directorates of NSF, but principally from the Biological Sciences Directorate (BIO) and the Geosciences Directorate (GEO). In the first decade, LTER funding was exclusively from the NSF Division of Environmental Biology in the Biological Sciences Directorate (BIO/

Table 3.3 Changes in frequency of LTER review criteria in NSF rfps, 2010–2020

	2010	2012	2014	2016	2018	2020
Original criteria in 1980						
Information (data) management	3	13	14	8	9	9
Site (project) management, diversity	2	3	3	3	3	2
Long-term	3	15	18	17	17	5
Cross-site, network	10	28	7	8	8	2
Goal	0	8	5	5	5	1
Core area	0	3	2	2	7	6
Criteria added in 1997 or later						
Integrate, -s, -d, -ing	1	15	11	17	17	2
Conceptual framework	3	6	5	7	7	7
Social, socio-, human	0	12	6	13	11	2
Theory, -ies	0	3	1	2	2	2
Model, -ing	2	4	8	6	5	4
Publications	4	0	1	1	7	6
Predict	0	1	1	7	3	3
Total words in instructions	1230	2205	2546	2664	3034	3186
Turnover in instructions (%)	1	100	37	30	26	31

Total words and turnover were calculated for Section V of the RFP: proposal preparation
Turnover = (number of words in in year t that do not appear in year t−1 + number of words in year t−1 that do not appear in year t)/(total words in year t and year t−1). Turnover from 1980 to 2010 was <21%

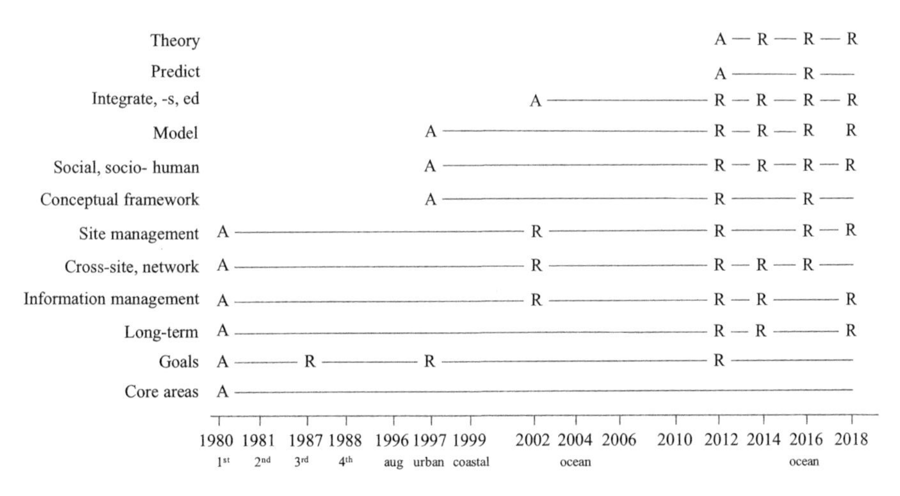

Fig. 3.4 First appearance (A) and subsequent revisions (R) of key terms and concepts in NSF LTER review criteria in NSF solicitations (RFPs) issued on dates shown on x-axis. NSF competitions for new or augmented ("aug") LTER sites (from Table 3.2) are shown below the dates

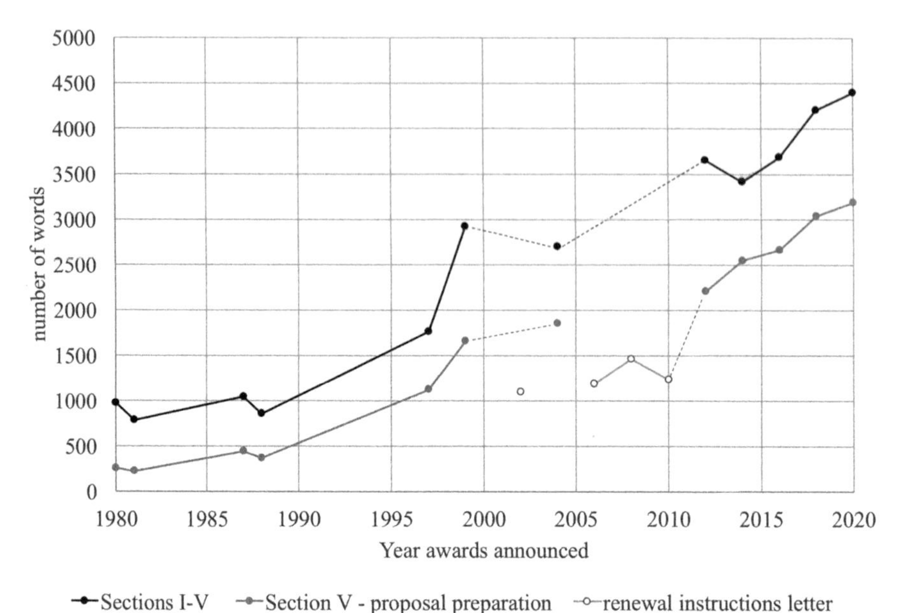

Fig. 3.5 Trends in numbers of words in NSF RFPs and renewal instruction letters for the LTER program. Section V is guidelines for proposal submission, including review criteria. Numbers for 2012–2018 refer to renewal RFPs only (see Table 3.2)

DEB), and focused on terrestrial sites, such as forest, grassland (including agricultural), and desert (Table 3.2, Fig. 3.3a). Two of three freshwater sites were terminated by 1990.

In the second decade, ecosystem types of LTER sites expanded to include polar, urban, and coastal/land margin ecosystems. Polar sites were funded by the Office of Polar Programs of the Geosciences Directorate (GEO/OPP). Urban sites were funded by BIO/DEB with contributions for the first few years from the Directorate of Social, Behavioral, and Economic Sciences (SBE), Education and Human Resources (EHR), and the Engineering Directorates (S.L. Collins, personal communication). Coastal/land margin sites were funded by the Division of Ocean Sciences of the Geosciences Directorate (GEO/OCE) (Table 3.2, Fig. 3.3a).

In the third and fourth decades, ecosystem types of LTER sites expanded to include coastal ocean ecosystems (funded by GEO/OCE and GEO/OPP). No new terrestrial sites have been initiated with BIO/DEB funding since 1988. BIO/DEB provides all the funding for the urban sites, which began in the 1990s (S.L. Collins, personal communication). All new LTER sites since 1988 have involved funding from the GEO (and briefly, from the SBE) Directorates of NSF. All seven of the sites that have been terminated were funded by BIO/DEB at the time of termination. Thus, as of 2018, 13 of the 26 sites that retain funding are funded by BIO/DEB, and 13 sites are funded by GEO/OPP or GEO/OCE, with a few sites receiving co-funding from two or more directorates (Table 3.2, Fig. 3.3a).

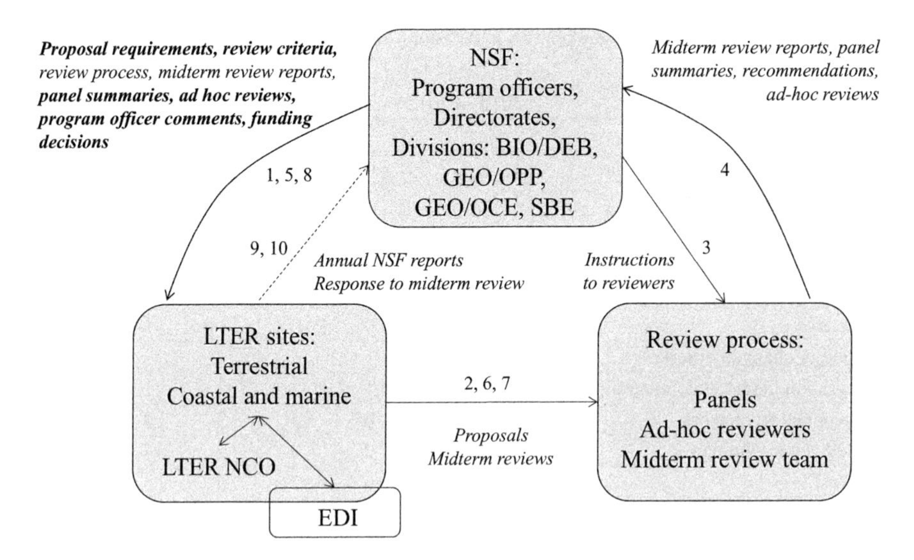

Fig. 3.6 Structure of the LTER community, and information and communication flows that influence binding decisions on the structure and composition of the LTER program over time. This study analyzed the items in bold font. *NCO* LTER Network Communication Office, *EDI* Environmental Data Initiative. NSF Directorates and Divisions are defined in Table 3.2. Numbers are explained in the text

As a result of these funding patterns, the total years of LTER funding also vary by ecosystem type and decade (Fig. 3.3b). Years of funding (and associated data and long-term analyses) are dominated by forest sites, then grassland, alpine/polar, and coastal sites. Desert, freshwater, ocean, and urban sites have the least total years of funding. The average years of funding per site (which are a measure of the expected length of longest records) also vary by ecosystem type and decade (Fig. 3.3c). Average funding periods of forest, grassland, and desert exceed 35 years for the entire program history, as does the funding period for the single freshwater lake site. Average funded periods for coastal and urban sites exceed 20 years. The average funding period is shortest for ocean sites (Table 3.2, Fig. 3.3c).

3.3.2 LTER Site Renewal Process and Review Criteria

The review process for LTER site renewal involves multiple participants, and is directed by NSF. The renewal process includes institutions (NSF, LTER sites) and individuals (program officers, PIs, reviewers, and panelists) (Fig. 3.6). The principal flows of information that control decision-making are: (1) requests for proposals issued by NSF, (2) submission of research proposals from LTER sites, (3) instructions to reviewers from NSF, (4) reviews, summaries and panel recommendations to NSF, and (5) funding decisions from NSF to LTER sites (Fig. 3.6). Proposals are

Table 3.4 Frequency of use of key phrases in decisions on probation and termination from 2010 to 2018 (n = 12)

Review criteria	2010	2012	2014	2016	2018	Total	%
Original criteria							
Information (data) management	1	2	0	3	1	7	58
Site (project) management	2	2	1	1	0	6	50
Long-term	1	1	1	1	1	5	42
Cross-site, network	2	1	0	0	0	3	25
Goal	0	1	0	1	0	2	17
Core area	0	0	0	0	1	1	8
Added or changed criteria							
Integration, integrated, integrating	2	2	1	3	2	10	83
Conceptual framework	1	2	0	4	2	9	75
Social, socio-, human	1	0	0	4	2	7	58
Theory	1	2	0	2	0	5	42
Model	1	1	0	2	0	4	33
Publications	1	1	0	0	1	3	25
Predict	0	0	0	1	0	1	8
Total decisions	2	2	1	4	3	12	
Total criteria used	13	15	4	22	12	66	
Average no. criteria used per decision	6.5	7.5	4	5.5	4	5.5	46

Analysis included documents from 12 of 14 panel reviews affecting 11 sites; 11 were probation decisions and 3 were termination decisions. % = percent of times used in probation or termination

Table 3.5 Cumulative numbers of LTER site proposals by outcome and proportions of proposals leading to probation and termination, at the end of each of four decades of the NSF LTER program

	1990	2000	2010	2018
Cumulative sites funded	20	27	29	33
Net sites with continued long-term funding	17	24	25	26
Cumulative renewal proposals reviewed	20	50	92	133
Cumulative renewal proposals funded	14	40	77	107
Cumulative renewal proposals leading to probation	3	7	11	19
Cumulative site terminations	3	3	4	7
Cumulative terminations/cumulative funded	0.15	0.11	0.14	0.21
Cumulative probations/renewal proposals	–[a]	0.14	0.12	0.14
Terminations/probations in that decade	–[a]	0	0.25	0.38

A total of 33 funding decisions has been made to fund a total of 32 LTER sites from 1980 to 2018 (Sevilleta was funded twice)

[a]There are no records of probation before 1990

reviewed by the panelists, and (since 2012) additional ad-hoc reviews are solicited from other scientists prior to the meeting of the panel; often one or more of the ad-hoc or panel reviewers are drawn from within the LTER network. Supplemental flows of information include (6) midterm site reviews by NSF, (7) midterm review reports by review teams to NSF, (8) NSF midterm review evaluation sent to LTER

Table 3.6 Numbers and fates of LTER renewal proposals submitted by each site over its entire funded period, and effects on site-level probation and termination

	Numbers of renewal proposals submitted								
	0	1	2	3	4	5	6	7	All
N of sites	3	2	3	3	5	4	8	4	32
N of proposals reviewed	–	2	6	9	20	20	48	28	133
N of proposals leading to probation or termination	–	2	2	1	4	2	9	6	26
N of proposals leading to probation	–	0	1	1	3	2	7	5	19
N of sites on probation	–	0	1	1	3	1	5	4	15
N of sites terminated	–	*2*	*1*	0	1	0	2	1	7
Proposals leading to probation or termination (%)	–	*100*	*33*	11	20	10	19	21	31
Sites on probation or termination (%)	–	0	*33*	33	60	25	63	100	45
Sites terminated (%)	–	*100*	*33*	0	20	0	25	25	29
Sites terminated given probation (%)	–	–	*100*	0	33	0	40	25	33

Italics indicates this figure includes termination of three sites in the first decade of the program, 1980–1990, before the first documented instance of probation identified in this study

sites, and (9) response to midterm review from LTER sites. Finally, (10) LTER sites submit annual reports to NSF.

From 1980 to 2020, there has been expansion and turnover of program officers and directorates involved in LTER (Table 3.2). In the 1980s, the LTER program was initiated by a single program officer in one NSF directorate/division (BIO/DEB). In the 1990s, LTER had expanded to at least three directorates (BIO, GEO, SBE), and four new program officers were added. In the 2000s (2002 to 2010), eight program officers from three NSF directorates and four NSF divisions were listed on LTER RFPs. Of these eight program officers, seven were new, and one (Taylor) had been listed on an LTER RFP in 1999. In the 2010s (2012 to 2020), thirteen program officers from three NSF directorates and four NSF divisions were listed on LTER RFPs. Of these, twelve were new, and one (Garrison) had been listed on an LTER RFP in the previous decade. Over the history of the program, some LTER program officers had prior experience as LTER researchers, and others did not.

The length of RFPs and number of review criteria have increased steadily since 1980. From 1980 to 2020, the total length of the RFP increased by four times, and the length of Section V (instructions for proposal submission) increased by more than ten times (Fig. 3.5, Table 3.3).

The original RFPs (issued in 1980 and 1981) established the main goals and essential features of LTER: core areas, long-term questions, cross-site research and network participation, information management/data availability, and continuity of leadership (Table 3.2, Fig. 3.4). Criteria involving "conceptual framework," modeling, and social factors (for urban sites) were added in the late 1990s. In the 2012 RFP, the language changed for solicitation-specific review criteria including integration, conceptual frameworks, modeling, social science, and the terms "theory," and "predict" were added. The 2012 RFP changed the language regarding cross-site research and network participation, information management, and site management. From 2012 to 2018 there were a number of important changes in the wording of these new criteria (Fig. 3.4). Over the period 2012–2018, all but two of the criteria used in the early decades of the LTER program decreased in frequency, while terms involving integration ("integrate, -ed, -ing") and social processes ("social, socio-ecological, social-ecological, human") increased in frequency. In 2020, these latter terms decreased in frequency (Table 3.3, NSF announcement for 2020).

3.3.3 Changes Over Time in the Wording of Original Review Criteria for LTER

This section summarizes major changes from 1980 to 2020 in the wording used to describe the original LTER review criteria. NSF's program announcements and calls for proposals are listed in Table 3.2: references in brackets in this and subsequent sections are abbreviated as "NSF" and year, and refer to the announcement in Table 3.2.

Overall Goals and Mission The overall goals of LTER were defined in the 1980 RFP and have not changed. In 1980, program goals were to *"(1) initiate the collection of comparative data at a network of sites representing major biotic regions of North America and (2) evaluate the scientific, technical and managerial problems associated with such long-term comparative research"* (NSF 1980). Key phrases were added to the goals in 2012 to 2014; these include *"mechanistic understanding," "multiple scales," "predict ... responses to future environmental change,"* and *"social responses"* (NSF 2012, 2014).

Core Areas The core areas first appeared in the 1980 RFP: *"Investigators must focus on a series of core research topics, coordinate their studies across sites, utilize documented and comparable methods, and be committed to continuation of work for the required time. The core research areas are: (1) pattern and control of primary production, (2) dynamics of populations of organisms selected to represent trophic structure, (3) pattern and control of organic matter accumulation in surface layers and sediments, (4) patterns of inorganic inputs and movement of nutrients through soils, groundwater, and surface waters, and (5) patterns and frequency of disturbances"* (NSF 1980). The description of the core areas has not changed since 1980, although the 2018 RFP added core areas to the review criteria (Fig. 3.4, Table 3.3).

Long-Term Data and Long-Term Research In 1980, the RFP had no specific wording about long-term data and research, although emphasis was placed on the need for a long-term commitment by the investigators. From 2012 to 2018, "long-term" appeared frequently in the RFP (Table 3.3), and the wording was changed in each successive RFP (Fig. 3.4), including *"questions that uniquely demand study on* *decadal time scales," "justify the need for long-term support," "questions that arise from the analysis of long-term data," "questions that ... require uninterrupted, long-term collection, analysis, and interpretation of environmental data"* (NSF 2012, 2014, 2016b, 2018). The frequency of "long-term" increased sharply from 2012 to 2018, but decreased in the 2020 RFP (Table 3.3, NSF 2020).

Information Management and Data Availability Information management and data availability have been a hallmark of LTER since its inception, and requirements have become more numerous and specific over time. *"Data storage and retrieval"* was required in the 1980 RFP (NSF 1980). Starting in 1997, the RFP required proposals to explain data accessibility, data completeness, and how data from LTER research were provided to LTER information managers (NSF 1997). In 2012, the RFP required reporting of specific timelines for data release for "core" datasets, and documentation of data use, but the latter requirement was dropped in 2014 (NSF 2012, 2014). Starting in 2012, the RFP required sites to comply with *"LTER Network Access goals,"* and in 2018 the RFP required sites to comply with the *"LTER Network's Information Management Policy."* The 2018 RFP also lists specific requirements for depositing data in public data repositories and reporting on the data deposited at these sites. In addition, in 2018, the solicitation-specific review criteria require *"comprehensive availability of data previously collected by the site"* (NSF 2018), which might refer to data collected with NSF LTER funding, as well as other funding.

Site Management and Leadership, Including Diversity Continuity of leadership has been a key component of LTER since it began. In 1980, the RFP stated *"The principal investigators must be prepared to make long-term time commitments and should consider ... continuity of site leadership"* (NSF 1980), and continuity of site leadership remains a criterion in the 2020 RFP (NSF 2020). Originally the emphasis was on continuity of leadership within a single LTER award (5 or 6 years). However, many sites were led by their founding PI for several grant cycles, and therefore, starting in 2012, the RFP specifically requires a description of how site leadership transitions are planned and managed (NSF 2012, 2014, 2016b, 2018, 2020). Starting in 2002, the RFP also required sites to explain how they encourage participation in LTER by *"non LTER scientists"* and how site management *"enhance[s] ... diversity of scientists"* (NSF 2002). In 2018, wording was added, *"New participants bring new ideas and fresh perspectives, which are likely to enrich the development of research at the site"* (NSF 2018).

Cross-Site Research and Network Participation The early phases of LTER strongly emphasized cross-site research and network participation, but RFPs have de-emphasized it since 2012. In 1980, the RFP stated, *"Investigators must ... coordinate their studies across sites"* (NSF 1980). Cross-site analysis and network coordination was specified in all RFPs and included as a specific review criterion until 2014, when it was made optional. In the 2012 and 2014 RFPs, *"sites are encouraged to develop network-level interactions"*, and *"proposals are encouraged to* *broaden the spatial scale ... through comparative research with other LTER sites or studies outside of the LTER network"* (NSF 2012, 2014). However, in 2016, cross-site work became optional: *"where appropriate, projects among sites or with collaborators outside of the LTER network may be included"* (NSF 2016b).

3.3.4 Additional Review Criteria Added Starting in the Mid-1990s

This section describes additional solicitation-specific review criteria that were included starting in the mid-1990s, and changes in these criteria to 2018.

Integration The first appearance of "integration" in instructions for LTER proposal preparation was in the renewal guidelines in 2002: *"describe the methods and planned analyses in detail and ... conceptually integrate these efforts to your long-term studies ... [C]lose ... with a synthesis that shows how your major activities will be integrated"* (NSF 2002). Starting in 2012, the term "integrate, -ed, -ive" appeared frequently in the RFP linked to many different review criteria, and wording using this term was expanded or modified in each successive RFP through 2020 (Table 3.3, Fig. 3.4). Lack of integration was cited as a reason for 83% of probation and termination decisions from 2010 to 2018 (Table 3.4). In 2020, the term had largely disappeared from the RFP (Table 3.3, NSF 2020).

Conceptual Framework Conceptual frameworks were added to NSF RFPs for LTER in 1997: *"LTER research should be developed around a site-specific conceptual framework that generates questions requiring experiments and observations over long time frames and broad spatial scales"* (NSF 1997). This definition was omitted starting with the renewal instructions in 2002, which merely stated, *"Develop and explain the conceptual framework that provides the unifying ecological theme for your site"* (NSF 2002). The 2012 RFP refers to, but does not define, conceptual frameworks (NSF 2012). In 2014, sites were required to *"extend"* or develop *"new"* conceptual frameworks (NSF 2014). In 2016, key new phrases were added requiring a conceptual framework that *"examines and predicts," "produce[s] a comprehensive understanding," "integrates across populations, communities, and ecosystems,"* and *"develop[s] predictions"* (NSF 2016b). Issues with the conceptual framework were cited as a reason for 75% of probation and termination decisions from 2010 to 2018 (Table 3.4). The 2020 RFP re-defined the conceptual framework as something that, *"motivates questions requiring experiments and observations over long time frames. The conceptual framework should explicitly justify the long-term question(s) posited by the research and it should identify how data in LTER core areas and any experimental work contribute to an understanding of the question(s) while testing major ecological theories or concepts. The framework should provide the justification for all studies outlined in the proposal and should be informed by ongoing analyses of long-term data"* (NSF 2020).

Social Science (Social, Socio-, Human) The term "social" did not appear in LTER RFPs until 2012, with two exceptions. In 1994, the term "social" was introduced into LTER RFPs in a supplemental competition for expanded research (NSF 1994, Table 3.1). In 1997, a competition for urban LTER sites brought social factors to prominence and added them as a required review criterion for urban sites (NSF 1997). The word "social" was absent from subsequent RFPs and renewal guidelines until 2012. In 2012, the RFP added references to *"social scientists," "socio-ecological connections," "social responses,"* and *"social strategies"* (NSF 2012). In the 2012 RFP, if social science was proposed, it was to be evaluated based on *"the extent to which the research draws from and contributes to social science theory and understanding."* The 2012 RFP stated that all LTER sites "may elect to" include social science *"if there are key, conceptually motivated social science questions."* In 2014, the RFP language was qualified by the addition of *"if appropriate"* preceding *"social factors,"* and the recommendation was qualified as *"[sites] may elect to include social science research" ... "if it helps to advance or to understand key, conceptually motivated ecological questions"* (NSF 2014). In 2018, this wording was changed to simply, *"The disciplinary breadth of LTER research includes ... in some cases, social and economic science"* (NSF 2018). The terms "social, socio-, human" etc. appeared frequently in the RFP from 2012 to 2018 (Table 3.3), and the wording associated with these terms was modified in each successive RFP through 2020. Issues with these concepts were cited as reasons for probation and termination in 58% of cases from 2010 to 2018 (Table 3.4). In the 2020 RFP these terms rarely appeared (Table 3.3).

Models The word "model" or "modeling" did not appear in RFPs from 1980 to 1996. From 1997 to 2012, RFPs noted that modeling was important and required modeling efforts to be *"discussed in detail as appropriate"* (NSF 1997, 2010). In 2012, the RFP included new wording that required use of models, or development of models, and mentioned specific categories of models. In 2014, this wording was modified to include *"refinement"* of models to *"incorporate sources of uncertainty"* and *"model-data assimilation"* (NSF 2014); this wording was dropped in the 2016 RFP. The 2018 RFP stated that proposals must include *"development, refinement, and testing of quantitative models that provide a mechanistic understanding of ecological processes fundamental to the conceptual framework and inform future work"* (NSF 2016b, 2018). The term "model, –ing" increased in frequency in RFPs starting in 2012 (Table 3.3), and wording was modified in each successive RPF (Fig. 3.4). Issues with models were cited in 33% of decisions for probation and termination from 2010 to 2018 (Table 3.4).

Theory The term "theory" was absent from RFPs until 2012 although *"general systems theory"* appeared in RFPs of 1987 and 1988; and *"theoretical efforts"* appeared in the RFP of 1997 (NSF 1980, 1981, 1987, 1988, 1997). In 2012, new wording and solicitation-specific review criteria were added requiring *"test[ing] of ecological or ecosystem theories"* (NSF 2012). Wording associated with the term "theory" was modified in each of the successive RFPs (Fig. 3.4, NSF 2014, 2016b,

2018, 2020). Issues with theory were cited in 42% of decisions for probation or termination from 2010 to 2018 (Table 3.4).

Predict From 1980 to 2011 the word "predict" was not mentioned in the RFPs. Starting in 2012, the RFP required research to *"predict ecological, evolutionary, and social responses"* (NSF 2012). In 2016, revised wording requires *"a conceptual framework that describes or predicts"* and *"testing of predictive models"* (NSF 2016b). The 2018 RFP includes the evaluation criterion, *"develop predictions that link processes and observations across levels of organization or across temporal or spatial scales"* and *"[predict] how populations, communities, and other ecosystem components interact"* (NSF 2018). The 2020 RFP includes the evaluation criterion, *"Conceptually-based predictions that link processes and observations across levels of organization (population, community, and ecosystem) or across temporal or spatial scales"* (NSF 2020).

3.3.5 Outcomes of LTER Site Renewal

The outcomes of the LTER renewal process vary by decade. Three of the first eleven sites were terminated in first decade of the program. The fraction of renewal proposals that led to probation increased in the fourth decade. Four sites were terminated in the fourth decade after a period of two decades with no terminations (Table 3.5, Fig. 3.2).

During the fourth decade of the LTER program, review criteria that had been added or changed since the mid-1990s were identified as the basis for decisions for probation and termination more frequently than the original review criteria for the LTER program (Table 3.4). Integration and conceptual framework were the most commonly cited review criteria for decisions for probation and termination in the fourth decade.

The number of renewal proposals a site has submitted is associated with the outcome of renewal decisions. There is an increased likelihood of probation with an increased number of renewal proposals submitted (Table 3.6). Four sites that were terminated in the fourth decade had submitted 4 to 7 renewal proposals. Of the 21 sites submitting their 4th to 7th renewal proposal, 11 have been placed on probation and 4 have been terminated in the 4th decade.

3.4 Discussion

The LTER program has transitioned from 1980 to 2020. In the 1980s, twenty terrestrial and freshwater ecosystems received LTER funding from the BIO Directorate. As of 2018, there are 26 sites with continued funding, equally divided between the BIO and GEO Directorates. From 1980 to 2018, the LTER program had three distinct periods: an initial period of rapid growth with some terminations (1980s), a middle period of slower growth with no terminations (1990s and 2000s), and a third period of no net growth, with added and terminated sites (2010s). As the result of new funding and terminations, the character and composition of LTER sites changed, especially in the 1980s and the most recent decade (2010–2018).

The early period (1980s) had a single NSF program officer and funding from one NSF directorate (BIO). LTER sites in terrestrial and aquatic ecosystems were funded in this first period. Three aquatic ecosystem sites were terminated. The second period (1990s and 2000s) had an increasing number and turnover of program officers. In this period, the original Directorate (BIO) funded no new sites, and at least three additional directorates in NSF (including GEO, SBE, and EHR) contributed funding for LTER sites. LTER funding was extended to include urban, polar, land margin, and coastal ocean ecosystems. In this second period, a number of sites were placed on probation, but none were terminated. The most recent decade (2010s) had very high turnover of program officers, a complete revision of review guidelines, additions and revisions of review criteria, shift in emphasis of review criteria in panel summaries, increased frequency of probation, and multiple terminations of sites. All new sites in this period were marine sites funded by the GEO Directorate, and all sites terminated in this period (grassland, desert, forest, urban) had been funded by the BIO Directorate for several decades.

Changes in the character and composition of LTER sites are associated with many factors, both at LTER sites and in the review process. LTER sites and the review process are simultaneously attempting to respond to changes in the science community and society. LTER sites face both scientific and social challenges in maintaining a long-term research program. A number of factors associated with individual LTER sites may be responsible for their ability to achieve continued funding in successive renewal proposals. These include: (1) ongoing changes in science and the expectations of the science community, (2) changes in site leadership, coordination, and communication, and (3) changing relationships with host institutions and partner institutions. These are described below.

Over time, as science and the scientific community change, LTER site research programs experience an evolution of scientific understanding and priorities. There is a simultaneous evolution in LTER research as expressed in renewal proposals, the renewal criteria, and the scientific community which provides panelists and reviewers. Thus, one explanation for the increased frequency of probation and termination in the 2010s is that sites are unable to adapt to the changing expectations of the RFPs and reviewers. Alternatively, it may become increasingly difficult to propose novel and innovative research at the same site after several decades of continued funding.

Site leadership, coordination, and communication are also important. Creating and maintaining an "integrated conceptual framework" in an LTER project requires prolonged intensive communication and coordination among a large set of project elements and participants. Changes within individual LTER sites may weaken communication and coordination. Changing leadership and composition of researchers within an LTER site, including loss of experienced leaders, varying availability of interested researchers within participating institutions, and lack of budget incentives or effective mentoring of researchers new to LTER may all contribute to an apparent lack of integration, or a conceptual framework which reviewers perceive as inadequate. Such changes may explain why some LTER sites, despite two or more decades of success in long-term research, were perceived as lacking integration or having inadequate conceptual frameworks, leading to probation and termination over the past decade.

Changing institutional understanding and support likely also affect the ability of sites to meet LTER review criteria for renewal. For example, partner or lead institutions may be essential to making key data available. Lead and partner institutions also may provide administrative assistance, cost-sharing or PI salaries, and reduced overhead, all of which may effectively expand the LTER budget and the corresponding scope of the LTER project. Changes in these relationships may undermine the ability of an LTER site to sustain an innovative program.

A number of factors associated with the review and funding process at NSF may also be responsible for the changes in the LTER program. These include: (1) changes in review criteria, (2) continuity of leadership within NSF, (3) experience and expertise of LTER program officers, (4) changes in NSF procedures for managing the renewal process, and (5) continuity and commitment to funding.

Changes in review criteria for LTER site establishment and renewal, and changes in reviewer attention to these criteria, have shaped the LTER program over time. In the early years of the program, LTER review criteria focused on program goals and core areas, establishing and maintaining long-term research, and making data available. However, in the fourth decade, RFPs, reviewers and panelists focused increasingly on conceptual frameworks and integration in renewal decisions, especially for long-running LTER sites. The requirement for a "conceptual framework," introduced in the 1990s, was initially a request for some kind of depiction of how things fit together in an LTER program (S.L. Collins, personal communication). Initially, sites had considerable latitude in how they chose to present the conceptual framework motivating their LTER research, but the conceptual framework has attracted increasing attention from reviewers in the past decade.

Another example is the increased attention to the concept of "integration," since its first appearance in the 2002 RFP, in renewal decisions for LTER sites. The LTER program has always emphasized integration of studies under a common overarching research theme or question(s). Nevertheless, in the past decade the term "integrate" proliferated in successive versions of the RFP, and lack of integration perceived by reviewers was a criterion in almost all (83%) of probation and termination decisions. Although the term "integrate" was largely erased from the 2020 RFP, increased attention of reviewers to conceptual frameworks and integration in the past decade is associated with declining use of the original review criteria for LTER sites in renewal decisions.

Continuity of leadership at NSF also appears to be an important factor influencing the evolution of the LTER program. During periods of sustained leadership lasting more than a decade, individual program officers in BIO/DEB (Tom Callahan, program officer from 1980–1994) and GEO/OCE (Dave Garrison, program officer from 2004--2019) were associated with expanded numbers of LTER sites in these directorates. In contrast, a period of leadership transition with high turnover and multiple, short-term program officers in the past decade (2010–2018) coincided with multiple site probations and terminations of sites, all funded by BIO/DEB. The roles of NSF Division Directors may also have been crucial, but no data on this was available for this study. The LTER experience since 1990 suggests that sustained leadership of LTER programs within NSF is associated with continuity and expansion of the LTER program and LTER sites, while turnover is associated with site probations, terminations, and no net growth.

The research experiences and areas of expertise of individual LTER program officers may also have affected how the LTER program has changed over time. Program officers are responsible for establishing the wording of the RFPs, organizing review panels, evaluating panel summaries and midterm reviews, and ultimately making funding and renewal decisions (Fig. 3.6). Modification and updating of review criteria is part of due diligence on the part of program officers in response to evolving science or societal needs. Program officers' experience in, and attitudes toward, long-term ecological research influence their interpretation of the evolution of scientific understanding and priorities for the LTER program.

Changes in NSF procedures for the LTER renewal process also may have played a role in how the LTER program has evolved. In the first decade of the program, three sites were terminated without probation. The probation process was instituted in the second decade of the program as a means of protecting NSF's investment in long-term research. In multiple instances, sites corrected perceived problems and were renewed after probation, thus preserving and extending the long-term research at these sites. However, the probationary process requires submission of a second renewal proposal 2 years later to a new panel (and in recent years, responding to a significantly revised RFP). Data presented above demonstrate that the more renewal proposals a site submits, the higher is the chance of probation and termination. At some times in the past, when renewal proposals had minor deficiencies, NSF program officers have requested "addenda" (explanatory documents submitted by LTER PIs to program officers) as a means of clarifying issues with renewal proposals rather than placing the site on probation. However, this process has been little used in the past decade, when the rate of probation was higher than in previous decades.

Continuity and commitment to funding is an obvious factor influencing the changes over time in the LTER program. Decisions to allocate funding to long-term ecological research made at the level of NSF Directorates clearly influenced the composition of the LTER program over time. No data were available to explain why LTER funding expanded in the GEO Directorate after 1990, why funded LTER sites in the BIO Directorate decreased from 18 in the late 1980s to 13 as of 2018, or whether funding issues affected LTER sites (i.e., urban sites) with shared funding from several Directorates.

3.5 Conclusions and Implications for the Future of LTER

The history of the LTER program suggests that programs to support long-term ecological research must strike a balance between continuity and change. Continuity in science leadership, administration, and expectations is important, and at the same time research and administration must respond to ongoing changes in science, in ecosystems, and in the long-term study sites. Long-term ecological research spans periods of evolving scientific understanding and priorities. As long-term ecological research matures, the expectations for that research will also grow. If LTER continues into the future, there will continue to be turnover of LTER researchers and NSF personnel as well as evolution of research topics, conceptual frameworks, research methods, and information management technologies. In this changing environment, continuity in administration and expectations can be fostered through communication at four scales: long-term leadership and mentoring of leadership transitions within each LTER site; communication, collective memory, and mentoring within the LTER network; open discussion of LTER program history and management between LTER sites and NSF; and attention by NSF to continuity and experience of LTER program officers, LTER review criteria, and review processes.

At current (4th decade) rates of termination, especially of sites with long-term funding, the LTER program is on a path to lose many of its longest terrestrial ecosystem sites in the next few funding cycles. How will this affect the future of the LTER program and long-term ecological research in general? Alternatively, might NSF and LTER sites make adjustments that could protect the science community's investment in long-term research, at a time of heightened awareness of its value (Kuebbing et al. 2018) and influence (Hughes et al. 2017)?

Given the accelerated pace of environmental change, the need for long-term ecological research is even more urgent today than when NSF established the pioneering LTER program in 1980. LTER sites are valued not only for their novel research, which is critically evaluated in every proposal, but also for their long-term data on core research topics. Both are essential for documenting ecosystem responses to environmental change, for providing a means of predicting responses to future change, and as a basis for environmental policy. The context for, and lessons from, long-term ecological research continue to shift in response to environmental, social, and technological change. The NSF LTER program is designed to be able to respond to all of these drivers of change, through a process of 6-year grants that are renewed based on peer review. The LTER Program history reveals important lessons for how to structure and manage long-term ecological research.

Acknowledgements This work was supported by funding to the Andrews Forest LTER program (NSF 1440409). We thank J. Blair, S.L. Collins, L. M. DiGregorio, C.T. Driscoll, D.R. Foster, S.K. Hamilton, D. Reed, G.P. Robertson, E. Seabloom, F.J. Swanson, J. Van de Castle, and R.B. Waide for information and helpful comments on earlier drafts of this manuscript.

References

Callahan, J.T. 1984. Long-term ecological research. *Bioscience* 34: 363–367.

Foster, David. 2012. Expanding the integration and application of long-term ecological research. *Bioscience* 62: 323.

Gosz, J.R., R.B. Waide, and J.J. Magnuson. 2010. Twenty-eight years of the US-LTER program: Experience, results, and research questions. In *Long-term ecological research: Between theory and application*, ed. F. Müller, C. Baessler, H. Schubert, and S. Klotz, 59-74. Dordrecht: Springer.

Hughes, B.B., R. Beas-Luna, A.K. Barner, K. Brewitt, D.R. Brumbaugh, E.B. Cerny-Chipman, S.L. Close, K.E. Coblentz, K.L. De Nesnera, S.T. Drobnitch, J.D. Figurski, B. Focht, M. Friedman, J. Freiwald, K.K. Heady, W.N. Heady, A. Hettinger, A. Johnson, K.A. Karr, B. Mahoney, M.M. Moritsch, A.K. Osterback, J. Reimer, J. Robinson, T. Rohrer, J.M. Rose, M. Sabal, L.M. Segui, C. Shen, J. Sullivan, R. Zuercher, P.T. Raimondi, B.A. Menge, K. Grorud-Colvert, M. Novak, and M.H. Carr. 2017. Long-term studies contribute disproportionately to ecology and policy. *Bioscience* 67: 271–281.

Kratz, T.K., L.A. Deegan, M.E. Harmon, and W.K. Lauenroth. 2003. Ecological variability in space and time: Insights gained from the US LTER program. *BioScience* 53(1): 57-67.

Kuebbing, S.E., A.P. Reimer, S.A. Rosenthal, G. Feinberg, A. Leiserowitz, J.A. Lau, and M.A. Bradford. 2018. Long-term research in ecology and evolution: A survey of challenges and opportunities. *Ecological Monographs* 88: 245–258.

Chapter 4
Sustaining Long-Term Ecological Research: Perspectives from Inside the LTER Program

Merryl Alber, John Blair, Charles T. Driscoll, Hugh Ducklow, Timothy Fahey, William R. Fraser, John E. Hobbie, David M. Karl, Sharon E. Kingsland, Alan Knapp, Edward B. Rastetter, Timothy Seastedt, Gaius R. Shaver, and Robert B. Waide

Abstract Principal Investigators from several sites within the Long Term Ecological Research (LTER) program offer their insights about how long-term research has been effectively sustained from periods ranging from 20 to 40 years. The sites are: Hubbard Brook (New Hampshire), Konza Prairie (Kansas), Niwot Ridge (Colorado), Arctic (Alaska), Palmer Station (Antarctica), and Georgia Coastal Ecosystems (Georgia). The main themes discussed include: the importance of a strong foundation and common vision, creating a culture of collaboration and cooperation, showing the relevance of research to societal needs, managing conflict resolution, encouraging innovation, facilitating an exchange of ideas, working to build collaborations, willingness to adopt new management structures, and careful attention to transitions in leadership. The conclusion summarizes themes based on this chapter as well as other chapters in the book.

M. Alber
Department of Marine Sciences, University of Georgia, Athens, GA, USA
e-mail: malber@uga.edu

J. Blair
Division of Biology, Kansas State University, Manhattan, KS, USA
e-mail: jblair@ksu.edu

C. T. Driscoll
Department of Civil and Environmental Engineering, Syracuse University, Syracuse, NY, USA
e-mail: ctdrisco@syr.edu

H. Ducklow
Earth and Environmental Sciences Department, Lamont-Doherty Earth Observatory, Columbia University, Palisades, NY, USA
e-mail: hducklow@ldeo.columbia.edu

R. B. Waide, S. E. Kingsland (eds.), *The Challenges of Long Term Ecological Research: A Historical Analysis*, Archimedes 59,
https://doi.org/10.1007/978-3-030-66933-1_4

Keywords LTER Program · Long-term ecological research · Scientific collaboration · Scientific culture · Scientific leadership · Team science · Program sustainability

4.1 Introduction

When the Long Term Ecological Research Program was created in 1980, investigators were told that they would be expected to make long-term commitments to the research program, although nobody knew just how long that meant. The original funding commitment was only for 5 years, but in 2020 the program reached its fortieth anniversary, with four of the original six sites still part of the Network. In addition to a long-term commitment, sites were expected to form a "network" by engaging in cross-site comparisons and collaborations. Moreover, sites were expected to implement a multi-disciplinary approach involving teams of researchers.

T. Fahey
Department of Natural Resources, Cornell University, Ithaca, NY, USA
e-mail: tjf5@cornell.edu

W. R. Fraser
Polar Oceans Research Group, Sheridan, MT, USA
e-mail: bfraser@3rivers.net

J. E. Hobbie · E. B. Rastetter · G. R. Shaver
The Ecosystems Center, Marine Biological Laboratory, Woods Hole, MA, USA
e-mail: jhobbie@mbl.edu; erastetter@mbl.edu; gshaver@mbl.edu

D. M. Karl
Center for Microbial Ecology: Research and Education, School of Ocean and Earth Science and Technology, University of Hawai'i, Manoa, HI, USA
e-mail: dkarl@hawaii.edu

S. E. Kingsland
Department of History of Science and Technology, Johns Hopkins University, Baltimore, MD, USA
e-mail: sharon@jhu.edu

A. Knapp
Department of Biology and Graduate Degree Program in Ecology, Colorado State University, Fort Collins, CO, USA
e-mail: aknapp@colostate.edu

T. Seastedt
Department of Ecology and Evolutionary Biology and Institute of Arctic and Alpine Research, University of Colorado, Boulder, CO, USA
e-mail: timothy.seastedt@colorado.edu

R. B. Waide (✉)
Department of Biology, University of New Mexico, Albuquerque, NM, USA
e-mail: rbwaide@unm.edu

All of these expectations resulted in significant social and cultural changes within the discipline of ecology, as discussed by Waide (2016). At first, it was necessary for the sites to improvise approaches to long-term research. For cross-site comparative work, sites had to come up with more broad-based research questions, which in turn required new standards for data, coordination of approaches, and recruitment of experts in information management. How the sites met the challenge of conducting cross-site research is discussed in Chap. 12 by Magnuson and Waide. In this chapter, we take up another problem, namely the social organization required to manage these complex, multi-disciplinary, long-term projects. As Waide (2016) has explained, understanding long-term processes requires constant supervision, close collaboration, and careful budgeting of time and resources.

The challenges faced by long-term research projects are considerable (Gosz et al. 2010). Sites must focus on long-term objectives, while also addressing new research opportunities that arise as knowledge expands about the ecosystem under study. Sites must foster a culture of collaboration and cooperation, which can be counter to the usual competitive ethos of science. Over time, sites are expected to accomplish more, while budget increases are constrained. As research initiatives accumulate over time, and as the scientific community makes increasing demands to use data or work at the sites, the support for research has not kept up with these demands. Sites must apportion funds carefully, and support projects with other sources of funding to accomplish goals. This site research, beyond the direct LTER research, brings in on average 2.8 times the support provided by the LTER program. Managing and coordinating these projects increases the importance of strong leadership (Waide 2016). Moreover, working on long-term research requires a certain level of altruism. Not only are investigators working as part of multi-disciplinary teams, but answers to questions can take decades and extend beyond an individual's career.

In this chapter, past and present Principal Investigators at a variety of sites, located in very different types of ecosystems, offer their perspectives on the social and cultural aspects of sustaining long-term ecological research. The sites are Hubbard Brook (New Hampshire), Konza Prairie (Kansas), Niwot Ridge (Colorado), Arctic (Alaska), Palmer Station (Antarctica) and Georgia Coastal Ecosystems (Georgia). Each site has successfully maintained its research program for between two and four decades. In this chapter, we offer reasons for these successes, ideas about what might have been done differently, and evaluations of continuing challenges. The main themes discussed include the following: the importance of a strong foundation and common vision, fostering a culture of collaboration and cooperation, showing the relevance of research to societal needs, managing conflict resolution, encouraging innovation, facilitating an exchange of ideas, working to build collaborations, willingness to adopt new management structures, and careful attention to transitions in leadership. We end by summarizing these themes and expanding on them based on our knowledge of the full range of sites in the LTER Network as well as conclusions from other chapters in this book.

4.2 A Strong Foundation and a Common Vision: Hubbard Brook LTER, New Hampshire

The Hubbard Brook LTER project in New Hampshire joined the LTER Network in 1987, building on a strong foundation of small-watershed-based ecosystem research going back to 1963. Timothy J. Fahey and Charles T. Driscoll were both early career scientists at the time they assumed leadership of the new LTER in 1987, and they continued in that role for 30 years until 2016, when they both decided to step down. This LTER project benefited from an earlier founding vision that produced important scientific results well before the start of the LTER program. Its scientific accomplishments, discussed in a recent book by R. T. Holmes and G. E. Likens (2016) have long been used to support the value of long-term research. The initial vision has been further sustained by careful cultivation of a research community that supports it. Equally important has been continuity of leadership. What follows is their description of the organizational structure of this LTER and their assessment of the reasons for the program's success.

4.2.1 Tim Fahey and Charles Driscoll on the Hubbard Brook LTER

The Hubbard Brook Ecosystem Study (HBES) was the progenitor of the Hubbard Brook LTER project and in some respects for the entire LTER Program. The HBES originated in 1963 as a cooperative endeavor between university scientists, F. Herbert Bormann, Gene E. Likens, both ecologists, and N. M. Johnson, a geochemist, and the U.S. Forest Service (USFS). Robert S. Pierce, a USFS scientist and the lead scientist for the Hubbard Brook Experimental Forest (HBEF), had sought academic involvement to increase the scientific capacity and scope of the HBEF. Building upon the insight of Bormann, the HBES applied the small watershed approach to quantify forest biogeochemistry, constructing budgets for water and mineral elements, their response to experimental manipulations and long-term changes. Long-term measurements of meteorological, hydrological and biogeochemical parameters were conducted, and Richard T. Holmes was monitoring breeding bird populations. This work was supported by U.S. Forest Service and a series of competitive research grants from the National Science Foundation (NSF). Over its first decade this research led to the discovery of acid rain in North America and its impact on terrestrial and aquatic systems and showed that clear-cutting and other natural disturbances in forests severely disrupted the nitrogen cycle. These results validated the small-watershed approach to the study of ecosystem processes. Well before the start of the LTER program, therefore, the HBES was recognized as an important long-term research study.

The HBES was coordinated by a Scientific Advisory Committee (SAC) that served as an intellectual advisory group and clearinghouse for proposed research at the HBEF. The Scientific Advisory Committee consisted of the original lead principal investigators, including the U.S. Forest Service project leader who had ultimate responsibility for approving research at the site. For many years the overall HBES was small enough that this management approach worked well. Graduate students and research associates of the HBES were housed at a common facility, Pleasant View Farm (PVF), which served as an intellectual center for the science conducted.

In 1987 the HBES joined the LTER network under the leadership of Timothy J. Fahey from Cornell University, and Charles T. Driscoll from Syracuse University. Both were early career scientists who had been conducting research at the site. The original LTER proposal advanced the theme of the response of northern hardwood ecosystems to disturbance, which has remained the central organizing concept of the project ever since. Participating institutions included Cornell University, Dartmouth College, the Institute of Ecosystem Studies (now the Cary Institute of Ecosystem Studies), Syracuse University, Yale University and U.S. Forest Service. The pre-existing Science Advisory Committee continued to coordinate the overall HBES and incorporated the two Principal Investigators of the HBR LTER project. Several new long-term measurement activities were added to the HBES, aimed at "taking apart the black box" of the small watersheds and scaling this understanding up to the broader Hubbard Brook valley and White Mountain and Northern Forest regions.

In addition, long-term monitoring of breeding birds and their food sources was maintained with LTER funding. Notably, funding for the breeding bird research was eliminated from the LTER project for one round of funding at the direction of an NSF panel; the bird research was continued under the auspices of a different program, Long Term Research in Environmental Biology (LTREB), which was created in 1980 to complement the LTER program, and was intended to support research in ecology, systematic biology, population biology, physiological ecology, and ecosystem studies. The bird-monitoring studies at Hubbard Brook have since been returned to the core LTER. For many years another LTREB helped to maintain long-term biogeochemistry measurements for which LTER funding was insufficient; most recently funding limitations have forced the discontinuation of some long-term measurements (e.g., reducing the stream monitoring network).

In 1996, in response to a marked increase in the complexity of the HBES and a steady increase in the number of participating researchers, the governance structure of the HBES and HBR LTER was reorganized. At about the same time the SAC had initiated a "Friends' group" and science-interface organization, the Hubbard Brook Research Foundation (HBRF), to help coordinate and secure funding for education and outreach activities as well as housing and facilities management challenges. A Committee of Scientists (COS) was formed, consisting of all the principal investigators participating in the overall HBES (at the time about 50 scientists). The COS acts as the center of the current governance structure of the HBES and meets quarterly to discuss scientific issues of interest. The July COS meeting occurs in tandem with the Annual Cooperators' meeting, which includes a community outreach event with a general interest speaker and field trips; the formal Cooperators meeting where all scientists working at the HBEF (undergraduate, graduate students, post-doctoral associates, senior scientists) give short research presentations; and a suite of committee and subproject meetings. A smaller executive group, the Scientific Coordinating Committee (SCC) provides leadership for the COS fostering integration, synthesis, encouraging new scientists to work at the HBEF and promoting interaction, communication and outreach. The SCC includes eight members, four elected by the COS, an HBR LTER lead PI, the U.S. Forest Service lead scientist, the Director of the HBRF and an external advisor. Although not all members of the COS are supported financially by the HBR LTER project, a rather seamless structure of the HBES and HBR LTER has developed over the years under this governance structure.

The Scientific Coordinating Committee and Committee of Scientists of the HBES support three active sub-committees. The Research Approvals Committee evaluates and approves proposed research at the HBEF and prevents conflicts. The Committee is chaired by a Forest Service lead scientist, since the Forest Service has ultimate responsibility for approving research at the HBEF. The Information Oversight Committee maintains the joint HBES and HBRF website (https://hubbardbrook.org/) and is responsible for organizing data management and the sample and document archives of the project. The Education and Outreach Committee works closely with the HBRF on various activities that fulfill the commitment of the HBES to Broader Impacts (part of the evaluation criteria used by NSF). The HBRF itself manages facilities for housing researchers and visitors to the site as well as some laboratory space adjacent to the PVF dormitory. With the growth of the project during the 1980s and 1990s, the PVF dormitory was inadequate to house the large population of researchers during the field season, and the HBRF conducted a major capital improvement project, based on private donations and Federal government support, that resulted in the purchase and improvement of the Mirror Lake campus; the campus provides housing for about twenty-five researchers as well as some dry laboratory space.

The HBRF has also supported a variety of major education and outreach projects, most notably the Science Links series that has covered such topics as Acid Rain, Nitrogen Deposition, Mercury Pollution, and Carbon Management with the aim of actively informing key policy makers and the public on issues of environmental concern. Most recently, the HBRF was funded by the NSF Education directorate to support an investigation of public engagement in science, evaluating a model of community-scientist partnerships in guiding the research endeavors of the HBES. COS members are active participants in all the endeavors coordinated by the HBRF.

The two lead PIs of the HBR LTER continued in this role for 30 years from 1987 until 2016 when they decided to step down. The decision on transferring LTER project leadership was somewhat informal, as the lead PIs worked with several senior co-PIs to identify the ideal new leadership team (Gary M. Lovett and Peter M. Groffman from the Cary Institute) who agreed to serve for at least one 6-year round of LTER funding. Shifts in the responsibility for core research activities have not, in general, been the subject of COS group deliberations but rather have been decided on a case by case basis by the researchers involved in consultation with the lead PIs. Several such major changes have occurred in recent years, including monitoring of precipitation and stream chemistry, breeding bird populations and forest vegetation.

The HBES and HBR LTER have remained a vigorous research program through various challenges partly because of the cordial interactions among the scientists and U.S. Forest Service, but especially because of a common vision to understand all aspects of an iconic ecosystem. This vision has also played a key role in the recruitment of young scientists to join the project. Rather than actively identifying and recruiting new participants, we have been successful in attracting young scientists who express interest because they recognize how their research can contribute to better understanding the Hubbard Brook ecosystem, the value of conducting their research in the context of the long-term measurements of ecosystem structure and function, and the collegial environment for conducting ecosystem science. Typically, these scientists are encouraged to contribute a talk or participate at one of the COS meetings to assure a good fit with ongoing research and to integrate ideas into the overall HBES.

One feature of the HBES and HBR LTER that might be predicted to be a curse has proven to be a blessing; the home institutions of the project participants are exceptionally widely dispersed. For example, at the time of this writing 23 different institutions are represented among the active members of the HBES Committee of Scientists. The upshot is that special effort has been continuously expended to assure a high level of communication among project participants. The fall-winter-spring quarterly meetings of the COS are very well attended (averaging roughly 40) and the summer Cooperators' meeting attracts over 100 participants who are fully committed by being away from responsibilities at their home institutions.

These meetings usually include a thematic session that explores new directions for the project, and initiates planning for competitive grant proposals and for synthesis efforts. Coalescence of research teams within the COS has been fostered during informal interactions at COS meetings. Moreover, non-attendance at these meetings serves as a clear signal that particular team members have moved away from active involvement in the HBES, serving to prune the project without acrimony. Ultimately, the success of the HBES and HBR LTER is measured by its contributions to ecosystem science and by maintaining a steady stream of funding to facilitate those contributions. Our ability to remain a productive research operation is a testament to the vision of our founders, to the iconic ecosystem study they initiated and to the hundreds of dedicated scientists, staff and students who have been inspired to carry on the tradition of the HBES.

4.3 Collaboration and Cooperation: Konza Prairie LTER, Kansas

Konza Prairie was part of the original cohort of six sites that began the LTER program in 1980 and is still in the Network in 2020, along with three other original sites (Andrews Forest, North Temperate Lakes, and Niwot Ridge). Located in northeastern Kansas, it consists mostly of native tallgrass prairie vegetation. Alan Knapp and John Blair were lead Principal Investigators in 1991–1999 and 1999–2017 respectively. They discuss the particular challenges faced by this site, which, unlike Hubbard Brook and many of the other sites, did not have a long history of scientific accomplishments to draw on. But it did have an important institutional asset that served as the focal research site, the Konza Prairie Biological Station, named for the indigenous Kansa people. Founded in 1971 by Lloyd C. Hulbert, the biological station supported ecological research by faculty at Kansas State University, and in partnership with The Nature Conservancy expanded in size during the 1970s. As Knapp and Blair discuss in the following section, support from Kansas State University was crucial to the success of the Konza Prairie LTER, but efforts had to be made to ensure that the level of support did not fall victim to complacency. Also important was development of a collaborative leadership model, cultivation of a cooperative and congenial group of investigators, and maintaining effective communication within a large, diverse, and geographically separated group of investigators.

4.3.1 Alan Knapp and John Blair on the Konza Prairie LTER

The Konza Prairie (KNZ) LTER program was initially funded in 1980 as one of the six original LTER sites. KNZ was unique among most of this first cohort, as the Konza Prairie Biological Station (KPBS) was still a relatively new research site in 1980 and lacked a long history of sampling protocols, datasets, and research accomplishments to build upon. The original lead Principal Investigator (PI) was Dick Marzolf, a limnologist (and also the first LTER Network Chair). Marzolf left Kansas State University (KSU) in the mid-1980s and was followed as KNZ lead PI by Don Kaufmann (a mammalogist) in 1986, and then by Tim Seastedt (an ecosystem ecologist) in 1988. Seastedt left KSU in 1991, and Alan Knapp (trained as a plant ecophysiologist) became lead PI and continued in the role until 1999. John Blair (an ecosystem ecologist) joined KSU in 1992, and was recruited with the intention that he would eventually assume the role of lead PI. Blair took over following the KNZ mid-term review in 1999, and led his first KNZ renewal proposal in 2002. Blair continued as lead PI through two additional KNZ renewals until the 2017 mid-term review when the program transitioned to Jesse Nippert (a plant ecologist who earned his Ph.D. at Colorado State University with Knapp) as lead PI.

The rapid succession of leadership during the establishment years of the KNZ program could be viewed as a time of risk, as termination of some well-established LTER programs has coincided with a disruption in leadership. However, in many ways these leadership transitions shaped the evolution of the KNZ LTER management model and the resilience of the program. Management and leadership of KNZ evolved from a more traditional single PI/Director model to a broader, more collaborative model where the co-PIs provide much greater input and play a more significant role in directing and managing all aspects of the program. This collaborative "group leadership" model began when Knapp, a relatively young PI and Assistant Professor, was charged with leading the program following the departure of Seastedt. Knapp successfully led the KNZ program through this time of transition with strong support from co-PIs John Briggs, Dave Hartnett, and Don Kaufmann (who were also co-PIs on earlier LTER grants). This mode of collaborative leadership continued throughout subsequent leadership changes and has served KNZ well for over 30 years.

In this model, the PI serves as the primary point-of-contact at the university, network and NSF levels and coordinates project activities. However, the co-PIs provide regular significant and meaningful input on issues ranging from research foci to personnel and budget decisions. KNZ LTER renewal proposals are a collaborative effort, with input from all co-PIs and senior investigators and the PI providing integration and a "single voice" for the final product. In many ways, the success of the KNZ LTER program is built on this shared intellectual leadership, as well as continuity of former PIs as co-PIs with substantial roles. As a result, KNZ has been a successful test of the LTER vision of supporting long-term, site-based research managed in such a way that turnover of individual investigators and/or completion of scientific careers is not detrimental to the research program.

Research Strategies and Challenges To a large extent, the KNZ LTER program and the Konza Prairie Biological Station (KPBS) site evolved together. The KPBS site was established under the guidance of Lloyd Hulbert, a Kansas State University plant ecologist and early KNZ investigator, to conduct research on the structure, composition and functioning of tallgrass prairies, with a particular focus on three key factors – fire, grazing and climate – that have become widely recognized as essential to the formation and maintenance of tallgrass prairie and other mesic grasslands worldwide. Hulbert's vision for the KPBS site included whole watershed manipulations of a range of fire return intervals and the re-introduction of native grazers (bison) as part of the site-based experimental design, along with studies that were sufficiently long-term to assess and document responses to a variable continental climate. This focus on fire, grazing and climate as essential and interactive drivers of ecological processes became the core of the KNZ LTER research program. Of course, the KNZ LTER program expanded upon this theme over time including adding new treatments to the site-based experimental design and developing numerous smaller-scale experiments that complement the watershed studies. However, having a recognizable and consistent theme – "the Konza brand" – has provided a unique identity for the KNZ program, and has been an important framework around which the program has developed over multiple funding cycles. In fact, a key to the growth and success of the KNZ LTER research program has been a highly cooperative relationship between KPBS as a site and the KNZ LTER research program. While these independent entities have somewhat different missions and priorities, all KPBS Directors have also been co-investigators in the KNZ program, which has contributed to the success of both.

The initial KNZ LTER proposal focused on a few key questions and establishing long-term sampling sites, protocols and datasets. However, the breadth and complexity of the KNZ program has grown with each successive renewal. As the scope of research and number of experiments and datasets expanded, KNZ investigators have had to grapple with many new challenges, including the tension between maintaining long-term studies that increase in value with time versus initiating more "exciting" new research that is typically viewed as essential to the success of each renewal. Given limited funding, personnel, and time, difficult choices must be made about when to discontinue specific experiments or data sets in order to initiate new studies. This tension is real from both a fiscal and investigator perspective. In other words, it is a challenge to adequately fund ongoing research *and* allocate funds for new studies, and equally challenging to increase the number of new investigators to permit both. However, new investigators are the life-blood of vibrant and successful LTER programs. They inject fresh perspectives and ideas into each funding cycle for building on long-term data and studies, as well as initiating new experiments and research directions.

Konza Prairie LTER has dealt with the challenge of allocation of LTER funds in several ways. For example, the majority of funds are dedicated for core personnel (e.g., field and laboratory assistants, information managers) and core research activities (e.g., research facilities and experiments or datasets) that are used by multiple LTER investigators rather than individual investigator interests. This model works well with investigators that are somewhat altruistic and whose research interests overlap with, or closely complement, core LTER activities. KNZ also invests heavily in graduate students. This investment both supports core KNZ activities and provides incentive to encourage investigator involvement in the program. Many of these former graduate students remain active in Konza research today and they bring their own students to the site.

KNZ has also been successful at leveraging LTER funds with external grants to initiate new research. These projects have often been folded into KNZ LTER research after the initial high costs of set-up have been borne externally. For example, the significant grassland restoration component of the present KNZ LTER program, currently overseen by co-PI Sara Baer (University of Kansas and a former KSU and KNZ graduate student) was initiated with non-LTER funding in the 1990s. Knapp (PI), Blair, and Scott Collins initially led this grant. Baer has continued to leverage LTER data and generate outside funding for this research, but core support for this research and an expanded KNZ focus on restoration ecology has been provided by the KNZ LTER program. A similar strategy was used to initiate KNZ climate manipulation experiments that required significant infrastructure at the onset, such as the Rainfall Manipulation Plots (RaMPs) project, which Knapp (PI) and Blair initially established with external support from multiple agencies. The ongoing Climate Extremes Experiment was established in a similar manner by co-PI Melinda Smith (Colorado State University) and another former KSU and KNZ graduate student.

Long-term data and studies *define* successful LTER programs and while many new questions emerge naturally and obviously from long-term studies, others are unanticipated and serendipitous (e.g., taking advantage of long-term data to better understand unique events such as wildfires, extreme climate events, etc.). LTER sites are uniquely poised to capitalize on these unexpected events when they occur, and KNZ has taken advantage of this several times, as have other LTER sites.

Management Strategies and Challenges Initiating an LTER program at a new research site provided obvious challenges and early KNZ investigators put a great deal of effort into "catching up" with sites that had a much longer and stronger history of research activities and accomplishments. This motivated KNZ investigators to work diligently to meet LTER program expectations, and to garner significant institutional support for the developing program. For example, KNZ played a significant role in development of Information Management policies under the direction of Briggs, and KNZ was the first to publish a volume in the LTER synthesis book series in 1998 (*Grassland Dynamics*, edited by Knapp, Briggs, David Hartnett, and Scott Collins). Over time KNZ transitioned from a "new" site and program in the 1980s to one of the most successful LTER programs.

One obvious and perhaps unavoidable byproduct of this transition is institutional fatigue or complacency in providing support for the program. Strong institutional support from Kansas State University was essential to the early success of a young program and its emergence as a mature LTER site. This support has included reduced indirect cost recovery to allow greater investment into building the program, as well as financial support for the developing KPBS site, prioritizing targeted replacements for key LTER scientists that left, and hiring new scientists with research interests that would benefit KNZ. However, motivation for continuing this kind of support can be difficult to maintain, particularly as university administrators turn over, and present-day administrators view KNZ as a well-established and successful program and, consequently, turn their attention to supporting newer programs. As a result, KNZ PIs today must continually work to battle complacency and maintain a high-level of institutional awareness and support in order to ensure continued success of the program.

KNZ PIs encourage use of the KPBS site and KNZ LTER data by new investigators through a variety of mechanisms, including: on-site research support, support for new graduate students, opening the application for LTER supplements to new researchers, providing LTER "seed money" to researchers that get involved during the course of LTER VII (i.e., the seventh funding cycle) and, if feasible, more formal subcontracts. The KNZ program also strives to engage junior faculty and to promote institutional diversity and gender equity in KNZ leadership positions. The

KNZ program has been very successful at attracting new investigators, from both KSU and other institutions.

However, a by-product of this growth is the challenge of effective communication and cohesion among a large group of diverse and geographically separated investigators. Initially, the KNZ program was comprised almost exclusively of KSU faculty, with most housed in the same building on campus. Frequent hallway conversations were an effective communication mode. But now, as the program has become more diverse disciplinarily and geographically, communication between on and off-campus co-PIs and among scientists in disciplines with different cultures and jargon has become a challenge.

Under the leadership of Blair, KNZ has dealt with this challenge in several ways. All investigators (at KSU and at other campuses) are included on e-mail lists through which information and requests are distributed. In addition, a Konza listserv (Konza-l) provides a means of broadcasting announcements and disseminating information to all KNZ and KPBS researchers. During the academic year, monthly meetings for all KNZ scientists and students are held, with off-campus participants joining by teleconference (*e.g.*, Zoom). On an annual basis, a Konza Prairie LTER Workshop is organized. This meeting is attended by researchers, students, KNZ support staff, K-12 educators, and docents. These workshops include oral and poster presentations, with ample opportunities for informal interactions, as well as a formal planning meeting.

A challenge faced by LTER sites that have been funded over long periods is coping with the shifting landscape of NSF and programmatic expectations. For example, there have been changes in LTER renewal guidelines over time, and changes in the expectations regarding allocation of time and effort towards site-based research activities vs. broader cross-site and LTER network activities. It has been important to have a management model that is adaptive and responsive to these changes. In the case of KNZ, the organization of major research groups has been adjusted to accommodate changes in emphasis on site-based and cross-site research. In addition, KNZ has endeavored to maintain a high level of activity in LTER Network activities among both younger and more senior investigators.

Final Thoughts Although there may be multiple models for a successful LTER program, the key to the success of the KNZ LTER effort has been a highly collaborative and congenial group of investigators who enjoy working together. This is due in large part to a history of PIs and co-PIs that value and encourage collaboration, think broadly and inclusively about the discipline of ecology, appreciate the value of long-term research, and work to cultivate these characteristics in young investigators so that they can assume future leadership roles. The payoff from this high level of collaboration and comradery is evident in the number of former KNZ investigators, former graduate students, and second and third generation KNZ scientists that continue to work at the KPBS site and contribute to KNZ's success.

4.4 Addressing Societal Needs: Niwot Ridge LTER, Colorado

The Niwot Ridge LTER site in Colorado was also part of the first cohort of LTER sites established in 1980. Located in the high mountains of the Rockies, the landscape includes forest, tundra, and lakes, and is the sole alpine LTER site. Timothy Seastedt became Principal Investigator at Niwot Ridge in 1992, after serving as Principal Investigator at Konza Prairie LTER from 1988 to 1991. As he explains in the following section, when he joined the Niwot Ridge LTER there was increasing emphasis on cross-site comparative research, in order to develop the "network" aspects of the LTER program (discussed in more detail in Chap. 12 by Magnuson and Waide). But Niwot Ridge was in some respects unique because it was an alpine site, and cross-site research would not be as easy there as at grassland sites like Konza Prairie. Instead, Niwot Ridge's longevity was aided by the way its research findings helped explain the processes that affected the availability of water for human populations. Seastedt's account of the challenges faced by the Niwot Ridge LTER focuses on the question of how the longevity of LTER sites can be influenced by the relevance of their research to society's needs. However, the fact that portions of the LTER land are owned by the city of Boulder sets limits to where experimental manipulations can be undertaken.

4.4.1 Tim Seastedt on the Niwot Ridge LTER

The end of the first decade of the LTER coincided with new marching orders from NSF. We were to become a network, involving cross-site studies to expand the generality of our findings. The LTER network had by then a number of replicated grassland sites, but by 1992 I had become the Principal Investigator of the Niwot Ridge, the sole alpine LTER site with essentially no true cross-site partners. Nonetheless, a number of stand-alone inter-site research projects such as a long-term decomposition experiment (Parton et al. 2007) emerged where numerous sites could participate, and we could interact with the arctic LTER site (Walker et al. 1999) as well as become the 'cold regions outlier' to grassland studies (Knapp and Smith 2001). Clearly, however, these inter-site activities were not sufficient to justify our alpine research focus.

Instead, our success in the 1990s and into the early twenty-first century in part was based on the recognition that we were the sole high elevation system needed to understand an absolutely critical piece of the controls on western US water availability. By the late 1990s, climate models had developed more credibility, precision, and certainty regarding the changes in precipitation and water budgets in the coming decades. High elevation systems were responsible for scavenging 70% of western US water from the atmosphere, and the fate of that water in terms of the intensity and seasonality of water provisioning was certainly on the radar screens of policy makers. And, thanks to a few droughts, stakeholders recognized this as well.

Our position was strengthened by partnering with a very strong carbon flux (Ameriflux) study site developed by Russ Monson. Ameriflux is a network of principal-investigator managed sites that measure carbon dioxide, water, and energy fluxes in ecosystems in North, Central, and South America. Launched in 1996, Ameriflux started with a set of fifteen sites in 1997. Monson's site, which began in 1998, was in a subalpine forest ecosystem in Colorado. Partnering with that study site made our subalpine forest another benchmark data set showing relationships between water availability and long-term carbon dynamics.

That effort was subsequently expanded to include alpine flux towers which have documented substantial differences in the alpine and subalpine response to warming (Knowles et al. 2015; Knowles et al. 2019). Behind this solid "need to know" argument for energy and precipitation relationships, our alpine researchers continued to make inroads into understanding outcomes of biotic and abiotic interactions across a remarkably beautiful and heterogeneous landscape. As was done at Konza Prairie, our success involved having a diverse group of scientists actively publishing in a broad array of biogeochemical, ecosystem, and community ecology subjects.

Our within-site activities in the alpine LTER have been criticized for a lacking of coupling of our terrestrial and aquatic components. That criticism is perhaps universal for all LTER sites with large terrestrial and aquatic components to their research. We have continued the legacy of Hubbard Brook and Coweeta by publishing entire watershed-level findings. Unlike Konza Prairie, however, which was a site initially designed as an experiment and managed by a single LTER partner, our alpine site was owned by two different entities. Upper regions are owned and controlled by the U.S. Forest Service, an agency willing to host responsible experimental research. This has allowed the National Ecological Observatory Network (NEON) to colocate on our site, so we have a partner at last, even if our site footprints strongly overlap.

The lower regions of our LTER are owned by the City of Boulder, which views this drainage as a pristine water supply. Not surprisingly, this is not a welcoming group when it comes to large-scale manipulative research. Thus, while we could monitor all portions of our study area, we could not manipulate portions of it. Our efforts to incorporate the experimental and modeling approaches used to interpret terrestrial-aquatic couplings remain an ongoing challenge.

It's now been nearly four decades since the LTER program began its functions, and the watershed and plot manipulations of the original LTER programs are often as old – and sometimes older – than the LTER itself. The combination of experimental manipulations and length of observations is somewhat unique to ecology, at least in this country. Having survived as a research program for this long, our understanding of those parts of the ecosystem that change slowly relative to the rest of the ecosystem are now dominating much of the science. Our future and that of ecosystem ecology as a discipline in general is our ability to interface societal issues of conservation and restoration with sustainability science. Only a focus that concurrently addresses biological and physical processes as an integrated whole is likely to produce the insights needed to provide mitigation and adaptation strategies for dealing with our rapidly changing environment.

4.5 Collaboration, Conflict Resolution, Innovation: The Arctic LTER, Toolik Lake, Alaska

The Arctic LTER site is located in the foothills region of the Brooks Range, North Slope of Alaska, and the project is based at the University of Alaska's field station at Toolik Lake, where research began in 1975 (Hobbie and Kling 2014). It became part of the LTER Network in 1987. Institutional support is provided by The Ecosystems Center at the Marine Biological Laboratory in Woods Hole, Massachusetts. The LTER's links to longer traditions of ecological research in the Alaskan Arctic are discussed in Chap. 7. Edward B. Rastetter, Gaius R. Shaver, and John E. Hobbie, all based at The Ecosystems Center, have been involved in the LTER and in arctic research for many decades. Hobbie and Shaver have both served as lead Principal Investigators of the LTER, and Rastetter is the current lead Principal Investigator. Their analysis, presented in the next section, attributes the longevity of this LTER to fostering a culture that emphasizes a collaborative, team-based approach, but which at the same time is open to bringing in complementary projects that attract new researchers and therefore new ideas. They also discuss how potential conflicts are handled and resolved, and governance structures that emphasize consensus-building.

4.5.1 Ed Rastetter, Gus Shaver, and John Hobbie on the Arctic LTER

Since 1987, the Arctic LTER (ARC-LTER) has been a major driver in making the landscape around Toolik Lake the most intensively studied tundra ecosystem in the world (Metcalfe et al. 2018). In part to maintain closer parity between their logistical support for arctic and antarctic research, the NSF Office of Polar Programs (OPP) maintains infrastructure at the field station and provides user-day support and helicopter time through a cooperative agreement with the University of Alaska, Fairbanks, and a contract with Polar Field Services, Inc.

Research funding for other projects attracted to Toolik by the groundwork established by the ARC-LTER now exceeds direct funding to the LTER by several-fold. Thus, in addition to potential conflicts arising from funding needs within the LTER, this heavy use of the landscape also leads to potential conflicts, both spatial and intellectual, between the LTER and other projects.

One way to avoid conflict is to prioritize funding of collaborative research. Funding for the LTER covers 3 months per year for the principal investigator (PI),

0–2 months for the co-principal investigators (co-PI), a full-time senior research assistant in each of the four focus areas, summer research assistants and research experience for undergraduate students (REUs), our school-yard program that sponsors two to four K-12 teachers per year, travel to and from Toolik Field Station (TFS), an annual meeting of affiliated scientists, postdoctoral fellows and students, plus equipment and supplies. Most of the LTER funding is used to support maintenance of core experiments, long-term observations, network collaborations, and data management and synthesis, leaving only a small amount to expend on any new endeavor. ARC-LTER resources are allocated to obtain maximum leverage from collaborating projects and the LTER does not pay for *anybody's* individual research (except perhaps to help new investigators get started). Virtually all ARC-LTER funds are expended on creating and maintaining opportunities for long-term collaborations.

Land-use conflicts are avoided in large part because the land belongs to the Bureau of Land Management (BLM) and permits for research are required; the holder of a current permit could simply request that a conflicting permit not be granted. In addition, the Toolik Field Station Environmental Data Center (EDC) maintains maps of the location and use of past and current plots that have either been manipulated experimentally or sampled in a way that might interfere with future use. It is expected practice that any newly planned experiments or measurements first be reviewed by the Environmental Data Center. Care is taken not to interfere with long-term records monitored by the LTER or any other Toolik-based project. Any land-use conflicts that do arise are resolved through collegial negotiation. Toolik Field Station, as the holder of the NSF cooperative agreement for the field station, helps to mediate these land-use conflicts.

Intellectual conflicts, per se, are of course not a problem. They are what drives science and are appropriately aired in the peer-reviewed literature. The ARC-LTER also offers opportunities for public debate through presentations at our annual meetings and there are ample opportunities for intellectual discussion at the annual meeting and because of the close community and shared meals at the TFS. The more problematic issue is priority in use of data. The ARC-LTER data base includes not only core LTER data, but also data from most of the affiliated projects. Each data set is labeled with the name of the researcher responsible for the data. It is expected practice that anyone using the data first consults with that responsible person, both to avoid conflict and to assure proper interpretation and representation of the data. ARC-LTER core data are freely available and control of other data is left to the researcher listed as responsible for the data but under the guidelines set by NSF for data accessibility. When conflicts do arise, they are handled collegially, generally caught before publication, but occasionally require post hoc corrections or attributions. In the end, a reputation for either abusing the norms of intellectual property rights or for holding data too tightly or for too long does not serve a researcher well in the long-term and can be avoided by simply appealing to common sense.

Reflecting the collaborative, team-based legacy of the International Biological Program, participants in the ARC-LTER come from multiple disciplines and institutions throughout the US and the world. The LTER is governed by consensus of an executive committee that represents the full research domain of the project. This approach forces a broad ecosystems perspective that encompasses interactions among landscape components as well as a focus on component-specific questions. The executive committee has representatives from the four focus areas of the ARC-LTER: (1) terrestrial ecology, (2) stream ecology, (3) lake ecology, and (4) land-water interactions. Each of these representatives is a co-PI on the ARC-LTER grant, typically has a sub-contract on the grant, and, with the aid of a senior research assistant assigned to each of the four focus areas, is responsible for coordinating research within their specific domain. The members of the executive committee have primary responsibility for coordinating among the focus areas and for each of the focus areas making progress that advances knowledge aligned with the project's overall conceptual framework. Decisions on work within each focal area are also made by consensus, but ultimate distribution of resources, primarily in the form of logistical support, is made by the executive committee.

Transition of leadership, for both the PI and for the co-PIs in charge of each of the focus areas, is also done by consensus of the executive committee with solicited input from all affiliated scientists. In anticipation of this turnover, we ask candidate replacements to sit in on executive committee meetings at least 1 year ahead of the transition. This turnover inevitably results in changes to the research focus, but only after critical evaluation of the effects on the long-term data record and the LTER mission. These changes are of course reflected in the renewal proposal.

The executive committee meets formally at least twice a year, a tele-conference in the fall and a face-to-face meeting in the spring in an all-day session just before the annual project meeting. In the fall we resolve issues that arose over the course of the field season and begin to plan the annual meeting. In the spring executive committee meeting, priorities are set for the next field season and issues needing to be resolved during the annual meeting are identified.

The annual meeting itself is attended by between 40 and 70 people including researchers from affiliated projects, post-doctoral scholars, and students. The main goal of the meeting is communication among LTER and associated scientists and coordination for the upcoming field season. It begins with a full day of science talks and posters including syntheses by the executive committee on results from the prior field season in each of the four focus areas. The second day of the annual meeting is devoted to coordination for the upcoming field season. The attendees break up into the four focus areas of research to plan and coordinate efforts. The groups then reconvene in plenary to inform each other of plans and to coordinate logistics. On the last day of the meeting the group again meets in plenary to resolve remaining issues and then breaks up into ad hoc groups to discuss papers, spin-off projects, and any remaining issues related to the field season.

In addition to monitoring natural conditions, the ARC-LTER maintains multi-decade experiments on terrestrial, stream, and lake ecosystems. These experiments involve, for example, the addition of nutrients, warming, species removal, and exclusion of grazers. Rather than a standalone project, the ARC-LTER is viewed as a backbone providing long-term continuity that can be leveraged in support of other shorter-term projects. Such spinoff projects might not be affordable under the LTER directly or might not fit well under the current goals of the ARC-LTER. Nevertheless, each new study potentially magnifies the value of existing studies because it provides a new vantage point for comparison. Thus, the value of the LTER research increases exponentially with the addition of spinoff projects, not just additively, and these efforts can help drive the evolution of the ARC-LTER goals in successive renewal cycles. To help foster these spinoff studies, we set aside funds in the budget to bring new collaborators to Toolik every year. These potential new collaborators are selected based on the need to fill a perceived gap in knowledge or approach and are discussed by the executive committee. We are usually able to cover travel and, through the cooperative agreement, provide user-days and facilities at TFS.

In adherence to this philosophy of being broadly open to collaborators from outside the LTER project, we encourage complementary studies using our site and our experimental manipulations (e.g., Deslippe et al. 2011; Crump et al. 2012; Sistla et al. 2013; Cory et al. 2014, among many others). For example, it is our practice to set aside empty plots in our experimental blocks, not to serve as controls, but rather to provide opportunities for new treatments to be embedded in our design. These added treatments can leverage off existing controls, off the time series of data collected prior to establishing the new treatment, and off comparisons to other treatments in the experiment. With few exceptions, the benefits to the LTER derived from this openness far exceed the costs.

The long-term success of the ARC-LTER can be attributed to two characteristics of our project. First, the collaborative, team-based approach to both governance and research fosters a culture of collaboration that maintains a broad ecological perspective, promotes multi-disciplinary research, and helps keep us up-to-date on current questions driving the science. Second, the openness to adding complementary projects and new researchers continuously promotes new ideas and innovative approaches. Both these characteristics keep the ARC-LTER novel and relevant for the next renewal. Maintaining this freshness of outlook and approach is vital to long-term success. No LTER is, or should be, guaranteed renewal by simply meeting specific program requirements for education, “networking”, “broader impacts,” and data collection to address the five LTER core areas (primary production, population studies, movement of organic matter, movement of inorganic matter, and disturbance patterns). Thus, although we meet these specific program goals, our primary aim is to advance ecosystem science in a long-term context.

Over five renewal cycles, our long-term goal has been to develop a landscape understanding of ecological functioning in the Arctic based on the interactions among tundra terrestrial, stream, and lake ecosystems. Renewals always require a balance between maintaining the legacy of long-term monitoring efforts and experiments and pursuing the newest, up-and-coming ideas in ecology. Ecosystems respond so slowly that a mere 30 years barely scratches the surface of the long-term responses on which the LTER program was designed to focus (Rastetter 1996; Rastetter et al. 2003). It is therefore appropriate that each renewal should continue the long-term monitoring and experiments. However, these efforts can always be viewed in a new light emanating from advances in ecological thinking and some modifications to the LTER research design can be made, but only within the constraints of the budget. The spin-off projects help immensely in exploring avenues of new research and approach and, if they pay off, are sometimes subsumed within the long-term purview of the ARC-LTER, budget allowing. Our long-term success depends on this evolution in focus, which in turn is emergent from our team-based approach and our openness to new and complementary collaborations involving multiple institutions and disciplines.

4.6 Sustaining Research in a Hostile, Remote Environment: Palmer Station, Antarctica LTER

Palmer, Antarctica LTER (PAL) joined the LTER Network in 1990 and commenced regular time-series oceanographic observations in January, 1993. PAL was the first marine LTER, a biological oceanographic research program with an emphasis on conducting observations from large research vessels often working remote from land. Biological oceanography is closely allied with other oceanographic research disciplines and has strong ties to geophysical research; in the case of PAL, the main connections are with research on climate change, physical oceanography and sea ice processes. The fieldwork also focuses on nearshore processes, exemplified by the iconic, nearly 50-year long time series of Adélie penguin observations on Torgerson Island near Palmer Station. Hugh Ducklow was lead Principal Investigator for three cycles, 2002–2020, and William R. Fraser was co-Principal Investigator from the start of the Palmer Antarctica LTER project in 1990. David M. Karl, microbial ecologist at the University of Hawaii, has been involved in Antarctic research since the 1970s. In this section they investigate the institutional and conceptual roots of PAL, with emphasis on establishing and maintaining a sustainable long-term research program under physically harsh conditions in a remote ocean.

4.6.1 Hugh Ducklow, David Karl, and William Fraser on Palmer Antarctica LTER

Even before its scientific foundation, PAL required a prior commitment of extraordinary and sustained institutional support, including a new polar research vessel (with the addition of new polar marine sites, PAL is no longer unique in this regard). PAL was conceived in the late 1980s during conversations between the late Tom Callahan, program officer at NSF's Division of Environmental Biology (DEB) and Antarctic Sciences Program Director Polly Penhale about Antarctic LTER sites. As Penhale recalls, "[we] got talking and cooked up the idea ourselves and moved it up our chains of command for an agreement that Antarctic proposals could 'compete' in the DEB LTER network competitions (no separate announcement), meeting the criteria of the LTERs, and that OPP [Office of Polar Programs] would fund them.

We envisioned two Antarctic LTERs, based on the funding commitment I could get from our office." (P. Penhale, email to Ducklow, 8 May 2018).

The progress of the PAL time series can be seen best in the archived data and cruise records (http://pal.lternet.edu/data). Selected hydrographic and ecological observations are obtained over two spatial scales: the nearshore local scale (~50 km) characteristic of penguin foraging ranges, and the offshore regional scale (~500 km) over which many oceanographic processes are manifested. The former are performed twice weekly from October to April using small boats at Palmer Station. The regional observations are obtained from a sea-ice capable research vessel over the 200 × 700 km LTER grid (Waters and Smith 1992) every January (Austral summer). The first field season of local sampling was November 1991 to the end of February 1992, building on the continuing Adélie penguin program (1975-present) that joined PAL in 1990.

Scientific and Conceptual Roots Perhaps the most important single factor ensuring PAL's success over the long term was a viable conceptual framework developed right at the start. Sustained time-series observations are mandatory for successful LTER proposals; even so, the practical and conceptual foundations of time series approaches to scientific questions vary from field to field and site to site. The LTER Network of sites from sea ice to coral reefs and deserts to rainforests and the open sea is a valuable resource for exploring these foundations (Willig and Walker 2016).

The need for extended time series approaches to capture longer-period phenomena was explicit in the founding PAL aim to investigate the relationship between the 6–7 year sea ice cycle and diatom-krill-penguin recruitment. At this early juncture there was no recognition of trends in climate or ecosystem processes along the Peninsula. Krill biologists Robin Ross and Langdon Quetin, future lead PIs for the new site, had been working on krill physiology and recruitment in relation to sea ice in the Palmer region since 1982 (Quetin and Ross 1984). Coauthor and microbial ecologist David Karl worked in the Bransfield Strait on four seasonal cruises in the RACER Program in 1986–1987 (Huntley et al. 1991), cementing the idea of "time" for future studies in the region. At about the same time, the Joint Global Ocean Flux Study (JGOFS) established new oceanic time series stations at Bermuda and Hawaii, the latter under Karl's leadership. These "ocean observatories" included physical, plankton and biogeochemical properties and continue today (Karl and Church 2017).

The cross-fertilization of thought and practice between the Hawaii Ocean Time-series (Station ALOHA, 1987-present) and PAL were critical for shaping the PAL project. Karl melded JGOFS and LTER science with a successful proposal to add microbial biogeochemistry to PAL starting in 1991. The signature of the JGOFS time series is evident in the list of properties sampled at Hawaii and in PAL: time-series sediment traps, dissolved inorganic carbon (CO_2) and plankton and microbial processes, all of which continue at both sites today.

Meanwhile, time series observations of Adélie penguins and other seabirds and marine mammals continued near Palmer Station, as Bill Fraser took over from his doctoral mentor, David Parmelee, who first explored the PAL region in 1972

(Parmelee and Macdonald 1973). Thus began the observations that would later develop into the long time series that shaped PAL's early conceptual models.

An important and even critical catalyst in the development of these models was the start in 1983 of the Antarctic Marine Ecosystem Research at the Ice Edge Zone (AMERIEZ) program, which was uniquely focused on investigating and understanding the role of sea ice in the marine ecosystem of the eastern Weddell Sea (Sullivan and Ainley 1987). AMERIEZ was funded by NSF and led by Neil Sullivan and David Ainley working in conjunction with a large steering committee of experienced Antarctic researchers. Over the course of 5 years (1983–1988), AMERIEZ's multidisciplinary team conducted spring, fall, and winter research cruises using two vessels that sampled in tandem within and outside the pack ice of the marginal ice edge zone. Two outcomes were especially critical to PAL's development. The first was that AMERIEZ validated the hypothesis that the ice edge functioned as an oceanographic front that separated ecologically distinct communities whose life histories were strongly influenced by the presence or absence of sea ice (Ainley et al. 1986; Fraser and Ainley 1986; Ribic et al. 1991). The second was that by the end of AMERIEZ in 1988, program results led to a fundamental change in the scoping of U.S. Antarctic marine ecosystem research, ushering in a new perspective on the role of sea ice as a driver of processes linked to marine ecosystem structure and function. The first PAL proposal, funded in 1990, elegantly captured this progression of ideas in its central driving hypothesis, which states that "the annual advance and retreat of the pack ice is a major physical determinant of spatial and temporal changes in the structure and function of Antarctic marine communities, from total primary production to the breeding success of seabirds" (Proposal number NSF-DPP-9011927).

Another AMERIEZ outcome, more subtle and thus much less appreciated, is that the program by design was inclusive of researchers representing top predator components, namely seabirds and seals (Sullivan and Ainley 1987). Southern Ocean biological oceanography prior to AMERIEZ had for the most part been largely "krill-centric", with little thought given to the idea that it may be important to understand the role of apex species in an ecosystem context. One of the key observations to arise from AMERIEZ was that wintering populations of penguins in the Weddell Sea did not share the same habitats, with Adélie penguins preferring sea ice and Chinstrap and Gentoo penguins largely avoiding it. These differences in life history affinities to sea ice led to the hypothesis that recent changes in these penguin populations were linked to the then fragile and controversial idea that the Western Antarctic Peninsula was warming due to global climate change, and the loss of sea ice was benefiting ice-intolerant species over ice-dependent species (Fraser et al. 1992). This along with the further discovery that trends evident in penguin diets accurately tracked changes in krill population dynamics (Fraser and Trivelpiece 1995) ultimately consolidated the role that seabirds would play, giving PAL its sense of time and space. Defining the operational domain the program would consider to scale the diversity of processes was needed to holistically link the presence or absence of sea ice through the food web to seabird population responses. The recent work by Saba et al. (2014), integrates decades of work by the major PAL components to understand the processes influencing krill recruitment. Critically, its krill time-series was derived from Adélie penguin diet samples.

Logistical and Social Influences Our review has so far focused on the conceptual and historical origins of PAL, but we also need to recognize the logistical and social infrastructures needed to ensure the survival of high-level scientific research in one of the most remote and hostile regions on the planet (it takes 6 days to get to Palmer Station from the USA, as there is no air service beyond South America). PAL and soon thereafter, the McMurdo Dry Valleys LTER were started with strong commitments from the NSF Office of Polar Programs. PAL required intensive and extensive (and expensive) logistics support (Augustine 2012): operating a modern, well-equipped marine laboratory and a dedicated research and supply vessel capable of navigating safely in the world's roughest seas and through year-round sea ice, plus long, military-style supply lines for people, gear, food and fuel. PAL provided strong justification for adding two new research vessels to the Antarctic Program. When it started, PAL relied on a Norwegian charter vessel. The US military also played an important role in transporting these needs for the entire US Antarctic Program until the late 1990s. Military support for the Earth Sciences expanded greatly after WWII, reflecting growing recognition of the importance of oceanography and other disciplines for national security in the Cold War era (Doel 2003). Just the idea of a sustainable Antarctic marine LTER was inconceivable without the promise and fulfillment of this support commitment.

With respect to PAL's social infrastructure, we cannot divorce the science from the issue of doing the science, i.e., the collective social component of what we do, meaning the mix of people needed to do a dangerous job in a dangerous place while living with each other in isolation for months on end. The dangers of fieldwork in Antarctica, being far from immediate medical services, mean that all researchers must be in good physical shape. Everyone travelling to Antarctica must pass rigorous medical and dental exams annually. Since Antarctic research is to an important extent a young person's game, there develops a tradeoff between experience and capability/stamina. The hardest part of managing PAL science is finding people who can do the job under these conditions, and thus simply keep things going. We suggested above that PAL's remote location is fundamental to its identity. This isolation also sets up the unique leadership, logistical and social conditions that *must* be up and running annually and over the decades to make PAL work. Maintaining PAL has been a rolling process that is for the most part black or white; there is no grey zone, no latitude to correct an error in a place that does not forgive poor judgment, bad planning and narrow perspectives on how critically people and their programs depend on each other, and thus on the success of the key scientific objectives that have circumscribed PAL science.

Early failures in leadership and changes in the composition of program co-PIs seem to be a common theme in the early history of most if not all LTERs, including PAL. Retrospectively, the early personnel turnover in PAL reflected the difficulty in accepting the idea that PAL's isolation actually required a very different, highly elevated level of engagement and planning in the context of conducting collaborative research. This acceptance has to start at home, in the US, and engage everyone from technicians to students and PIs right from the start, so that no one is left with the idea that the parts are somehow greater than the whole. It took PAL over a decade to really achieve this goal.

Another noteworthy challenge was instilling the idea of a decadal research program in a research support structure entirely used to supporting typical 3-year grants with one or at most two field seasons (this was a challenge for JGOFS also). In the formative first decade, Polly Penhale spent some time at Palmer Station almost every year. She experienced the conflicts, the successes, what worked, what did not work, where the weak points were, where the strong points were, and so on. She left Palmer every year with a plan to have solutions to issues in place by the following field season. This turned out to be absolutely essential to how the "PAL experiment", was able to navigate the logistics and management issues of those early days, and set the stage for the continuing program once inevitable NSF turnover began to erase institutional memory.

In this last regard, Fraser's team of seabird scientists also provided important long-term memory in the day-to-day program from the beginning. Their tenure usefully linked the past to the present, thus providing PAL with a reference point, a means of addressing its science, logistics, and management with a lot less second-guessing compared to its early years. Finally PAL, from its first year interns and technicians to grad students, coPIs and team leaders, keep it going through love for science, dedication to an important mission and leadership at many levels. In this way, PAL is really not much different from any other site or any other science.

4.7 An Evolving Management Structure: Georgia Coastal Ecosystems LTER

The Georgia Coastal Ecosystems LTER is relatively new, having joined the LTER Network in 2000. Merryl Alber, Professor in the Department of Marine Sciences at the University of Georgia, is the Project Director for the LTER. Her account of the organization of this LTER analyzes particularly well the learning process that LTERs must go through to figure out how best to sustain the kind of collaborative research that is the hallmark of LTER projects. She offers insights into the process of shifting from a collection of independent projects into an integrated program. Her description and analysis of this LTER touches on many of problems inherent to LTER projects in general, and shows how these communities engage in a process of continual evaluation and adjustment, develop strategies to keep principal investigators engaged, and over the years create an organization that is tighter and more team-oriented. Through this case study, she shows how LTER projects worked toward and achieved a basic cultural shift within ecology, from an individualistic undertaking to a more collaborative, team-based and multidisciplinary undertaking.

4.7.1 Merryl Alber on the Georgia Coastal Ecosystems LTER

The Georgia Coastal Ecosystems (GCE) LTER project was started in 2000. Scientists involved in the GCE LTER study the marshes and estuaries of the Georgia coast in order to understand how these ecosystems function, to track how they change over time, and to predict how they might be affected by future variations in climate and human activities. University of Georgia (UGA) is the home Institution for the program, which uses the UGA Marine Institute on Sapelo Island as its base of field operations. The GCE currently has 22 Principal and 11 Affiliated Investigators from 9 Institutions. Below we discuss the GCE administrative structure, the strategies we use to optimize PI involvement, and some of the changes that have occurred over time.

4.7.1.1 Georgia Coastal Ecosystsms Administrative Structure

The administrative structure of GCE involves five key elements. We have not only a Project Director, but also a co-Project Director; we created a set of bylaws to describe roles and responsibilities of Principal Investigators and other project personnel; we have an Executive Committee; we organize annual project meetings; and we have an External Advisory Committee. Their roles are described in turn.

Co-Project Director Model All LTERs have a Project Director (PD) who serves as the main point of contact with NSF. However, the GCE also has a co-PD who shares responsibility for project management. Leading a large project involves multiple tasks, from establishing the overall conceptual goals of the research to administering the budget to writing project reports to contributing to network-level efforts. The workload can be overwhelming and requires multiple skills. Having two people involved automatically brings in a partner who can share in the effort and brings additional abilities to the table. It also means there is a built-in sounding board when difficult decisions have to be made. I have served as PD of the GCE since 2006, and Steve Pennings has been co-PD since the project began in 2000 (the first GCE lead PD was Tim Hollibaugh). I manage the budget and subcontracts; take the lead on annual reports and other program documents; organize and run project meetings; serve as the primary contact with NSF; and handle overall project coordination. Steve manages the field program (including hiring and supervising field technicians; setting priorities and organizing field work); has represented the GCE on the network executive committee; provides critical (and immediate) feedback on draft documents; and helps with many other tasks. Steve and I share responsibility for personnel decisions and coordinating education and outreach efforts, and we also work together to frame the scientific goals of the program. We are equals in terms of speaking on behalf of the project and our split makes it possible to accomplish more than either of us could alone. It is frankly hard for us to imagine how all of the work could get done with just a single PD.

Bylaws The GCE operated for the first few years without a formal management structure. This was noted during the initial NSF 3-year review of the project and the review team suggested that we would benefit from an Executive Committee. Not only did we establish an Executive Committee at that time, but we also adopted bylaws that describe the roles and responsibilities of PIs and other project personnel. The bylaws are modeled after those of an academic department and treat the Lead PI similarly to a department chair (e.g. the PD is "re-appointed" by a vote taken every 6 years). Having bylaws has both benefits and disadvantages: On the plus side there is transparency and democracy in the process (e.g., all members have a say when it comes to bringing in new PIs). On the minus side we are less nimble when it comes to project decisions (it can be difficult to get a quorum). Although the bylaws have been helpful for clearly spelling out roles and responsibilities, if we were starting over I would recommend a simpler document that describes how the group will function and provides more flexibility for the PD to make decisions.

Executive Committee The GCE Executive Committee is comprised of six members. This includes the PD and co-PD, the Information Manager, and 3 at-large PIs. At-large members serve 6-year terms and are elected at the regular GCE-LTER meeting that precedes the 6-year term of each NSF grant cycle. Having an Executive Committee helps to distribute decision-making and to engage additional PIs in project management. It also provides a way to train future PDs. The Committee stays on top of the business end of the program (comings and goings of field technicians; infrastructure needs), helps to set the agenda for annual meetings and other project events (e.g. 3-year review), and weighs in on potential collaborations and other opportunities. Executive Committee members are actively involved in drafting the proposal and they often take the lead on specific sub-projects. We have found that regular monthly meetings of the Executive Committee are best, with more frequent communication by e-mail. We strongly recommend an Executive Committee, particularly given the large number of PIs involved in most LTER projects. We have also benefitted immensely from having the Information Manager (Wade Sheldon) involved at this level of project management, as it gives him a perspective that informs our data management structure and allows him to keep the Executive Committee apprised of NSF data requirements and any data issues that arise.

Project Meetings The GCE has an annual all-hands meeting, which provides a way for all project personnel as well as many of our partners to get together. The meeting is the one time when we are all in the same place and we also use it as a vehicle to bring in prospective investigators (many of whom start out as Affiliated Investigators). Depending on where we are in the LTER cycle, the meeting is spent in a combination of reporting on research results, field planning, and brainstorming proposal ideas. The formal program is generally 1½ days, although the meeting is often extended with project workshops or other activities (e.g. paper-writing sessions). Over the years, the annual meeting has become much more coherent in that investigators generally work together on presentations and participate in the whole meeting. In addition, most sub-projects now have regular meetings (these range from monthly to quarterly). These meetings have improved communication by getting busy people together and helping the group stay on task (most people respond to a deadline). Sub-project meetings are also a good way to involve graduate students and technicians in decisions.

External Advisory Committee The GCE has an external Advisory Committee comprised of four or five of our peers. They are selected for their expertise and experience with large projects and generally include at least one representative from another LTER site. The members of the Advisory Committee attend the annual meeting and provide feedback. The Committee was essential to our success during the first 6–8 years of the project, pointing out gaps that needed addressing (e.g., they encouraged us to develop a hydrodynamic model, which has benefitted us immensely) and pushing us to improve our integration. During the first few years of the GCE the annual meeting often functioned as a report-out to the committee, and they often commented "this meeting is not for us". Thankfully, we are now at a point where we truly are talking to each other, and, although we still benefit from the wisdom of the committee, the annual meetings would be much the same whether or not they were in attendance.

4.7.1.2 What Makes Things Work

I have heard an LTER project described as a "coalition of the willing", and getting to a position where everyone in an LTER feels like they are part of a team that is all pulling in the same direction can be difficult. Although this is true for most research projects, working on an LTER can be even more challenging because an individual's research needs to fit into the larger context of the project and so requires additional coordination; the project provides only minimal funding to each investigator; there are generally multiple people involved at the site (let alone across the network), so it is hard to keep up with co-PIs; and the work often involves interdisciplinary teams (which can make communication challenging). Ideally, the "willing" LTER PI is someone who is informed about the larger questions being asked and knows how their work fits into the project; who recognizes that the funding goes to support the core program and that this is valuable to them (and seeks leveraged funding as appropriate); who builds long-term relationships at both the site and network level; and who finds interdisciplinary interactions rewarding. Although this means that LTER research is not for everyone, here are some of the strategies that the GCE uses that we think help to keep our PIs engaged.

Integrated Research An LTER is uniquely situated to ask large-scale, long-term questions, and feeling like you are a part of that makes the project intellectually exciting. However, at the outset the GCE functioned more as a collection of independent research projects rather than an integrated program. This is likely typical for a new project, and prompted a lot of feedback from our Advisory Committee (how do these pieces fit together?). Our organization has gotten tighter over the

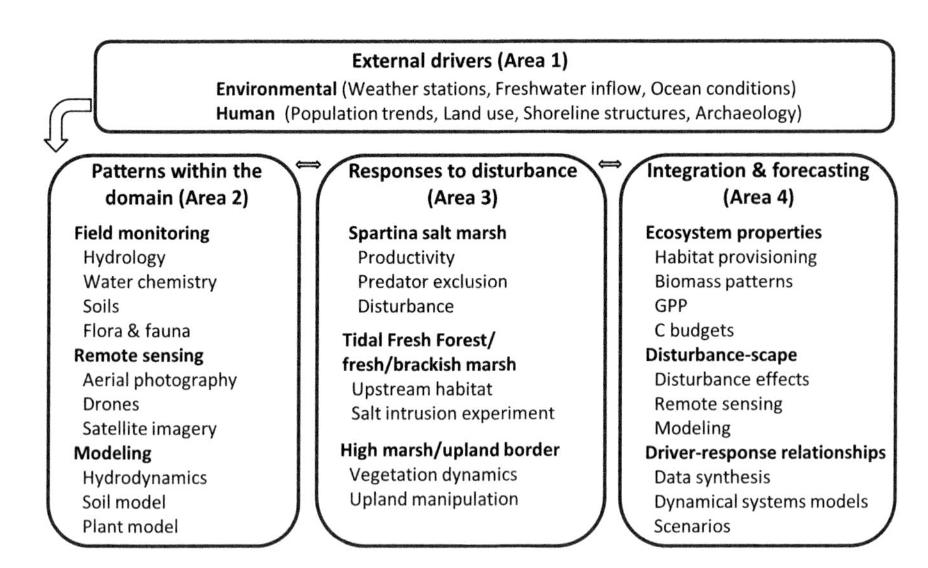

Fig. 4.1 GCE-IV Research Portfolio, showing major program components and how they are divided into "Areas". The PIs involved in each task are identified in the proposal

years, and we are now working in teams on almost every aspect of the research. We attribute much of this improvement to the time we spend on framing the introduction to the proposal and developing a corresponding conceptual model. This begins early in the proposal-writing cycle, with brainstorming at the annual meeting in or before year 4 of the current grant cycle. When people submit ideas for the new proposal we ask them to show how their proposed work will fit within the larger questions we are addressing. The Executive Committee works through these submissions, and we do not include projects (interesting though they may be) that do not fit the framework. This is an iterative process (the framework evolves over time, and we circulate the draft introduction before we write the bulk of the proposal). We also create a "research portfolio", which serves as a way to organize our major objectives (Fig. 4.1). This research portfolio flows through to the budget process, described below, and also makes it straightforward to lay out the project objectives in annual reports. An LTER is a difficult thing to see as a whole, and we spend time at every annual meeting presenting our conceptual model and our research portfolio (there are always newcomers and guests in the room). Our goal is to ensure that everyone involved in the project (including students and technicians) understands the main questions and can see how their work fits into the larger framework.

Funding Realities Although an LTER budget may seem large from the outside, it is difficult to stretch the resources to fund everything that is required by the project, and most PIs get relatively small individual budgets. One of the ways we address this issue is by explaining how funds are allocated and checking in with each PI during the proposal-writing process to make sure they are on board. We actually wait to do the budgeting until there is a complete first draft of the proposal that we can use to make a realistic assessment of budget requirements for each task. At that point the co-PDs write individualized memos to each investigator that describes the support that GCE provides for baseline infrastructure and core project personnel (field technicians and information managers), which together comprise about 1/3 of the budget, and details how these activities benefit the entire group. The memo includes an individualized list of tasks that are associated with the PI as described in the draft proposal, and the amount of funding we have available for those tasks. It also points out the benefits of participation in an LTER that go beyond money, such as participation in network activities, publicity for high profile papers, and training opportunities. Finally, we give people an opportunity to modify or opt out, telling them that "We know that these budget allocations are small and that some of you may not feel it is worth staying in the group for this modest amount of money. We truly hope that you won't choose to leave… If you want to suggest revisions to your budget and task list, please let us know what you think would work in the context of the proposal…." The memo serves as an opportunity to remind people that we are all in the same boat (i.e., no one is getting a lot of money from this), and that they benefit from the project as a whole. Although there has been some grumbling we have not lost anyone at this point in the process, which I attribute to having PI's who have bought into the program and are intellectually engaged in the science outlined in the proposal.

Bonding Opportunities PIs working on an LTER site are generally juggling multiple research projects, teaching classes, advising students, etc. It takes time and commitment to do collaborative research, which involves understanding what others bring to the table and ceding some autonomy. Some of this comes naturally over time: after nearly 20 years we understand each other's work habits and have deeper ties. There is no magic bullet for building trust, but the GCE field site at the UGA Marine Institute on Sapelo Island serves as a way to bring people together from across the different Institutions. The summer field season is particularly busy, and there are lots of informal opportunities for interactions. We also hold weekly summer seminars that include the interns and the summer Schoolyard teachers. Other opportunities to bring together different groups come through intense field campaigns surrounding research cruises and our annual fall monitoring, and we also have focused field trips with representatives from different laboratories (e.g., to sample experiments). Preparing for the 3-year review and writing the proposal are also times when we bring in the whole group, and the annual meeting always includes an informal poster session followed by group dinners. The All Scientists Meeting is also an excellent chance to interact with people from across the network, and is essentially the only time that most of our personnel have that opportunity. In this regard, being part of a national network creates incredible collaborative opportunities for LTER scientists.

4.7.1.3 Transitions

Project Director Transitions We have only had one PD transition to date, when I took over for Tim Hollibaugh at the end of GCE-I. Tim was always clear that his intention was to stay for only one funding cycle in order to get the project launched, and I had committed to the new role (and been approved by a vote of the PIs) well before the renewal proposal was due. The transition was straightforward, as Tim remained on the Executive Committee and was always just down the hall when I needed advice. We are now looking ahead to another transition (likely in 2024), and have recently hired a new faculty member at UGA with this in mind. Recruiting a new PD was strongly encouraged by the review team during our most recent 3-year review, and these comments were extremely helpful in convincing the University administration to allocate the position. The new hire will join the Executive Committee immediately, allowing us 3 or 4 years to work together before committing to a final decision about the leadership transition. Assuming that all goes forward, Steve and I anticipate staying on the NSF coversheet as co-PDs for the next funding cycle in order to facilitate a smooth transition.

Principal Investigator Transitions The GCE PI list has turned over considerably since the project began in 2000. Only 5 out of the original 16 PIs are still on the roster and we have added between 6 and 9 new PIs during each renewal. Most of the PIs we have added began as affiliated investigators and so were already familiar with the GCE and had attended our annual meetings. Two of our new PIs were first introduced to the project through an ROA (Research Opportunity Award), which targets faculty at primarily undergraduate institutions; two others were originally graduate students in the project and so came up through the ranks. The PIs that have stepped down have left for numerous reasons, including retirement, a shift in the research focus of the PI or of the project, and a lack of engagement. A one-page letter of intent is required as part of the proposal writing process, and that provides an easy mechanism for PIs to step away if they are not a good fit for the upcoming proposal.

Changing Research Questions At the 50,000 foot level the overarching question of the GCE has remained the same since the project began: We are interested in long-term change in coastal ecosystems as the result of human and climate drivers. We are also committed to tracking external drivers and long-term patterns as part of our core monitoring program, and have several large-scale, long-term manipulations in place that continue from one funding cycle to the next. However, we have shifted our focus over time: During GCE-I (i.e., the first funding cycle) we began to describe the patterns of variability in the system with an emphasis on the marked spatial variation in freshwater inflow as a primary environmental forcing function in our domain. In GCE-II we added a more detailed understanding of the movement of water, taking into account freshwater-marine gradients along the longitudinal axes of the estuaries as well as lateral gradients including tidal exchange on and off the marsh platform. In GCE-III we focused on salinity and inundation as the major structuring variables for intertidal wetland ecosystems. In GCE-IV, which is just beginning, we are examining disturbance responses at the landscape scale. These transitions are due in part to a natural progression wherein an investigation raises additional questions or an incoming PI brings new expertise. Improvements in technology such as drones allow us to address topics that were logistically impossible just a few years ago. Finally, some of the shifts are due to feedback from NSF and the LTER community. For example, guidance in the RFP led us to emphasize ecological theory in our current proposal.

4.7.1.4 Final Thoughts

Although I consider the GCE a successful project, there are always things that we can do better. Our communication has improved over time but there are still sub-projects that do not meet regularly and not everyone has a complete picture of the research. This is particularly true for graduate students and technicians, who turn over at a high rate. It is also difficult to foster connections among students from different Institutions. We have listservs for most of our major sub-projects as well as a program-wide weekly e-mail newsletter, and this has helped to improve communication and coordination. Another challenge has to do with ensuring the integrity of data sets, particularly when personnel change or the measurement shifts from one lab to another. We have had a few distressing occasions where we discovered that protocols had been inadvertently altered over time. A related issue is that when there is a problem (e.g. a piece of equipment is failing) it may take some time for the problem to be discovered, resulting in lost data. Although we have automatic quality controls and real-time output for many of our data streams, it is still important for someone to be paying attention. We are currently in the process of instituting regular protocol audits and a data sign-off process as part of our work flow.

Finally, the GCE is not always able to identify someone with the appropriate interest or expertise to participate in every network-level committee, and as a result there are potential opportunities that get overlooked (e.g., we do not currently have a contact for arts and humanities). We do work fairly closely with the other East Coast coastal LTER sites (PIE, VCR, FCE), sending representatives to each other's annual meetings and collaborating on research projects. However, we could be more proactive in terms of making GCE investigators and students aware of cross-site activities and options for increased engagement at the network level.

The administrative responsibilities associated with running an LTER project are substantial, and can be both stressful and demanding. The PDs spend a large amount of time and effort facilitating other people's research and organizing meetings rather than doing science, which can leave you asking yourself why you agreed to take this on. On the other hand, having a thriving LTER is very rewarding and provides opportunities that are well beyond the scope of most scientists. There are few other projects with the type of long-term support offered by an LTER, and that changes the scope of the questions that can be asked and encourages large-scale thinking. In addition to the multiple interactions within the site, being part of the larger LTER network expands your horizons even further. The LTER PIs interact regularly at the Science Council meeting, and that provides a built-in group of collaborators and connections to some of the leaders in the field. My experiences with the LTER have been extremely rewarding, and I feel fortunate to have had the opportunity to serve as the PD of the GCE.

4.8 Common Themes

As evident from the six examples above, LTER projects employ a broad range of approaches to sustain long-term research at their sites. All these approaches have the same aims: to develop a strong, integrated research program that addresses key ecological questions; to build a research organization that can sustain the research program over decades or centuries; and to ensure sufficient flexibility for their research program to adapt to changes in the scientific or funding landscapes. Sites that realize these aims can attain their long-term research goals; those that fail face probation and termination. To avoid this fate, LTER projects must also navigate an evaluation process that includes an onsite mid-term peer review and a closed, external review of research proposals every 6 years (Jones and Nelson, Chap. 3). There are four possible results from this evaluation process. Proposals can be approved as submitted. Project leaders can be asked to submit an addendum to their proposal clarifying points raised by external reviewers, the NSF review panel, or program officers. Projects can be placed on probation for 2 years during which they are tasked with writing a new proposal that addresses any weaknesses perceived by the reviewers or panel. Projects can also be terminated if this second proposal fails to correct deficiencies in science, project management, or data management. Because seven of the 20 LTER projects funded by the Division of Environmental Biology (35%) have been terminated (Jones and Nelson, Chap. 3, this volume), this outcome is a real danger that adds tension and suspense to the renewal process.

Research at LTER sites has often been compared to a marathon, but a better metaphor would be a steeplechase in which obstacles become higher as the race progresses. As LTER sites mature, increasing numbers of investigators, increasing disciplinary breadth, increasing numbers of measurements and experiments, and increasing expectations add to the burden of sustaining an LTER project (Risser et al. 1993). Success under these circumstances depends on how well LTER scientists manage their projects, integrate their research efforts, and adapt to changing expectations.

4.8.1 *Project Management: Leadership, Coordination, and Communication*

LTER projects employ a variety of management models, but all models must address three key objectives. Each successful approach ensures continuity of leadership, develops mechanisms to coordinate diverse research interests, and identifies strategies to inform and engage investigators over the long-term. Projects that fail to achieve any one of these objectives put themselves at risk of probation.

Leadership Lead scientists are their projects' representatives to the LTER Network, NSF, and the scientific community. They are guardians of the project's institutional memories and often serve as arbiters of protocols and processes. They manage diverse and changing personnel and recruit new participants to fill gaps in project expertise. Lead scientists combat institutional fatigue or complacency to protect resources allocated to the project. They share ideas from other LTER sites and communicate shifting expectations or changing personnel at NSF. They continually evaluate and adjust project activities to improve outcomes and assure stability.

Changes in leadership or key senior personnel are important events in the LTER universe and as such are carried out with deliberation. Replacement of a project's lead scientist, for whatever reason, disturbs the relationship between the site and NSF and may result in closer scrutiny of renewal proposals. As a result, there is a correlation between changing leadership and going on probation (Jones and Nelson, Chap. 3). To counteract this trend, sites often announce leadership changes in advance and plan a transition period of a year or more. Most sites have developed formal mechanisms to select and prepare new leaders. Some sites share leadership responsibilities between two people, spreading the workload and ensuring continuity in project objectives.

Most sites employ a small management team to ensure inclusivity in decision making, and some sites make major decisions, such as selecting new leaders, as a community. Many sites have written bylaws or protocols to guarantee consistency and fairness in policy decisions. Management teams often include students, technicians, educators, or information managers to bring a broad perspective to bear on project decisions.

Coordination LTER projects are complex organizations best served by open and inclusive leadership that fosters a culture of collaboration. A major challenge to leadership is the alignment of multiple individual researchers with a set of common goals and procedures. This challenge is particularly acute for isolated sites with difficult logistics or restricted field seasons. Under these conditions, a well-disciplined, team-oriented organization is required for maximum efficiency. Even under less stringent conditions, coordination of effort is a requirement for a successful LTER project.

Communication A well-coordinated LTER project requires frequent and effective communication among a large group of diverse and often geographically separated investigators. Communication under these circumstances can be challenging. Twitter, e-mail, and newsletters have a role in disseminating information but cannot replace real-time dialogue as a means of exchanging ideas. Many sites choose to have weekly or monthly project meetings, especially if project investigators are from the same or nearby institutions. For sites whose investigators are more geographically dispersed, online meetings are popular in lieu of physical proximity. Most sites also schedule an annual meeting where a detailed discussion of research results takes place, and which is open to all investigators. Because coordination is impossible without good communication, most LTER projects have developed strategies to engage investigators and keep them apace of the activities of their colleagues.

4.8.2 Research Integration

Successful LTER projects can sustain their research for decades or centuries to allow long-term observations and experiments to reach their conclusion. Most projects have begun studies designed to last for decades; the longest planned LTER study is a 200-year log decomposition experiment at the H.J. Andrews LTER site. Achieving success in these efforts depends on many factors, some of which are outside the control of LTER. The projects that have sustained their research for four decades have several common characteristics. Their research themes are stable over time, providing them with a recognizable identity. The research questions that they pose flow naturally from previous results and show a clear connection between data and understanding. Most importantly, these projects have been able to integrate disparate research ideas into a shared conceptual framework and to communicate that framework unambiguously. The inability to demonstrate integration and a clear conceptual framework are the most common reasons why LTER sites fail (Jones and Nelson, Chap. 3).

Successful sites can balance the tension between preserving their long-term research plans and initiating new, leading-edge studies. The expectation that LTER projects will include new research ideas in their renewal proposals has been a quandary for sites since the beginning and has been discussed in every decadal review of the program. Practically speaking, sites cannot accommodate much new work in their budgets without cutting existing studies. However, fresh science is alluring to reviewers and external advisory committees, and thus successful sites constantly evaluate and update their priorities, often leading to difficult choices. Research teams at most sites expand over time as complementary projects are added, but funding for these projects often comes from outside LTER.

A potential stumbling block in any multi-investigator project is the allocation of resources. Because many of the resources in an LTER project are committed to long-term observations and experiments, very little discretionary money is allocated to individual scientists. Moreover, because budgets for LTER sites remain constant between renewal proposals every 6 years, adding new science requires other sources of funding. LTER sites attract an average of nearly $3 in other funds for each LTER dollar, but even successful fund-raising can have its drawbacks. Studies funded from outside the LTER program are often short-term, and there may be strong pressure on LTER sites to adopt and continue these studies once funding has ended. This in turn may lead to mission creep and difficulties in disentangling project finances for reviewers. The development of comradery and trust among team members is a critical tool in avoiding disputes about resource allocation.

4.8.3 Adapting to Change

The development of an LTER renewal proposal is a very complex and arduous process, littered with potential pitfalls. On one hand, it may be difficult for scientists to understand why a long-term research plan that was deemed excellent 6 years ago must be modified at all. On the other hand, much has changed since that last proposal, including external reviewers, panel members, NSF program officers, and their superiors. Membership in the research team may have changed. NSF's requirements for renewal proposals almost certainly will change (Jones and Nelson, Chap. 3), as may funding for NSF programs as well as competition for that funding. In 6 years, attitudes in the whole scientific community may shift significantly. All these potential changes must be borne in mind in developing a renewal proposal.

Renewal proposals themselves are challenging. In addition to presenting a rigorous justification for the proposed science, LTER renewals must:

- Demonstrate how the proposed research links to previous results
- Communicate and justify an integrative conceptual framework
- Chart ongoing long-term observations and experiments and justify their continuation
- Propose and defend additions to the research plan
- Justify a $6 million budget
- Explain and defend data management and accessibility plans
- Document research productivity
- Report network activities
- Address any new renewal requirements

A successful proposal must integrate these tasks seamlessly in a document which is clear, concise, and informative. Failure to accomplish these tasks can result in probation, the beginning of a slippery slope. With this in mind, proposal preparation may begin as much as 2 years before the due date while recommendations of the mid-term review are still fresh. Ideas are gathered from the research team and evaluated against the existing conceptual framework, which may have to be tweaked to accommodate new results or approaches. The lead investigator or the management team must decide what new ideas to include. Small teams or individuals gather information for specific sections of the proposal. Budgets must be negotiated, and institutional support justified. Investigators must document and submit data and publications. Once preparations are completed, a writing team writes and circulates as many drafts as possible before the due date.

Project leadership must ensure that the science agenda in the proposal adapts both to new research results and changes in external expectations. The appropriate balance of ongoing and fresh research studies demonstrates that the research team has maintained its focus and creativity even in the face of internal and external changes. Attaining this balance is one of the most important challenges facing an LTER site and may determine whether the site can be sustained over the long-term.

References

Ainley, D.G., W.O. Smith, C.W. Sullivan, J.J. Torres, and T.L. Hopkins. 1986. Antarctic mesopelagic micronekton: Evidence from seabirds that pack ice affects community structure. *Science* 232: 847–850.

Augustine, N. ed. 2012. *More and better science through increased logistical effectiveness.* Report of the US Antarctic Program Blue Ribbon Panel. Washington, DC: National Science Foundation.

Cory, R.M., C.P. Ward, B.C. Crump, and G.W. Kling. 2014. Sunlight controls water column processing of carbon in arctic fresh waters. *Science* 345: 925–928.

Crump, B.C., L.A. Amaral-Zettler, and G.W. Kling. 2012. Microbial diversity in arctic freshwaters is structured by inoculation of microbes from soils. *ISME Journal* 6: 1629–1639.

Deslippe, J.R., M. Hartmann, W.W. Mohn, and S.W. Simard. 2011. Long-term experimental manipulation of climate alters the ectomycorrhizal community of Betula nana in Arctic tundra. *Global Change Biology* 17: 1625–1636.

Doel, R.E. 2003. Constituting the postwar earth sciences: The military's influence on the environmental sciences in the USA after 1945. *Social Studies of Science* 33: 635–666.

Fraser, W.R., and D.G. Ainley. 1986. Ice edges and seabird occurrence in Antarctica. *Bioscience* 36: 258–263.

Fraser, W.R., and W.Z. Trivelpiece. 1995. Palmer LTER: Relationships between variability in sea-ice coverage, krill recruitment, and the foraging ecology of Adélie penguins. *Antarctic Journal of the United States* 30: 271–272.

Fraser, W.R., W.Z. Trivelpiece, D.G. Ainley, and S.G. Trivelpiece. 1992. Increases in Antarctic penguin populations: Reduced competition with whales or a loss of sea ice due to global warming? *Polar Biology* 11: 525–531.

Gosz, James R., Robert B. Waide, and John J. Magnuson. 2010. Twenty-eight years of the US-LTER Program: Experience, results, and research questions. In *Long-term ecological research*, ed. F. Muller, C. Baessler, H. Schubert, and S. Klotz, 59–74. Dordrecht: Springer. https://doi.org/10.1007/978-90-481-8782-9_5.

Hobbie, J.E., and G.W. Kling, eds. 2014. *Alaska's changing Arctic: Ecological consequences for tundra, streams, and lakes*. New York: Oxford University Press.

Holmes, R.T., and G.E. Likens. 2016. *Hubbard Brook: The story of a forest ecosystem.* New Haven and London: Yale University Press.

Huntley, M.E., D.M. Karl, P.P. Niiler, and O. Holm-Hansen. 1991. Research on Antarctic coastal ecosystem rates (RACER): An interdisciplinary field experiment. *Deep-Sea Research* 38: 911–941.

Karl, D.M., and M.J. Church. 2017. Ecosystem structure and dynamics in the North Pacific Subtropical Gyre: New views of an old ocean. *Ecosystems* 20: 433–457.

Knapp, A.K., and M.D. Smith. 2001. Variation among biomes in temporal dynamics of aboveground primary production. *Science* 291 (5503): 481–484.

Knowles, J.F., S.P. Burns, P.D. Blanken, and R.K. Monson. 2015. Fluxes of energy, water, and carbon dioxide from mountain ecosystems at Niwot Ridge, Colorado. *Plant Ecology & Diversity* 8 (5–6): 663–676.

Knowles, J.F., P.D. Blanken, C.R. Lawrence, and M.W. Williams. 2019. Evidence for non-steady-state carbon emissions from snow-scoured alpine tundra. *Nature Communications* 10 (1): 1–9.

Metcalfe, D.B., T.D.G. Hermans, J. Ahlstrand, M. Becker, M. Berggren, R.G. Bjork, M.P. Bjorkman, D. Blok, N. Chaudhary, C. Chisholm, A.T. Classen, N.J. Hasselquist, M. Jonsson, J.A. Kristensen, B.B. Kumordzi, H. Lee, J.R. Mayor, J. Prevey, K. Pantazatou, J. Rousk, R.A. Sponseller, M.K. Sundqvist, J. Tang, J. Uddling, G. Wallin, W. Zhang, A. Ahlstrom, D.E. Tenenbaum, and A.M. Abdi. 2018. Patchy field sampling biases understanding of climate change impacts across the Arctic. *Nature Ecology & Evolution* 2: 1443–1448.

Parmelee, D.F., and S.D. Macdonald. 1973. Birds of the Antarctic ice pack. *Antarctic Journal of the United States* 8: 150.

Parton, W., W.L. Silver, I.C. Burke, L. Grassens, M.E. Harmon, W.S. Currie, and B. Fasth. 2007. Global-scale similarities in nitrogen release patterns during long-term decomposition. *Science* 315 (5810): 361–364.

Quetin, L.B., and R.M. Ross. 1984. School composition of the Antarctic krill *Euphausia superba* in the waters west of the Antarctic Peninsula in the Austral summer of 1982. *Journal of Crustacean Biology* 4: 96–106.

Rastetter, E.B. 1996. Validating models of ecosystem response to global change. *Bioscience* 46: 190–198.

Rastetter, E.B., J.D. Aber, D.P.C. Peters, D.S. Ojima, and I.C. Burke. 2003. Using mechanistic models to scale ecological processes across space and time. *Bioscience* 53: 68–76.

Ribic, C.A., W.R. Fraser, and D.G. Ainley. 1991. Habitat selection by marine mammals in the marginal ice zone. *Antarctic Science* 3: 181–186.

Risser, Paul G., Jane Lubchenco, Norman L. Christensen, Philip L. Johnson, Peter J. Dillon, Pamela Matson, Luis Diego Gomez, Nancy A. Moran, Daniel J. Jacob, Thomas Rosswall, and Michael Wright. 1993. *Ten-year review of the National Science Foundation's Long-Term Ecological Research program.* https://lternet.edu/wp-content/uploads/2010/12/ten-year-review-of-LTER.pdf. Accessed 4 Aug 2020.

Saba, G.K., W.R. Fraser, V.S. Saba, R.A. Iannuzzi, K.E. Coleman, S.C. Doney, H.W. Ducklow, D.G. Martinson, T.N. Miles, D.L. Patterson-Fraser, S.E. Stammerjohn, D.K. Steinberg, and O.M. Schofield. 2014. Winter and spring controls on the summer food web of the coastal West Antarctic Peninsula. *Nature Communications* 5: 4318.

Sistla, S.A., J.C. Moore, R.T. Simpson, L. Gough, G.R. Shaver, and J.P. Schimel. 2013. Long-term warming restructures arctic tundra without changing net soil carbon storage. *Nature* 497: 615–618.

Sullivan, C.W., and D.G. Ainley. 1987. AMERIEZ 1986: A summary of activities on board the R/V Melville and USCGC Glacier. *Antarctic Journal of the United States* 22 (5): 167–169.

Waide, Robert B. 2016. Sustaining long-term research: Collaboration, multidisciplinarity and synthesis in the long-term ecological research program. In *Long-term ecological research: Changing the nature of scientists*, ed. M.R. Willig and L.R. Walker, 29–41. New York: Oxford University Press.

Walker, M.D., D.A. Walker, J.M. Welker, A.M. Arft, T. Bardsley, P.D. Brooks, J.T. Fahnestock, M.H. Jones, M. Losleben, A.N. Parsons, T.R. Seastedt, and P.L. Turner. 1999. Long-term experimental manipulation of winter snow regime and summer temperature in arctic and alpine tundra. *Hydrological Processes* 13: 2315–2330.

Waters, K.J., and R.C. Smith. 1992. Palmer LTER: A sampling grid for the Palmer LTER program. *Antarctic Journal of the United States* 27: 236–239.

Willig, M.R., and L.R. Walker, eds. 2016. *Long-term ecological research: Changing the nature of scientists*. New York: Oxford University Press.

Part II
An In-Depth Perspective on Selected Sites: History, Foundations, and Partnerships

Chapter 5
The Luquillo Experimental Forest: A Neotropical Example of the Interaction Between Forest Conservation and Long-Term Ecological Research

Ariel E. Lugo

Abstract The origins of the Luquillo Long Term Ecological Research Program are traced through four historical trends that still influence research activity in Puerto Rico's Luquillo Mountains: (1) A history of identifying lands for protection and their designation for public uses; (2) A history of governmental and non-governmental institutions acting with the foresight to pursue scientific research for the benefit of economic development and the will to support scientific activity; (3) The uninterrupted progression of scientific activity through projects and programs that cumulatively developed a knowledge base that supported succeeding projects and programs; (4) The excellence of the individual scientific talent that participated in research in this location over the last 200 years. These historical trends took place within the context of four time intervals: (1) Discovery and colonization of Puerto Rico (1493 to 1898); (2) The negotiation of the Paris Treaty in 1898, which transferred Puerto Rico to the United States of America and led to the Scientific Survey of Puerto Rico (1913 to 1957); (3) Development of tropical forestry and ecosystem-level research through the establishment of the USDA Forest Service Tropical Forest Experiment Station in 1939 and the funding of radio-ecological and ecological research by the Atomic Energy Commission and the Department of Energy from 1963 to 1988 (1939 to 1988); (4) Funding of multiple research programs, including the Luquillo LTER, by the National Science Foundation (1989 to the present). This history demonstrates how advancing scientific understanding of tropical ecosystems benefited science, society, and the conservation of Neotropical natural resources.

Keywords LTER program · Long-term ecological research · Disturbance ecology · Tropical ecology · Radiation ecology · USDA Forest Service · Social-ecological research · Puerto Rico · International Institute of Tropical Forestry

---

A. E. Lugo (✉)
International Institute of Tropical Forestry, USDA Forest Service, Río Piedras, Puerto Rico
e-mail: ariel.lugo@usda.gov

R. B. Waide, S. E. Kingsland (eds.), *The Challenges of Long Term Ecological Research: A Historical Analysis*, Archimedes 59,
https://doi.org/10.1007/978-3-030-66933-1_5

5.1 Introduction

In this essay, I explore the historical roots of the Luquillo Experimental Forest Long -Term Ecological Research Program (LUQ LTER) to illustrate a successful example of the integration of local ecological research and the conservation[1] of a complex Neotropical forest. The historical roots of the LUQ LTER are as many and as intertwined as are the roots of tabonuco (*Dacryodes excelsa*) tree unions in the Luquillo Mountains (Fig. 4.6 in Lugo and Scatena 1995). These historical roots connect the Luquillo Mountains and the LUQ LTER to the early human inhabitants of Puerto Rico,[2] Spanish conquistadors, early natural history research scientists, the emergence of radiation ecology and ecosystem-level ecology in the tropics, the integration of the subfields of population and community ecology, and the systematization of long-term ecological research. Moreover, the roots of the LUQ LTER are solidly connected with the emergence of the New York Botanical Garden as a premier scientific organization in the United States of America, with the origins of the United States Department of Agriculture (USDA) Forest Service, and even the early history of the Ecological Society of America. Much of this institution building activity early in the twentieth century occurred simultaneously with the development of the field of modern ecology (Golley 1993; Kingsland 2005). Those involved in the new research and conservation organizations and emerging scientific fields of study accumulated a knowledge base upon which LUQ LTER scientists could develop new paradigms about tropical forest resilience and restoration management (Lugo et al. 2012; Brokaw et al. 2012a).

None of the rich history of ecological research and conservation in the Luquillo Mountains would have been possible without the talent and collaboration of pioneering educators at the University of Puerto Rico, and the support of the colonial government of Puerto Rico and its legislature. The key players in this history were Nathaniel Lord Britton and his wife Elizabeth Gertrude Britton, Raphael Zon, Carlos Eugenio Chardón Palacios, Howard Thomas Odum, and Frank H. Wadsworth. These main actors were associated with a distinguished and substantial cast of supporters that included Gifford Pinchot, Henry Allan Gleason, the Vanderbilt family, Eugene P. Odum, José Marrero, Leslie R. Holdridge, and many others. Key institutions in this history include the New York Botanical Garden, the University of Puerto Rico, the USDA Forest Service, the government of Puerto Rico and its legislature, and the US Atomic Energy Commission. The roles that these actors and organizations played in the history of the LUQ LTER will become evident in the narrative that follows.

---

[1] I use 'conservation' and 'management' interchangeably, as both concepts involve notions of 'use' and 'preservation' based on scientific information (Lugo 1989).

[2] For several decades after the United States of America invaded Puerto Rico, the island was identified as Porto Rico. I use Puerto Rico throughout, except when citing official documents or publications.

5.2 Legacies Before the *Scientific Survey of Porto Rico and the Virgin Islands*

The Luquillo Mountains attracted the attention of humans for millennia. The Taínos considered the mountains a "sacred and spiritual landscape" (Walker 2014, p. 26) and the inspiration for their *cemi*, which portrays the mountains in the form of an animal (Walker 2014, p. 18). The earliest date given by Wadsworth (1970) for the exploration of the Luquillo Mountains was 1513, when Diego Columbus began searching for gold. Spanish conquistadors harvested tropical timbers and contributed to the deforestation of the island during the eighteenth and nineteenth centuries. They mined the Mameyes River in the Luquillo Mountains during the first half of the nineteenth century. Formal management on these mountains began as a result of forest policies originating in Spain and applied to the Caribbean during the second half of the nineteenth century (Domínguez Cristóbal 2000). In 1876 King Alphonse XII of Spain promulgated *La Ley de Montes*, a law that established forest management on all Spanish colonies in the Caribbean (Domínguez Cristóbal 2000). At this time not only was the *Inspección de Montes de Puerto Rico* (Puerto Rico's Forest Service) established, but also portions of the Luquillo Mountains where the LUQ LTER was to be established were designated as a reserve dedicated to the conservation of its timbers and waters. Professional foresters trained in Spain were assigned the conservation of the reserve, and the inventory of forests began (Domínguez Cristóbal 2000).

While all this forestry activity was brewing on the island, the transfer of Puerto Rico from the Spanish empire to the emergent American empire was imminent, the result of the Paris Treaty of December 10, 1898 between the United States of America and the government of Spain. On the mainland United States, New York City in particular, a group of influential botanists were working to establish a research organization in the city (Mickulas 2007). The group was led by a husband-wife team (Nathaniel and Elizabeth Britton) that later became the first scientific partnership to conduct research in the Luquillo Mountains. However, before the dawn of the twentieth century, they were busy designing and establishing the New York Botanical Garden as a scientific organization meant to be the United States of America's version of the Royal Botanic Gardens at Kew, London, England (Kingsland 2005; Mickulas 2007). The New York Botanical Garden was incorporated in 1891 and in 1896 began operations. In 1898 and 1899, the Garden sent plant collectors to Puerto Rico, and they returned with 8000 specimens (Kingsland 2005). Amos Arthur Heller, who returned to Puerto Rico in 1900, and 1902 to 1903, led these collecting trips (Santiago Valentín 2005). Heller collected island-wide, including in the Luquillo Mountains.

At the turn of the twentieth century, Puerto Rico lacked self-government and the President of the United States of America appointed most of its government officials. All the activities in the island by the USDA Forest Service and its predecessor agencies and by Britton's Scientific Survey were performed with the knowledge and strong collaboration of the Insular Government and Legislature. An example of local government influence on the federal government was through the relationship between Governor Rexford G. Tugwell and the President of the United States of America. The appointment of USDA Forest Service Chief Ferdinand Silcox benefitted from the friendship with Governor Tugwell, who recommended him to the President for the position (Tugwell 1947).

The American colonization of Puerto Rico led to a burst of scientific activity, including the botanical collectors from the New York Botanical Garden, and botanical and zoological collectors from Puerto Rico and Europe. The identity and contributions of these collectors and natural historians are itemized by Wadsworth (1970), and in every one of the 19 volumes of the *Scientific Survey of Porto Rico and the Virgin Islands*. The United States of America Fisheries Commission sponsored the first organized scientific exploration of the Luquillo Mountains. They sent the steamer *Fish Hawk* to Puerto Rico to study the aquatic resources of the island. The steamer sailed the same month that the Paris Treaty was signed (December 1898), arrived in San Juan on January 2, 1899, and remained in Puerto Rico until late February 1899. This ship spent 52 days circumnavigating the island and adjacent cays and 38 days of actual fieldwork while visiting all available ports (Evermann 1899). They produced detailed maps of the San Juan and Guánica Bays, studied the mollusks of the island, produced beautiful color plates of fish species, and visited with island naturalist and physician Agustín Stahl in Bayamón. Stahl was the island's first internationally known scientist and the first to write an insular flora (Santiago Valentín et al. 2015).

The *Fish Hawk* expedition also surveyed fisherman and described in some detail the fisheries industry of Puerto Rico, including the techniques used for fishing. Among the inland sites they visited, was El Yunque (the Luquillo Mountains), thought to be the tallest mountain in Puerto Rico. There, they studied what they identified as the Luquillo River, which I surmise was the Mameyes River, and described the pool and ripple habitats and the shrimp that the LUQ LTER has studied continuously since 1988 (Crowl et al. 2012).

Robert T. Hill, an employee of the United States Geological Survey, spent a month in Puerto Rico to examine the forest conditions of the island (Hill 1899), including those of the Luquillo Mountains. He also arrived in January 1899, but probably after the arrival of the *Fish Hawk*. Gifford Pinchot transmitted Hill's report to the Secretary of Agriculture to be published as a Bulletin of the Forestry Division that he headed. As a geologist, Hill noticed the importance of soils and geological formations to the flora of the island, and described problems of erosion and deforestation. In 1905, John C. Gifford, working as a consultant for the USDA Bureau of Forestry (previously known as the Forestry Division), visited the Luquillo Mountains for the purpose of studying the forestry situation of the island and making forest management recommendations. His nine recommendations represent positive steps towards professional conservation of forestlands in Puerto Rico, the Virgin Islands, and the Luquillo Mountains (Gifford 1905).

The University of Puerto Rico was established in 1903, and in 1908 became a land grant college, which allowed for the establishment of the College of Agriculture in Mayagüez in 1911. Both campuses were engaged with the Scientific Survey of

Britton. Several stations of the Universities' Agriculture Experiment Station were strong collaborators of the Scientific Survey. The Río Piedras Station in particular supported the ecological survey by Henry A. Gleason and provided support to Mel T. Cook, the co-author of the ecological study. Early in the twentieth century the University of Puerto Rico hosted offices and nurseries of the Puerto Rico Forest Service and the USDA Forest Service from which reforestation programs were managed. Today, the USDA Forest Service International Institute of Tropical Forestry is located on the grounds of the Agriculture Experiment Station of Río Piedras.

Between November 1911 and May 1912, Louis S. Murphy (1916), Forest Examiner of the USDA Forest Service (previously known as the Bureau of Forestry), studied the forestry problems of Puerto Rico under an agreement between the USDA and the governor and Board of Commissioners of Agriculture of Puerto Rico. His report asked for an expansion of the authority of the Commissioners of Agriculture so that the management of forests and the establishment of an Insular Forest Service be delegated to them.[3]

Around the same time in Washington, D.C., Raphael Zon was leading the Office of Special Investigations in Forest Economics of the USDA Forest Service. Zon had left his native Russia in 1896, and after studying for two years in Belgium and England he emigrated to the USDA in 1898, coincident with the acquisition of Puerto Rico by the United States (Rudolf 1957). After completing a degree in forestry at Cornell University in 1901, he joined Gifford Pinchot's Bureau of Forestry (later the USDA Forest Service) in the U. S. Department of Agriculture. He was among those who established the *Journal of Forestry* and served as its editor-in-chief for six years. Zon was Chief of the Office of Silvics (1907), led the Office of Special Investigations in Forest Economics (1920) and was Director of the Lakes States Forest Experiment Station in St. Paul, Minnesota (1923). He was the right hand of Gifford Pinchot at the USDA Forest Service, and was responsible for recommending the establishment of the Forest Research Experiment Stations and Experimental Forests and Ranges, which Pinchot approved on May 6, 1908. The Luquillo LTER would later benefit from both designations.

Zon collaborated with William N. Sparhawk, Senior Forest Economist of the USDA Forest Service, in the compilation of the first global report of the status of forests, including those of Puerto Rico (Zon and Sparhawk 1923). This publication summarized available information about the forests of the island and was foundational for developing local forestry policy. All these events resulted in the establishment of a strong coalition between federal and state governments, including the University of Puerto Rico. This coalition has remained strong into the present in its support of the LUQ LTER.

Britton played a dual role in Puerto Rico. On the one hand, he directed and played a scientific role in the *Scientific Survey of Porto Rico and the Virgin Islands*

[3] By this time, the Puerto Rico Forest Service established by the Spanish Government was no longer operating. In 1917, the Puerto Rico Legislature established the current Puerto Rico Forest Service.

(from now on Scientific Survey), and on the other hand he promoted forestry, conservation, and education in Puerto Rico, taking advantage of his position and opportunities to express his opinions to the public. Britton's proposals (discussed in more detail below) for conserving forest cover in Puerto Rico, increasing the area of protected lands, expanding research activity, and incorporating professional forest management practices were a synthesis of all the knowledge and practices that had accumulated by the procession of technical visitors and interventions since the Spanish era, augmented by excursions by professionals affiliated with the US Government and American universities and institutions (Britton 1919a). This wealth of information was about to be greatly multiplied by the Scientific Survey and together formed one part of the technical basis upon which the LUQ LTER was conceived and eventually established. The other part includes the studies that took place after the Scientific Survey. Both parts of the technical roots of the LUQ LTER are discussed next.

5.3 The *Scientific Survey of Porto Rico and the Virgin Islands*

History and Scientific Content of the Survey On the evening of November 19, 1912, the Executive Council of the New York Academy of Sciences received a proposal from Britton to conduct a Scientific Survey of Puerto Rico (Baatz 1996). The Survey was designed to contribute to the economic development of Puerto Rico in terms of forestry, agriculture, and mineral exploration (Fig. 5.1) The *Scientific Survey of Porto Rico and the Virgin Islands* was a seminal research program of the New York Botanical Garden, as it gained a national and international reputation in the sciences. Top scientific talent from the New York Botanical Garden and other American institutions participated in the Scientific Survey.

To the promise of this science-based development, the Puerto Rico legislature contributed funding and in-kind support such as security by the Insular Police, and staff and facilities from the University of Puerto Rico, including its Agriculture Experiment Station (Britton 1930). Examples of participating Agriculture Experiment Station staff included Mel T. Cook on the ecology of the island and Carlos Eugenio Chardón Palacios on the fungi.

Chardón was born in Ponce and educated in Ponce, Mayagüez, and Cornell University. He is considered the first Puerto Rican mycologist, and began his career at the Agriculture Experiment Station in Río Piedras. He was the appointed Commissioner of Agriculture and Labor of Puerto Rico between 1923 and 1931. In 1931, Governor of Puerto Rico, Theodore Roosevelt III, appointed Chardón Chancellor of the University of Puerto Rico. His Plan Chardón resulted in the establishment of the Puerto Rico Reconstruction Administration, which he headed. This organization was a critical contributor to the training and development of agricultural technicians and the construction of conservation infrastructure in public forest reserves, including the Luquillo National Forest. After a period of time in exile from

Fig. 5.1 Scientists having lunch on the Guaynabo road leading to the San Juan Aqueduct early in 1924. Left to right: Nathaniel and Elizabeth Britton, Carlos Chardón (Agriculture Commissioner), W. P. Kramer (Puerto Rico Forest Service and Supervisor of the Luquillo National Forest), E. E. Dale and John S. Dexter (professors at the University of Puerto Rico). (From Kramer 1924)

Puerto Rico, he returned to direct the Puerto Rico Land Authority (1940) and the Tropical Agricultural Institute in Mayagüez (1942). His partnership with Britton began while he was a scientist at the Agriculture Experiment Station, and continued through his career as Commissioner of Agriculture and Labor and Chancellor of the University of Puerto Rico.

Chardón used his influence to secure funding for the Scientific Survey from the legislature of Puerto Rico (Baatz 1996). Among the original group of collaborators of the Scientific Survey (a list that was augmented dramatically over the subsequent decades) were the federal and insular Agriculture Experiment Stations, the American Museum of Natural History, the New York Botanical Garden, scientific departments of Columbia University, and the government of Puerto Rico.

The Scientific Survey covered all groups of plants and animals including algae, as well as the paleontology, meteorology, archeology, botany, plant ecology, and geology of the island. Among the prominent scientists that contributed to the Scientific Survey were Howard Augustus Meyerhoff (geologist), Arthur Hollick (paleobotanist), Harold Elmer Anthony (mammalogist), Alexander Wetmore (ornithologist), Percy Wilson (botanist), Karl Patterson Schmidt (herpetologist), John Treadwell Nichols (ichthyologist), John Alden Mason (archaeological anthropologist), Fred J. Seavers (mycologist), and many others. Elizabeth Britton also participated in the Scientific Survey. Crum and Steere (1957) recognized her collections of mosses in Puerto Rico as the largest in existence at the New York Botanical Garden. Scientists of the Scientific Survey developed tools such as geologic maps, and maps of suggested geographic locations for ports, mineral explorations, and natural resources assessments, to name a few. The results of the Scientific Survey are published in a 19-volume publication series (each with up to four parts) of the New York Academy of Sciences.

The Scientific Survey and the USDA Forest Service When Britton was elected to the National Academy of Sciences in 1917, his inaugural speech was on the subject of Forestry in Puerto Rico. For that speech, he developed a proposed forest management policy for the island, a proposal that he sent to the governor of Puerto Rico. At the time, Emory Murray Bruner, who was the Forest Supervisor of the USDA Forest Service Luquillo Forest Reserve, also led the Puerto Rico Forest Service as Chief Forester. The incumbent USDA Forest Service Forest Supervisor supervised the activities in both insular and federal forests (Wadsworth 2014). Britton's proposal was successful, and from then on, he worked closely with Bruner and those who followed him.[4] Britton also provided botanical advice to USDA Forest Service employees in Puerto Rico.

Britton used his influence with Gifford Pinchot, USDA Chief Forester, to support the designation of the Luquillo Forest Reserve as the Luquillo National Forest in 1907. Britton and Pinchot had known each other for many years, and both had strong ambitions to develop botany as a science and to develop educational institutions that would provide training for future botanists. In the early twentieth century, Britton and Pinchot were part of an elite group of scientific leaders who not only advocated for the expansion of American science, but with their personal wealth and connections were in a position to see their ambitious visions realized. As the New York Botanical Garden opened its doors in 1900, Pinchot was busy helping to establish, with his family's financial support, a school of forestry at Yale University. When the Carnegie Institution of Washington was founded in late 1901, it turned to Britton and Pinchot for advice on how it should develop its own programs in botanical research. During President Theodore Roosevelt's administration, Pinchot established the Forest Service in 1905, served as the first USDA Chief Forester until 1910, and became known as a prominent conservationist and advocate of scientific management of forests (Pinchot 1947).

The Luquillo National Forest was renamed Caribbean National Forest in 1935 and El Yunque National Forest in 2007. Since the pioneering work of Zon at the turn of the twentieth century, the USDA Forest Service supported the close relationship between science and conservation. In 1956, the USDA Forest Service proclaimed the totality of the Caribbean National Forest in Puerto Rico as an Experimental Forest, recognizing that management activities in such a complex environment by necessity are experiments that require a research approach. For the next 18 years starting in 1955, Wadsworth, Director of the Institute of Tropical Forestry, led both the management and research programs of the USDA Forest Service in Puerto Rico. Today, El Yunque National Forest still has the same boundary as the Luquillo

[4] William P. Kramer (1922–1931), Thomas R. Barbour (1931–1935), E. Worth Hadley (1935–1943), Arthur Upson (1943–1951), and Henry B Bosworth (19551–1953) followed Bruner over the next 36 years.

Experimental Forest (LEF), but the conservation of the forest is the dual responsibility of the management (National Forest System) and research (Research and Development) arms of the USDA Forest Service.

Britton also advocated for more protected lands on the central mountains of Puerto Rico (Mickulas 2007). In response to his advice, in 1924 the Puerto Rico government gave one of the first indications of official interest in forest conservation with the recommendation that trees in upper watersheds be protected (Little and Wadsworth 1964).

Henry A. Gleason and the Ecology of Puerto Rico In 1919, Britton announced the appointment of Henry Gleason as the first assistant to the Director in Chief of the New York Botanical Garden (Britton 1919b). Gleason obtained his PhD from Columbia University in 1906, based on research done at the New York Botanical Garden, which served as Columbia's botany department. Gleason gained fame as an ecologist for his plant succession work that led him to challenge the prevailing views of Frederic Clements (Clements 1928).

Clements had advanced the idea that plant communities developed toward a single climax community, with a distinctive and predictable species composition that ultimately depended on the climate. His conception of the plant community was that it was like a "complex organism", developing, as a single organism would, toward a given endpoint. Gleason, in a now-classic article published in 1926 called "The individualistic concept of the plant association," proposed to the contrary that species behaved individually, not as parts of a larger "complex organism." They arrived in a region individually, according to their tolerance of the environmental conditions, and assembled into communities composed of species. The pattern of assembly could occur in different ways, depending on local conditions, and as a result, he concluded, two regions that were similar in climate and physiography might have very different plant associations. Gleason defended his individualistic concept from an attack by George E. Nichols of Yale University (Egerton 2015) during a meeting of American ecologists attending the Fourth International Congress of Plant Sciences at Cornell University. A famous picture of attendants to that conference shows Zon and Gleason standing near each other. Eighty-seven years after Gleason's paper, scientists at the LUQ LTER were still debating the relative importance of Clementsian vs. Gleasonian patterns of community organization up the elevational gradient of the Luquillo Mountains (Willig et al. 2013).

The same year that Gleason published his individual species classic he also published *Plant Ecology of Porto Rico* (Gleason and Cook 1926). The collaborators for this work were the New York Botanical Garden, the University of Puerto Rico Agriculture Experiment Station, and the Puerto Rico Agriculture and Labor Department. Chardón, then Commissioner of Agriculture and Labor, Francisco López Domínguez, Director of the Insular Experiment Station, and Britton, Director of the New York Botanical Garden, provided leadership support. Clara Livingston, Puerto Rico's first female aviator and friend of Amelia Earhart, was among a large group of local cooperators of the study. William P. Kramer and Charles Z. Bates of the USDA Forest Service helped in the selection of sites to visit and provided personal guidance to the authors. Fieldwork was from January 18 to April 30, 1926. Over 90 years later, Gleason and Cook's work remains the most comprehensive description of Puerto Rican native plant communities (Lugo 1996).

Mentoring and Collaboration as Keys to Scientific Success Britton used a mentoring process in which he paired continental and insular scientists in collaborative work that strengthened the insular institutions and augmented human scientific power for the Scientific Survey. An example is the visit of Fred Jay Seaver, a New York Botanical Garden mycologist and editor of the journal *Mycologia* between 1909 and 1947. At the suggestion of Britton in 1923, the government of Puerto Rico invited Seaver to visit the island and spend time exploring with Elizabeth and Nathaniel Britton. Simultaneously, Chardón, a Puerto Rican mycologist, was teamed with Seaver to write the definitive work on the fungi of Puerto Rico as part of the Scientific Survey (Seaver and Chardón 1926).

Collaboration was an asset that Britton modeled and which contributed to the development of science and forest conservation in Puerto Rico. This collaborative style was in display in his address at the Muñoz Rivera Park in San Juan on November 1929. The occasion was the planting of a *Stahlia monosperma* tree, an endemic tree named after Stahl, Puerto Rico's first natural scientist, who had died in 1917 (Britton 1930). In short three pages, Britton acknowledged Kramer, who in 1922 had replaced Bruner as Forest Supervisor and Chief Forester and had grown the tree from seed, reviewed the ecology of *Stahlia* and other native tree species, advised on the proper upkeep of the growing seedling, highlighted the work of Stahl, mentioned the previous planting of a mahogany tree in the Bayamón Plaza to also honor Stahl's work, acknowledged the work of Chardón, and used the occasion to review the problems of water and soil quality associated with island deforestation and the need for reforestation and a forestry policy. Britton also acknowledged the work of reforestation of the USDA Forest Service and asked that everyone collaborate with those efforts. Britton finally thanked and credited the insular governor for his support for reforestation.

Final Outcome of the Survey At the end of the Scientific Survey in the mid-1940s, Puerto Rico had available a solid base of scientific information and expertise to manage its natural resources and advance many scientific fields.[5] Puerto Rico was also notable as the Neotropical location with the most comprehensive foundational knowledge about its natural resources. Many decades later, the Scientific Survey continued to inspire as the island's scientists celebrated its 80th anniversary by updating knowledge of its various original topics (Figueroa Colón 1996). Studies following the Scientific Survey, such as those discussed below, were poised to use available scientific knowledge for more effective research because they did not have to develop basic information and could directly address pressing conservation and research issues.

[5] The last publication was dated 1957, but was part of volume 7.

5.4 Two Evolving Lines of Research: Tropical Forestry and Radio-Ecological Research

After the Scientific Survey, the historical path to the LUQ LTER bifurcated into two independent but interacting pathways: that of the USDA Forest Service and other federal agencies in Puerto Rico, and the establishment of radio-ecological research at the University of Puerto Rico's Nuclear Center. I will discuss these two paths independently.

5.4.1 USDA Forest Service Tropical Forest Experiment Station

In 1939, research activity in the Luquillo Mountains was formalized through the establishment of the USDA Forest Service Tropical Forest Experiment Station, a unit that was designated in 1955 as the Institute of Tropical Forestry, and in 1992 as the International Institute of Tropical Forestry (from now on the Institute). The Institute was established because land restoration efforts by the USDA Forest Service were failing throughout the island. Officials recognized that forestry knowledge based on temperate zone experience was not sufficient to address the ecological issues in the tropics, where scientific understanding of forests was insufficient (Wadsworth 1995). Over the next 50 years the Institute developed successful research, application, and dissemination programs in tree dendrology, wood products, tropical silviculture, and tree propagation. Institute scientists applied newly acquired knowledge in reforestation projects and tree plantation establishment. They screened hundreds of native and introduced tree species for suitability for planting, and developed demonstration projects on land rehabilitation under a variety of tropical climates and soils. In the absence of tropical forestry schools in the Neotropics, the Institute conducted many tropical forestry trainings with regional participation, helped organize forestry schools in several countries, and edited and published a regional journal (the *Caribbean Forester*) to disseminate research findings in support of forest conservation.

Since its establishment, the Institute has been a consistent collaborator with all significant scientific activity in the Luquillo Mountains. Its scientists have continuously participated in the international science arena, particularly in the Neotropics, where they have influenced tropical forestry activities as well as the development of knowledge in support of forest conservation. Leslie R. Holdridge was the first scientist of the Institute, but by 1931 he had already started research activity in the Luquillo Mountains. He initiated a tree dendrology project, conducted fieldwork, supervised part of the initial botanical work, and wrote two preliminary volumes of the trees of Puerto Rico. Holdridge later became known for his Life Zone System for the delimitation of world plant formations (Holdridge 1967), a system used by scientists of the LUQ LTER and countless others around the world to classify the climates of the world.

Frank Wadsworth (Fig. 5.2) was another Institute scientist. His USDA Forest Service career epitomizes the merging of forest research and forest management activities with the objective of advancing the conservation of a complex tropical forest. At the beginning of his Forest Service career in 1938, Wadsworth worked under Gustav Adolph Pearson, Chief Silvicultural Scientist at the Southwestern Forest and Range Experiment Station at Fort Valley, Arizona. Wadsworth was transferred to Puerto Rico when he married Pearson's daughter, Margaret "Peggy" Pearson, and they arrived on the freighter *S.S. Maiden Creek* in January 1942. Wadsworth was appointed to the Tropical Forest Experiment Station as a replacement for Leslie Holdridge, who was taking an assignment in Haiti. Wadsworth, a silviculturist, faced many forestry challenges in Puerto Rico, called "the stricken land" by governor Rexford G. Tugwell (1947). His six-decade career as a tropical forester is legendary and described from his own perspective in Wadsworth (2014).

During a field campaign, Wadsworth was the cabin mate of Elbert Luther Little, Jr. Little became the Leading and National Dendrologist of the USDA Forest Service and in the 1950s they teamed up to describe the trees of Puerto Rico. The

Fig. 5.2 Frank Wadsworth (standing) circa 1960s, teaching to an international tropical forestry course at the Institute of Tropical Forestry of the USDA Forest Service. (From the historical archives of the International Institute of Tropical Forestry, USDA Forest Service)

completion of *Common Trees of Puerto Rico and the Virgin Islands* (Little and Wadsworth 1964), took five years from 1950 to 1955. Before publication in 1964, the manuscript was translated to Spanish and revised slightly in 1962. One of their main sources of information was the flora of Puerto Rico that was published by the Scientific Survey (Britton and Wilson 1923–1930). The information gathered by the Scientific Survey was critical to the systematization of forestry and long-term ecological research in the region. In addition, the New York Botanical Garden and the United States National Museum collaborated, identifying species and sharing herbarium specimens. Other important collaborators were the Division of Forests, Fisheries, and Wildlife of the Puerto Rico Commonwealth, and the Agriculture Experiment Station of the University of Puerto Rico.

Marrero translated to Spanish the *Common Trees of Puerto Rico and the Virgin Islands* (Little et al. 1967), a project sponsored by the College of Agriculture and Mechanical Arts of the University of Puerto Rico and the University's Agriculture Experiment Station and Agriculture Extension Service. The volume was elegantly produced with the addition of the beautiful and botanically accurate watercolors of Mrs. Frances W. Horne. Mrs. Horne, wife of University of Puerto Rico at Mayagüez professor Charles E. Horne, attended field trips with Britton, learned to identify the plants, and used to accompany herbarium specimens sent to the New York Botanical Garden with her watercolors. Roy O. Woodbury, plant taxonomist at the Agriculture Experiment Station, became a co-author in the second volume of the *Trees of Puerto Rico and the Virgin Islands* (Little et al. 1974), which contained the tree species not included in the first volume by Little and Wadsworth (1964) and was published with a new title. Many individuals and institutions collaborated with the elaboration of the second volume of the work and are acknowledged in the book. These two volumes describe all tree species in the island (native and non-native) and provide information essential for the conservation of forests and tree species, information that was foundational for the LUQ LTER.

The publication of *Trees of Puerto Rico and the Virgin Islands* is a milestone of federal government activities in Puerto Rico. Other federal agencies in Puerto Rico added to the wealth of scientific materials that contribute to the conservation of all the island's natural resources. For example, the Soil Conservation Service (now Natural Resources Conservation Service) described the soils of the island (Roberts 1942); the US Geological Survey did the same with the geology and geomorphology by developing geological and topographical quadrangles at a scale of 1:250,000 for the whole island, and gauging rivers and aquifers island-wide; and the Weather Bureau (now the National Weather Service) did so with the climate (e.g., Colón Torres 2009).

A significant technical problem facing Wadsworth as he focused on tropical silviculture was dealing with tropical trees that did not produce annual tree growth rings, as they did in Arizona where he had started his professional career in forestry. Without the benefit of annual tree growth rings, it was impossible to determine the age of trees and their rates of growth. He understood that he needed repetitive measurement of tree allometry over long time periods to estimate tree growth rates. Much as he had done under the supervision of Pearson in Arizona, Wadsworth tagged some 15,000 trees between 1943 and 1946. Wadsworth grouped the trees into permanent plots that by themselves alone, established long-term ecological research in the Luquillo Mountains. The tree growth plots were representative of the major forest types in the Luquillo Mountains and provided the first insights into the productivity of mature Neotropical forests (Briscoe and Wadsworth 1970) and on how hurricanes affected forest structure (Crow 1980). They were also the first step towards long-term measurements of forest structure and functioning in the Neotropics. When coupled to the ecosystem level work of Odum that I discuss below, the long-term behavior of the tree growth plots of Wadsworth provided the intellectual foundation for the first LUQ LTER proposal in 1980.

Through his activities as both manager and scientist, Wadsworth was responsible for the development of research infrastructure within the LEF, infrastructure that was vital to the success of the LUQ LTER. A few examples include the 1949 designation of the Baño de Oro Research Natural Area, containing the best examples of the historical primary forests of Puerto Rico (Weaver 1994); development of wood volume tables for critical species and forests, the first step towards the carbon cycling work advanced later by Odum and the LUQ LTER; development of road and trail infrastructure used by scientists today; establishment of tree plantations and an arboretum with iconic tropical timber species such as teak, mahogany, and Caribbean pine; management of forest stands; rehabilitation of degraded lands; and many other tropical forestry research initiatives. Marrero, a silviculturist at the Institute, was Wadsworth's partner and collaborator in many of these activities beginning with their overnight stays at what was to become the El Verde Field Station, establishment of long-term tree growth plots, and planting of millions of trees throughout the Luquillo Mountains and Puerto Rico (Marrero 1950).

Wadsworth signed the Special Use Permits to the US Atomic Energy Commission (AEC) to run the El Verde Radiation Experiment (Radiation Experiment from now on). He also permitted the Federal Agricultural Research Station to test defoliants on tropical forests,[6] Harvard University's Arnold Arboretum to conduct a seminal ecological and taxonomic study of the elfin forest at *Pico del Oeste* (Howard 1968), and the University of Puerto Rico to continue research at El Verde post-Radiation Experiment, a permit that enabled the LUQ LTER. Wadsworth was also responsible for the dedication of the El Verde Field Station, used later by Odum for the Radiation Experiment and by the LUQ LTER as its field headquarters. He facilitated the work of Odum and collaborated with the research through site selection, research history (Wadsworth 1970), and contribution to the understanding of tree growth and yield in the tabonuco forest (Briscoe and Wadsworth 1970).

[6] This permit was terminated when the site was found to be unsuitable.

5.4.2 The El Verde Radiation Experiment

In 1962, Howard Teas, a scientist at the Puerto Rico Nuclear Center, and Howard T. Odum (Fig. 5.3) wrote the original proposal for the Radiation Experiment. Odum was one of the most influential ecologists of the twentieth century. His father Howard W. Odum, a sociologist at the University of North Carolina at Chapel Hill, influenced him, as did his brother Eugene P. Odum, and his doctoral mentor, G. Evelyn Hutchinson (under whom he did his doctoral dissertation on the biogeochemical cycle of strontium at Yale University). Odum applied a holistic approach to all the ecosystems that he studied, including Texas bays, tropical coral reefs, mangroves, freshwater springs, cypress wetlands, and tropical forests. During the Second World War, Odum visited Puerto Rico for the first time as a meteorologist, and in the 1950s he conducted studies in the mangroves of Puerto Rico and the forests of the Luquillo Mountains. His contributions to tropical ecology are summarized in Lugo (1995a, 2003, 2004).

Under the Atoms for Peace Program of the Administration of President Dwight D. Eisenhower, the University of Puerto Rico had established the Puerto Rico Nuclear Center in 1957 with facilities in Río Piedras and Mayagüez. The University of Puerto Rico became a leader in Latin America in the field of nuclear energy both for medical and ecological applications. The Radiation Experiment at El Verde (1963–1967) was a collaboration between the University of Puerto Rico, the USDA Forest Service, and the Atomic Energy Commission. Oak Ridge National Laboratory in Tennessee was particularly influential, as it administered the funds for the Radiation Experiment for the AEC and provided scientific and technical support to the study.

Fig. 5.3 Howard Thomas Odum in 1963, when he was leading the radiation experiment at El Verde, Puerto Rico. (From the historical archives of the International Institute of Tropical Forestry, USDA Forest Service)

Teas and Odum's proposal was to the Division of Biology and Medicine of the AEC. In developing the proposal, Teas and Odum used the results of the Scientific Survey and the research of the Institute. The Radiation Experiment started in 1963 under Odum's leadership, who became part of the scientific staff of the Puerto Rico Nuclear Center. His office was located in the small space under the stairs leading to the second floor of the Nuclear Center's Río Piedras building at the Medical Center campus.

The Radiation Experiment (Fig. 5.4) attracted many scientists from mainland United States and European Universities and, from National Laboratories, as well as students, faculty, and staff from Puerto Rican universities and government agencies. Even the Puerto Rico Electric Power Authority (known then as *Autoridad de las Fuentes Fluviales*) collaborated by providing the helicopter that lowered the radiation source to the concrete pad from which it irradiated the forest for nearly three months. One participant scientist was William C. Steere, who was the only scientist to have participated in both the Scientific Survey (Crum and Steere 1957) and the Radiation Experiment (Steere 1970). At the time that he studied the effects of radiation on Bryophytes, Steere was Director of the New York Botanical Garden, the position held by Britton at the turn of the century.

The Radiation Experiment was conceived at a time when scientists and engineers were making rapid strides in the development of nuclear energy but the consequences of radiation exposure for organisms and communities was not clear (Bugher 1970). Research was needed to help clarify how nuclides moved through the environment and learn how sensitive organisms and ecosystems were to radiation exposure. Moreover, John N. Wolfe of the AEC's Division of Biology and Medicine was calling for greater emphasis on human ecology at a landscape scale with a focus on the interaction of whole ecosystems with technologically oriented humans (Wolfe 1970). Available research in the arctic and temperate zone was demonstrating high sensitivity to radiation in arctic communities and pine forests in the eastern United States (Wolfe 1970). Wolfe proved to be a visionary as he visualized the need for ecosystem-oriented, long-term studies through networks of sites as the way to develop sufficient understanding of ecological phenomena in a changing world. In his own words (p iv): "It does not seem unreasonable to suggest that massive ecological studies such as those at El Verde be federally funded and carried out in every biotic province of the continent on a continuing basis. Such endeavors would provide base lines and would answer many questions before disturbances occurred." In short, he anticipated the LTER program.[7]

Odum was among the first to ask ecosystem-level questions about the vulnerability of tropical forests to gamma radiation and suggested that the Radiation Experiment would begin to answer fundamental questions about what we call today the resilience of tropical forests. Dealing with this issue of resilience to disturbances

[7]In 1908, Zon established in the USDA Forest Service the first of a network of Experimental Forests and Ranges with similar national scope and scientific objectives (Lugo et al. 2006).

Fig. 5.4 The radiation source used to irradiate the wet forest at the El Verde sector of the Luquillo Experimental Forest. The two-ton lead container at the center of the concrete base contains the radioactive Cesium source, which when in operation, would be magnetically hoisted to below the oval metal structure at the top of the aluminum shaft. (From Odum 1970c)

was one of the themes that found its way to the LUQ LTER and became the central research topic of LUQ LTER for decades.

The funding of the Radiation Experiment benefitted from several institutional circumstances that made Puerto Rico attractive for hosting the study (Bugher 1970). At the national level, the AEC had been tasked with all matters related to peaceful uses of atomic energy. At the insular level, the University of Puerto Rico established the Puerto Rico Nuclear Center, which was tasked with developing nuclear energy applications for the benefit of Puerto Rico's economy. In addition to the Nuclear Center, the Commonwealth trained a new cadre of scientists and technicians for dealing with atomic energy and also established an atomic reactor in Rincón, Puerto Rico, to explore the potential for the generation of electricity for the island. The

Puerto Rico Nuclear Center focused on public health applications of nuclear energy, nuclear energy development, and marine and terrestrial environmental issues.

The Radiation Experiment was located in the terrestrial division of the Puerto Rico Nuclear Center. This institutional investment by the Commonwealth made it feasible for the AEC to locate the Radiation Experiment in Puerto Rico. Also, the intention of using nuclear explosions to open a new transoceanic canal in the Republic of Panama (Golley et al. 1975), favored Puerto Rico as a venue for this research because of its tropical location and the need to understand how tropical environments would respond to radiation effects as well as how nuclides would move in wet tropical conditions. Moreover, Puerto Rico had a unique level of foundational knowledge on tropical systems given the information about natural resources accumulated before, during, and after the Scientific Survey.

The timing of the funding for ecological research in the Puerto Rico Nuclear Center by the AEC also coincided with the development of new ways of thinking in the ecological field, a field that had its early development during the time the New York Botanical Garden was exerting scientific leadership in the United States of America (Kingsland 2005). The two ecological fields that were developing nationally in the 1950s and 1960s were the field of radiation ecology and the field of ecosystem-level ecology, neither of which had a tropical focus at their outset. Golley (1993) and Kingsland (2005) provide details on the connections among the AEC, the emergence of modern ecology, and the development of the ecosystem concept. In all these matters, Odum and his brother E. P. Odum were active participants. Of particular note in this regard was their joint study of a Pacific coral reef located in the Eniwetok Atoll that had been subjected to radionuclide fallout (Odum and Odum 1955). This study was an early example of ecosystem-level research with many lessons regarding the functioning of an ecosystem studied from both the population and community levels of complexity.

Odum's involvement in the Radiation Experiment assured that the El Verde Field Station would become a facility that hosted the visits of many scientists active in the discussion of the major ecological issues of the time. The Odum brothers and collaborators such as Golley, Richard Wiegert, and J. Frank McCormick were leaders within the Ecological Society of America (Egerton 2015). Their joint efforts at El Verde were reminiscent of the collaborative activities at the beginning of the century between scientists of the New York Botanical Garden and the USDA Forest Service during the origins of the Scientific Survey. All these historical developments would prove to be critical for the success of the LUQ LTER because the Luquillo Mountains in general, and the El Verde sector in particular, became known as centers of cutting edge research within the scientific community, both nationally and internationally.

The work at El Verde also transcended tropical ecological research, achieving international interest, including historical threads that lead to the projection of United States of America power over Latin America (Baldivieso 2016).

Odum's work and approach to scientific research was similar to that of Britton and Wadsworth, as these men had a strong sense of mission, welcomed collaboration, mentored scientists, and they all attracted dozens of scientists to Puerto Rico to advance the study and understanding of its natural resources. The notion of coupling forest research to forest conservation was also shared by Odum, as evidenced in his early writings (e.g., Odum 1962) and the strong presence of management issues in the Odum and Pigeon (1970) volume, which contains all the results of the Radiation Experiment between 1963 and 1970.

To the chagrin of some scientists, Odum prevented participants in the Radiation Experiment from publishing the results of their studies in journals. Instead, all contributions were published in a single volume edited by Odum and Robert F. Pigeon, Technical Editor at Oak Ridge National Laboratory. The book, titled *A Tropical Rain Forest: A Study of Irradiation and Ecology at El Verde, Puerto Rico*, is known among its users as *the rain forest book*, *the green giant,* or the *green bible*. Its impressive size (4.9 kg, 1644 pages, and 111 chapters) either encourages or discourages its use, depending on the user's intentions. The AEC published the volume, which received broad recognition for its comprehensive analysis of a tropical ecosystem, including positive reviews in major research journals. Its contents established a model for ecosystem-level studies of complex tropical forests. The volume also serves as an archive of field data and location of measurements for future reference.

The Radiation Experiment yielded numerous insights about the structure and functioning of tropical forests, insights that were useful to LUQ LTER scientists when developing the LTER proposal. A few examples suffice. The carbon cycle of the tabonuco forest was documented in detail during the Radiation Experiment. Part of the innovation involved was the early measurements of gas exchange rates of plants and soil in the field (Odum et al. 1970a), including the use of a giant cylinder to measure the metabolism and carbon balance of the whole forest (Odum and Jordan 1970). Odum and Lugo (1970) used forest floor microcosms to explore the issue of carbon enrichment of the atmosphere when respiration processes exceed photosynthesis rates. The coupling of the El Verde forest with global phenomena was done through the study of nuclide cycling and accumulation in forest epiphytes (Odum et al. 1970b). Microbial ecology studies drew attention to their high turnover and critical role in decomposition processes and nutrient cycling (Odum 1970a). Autecological studies of tropical forest species showed fundamental differences in life history strategies of pioneer and non-pioneer tree species (McCormick 1995). These population studies, conducted in the context of whole forest functioning, contributed to breaking down the false dichotomies between population and community-level approaches to ecology.

In his memoirs, Wadsworth (2014) recounts a lunch encounter with Odum. They were sitting on a log on the forest floor within tabonuco forest at El Verde. Their lunch discussion focused on all the perils encountered by tree seedlings on their way to canopy domination. This conversation epitomizes the fundamental research and management question that both individuals dedicated a lot of creative time to answering through research and application. Wadsworth's interest was in the development of seedlings into dominant trees with high quality wood volume, while Odum's interest focused on the ecological factors regulating the self-organization involved as seedlings developed into dominant members of the mature forest. This encounter of two of the most influential scientists on the island at that time, symbolizes the merging of science and conservation in the Luquillo Mountains.

When the Radiation Experiment ended with Odum's departure to the University of North Carolina at Chapel Hill in 1967, research continued on a second phase dealing with the movement of radionuclides through the forested landscape. Jerry R. Kline (1967–1968) and Carl Jordan (1968–1969), assisted by George E. Drewry led this phase of the study. Drewry had started with Odum in 1963 and acted as head of the El Verde Field Station. By 1970, Richard G. Clements and Drewry provided leadership to the program through its transition due to changes in the national funding landscape. In 1974, the AEC was abolished and substituted in part by the Energy Research and Development Administration, which continued to fund ecological research at El Verde. The Puerto Rico Nuclear Center was converted into the Center for Energy and Environment Research in 1976. In 1977, the Department of Energy was established and continued funding of ecological research at El Verde until the site was designated an LTER site by the National Science Foundation (NSF) in 1988. The transition under Clements culminated by 1980, when the research evolved into the Tropical Rainforest Cycling and Transport Program. This program focused on studies of succession, nutrient cycling, and animal ecology. After Clements, Laurence J. Tilly and Douglas P. Reagan led the research program followed by Robert B. Waide, who led the transition to the LUQ LTER.

5.5 The Luquillo Long-Term Ecological Research Program

Before the establishment of the LUQ LTER, Gerald Bauer, a USDA Forest Service forester, and Institute Project Leader Ariel E. Lugo spent several days in the National Forest searching for suitable locations to establish a watershed research program. They settled on three adjacent watersheds along the Bisley road. The USDA Forest Service constructed the road in the 1930s with labor from the Civilian Conservation Corps who followed an existing oxen trail (Scatena 1989). The oxen trail and road were used for a variety of purposes including transporting people and forest products to Sabana, the site where the National Forest Ranger had his home and office. Today, the Institute's Sabana Field Research Station is located at this site where it provides support to the LUQ LTER. Odum used part of the Bisley road to access the site of his 1950s biomass and metabolism study (Odum et al. 1970c). The three watersheds became the Bisley Experimental Watersheds, a long-term research program initially led by Institute scientist Frederick N. Scatena (Scatena 1989). The watersheds became the windward site that complemented the leeward El Verde site as the two main LUQ LTER study sites.

Funding for the LUQ LTER was secured in 1988 through a grant from the NSF to the University of Puerto Rico and the Institute. While funding from the NSF was significant and long-term, the LUQ LTER has always matched the NSF funding with substantial contributions from the University of Puerto Rico and the USDA Forest Service. The LUQ LTER has had functional similarities with the Scientific

Survey in that it is a multiagency collaboration where all participants contribute funds and scientists to the research program. Moreover, like the Scientific Survey, the LUQ LTER attracts the additional collaboration of many scientists and institutions from Puerto Rico, mainland United States, and internationally. As discussed above, the Institute's research program also operated on a similar philosophy, as did the Radiation Experiment. The high level of collaboration of the major ecological and forestry research and management programs in the Luquillo Mountains is one of the fundamental reasons why science and forest management has advanced so much in Puerto Rico and why the research and management programs have fed from each other to achieve excellence individually and through their synergy. Moreover, the LUQ LTER is connected to a national network of similar sites with similar missions: "to learn to protect and manage ecosystems, their diversity, and services" (Egerton 2015, p. 121).

The outcomes of the Radiation Experiment provided a framework for ecosystem-level studies in the LUQ LTER, a focus rare in other tropical sites at the time. The LUQ LTER was therefore in a good position to advance the understanding of tropical forests from ecosystem and community-level perspectives. For example, Reagan and Waide (1996) compiled the most detailed available analysis of the food web of an entire working tropical community because they had more than a decade of field research at El Verde to allow their comprehensive description and analysis of the complex community, and because the taxonomy and biology of the organisms were well known and dating back to results published in the Scientific Survey. Another synthesis effort used to boost research interest in the LEF and to support the drafting of NSF proposals was the summary of the research history and opportunities in the LEF (Brown et al. 1983). This publication became highly cited by LUQ LTER scientists and was latter updated (Harris et al. 2012).

The LTER program started in 1980 with James Callahan directing the program at the NSF. Luquillo scientists were unsuccessful in obtaining funding from NSF in 1980 and 1981. In 1987, when the NSF call for proposals invited submissions from tropical sites, the Luquillo proposal was partially funded. They received full funding in the 1988 competition. During the time the LUQ LTER proposals were being developed in the early 1980s, it was common for ecologists to identify themselves as population or community-level ecologists. This dichotomy was a source of tension among those scientists who assumed that these two parallel approaches to ecology had independent contributions to make to the understanding of tropical forests. This notion was initially accepted by the LUQ LTER, but was abandoned as participating scientists became more familiar with each other's work.

The LUQ LTER strategy for developing the research program at the Luquillo Mountains was to be inclusive and incorporate as much help as possible in the design of the research approach. They insured breadth of ideas by establishing an outside advisory committee that over time included such scientists as Odum, Wiegert, Jordan, Mathew Larsen, Gene E. Likens, Frank H. Bormann, Sandra Brown, and others. Scientists from other LTER programs were also invited to help in the design of the LUQ LTER. These scientists included Frederick J. Swanson

(Andrews LTER), David R. Foster (Harvard Forest LTER), Bruce Haines (Coweeta LTER), and Kristiina and Daniel Vogt (Andrews LTER).

To build strong bonds among all LUQ LTER scientists, weeklong meetings were held in natural settings suitable to scientific discourse and social interactions. One such meeting at a rustic facility at El Verde looking out over the forest was a memorable experience to all participants, and resulted in a tightly knit and scientifically unified group of scientists who understood each other's research and collaborated for decades to come. By the time that LUQ LTER published the book *A Caribbean Forest Tapestry; The Multidimensional Nature of Disturbance and Response* (Brokaw et al. 2012a), the focus for understanding tropical forests had evolved from a population and community ecology approach to an integrated approach based on the multidimensional nature of ecosystems. Such merging of the two sub-fields of ecology built on the results of the Radiation Experiment and represented one of many groundbreaking intellectual advances that resulted from the LUQ LTER.

Over a 30-year period the LUQ LTER has had numerous positive social and ecological outcomes. The passage of hurricane Hugo in 1989 over LUQ LTER sites led to better understanding of how disturbances affect tropical forest functioning (Lugo 2008) and exposed the accepted canopy gap-dynamics paradigm for explaining tropical forest regeneration as inadequate to explain the dynamics of forests exposed to hurricane winds (Lugo and Zimmerman 2002). Equally important to forest conservation and research, notions of ecosystem resilience, initially articulated by Odum (1970b), predominated during the LUQ LTER research synthesis (Waide and Willig 2012; Brokaw et al. 2012b). The results of this research falsified the prevailing paradigm that assumed tropical forests to be fragile (Farnworth and Golley 1974; Lugo 1995b; Sodhi et al. 2007).

In Puerto Rico, the LUQ LTER is a model for island-wide forest studies and a place where large numbers of underserved students (from Kindergarten to post-doctoral) train and learn about science, the tropics, and conservation. After the passage of hurricanes Irma and María in 2017, the research lessons after hurricane Hugo became critical to the design of management activities such as evaluating hurricane effects on trees, streams, and landslides. Luquillo LTER scientists had learned about the nature of forest response mechanisms through observation, measurement, and experimentation (Shiels and González 2014), and this knowledge proved useful for guiding post hurricane management actions.

Through the 200-year research history discussed here, Puerto Rico and the Neotropics have benefitted from the work of highly trained and talented scientists who spent time in the island and contributed to the development of new scientific talent, insular institutions, programs, and scientific and management activities. The progression of knowledge over time has followed a logical pathway starting with observation and spiritual expression among Taíno people, and ending with complex multi-scale research that spans from microbial to global scales with equal spiritual expression about the wonders of natural phenomena. The search for knowledge throughout this long period of research included exploration, inventory, and exploitation; taxonomic and natural history studies; forestry and silviculture research designed for assuring wise use of resources; and population and ecosystem-level studies involving experimentation to better understand and conduct forest conservation activities.

In the twelfth century, Bernard of Chartres introduced the notion of the progressive improvement of knowledge and how present generations benefit from the past as if standing on the shoulders of giants. Such is the case of the success of the LUQ LTER. Its accomplishments are rooted in 200 years of human investment in trying to understand the magic of the Luquillo Mountains. Therefore, in the temporal progression of events, the LUQ LTER is just another stage in the passing of the baton from the Taíno to tomorrow (Robinson et al. 2014).

5.6 Roads Taken and Not Taken: LUQ LTER in the Anthropocene

In his memoirs, Wadsworth (2014) included a section titled *frustrated visions* where he outlined missed opportunities at different times of his career. All careers and programs have such lost opportunities that in no way undermine accomplishments but that are important to learn from. The lost opportunities of LUQ LTER reflect a difference between the programs led by Britton, Wadsworth, and Odum, who were strong leaders operating independently under institutionally empowering missions, and the LTER Program, a national network of sites with strong top down constraints on participating sites. Many consider the fact that LTER scientists develop proposals to advance science as evidence of a bottom up approach. This is only partially true because top down control in the LTER is exerted through the framing of the required five core areas of research emphasis; required administrative and educational functions such as centralized data management and schoolyard programs; and the requirement to re-visit the research questions every six years as well as shifting objectives and approaches required by NSF.

Constraints imposed by NSF have benefits, such as advances in data management in ecological research, and comparative research across latitudes and ecosystem types. A strong justification for top down control in a long-term research network is the limitation of funds, which are barely sufficient to maintain core programs at participating sites, given that new research avenues require additional funding sources. The risk of a top down approach is in limiting creativity and individual initiatives when new avenues of research that conflict with program priorities are called for. As I discuss below, securing additional research funding sources outside the scope of LTER to study the Luquillo Mountains mitigates the limitations of the top down direction on ecological research of the LTER program.

Under the strong science leadership of Britton, Wadsworth, and Odum, past research programs in Puerto Rico were capable of exploring non-conventional topics such as Britton's exploration of forestry policy, Wadsworth's ecological approach to forestry, or Odum's application of computer simulations to advance system-level understanding of forests. In short, new research questions and new research directions naturally emerge from long-term ecological research. This also happened in the LUQ LTER. For example, scientists realized the importance of environmental gradients driving ecological processes in the Luquillo Mountains (González et al. 2013) and thus the need to study lowland forests, including secondary and emerging novel forest ecosystems (Lugo 2009). They also discovered the importance of past land use history to understanding present forest structure and species composition (Thompson et al. 2002), and the need to approach ecological systems from a social-ecological perspective (Muñoz Erickson et al. 2014). Moreover, new scientific frontiers were greatly expanding the ecological approach to include microbial, hydro-ecological, and geological scales that transcended traditional ecological scales of study. However in the context of the LTER, scientists did not have the same flexibility that Britton, Odum and Wadsworth had for exploring additional lines of research.

Given the above, LUQ LTER missed two opportunities: (1) not expanding research into lowland secondary and novel forests, and (2) not exploiting its momentum towards social-ecological research. These research topics conflicted with the traditional ecological focus of the LTER on mature forests and with the siloed-organization within NSF, which prevented a social-ecological approach to LTER research, with the exception of two urban ecology LTERs: the Central Arizona Phoenix, and the Baltimore Ecosystem Study.[8] In response to emerging research opportunities and LTER's resistance to the new topics of research, the LUQ LTER scientists were successful in articulating these new avenues of research activity through other funding sources. They received funding from other NSF programs to address new research questions about the ecosystems of the Luquillo Mountains, while the NSF and the USDA Forest Service funded social-ecological research and research on secondary novel forests outside the Luquillo Mountains. Unfortunately, the LTER Program was not designed for such dramatic expansion of research focus, and to save the LTER designation for the LUQ LTER, the scientific activity had to split according to other NSF Program objectives and research themes. By not addressing these developing lines of research and synthesis opportunities, the road was open to other organizations and research groups to fill the void, a process that strengthens ecological research island-wide, but does so outside the scope of LUQ LTER.

A positive outcome of roads not taken was that students and collaborators of senior LUQ LTER scientists are filling gaps left by the LTER program (e.g., Cusack 2013; Atkinson and Marín Spiotta 2015; Abelleira Martínez 2010; Muñoz Erickson et al. 2014) while some LUQ LTER scientists conduct significant research activity outside the scope of the LTER (e.g., María Uriarte, Grizelle González, Whendee Silver, Tana Wood, William McDowell). New research programs have arisen within the Luquillo Mountains with strong overlap with the LUQ LTER. Examples are the Tropical Responses to Altered Climate Experiments (TRACE; funded by the Department of Energy and the USDA Forest Service), the Luquillo Critical Zone

[8] NSF recently withdrew the Baltimore Ecosystem Study from the LTER network.

Observatory (LCZO, funded by NSF), and The Next Generation of Ecosystem Experiments (NGEE-Tropics, funded by the Department of Energy).

The LUQ LTER initially attempted to develop social-ecological research in the Luquillo Mountains and surrounding areas and it intended to expand this research focus in an upcoming renewal proposal. However, during the review process, the NSF LTER Program Manager strongly advised the LUQ-LTER principal investigators not to pursue social-ecological research within the scope of the LTER. At that time, the USDA Forest Service hired Tischa Muñoz Erickson, one of the pioneer scientists that were pursuing this research in collaboration with the LUQ LTER.[9] Muñoz Erickson assembled a transdisciplinary group that was funded through the Urban Long-Term Research Area (ULTRA) program with funds from NSF and USDA Forest Service. They focused on an urban watershed, the Río Piedras River Watershed (sanjuanultra.org), and the Institute established an International Urban Field Station in Río Piedras to support this research. The San Juan ULTRA evolved into a network of cities throughout the Neotropics and the temperate zones under the Urban Resilience to Extreme Events Sustainability Research Network managed by the Arizona State University. The NSF funded this network of cities, and the Institute's International Urban Field Station added cities in the Dominican Republic to further develop this line of research.

As research activity within the Luquillo Mountains multiplied, it became obvious that coordinating such a diversity of research programs, even in the small geographic area of Puerto Rico and the Luquillo Mountains, is a challenge. Coordination was necessary to ensure the forest conservation benefits from the results of new lines of research inquiry, and to prevent loss of information needed to advance knowledge about tropical forests. This has led to a third road not taken by the LTER, which is not integrating federal research in the Luquillo Mountains. Unfortunately, the different groups of scientists that participate in this diversity of research activity meet separately and have limited interaction with scientists from other groups studying the same forest stands. This results in loss of information for groups of scientists and loss of synergy among research groups that represent different NSF funding silos. The conservation benefits of the scientific synergy between different approaches to the study of the Luquillo Mountains are also lost.

Irrespective of the above, the LUQ LTER is blazing new trails regarding the study of Neotropical forests. These include the study through experimentation of the mechanisms of forest response to hurricanes, studies of the effects of droughts on wet forests, studies of gradients up the Luquillo Mountains, and studies of soil and microbial ecology. These are all understudied areas of research that are at present gaps in the understanding of tropical forest functioning. The long-term understanding of the forests of the Luquillo Mountains is particularly important in the Anthropocene Epoch, when changing environmental conditions challenge the adaptability of these forests. Moreover, the increased role and importance of people

---

[9] Others within LUQ LTER included Charles A.S. Hall, William H. McDowell, Jess K. Zimmerman, and Ariel E. Lugo.

in the Anthropocene also challenge forest managers that must deal day to day with uncertainty.

Although historical forests represent a small and shrinking proportion of the forests in Puerto Rico and the tropical world, they represent the forest type with the greatest complexity of all forest types, particularly more so than the simpler novel forests that predominate in Puerto Rico (Lugo and Helmer 2004) and are increasingly abundant in the world (Hobbs et al. 2013). Understanding and interpreting that complexity will require every advantage scientists can muster, and once again, the impressive information produced by the LUQ LTER and its predecessors will come into play to assure success and help the scientists support forest conservation in an age of environmental uncertainty. The remaining challenge is to integrate and make LTER studies relevant to the rest of the Anthropocene world.

Acknowledgments This work was conducted in collaboration with the University of Puerto Rico. Gisel Reyes and Jorge Morales assisted with the literature search. Mariela Ortiz is the graphic artist responsible for a timeline, and Sarah Stankavich, and Grizelle González contributed to the timeline of events. Carlos Domínguez Cristóbal, Kathleen McGinley, Tischa Muñoz Erickson, Robert B. Waide, and Sharon E. Kingsland reviewed and improved the manuscript. Patrick C. Kangas shared documents and a draft of the chapter on the El Verde Radiation Experiment of his upcoming book on the history of radiation ecology, and Helen Nunci helped with the production of the manuscript. Sylvia Zavala and Jorge Morales obtained the photos that illustrate the manuscript and their descriptions.

References

Abelleira Martínez, O.J. 2010. Invasion by native tree species prevents biotic homogenization in novel forests of Puerto Rico. *Plant Ecology* 211: 49–64.

Atkinson, E.E., and E. Marín Spiotta. 2015. Land use legacy effects on structure and composition of subtropical dry forests in St. Croix, US Virgin Islands. *Forest Ecology and Management* 335: 270–280.

Baatz, S. 1996. Imperial science and metropolitan ambition: The scientific survey of Puerto Rico, 1913-1934. *Annals of the New York Academy of Sciences* 776: 1–16.

Baldivieso, C. 2016. *Irradiating Eden: The El Verde experiment and the Atomic Energy Commission's nuclear prospecting in Latin America, 1954–1970*. M.A. thesis, University of Maryland, Baltimore County, Baltimore, Maryland.

Briscoe, C.B., and F.H. Wadsworth. 1970. Stand structure and yield in the tabonuco forest of Puerto Rico. In *A Tropical rain forest: A study of irradiation and ecology at El Verde, Puerto Rico*, ed. H.T. Odum and R.F. Pigeon, B79–B89. Oak Ridge: Division of Technical Information, Atomic Energy Commission.

Britton, N.L. 1919a. History of the survey. In *Scientific survey of Porto Rico and the Virgin Islands*, vol. 1, 1–10. New York: New York Academy of Sciences.

———. 1919b. Dr. Henry Allen Gleason, appointed first assistant. *Journal of the New York Botanical Garden* 20: 39–40.

———. 1930. The planting of a *Stahlia* on Arbor Day in Porto Rico. *Journal of the New York Botanical Garden* 31: 45–47.

Britton, N.L., and P. Wilson. 1923–1930. Descriptive flora-Spermatophyta. Botany of Porto Rico and the Virgin Islands, Vols. 5, 6. In *Scientific survey of Porto Rico and the Virgin Islands*. New York: New York Academy of Sciences.

Brokaw, N., T.A. Crowl, A.E. Lugo, W.H. McDowell, F.N. Scatena, R.B. Waide, and M.R. Willig, eds. 2012a. *A Caribbean forest tapestry: The multidimensional nature of disturbance and response*. New York: Oxford University Press.

Brokaw, N., J.K. Zimmerman, M.R. Willig, G.R. Camilo, A.P. Covich, T.A. Crowl, N. Fetcher, B.L. Haines, D.J. Lodge, A.E. Lugo, R.W. Myster, C.M. Pringle, J.M. Sharpe, F.N. Scatena, T.D. Schowalter, W.L. Silver, J. Thompson, D.J. Vogt, K.A. Vogt, R.B. Waide, L.R. Walker, L.L. Woolbright, J.M. Wunderle Jr., and X. Zou. 2012b. Response to disturbance. In *A Caribbean forest tapestry: The multidimensional nature of disturbance and response*, ed. N. Brokaw, T.A. Crowl, A.E. Lugo, W.H. McDowell, F.N. Scatena, R.B. Waide, and M.R. Willig, 201–271. New York: Oxford University Press.

Brown, S., A.E. Lugo, S. Silander, and L. Liegel. 1983. Research history and opportunities in the Luquillo experimental Forest. In *General technical report SO-44*. New Orleans: USDA Forest Service, Southern Forest Experiment Station.

Bugher, J.C. 1970. Project foreword. In *A tropical rain forest: A study of irradiation and ecology at El Verde, Puerto Rico*, ed. H.T. Odum and R.F. Pigeon, vii–viii. Oak Ridge: Division of Technical Information, Atomic Energy Commission.

Clements, F.E. 1928. *Plant succession and indicators: A definitive edition of plant succession and plant indicators*. New York: Hafner Press.

Colón Torres, J.A. 2009. *Climatología de Puerto Rico*. San Juan: La Editorial de la Universidad de Puerto Rico.

Crow, T.R. 1980. A rainforest chronicle: A 30-year record of change in structure and composition at El Verde, Puerto Rico. *Biotropica* 12: 42–55.

Crowl, T.A., N. Brokaw, R.B. Waide, G. González, K.H. Beard, E.A. Greathouse, A.E. Lugo, A.P. Covich, D.J. Lodge, C.M. Pringle, J. Thompson, and G.E. Belovsky. 2012. When and where biota matter, linking disturbance regime, species characteristics, and dynamics of communities and ecosystems. In *A Caribbean forest tapestry: the multidimensional nature of disturbance and response*, ed. N. Brokaw, T.A. Crowl, A.E. Lugo, W.H. McDowell, F.N. Scatena, R.B. Waide, and M.R. Willig, 272–304. New York: Oxford University Press.

Crum, H.A., and W.C. Steere. 1957. The mosses of Porto Rico and the Virgin Islands. In *Scientific survey of Porto Rico and the Virgin Islands*, volume 7, part 4, 395–599. New York: New York Academy of Sciences.

Cusack, D.F. 2013. Soil nitrogen levels are linked to decomposition enzyme activities along an urban-remote tropical forest gradient. *Soil Biology & Biochemistry* 57: 192–203.

Domínguez Cristóbal, C.M. 2000. *Panorama histórico forestal de Puerto Rico*. San Juan: Editorial de la Universidad de Puerto Rico.

Egerton, F.N. 2015. *A centennial history of the ecological society of America*. Boca Raton: CRC Press, Taylor & Francis Group.

Evermann, B.W. 1899. Summary of the scientific results of the Fish Commission expedition to Porto Rico. In *Investigations of the aquatic resources and fisheries of Porto Rico by the United States Fish Commission Steamer Fish Hawk in 1899*, xi–xv. Washington, DC: United States Fisheries Commission.

Farnworth, E.G., and F.B. Golley, eds. 1974. *Fragile eco-systems: Evaluation of research and applications in the Neotropics*. New York: Springer.

Figueroa Colón, J.C., ed. 1996. The scientific survey of Puerto Rico and the Virgin Islands: An eighty-year reassessment of the island's natural history. *Annals of the New York Academy of Sciences* 776: 1–273.

Gifford, J.C. 1905. The Luquillo Forest Reserve, Porto Rico. *Bulletin of the U.S. Department of Agriculture*, Bureau of Forestry, no. 54. Washington, DC: Government Printing Office.

Gleason, H.A. 1926. The individualistic concept of the plant association. *Bulletin of the Torrey Botanical Club* 53: 7–26.

Gleason, H.A., and M.T. Cook. 1926. Plant ecology of Porto Rico. In *Scientific survey of Porto Rico and the Virgin Islands*, vol. 7, 1–173. New York: New York Academy of Sciences.
Golley, F.B. 1993. *A history of the ecosystem concept in ecology: More than the sum of the parts*. New Haven: Yale University Press.
Golley, F.B., J.T. McGinnis, R.G. Clements, G.I. Child, and M.J. Duever. 1975. *Mineral cycling in a tropical moist forest ecosystem*. Athens: University of Georgia Press.
González, G.G., M.R. Willig, and R.B. Waide. 2013. Ecological gradient analyses in a tropical landscape. *Ecological Bulletins* 54: 7–250.
Harris, N.L., A.E. Lugo, S. Brown, and T. Heartsill-Scalley, eds. 2012. *Luquillo Experimental Forest: Research history and opportunities*. Washington, DC: U.S. Department of Agriculture, EFR-1.
Hill, R.T. 1899. Notes on the forest conditions of Porto Rico. *Bulletin of the Division of Forestry*, no. 25: Washington, DC: U.S. Department of Agriculture.
Hobbs, R.J., E.S. Higgs, and C.M. Hall, eds. 2013. *Novel ecosystems: Intervening in the new ecological world order*. West Sussex: Wiley-Blackwell.
Holdridge, L.R. 1967. *Life zone ecology*. San José: Tropical Science Center.
Howard, R.A. 1968. The ecology of an Elfin Forest in Puerto Rico, 1. Introduction and composition studies. *Journal of the Arnold Arboretum* 49: 381–418.
Kingsland, S.E. 2005. *The evolution of American ecology, 1890–2000*. Baltimore: Johns Hopkins University Press.
Kramer, W.P. 1924. El bosque del acueducto. *Revista de Agricultura de Puerto Rico* 12: 315–323.
Little, E.L., and F.H. Wadsworth. 1964. *Agriculture handbook 249: Common trees of Puerto Rico and the Virgin Islands*. Washington, DC: USDA Forest Service.
Little, E.L., F.H. Wadsworth, and J. Marrero. 1967. *Árboles comunes de Puerto Rico y las Islas Vírgenes*. San Juan: Editorial de la Universidad de Puerto Rico.
Little, E.L., R.O. Woodbury, and F.H. Wadsworth. 1974. *Trees of Puerto Rico and the Virgin Islands*, Agriculture handbook 449. Vol. 2. Washington, DC: USDA Forest Service.
Lugo, A.E. 1989. Biosphere reserves in the tropics: An opportunity for integrating wise use and preservation of biotic resources. In *Proceedings of the symposium on biosphere reserves, Fourth World Wilderness Congress*, ed. J.W.P. Gregg, S.L. Krugman, and J.D. Wood, 53–67. Atlanta: US Department of Interior, National Park Service.
———. 1995a. Contributions of H.T. Odum to tropical ecology. In *Maximum power: The ideas and applications of H.T. Odum*, ed. C.A.S. Hall, 23–24. Niwot: University Press of Colorado.
———. 1995b. Management of tropical biodiversity. *Ecological Applications* 5: 956–961.
———. 1996. Ninety years of plant ecology research in Puerto Rico. *Annals of the New York Academy of Sciences* 776: 73–88.
———. 2003. Obituary: Professor H.T. Odum (1924-2002). *Tropical Ecology* 44: 267–268.
———. 2004. H.T. Odum and the Luquillo Experimental Forest. *Ecological Modelling* 178: 65–74.
———. 2008. Visible and invisible effects of Hurricanes on forest ecosystems: An international review. *Austral Ecology* 33: 368–398.
———. 2009. The emerging era of novel tropical forests. *Biotropica* 41: 589–591.
Lugo, A.E., and E. Helmer. 2004. Emerging forests on abandoned land: Puerto Rico's new forests. *Forest Ecology and Management* 190: 145–161.
Lugo, A.E., and F.N. Scatena. 1995. Ecosystem-level properties of the Luquillo Experimental Forest with emphasis on the Tabonuco Forest. In *Tropical forests: Management and ecology*, ed. A.E. Lugo and C. Lowe, 59–108. New York: Springer.
Lugo, A.E., and J.K. Zimmerman. 2002. Ecological life histories. In *Agriculture handbook 721: Tropical tree seed manual*, ed. J.A. Vozzo, 191–213. Washington, DC: USDA Forest Service.
Lugo, A.E., F.J. Swanson, O. Ramos-González, M.B. Adams, B. Palik, R.E. Thill, D.G. Brockway, C. Kern, R. Woodsmith, and R. Musselman. 2006. Long-term research at USDA's Forest Service's experimental forests and ranges. *Bioscience* 56: 39–48.

Lugo, A.E., R.B. Waide, M.R. Willig, T.A. Crowl, F.N. Scatena, J. Thompson, W.L. Silver, W.H. McDowell, and N. Brokaw. 2012. Ecological paradigms for the tropics: Old questions and continuing challenges. In *A Caribbean forest tapestry: The multidimensional nature of disturbance and response*, ed. N. Brokaw, T.A. Crowl, A.E. Lugo, W.H. McDowell, F.N. Scatena, R.B. Waide, and M.R. Willig, 3–41. New York: Oxford University Press.

Marrero, J. 1950. La reforestación de tierras degradadas de Puerto Rico. *Caribbean Forester* 11: 16–24.

McCormick, J.F. 1995. A review of the population dynamics of selected tree species in the Luquillo Experimental Forest, Puerto Rico. In *Tropical forests: Management and ecology*, ed. A.E. Lugo and C.A. Lowe, 224–257. New York: Springer Verlag.

Mickulas, P. 2007. *Britton's botanical empire: The New York Botanical Garden and American botany, 1888–1929*. New York: The New York Botanical Garden Press.

Muñoz Erickson, T.A., A.E. Lugo, and B. Quintero. 2014. Emerging synthesis themes from the study of social-ecological systems of a tropical city. *Ecology and Society* 19: 23.

Murphy, L.S. 1916. Forests of Puerto Rico: Past, present, and future and their physical and economic environment. *Bulletin of the US Department of Agriculture*, no. 354. Washington, DC: U.S. Department of Agriculture.

Odum, H.T. 1962. Man and the ecosystem. In *Lockwood conference on the suburban forest and ecology*, 57–75. New Haven: The Connecticut Agricultural Experiment Station.

———. 1970a. Rain forest structure and mineral-cycling homeostasis. In *A tropical rain forest: A study of irradiation and ecology at El Verde, Puerto Rico*, ed. H.T. Odum and R.F. Pigeon, H3–H52. Oak Ridge: Division of Technical Information, Atomic Energy Commission.

———. 1970b. Summary: An emerging view of the ecological systems at El Verde. In *A tropical rain forest: A study of irradiation and ecology at El Verde, Puerto Rico*, ed. H.T. Odum and R.F. Pigeon, I191–I289. Oak Ridge: Division of Technical Information, Atomic Energy Commission.

———. 1970c. The AEC rain forest program. In *A tropical rain forest: A study of irradiation and ecology at El Verde, Puerto Rico*, ed. H.T. Odum and R.F. Pigeon, C3–C22. Oak Ridge: Division of Technical Information, Atomic Energy Commission.

Odum, H.T., and C.F. Jordan. 1970. Metabolism and evapotranspiration of the lower forest in a giant plastic cylinder. In *A tropical rain forest: A study of irradiation and ecology at El Verde, Puerto Rico*, ed. H.T. Odum and R.F. Pigeon, I165–I189. Oak Ridge: Division of Technical Information, Atomic Energy Commission.

Odum, H.T., and A. Lugo. 1970. Metabolism of forest-floor microcosms. In *A tropical rain forest: A study of irradiation and ecology at El Verde, Puerto Rico*, ed. H.T. Odum and R.F. Pigeon, I35–I56. Oak Ridge: Division of Technical Information, Atomic Energy Commission.

Odum, H.T., and E.P. Odum. 1955. Trophic structure and productivity of a windward coral reef community on Eniwetok Atoll. *Ecological Monographs* 25: 291–320.

Odum, H.T., and R.F. Pigeon, eds. 1970. *A tropical rain forest: A study of irradiation and ecology at El Verde, Puerto Rico*. Oak Ridge: Division of Technical Information, Atomic Energy Commission.

Odum, H.T., A. Lugo, G. Cintrón, and C.F. Jordan. 1970a. Metabolism and evapotranspiration of some rain forest plants and soil. In *A tropical rain forest: A Study of irradiation and ecology at El Verde, Puerto Rico*, ed. H.T. Odum and R.F. Pigeon, I103–I164. Oak Ridge: Division of Technical Information, Atomic Energy Commission.

Odum, H.T., G.A. Briscoe, and C.B. Briscoe. 1970b. Fallout radioactivity and epiphytes. In *A tropical rain forest. A study of irradiation and ecology at El Verde, Puerto Rico*, ed. H.T. Odum and R.F. Pigeon, H167–H176. Oak Ridge: Division of Technical Information, Atomic Energy Commission.

Odum, H.T., W. Abbott, R.K. Selander, F.B. Golley, and R.F. Wilson. 1970c. Estimates of chlorophyll and biomass of the Tabonuco Forest of Puerto Rico. In *A tropical rain forest: A study of irradiation and ecology at El Verde, Puerto Rico*, ed. H.T. Odum and R.F. Pigeon, I3–I19. Oak Ridge: Division of Technical Information, Atomic Energy Commission.

Pinchot, G. 1947. *Breaking new ground*. Washington, DC: Island Press.

Reagan, D.P., and R.B. Waide, eds. 1996. *The food web of a tropical rain forest*. Chicago: University of Chicago Press.

Roberts, R.C. 1942. USDA Series 1936, No. 8.: Soil Survey of Puerto Rico. Washington, DC: US. Government Printing Office.

Robinson, K., A.E. Lugo, and J. Bauer, eds. 2014. *Passing the baton from the Taínos to tomorrow: Forest conservation in Puerto Rico. FS-862*. San Juan: USDA Forest Service International Institute of Tropical Forestry.

Rudolf, P.O. 1957. R. Zon, pioneer in forest research. *Science* 125 (3261): 1283–1284.

Santiago Valentín, E. 2005. Amos Arthur Heller's Puerto Rico plant collecting itineraries of 1900 and 1902-1903 and their utility for the historical study of endangered plants. *Brittonia* 57: 292–294.

Santiago Valentín, E., L. Sánchez-Pinto, and J. Francisco-Ortega. 2015. Domingo Bello y Espinosa (1817-1884) and the new taxa published in his *Apuntes para la flora de Puerto-Rico*. *Taxon* 64: 323–349.

Scatena, F.N. 1989. *An introduction to the physiography and history of the Bisley experimental watersheds in the Luquillo Mountains of Puerto Rico*. General Technical Report SO-72. New Orleans: USDA Forest Service, Southern Forest Experiment Station.

Seaver, F.J., and C. Chardón. 1926. Mycology. In *Scientific survey of Porto Rico and the Virgin Islands*, vol. 8. New York: New York Academy of Sciences.

Shiels, A.B., and G.G. González. 2014. Tropical forest responses to large-scale experimental hurricane effects. *Forest Ecology and Management* 332: 1–135.

Sodhi, N.S., B.W. Brook, and C.J.A. Bradshaw. 2007. *Tropical conservation biology*. Malden: Blackwell Publishing.

Steere, W.C. 1970. Bryophyte studies on the irradiated and control sites in the rain forest at El Verde. In *A tropical rain forest: A study of irradiation and ecology at El Verde, Puerto Rico*, ed. H.T. Odum and R.F. Pigeon, D213–D225. Oak Ridge: Division of Technical Information, Atomic Energy Commission.

Thompson, J., N. Brokaw, J.K. Zimmerman, R.B. Waide, E.M. Everham III, D.J. Lodge, C.M. Taylor, D. García Montiel, and M. Fluet. 2002. Land use history, environment, and tree composition in a tropical forest. *Ecological Applications* 12: 1344–1363.

Tugwell, R.G. 1947. *The stricken land: The story of Puerto Rico*. Garden City: Doubleday.

Wadsworth, F.H. 1970. A review of past research in the Luquillo Mountains. In *A tropical rain forest: A study of irradiation and ecology at El Verde, Puerto Rico*, ed. H.T. Odum and R.F. Pigeon, B33–B46. Oak Ridge: Division of Technical Information, Atomic Energy Commission.

———. 1995. A forest research institution in the West Indies: The first 50 years. In *Tropical forests: Management and ecology*, ed. A.E. Lugo and C. Lowe, 33–56. New York: Springer Verlag.

Wadsworth, F. 2014. *A forestry assignment to Puerto Rico: Memoirs of Frank Wadsworth*. San Juan: Impresos Emmanuelli.

Waide, R.B., and M.R. Willig. 2012. Conceptual overview: Disturbance, gradients, and ecological response. In *A Caribbean forest tapestry: The multidimensional nature of disturbance and response*, ed. N. Brokaw, T.A. Crowl, A.E. Lugo, W.H. McDowell, F.N. Scatena, R.B. Waide, and M.R. Willig, 42–71. New York: Oxford University Press.

Walker, J. 2014. The prehistoric island to 1508. In *Passing the baton from the Taínos to tomorrow: Forest conservation in Puerto Rico*, ed. K. Robinson, J. Bauer, and A.E. Lugo, 17–27. FS–862. San Juan: USDA Forest Service International Institute of Tropical Forestry.

Weaver, P.L. 1994. *Baño de Oro natural area, Luquillo Mountains, Puerto Rico*, Station general technical report SO-111. New Orleans: USDA Forest Service, Southern Forest Experiment Station.

Willig, M.R., S.J. Presley, C.P. Bloch, and J. Alvarez. 2013. Population, community, and metacommunity dynamics of terrestrial gastropods in the Luquillo Mountains: A gradient perspective. *Ecological Bulletins* 54: 117–140.
Wolfe, J.N. 1970. Foreword. In *A tropical rain forest: A study of irradiation and ecology at El Verde, Puerto Rico*, ed. H.T. Odum and R.F. Pigeon, iii–v. Oak Ridge: Division of Technical Information, Atomic Energy Commission.
Zon, R., and W.N. Sparhawk. 1923. *Forest resources of the world*. New York: McGraw-Hill Book Co.

Chapter 6
Ecological Theory and Practice in Arid and Semiarid Ecosystems: A Tale of Two LTER Sites

Debra P. C. Peters

Abstract The chapter compares the conceptual history of two Long Term Ecological Research (LTER) programs at U.S. Department of Agriculture (USDA) research sites established following land overuse in the early 1900s. Early USDA studies were based on a Clementsian model, where vegetation following disturbance was expected to return to the climax determined by climate and soils. Management-based alternatives to the Clementsian model were developed after the dominant vegetation failed to recover and alternative states were observed. The Central Plains Experimental Range LTER (later Shortgrass Steppe LTER) program began in 1981 at a site in northern Colorado. Early studies focused on the inability of the dominant species, blue grama (*Bouteloua gracilis*), to recover on long-abandoned agricultural fields. These observations were in contrast to the resilience of blue grama-dominated ecosystems in response to livestock grazing and drought. The Jornada Basin LTER program began in 1982 at a Chihuahuan Desert site in southern New Mexico, where black grama-dominated *(Bouteloua eriopoda)* grasslands converted to shrublands following livestock overgrazing and drought. Early Jornada studies focused on the causes and consequences of this desertification. In collaboration with USDA scientists, the two LTER programs developed new research ideas and approaches which brought an improved understanding of historic dynamics and nuance to the Clementsian-based paradigm. New paradigms also reflected the importance of objective, long-term studies and the dynamic, multiscale nature of these arid and semiarid ecosystems. Comparing two LTER programs at research sites sharing a conceptual basis, but with different ecosystem properties, can provide insights into the development of US grassland ecology over the past century.

D. P. C. Peters (✉)
Agricultural Research Service, US Department of Agriculture, Washington, DC, USA

Jornada Experimental Range Unit, USDA ARS; Jornada Basin Long Term Ecological Research Program, New Mexico State University, Las Cruces, NM, USA
e-mail: Deb.peters@usda.gov

R. B. Waide, S. E. Kingsland (eds.), *The Challenges of Long Term Ecological Research: A Historical Analysis*, Archimedes 59,
https://doi.org/10.1007/978-3-030-66933-1_6

Keywords LTER program · Long-term ecological research · Grassland ecology · Arid ecosystems · Desertification · Ecological climax · International Biological Program · Livestock grazing · Drought

6.1 Introduction

In the US, in the western part of the central Great Plains and in the arid Southwest, the Homestead Act of 1862 and the construction of railroads and drilling of wells opened vast areas to settlement and economic development. Although crop failure and overgrazing were common in dry and average years, there were enough wet years and decades to sustain optimism for farming in the Great Plains and ranching in the Southwest; thus these land management practices continued in many places until the early 1900s. Because farmers, ranchers, and ecologists had a poor understanding of the physical and biological processes, and feedbacks constraining the recovery of grasses following large-scale disturbance, over-use of the land and its vegetation was common. This lack of understanding created large expanses of severely disturbed lands. Initial optimism for recovery remained high, since notable early ecologists suggested that removal of a disturbance, either cultivation or overgrazing, would allow a return to the previous grassland state in equilibrium with climate and soils (Clements 1916).

Beginning in the early 1900s in both geographic locations, large tracts of land were acquired by the federal government and placed in the public domain, often as national parks or as research sites protected from farming and livestock grazing to allow time for recovery. Protected areas in the Great Plains later became national grasslands under the USDA Forest Service (USFS), including the Pawnee National Grasslands (PNG) in northeastern Colorado (ca. 78,100 ha). A small area of about 6300 ha in the southwestern corner, the Central Plains Experimental Range (CPER), was established in 1937 and devoted to research; the land was transferred to the USDA Agricultural Research Service (ARS) in 1953. Funding for the Shortgrass Steppe (SGS) LTER program from the National Science Foundation (NSF) began in 1981, with a focus on the CPER that later expanded to other locations within the PNG (Lauenroth and Burke 2008).

In southern New Mexico in 1912, the Jornada Range Reserve (ca. 78,000 ha) was established from private and public lands and protected from overgrazing (Havstad 1996). Research began in 1915, when the Jornada was turned over to the USFS, and continued under the ARS beginning in 1954 as the Jornada Experimental Range (Havstad and Schlesinger 1996). In 1927, the adjoining property (ca. 26,000 ha) was given by the US Congress to New Mexico College of Agriculture and Mechanic Arts (currently New Mexico State University) for the purpose of conducting research, education, and demonstration projects on livestock, grazing methods, and range forage plants. This NMSU property, known locally as the College Ranch, became in 1982 the initial location of the Jornada Basin LTER program. LTER research expanded in 1989 to include both the NMSU property (now the Chihuahuan Desert Rangeland Research Center [CDRRC]) and the Jornada Experimental Range as the Jornada Basin LTER research site.

Here we compare the growth of science at the SGS and Jornada with a focus on the history of each research site that set up the central questions underlying early LTER proposals. We follow the development and change in research paradigms relative to these initial questions. Both research sites were established in response to over-use by human activities, and grounded in the Clementsian paradigm where the vegetation was expected to recover to grasslands given sufficient time. The inability of either native *Boueloua* grasses to recover within several decades provided early research questions that challenged this paradigm for research programs at both sites. These sites were influenced by the International Biological Programme (IBP) of the 1970s, when energy flow, systems ecology, and simulation modeling were brought to the forefront of ecology. The conceptual frameworks for both LTER research programs were modified to account for nonlinearities, feedbacks, and thresholds, in response to changing ecological dynamics. These changing dynamics either occurred naturally or were observed in response to experimental manipulation of environmental drivers. Comparing the scientific development of two LTER programs with a common conceptual beginning, but different ecosystem properties can provide insights into how grassland ecology in the US has developed through time.

6.2 Early History of the Shortgrass Steppe and Jornada Research Sites (USDA and IBP Research Before the 1980s)

Throughout much of the western US, including the locations where the CPER and Jornada would be established, a Clementsian view of succession prevailed as the country was settled from the east to west in the 1800s to mid-1900s. In this model, vegetation changes in a predictable way following disturbance until a climax state is reached that is in equilibrium with local climate and soils (Clements 1916). By the 1970s to 1980s, observations at both research sites led to questions about the validity of this conceptual model as thresholds had apparently been crossed and transitions had occurred leading to alternative vegetation states where the dominant grass species, either blue grama at the shortgrass steppe or black grama at the Jornada, had been replaced by other native species. Some challenges had to be overcome at each site before the new LTER programs in the 1980s could tackle these surprising dynamics and develop new conceptual models to replace the Clementsian model.

Shortgrass Steppe Grasslands dominated the rich soils of the Central Great Plains (CGP), which included the SGS, at the beginning of European settlement in the mid-1800s. As the country was settled by farmers and ranchers, there was a common belief that 'rain follows the plow'. However, the amount of rainfall received annually decreased along this east-west gradient, and the settlers who crossed the Mississippi River into the central Great Plains experienced a much different, and less forgiving, environment than their counterparts to the east. Livestock grazing was a major land use in the drier portions of the CGP where grazing by large herbivores, primarily bison, had occurred for 4000 to 5000y prior to their extinction in the late 1800s (McDonald 1981). This long evolutionary history of grazing by native herbivores, combined with semiaridity, resulted in native plants in the shortgrass steppe that were well-adapted to grazing by livestock and periodic drought that occurred throughout this time period (Milchunas et al. 1988).

The other major land use throughout the Central Great Plains was cultivation for crops where cotton, wheat, and corn were the primary crops grown in the region. By the 1920s, the high price of wheat, driven in part by the demand from World War I, encouraged farmers throughout the Central Great Plains to cultivate increasingly marginal land that included areas in the semiarid shortgrass steppe. The nearly decade-long drought of the 1930s combined with widespread cultivation led to high plant mortality on croplands, huge dust storms, and shifting sand dunes that collectively led to the 'Dust Bowl' (Savage 1937). An estimated five million acres of farmlands were abandoned and not returned to farming (Bement et al. 1965).

Cultivation and subsequent abandonment of land from farming was a novel disturbance, in particular in semiarid CGP grasslands with low annual rainfall and infertile soils, such that recovery of grasses did not occur immediately following abandonment. In response, the federal government purchased private land and created the National Grasslands within the USDA in the more arid regions of the West. One of the more devastated areas was northeastern Colorado where the Pawnee National Grasslands (PNG) were created. The CPER was established as a research site on the western side of the Pawnee National Grasslands to solve range management problems of the shortgrass prairie. The CPER originally consisted of a mosaic of native shortgrass steppe and abandoned farmland from the 1930s that was either left alone (locally called 'go back' land) or reseeded when taken over by the government, often with native grasses and non-native grasses (e.g., crested wheatgrass). Approximately 25–30% of eastern Colorado and the CPER was cultivated, abandoned, and reseeded by 1935 (Coffin et al. 1996).

Early USDA research focused on ways to improve range production through studies of livestock grazing, negative effects of rabbits on vegetation, and life history habits of the dominant grass, blue grama (Shoop et al. 1989). Many studies at the CPER, including construction of livestock grazing exclosures either with and without rabbits, charting of grass demography, and pastures with different livestock grazing intensities, were repeated at the more arid Jornada, located ca. 1200 km to the south, by different investigators, and this promoted comparisons of the two sites (Nelson 1934; Riegel 1941; Costello 1944; Klipple and Costello 1960). These studies were conducted under the Clementsian model of succession modified to account for livestock grazing (Dyksterhuis 1949, 1958). The Clementsian paradigm predicted a return to blue grama dominance in the shortgrass steppe following a series of successional stages (Judd and Jackson 1939; Costello 1944). These observations used the quadrat method (per m^2) to sample and measure variability in vegetation responses (Pound and Clements 1898). This scale-independent approach to sampling grassland vegetation uses the average cover of multiple, randomly located quadrats to represent an area.

Early studies showed that shortgrass steppe vegetation has very different responses to livestock grazing compared with abandoned farmland. Observations in the 1940s to 1950s showed that cattle grazing has little effect on shortgrass steppe vegetation (Klipple and Costello 1960). However, in the 1970s, vegetation on fields abandoned in the 1930s still had not recovered to dominance by blue grama (Hyder et al. 1975). These observations led USDA researchers to consider the Clementsian paradigm from two angles. *First,* blue grama was characterized as a relict species that had become established in a previous climate and had maintained dominance in the shortgrass steppe via traits, such as drought tolerance and high rooting density at shallow depths, that allowed it to out-compete other plants (Hyder et al. 1975). This paradigm allowed researchers to maintain a Clementsian model by invoking a change in climate. The paradigm was supported by field observations where: (a) reseeding of blue grama on disturbed areas often failed (Bement et al. 1965), (b) new blue grama seedlings were rarely observed, and (c) seedlings died by 6–10 weeks of age when adventitious roots failed to develop under field conditions (Hyder et al. 1971). A series of experiments and greenhouse trials confirmed the microenvironmental constraints on blue grama seedling establishment (Briske and Wilson 1977, 1978). *Second,* a new paradigm was forming, based on the concept of ecological thresholds—points where ecosystem function changes so dramatically that a return to the prior state is unlikely—and alternative stable states develop that are dominated by mid-successional grasses (e.g., *Aristida spp.*, *Sporobolus* spp.), where blue grama does not return to dominance (Laycock 1991). Evidence from repeat observations on abandoned fields and other disturbances, such as old roads, provided further support for this paradigm where recovery was not occurring more than 40 years following abandonment (Reichhardt 1982; Samuel 1985). This paradigm agreed with findings from other, more arid systems (Westoby et al. 1979) and complemented similar frameworks being developed at the Jornada and elsewhere in desert grasslands where alternative states had occurred and apparently were persistent through time (e.g., Archer 1989; Schlesinger et al. 1990).

Both paradigms predicted that blue grama was replaced as the dominant species, although the proximate cause was different. In the first paradigm, a change in climate led to a change in dominant species that was not blue grama but was in equilibrium with this new climate; thus this model still supports the Clementsian model. The second paradigm stated that any disturbance causing mortality of blue grama would shift the system to an alternative state dominated by different species of perennial grasses. Sometimes this state was referred to as a disclimax in keeping with Clementsian terminology. Testing these two models against the Clementsian model would require experiments of the underlying processes, long-term observations of different types and sizes of disturbances, and a synthesis of the experiments and observations within the entire grassland system. These studies would come later as part of LTER.

During this same time period (1972–1979), a group of researchers from Colorado State University in nearby Fort Collins, CO, started research at the CPER with funding from NSF as part of the IBP. Anticipating funding by NSF for the US IBP grasslands research (1968–1974), George Van Dyne led the establishment of the Natural Resource Ecology Laboratory (NREL) at CSU. As its first director (1967–1973), Van Dyne established a tradition of research at NREL based on the emerging field of systems ecology. He developed interdisciplinary teams of researchers with diverse expertise to address ecosystem-level problems facing land managers. Under his leadership, the Pawnee at the CPER was selected as the 'core' site for the Grasslands IBP. The aim was to study the whole grassland system with an emphasis on seasonal biomass and energy dynamics of grasslands across trophic levels, using ten grassland sites in the central and western US (Sims et al. 1978).

The impacts of computer simulation modeling and the systems approach developed at the CPER on the discipline of ecology cannot be overstated. The quadrat approach (/m^2) was used as a common metric among components of grassland ecosystems that provided baseline data for energetics-based grassland simulation models under development when the NSF call for LTER proposals occurred in 1980 (Innis 1978). In addition, postdoctoral fellows and graduate students trained under Van Dyne, George Innis, and others during the IBP (e.g., Robert Woodmansee, William Lauenroth) were at CSU and primed to become leaders in the LTER at the CPER/SGS.

Jornada In the arid Southwest, major naturally-occurring transitions between perennial grasslands and dominance by native woody plants occurred several times over the last 10,000 years (Van Devender 1995). The Chihuahuan Desert was exposed to livestock grazing as early as the 1500s, when Spanish explorers followed the Rio Grande northward along the Camino Real (Fredrickson et al. 1998). Throughout the American Southwest, many desert grasslands converted to dominance by woody plants by the early 1900s (Shreve 1917). Livestock grazing expanded in the late 1800s to 1900s, with a noticeable decline in range conditions accompanied by increases in shrub cover and density (Buffington and Herbel 1965; Gibbens et al. 2005). The Jornada research site was established in response to this region-wide decline in grasslands, and to the need for a large research area to develop a management plan that included livestock grazing under drought conditions (Wooton 1908).

Like the USDA studies at the shortgrass steppe site, early Jornada studies were conducted under a Clementsian model modified for rangelands, which included a focus on livestock grazing, shrub management, and grass recovery (Havstad 1996). Guidelines were developed for livestock grazing of black grama grasslands during drought (Campbell and Crafts 1938) as well as treatments for removing or reducing density of shrubs. Large exclosures (1 mile square; 260 ha) constructed to keep livestock and rabbits from affecting the vegetation, both with and without shrub treatments, were created in the 1930s as part of grass recovery efforts. Chemical treatments of shrubs, started after WWII, continued until the 1980s (Herbel 1983). Many other attempts were made to maintain or restore perennial grasses as the climax vegetation; most were unsuccessful (Herrick et al. 2006). By the 1950s, perennial grass cover was severely reduced throughout the Jornada. From 1951 to 1956, the most severe drought recorded during a 350-year period further reduced grass cover and increased the spread of shrubs, in particular mesquite and creosote-bush, which persisted after the drought (Fredrickson et al. 1998). Vegetation maps created by the USDA in 1915, 1926, and 1998 showed almost complete conversion of the Jornada and CDRRC from perennial grasslands to shrubs despite active management to limit their spread (Gibbens et al. 2005).

Thus, by the 1980s, the Clementsian paradigm had been called into question in the Chihuahuan Desert. Land managers recognized that long-term changes in vegetation from grass- to shrub-dominated at the Jornada had dramatically lowered rangeland forage productivity (Buffington and Herbel 1965). The cause of this change—climate change (Neilson 1986), fire suppression (Van Auken 2000), overgrazing (Archer 1989), or rising CO_2 (Idso 1992)—was hotly debated. Whatever the cause, it was obvious that the transition in vegetation had also caused a major reorganization of soil properties, from a relatively homogeneous pattern in grasslands to a patchy distribution of soil nutrients in shrublands (Schlesinger et al. 1996). In addition, black grama, the historical dominant perennial grass prior to the 1900s, had failed to recover following the cessation of grazing or with remediation treatments (Herrick et al. 2006). Black grama seedlings were rarely observed, leading some to speculate that this species was a relict species adapted to a past climate (Neilson 1986), similar to the speculations about blue grama at the CPER.

During the 1970s, the US IBP for the desert biome was led by Utah State University. The Jornada was one of the major validation sites under the leadership of Walt Whitford (1974–1979) at New Mexico State University (NMSU). Efforts in the Mojave Desert were located at Rock Valley on the Nevada Test Site. Nearly all desert sites organized their field studies around an individual shrub (e.g., *Larrea tridentata* or *Artemisia tridentata*) as the basic unit of study. While the IBP advanced a basic understanding of desert biology and recognized that shrub islands are hot spots of biological activity, its published works were not easy to assimilate into large-scale models of earth system function because the bare soils between shrubs were given little attention. It was nearly impossible to estimate basic ecosystem processes, such as net primary productivity, soil carbon content, and nitrogen mineralization, over large areas. This individual plant-based method is in contrast with the quadrat method developed and in use at the grassland sites where plant cover estimated as a proportion of total cover in a quadrat allowed scaling up to large spatial extents (Pound and Clements 1898). Like the Grassland IBP, the Desert IBP provided trained personnel in leadership positions (Whitford) and others with systems expertise in deserts (Gary Cunningham) ready to participate when the call for LTER proposals appeared from NSF in 1980.

Influence of the IBP on LTER Programs at Both Sites Both LTER Programs had a long history of research by USDA investigators dating to the 1910s (Jornada) and 1930s (CPER/SGS). Because of the history of overuse by livestock grazing (Jornada) and cultivation followed by abandonment (CPER/SGS), long-term treatments and observations existed at both research sites that were maintained through collaboration with the USDA. Each site also had a recent history of research from the IBP, either from the Desert Biome (Jornada) or the Grasslands Biome (CPER and Jornada) where large amounts of ecological data had been collected, archived, and made available from local university and federal sources, and used in simulation models. Each site was near a university where a group of ecologists was located who had experience with NSF as a result of the IBP. All of these factors likely played a role in the submission of LTER proposals from NMSU and CSU when the first call for proposals came out.

6.3 Early History of the LTER Research Programs (CPER from 1981 to 1985; Jornada from 1982 to 1987)

The first funded LTER proposals by both the CPER and Jornada LTER programs emphasized a catena or watershed as the organizing principle, concepts borrowed from the long-term watershed sites in forests (Bormann and Likens 1967; Swanson et al. 1988). For both sites, this conceptual model was useful in setting up long-term monitoring locations and providing baseline information about ecological patterns across relatively short environmental gradients. However, the conceptual model proved to be insufficient to explain ecological dynamics at multiple scales in these arid and semiarid systems at either research site, and a different conceptual model was needed for the second LTER proposal.

At the CPER, the catena concept was instrumental in characterizing patterns in plant species composition, soil nutrients, and related cattle behavior associated with the rolling hill topography of uplands and lowlands in the shortgrass steppe (e.g., Schimel et al. 1985; Senft et al. 1985, 1987). However, the aeolian reworking of surface soils combined with buried soils (paleosols) led to patterns in soil organic matter that did not follow the catena model of movement of material from upslope to downslope for many locations (Yonker et al. 1988). This redistribution of soil and nutrients has consequences for the movement of water and ecosystem dynamics. Thus, the catena selected for the first LTER proposal did not represent the multi-scale variability in patterns and dynamics across the CPER site, and a new model was needed for LTER II.

At the Jornada, a large-scale experiment was developed along a catena to determine if perennial bunchgrasses could return to dominance after soil resources are homogenized. A treatment and an adjoining control transect were selected along a change in topography from a black grama-dominated grassland at the base of a small mountain (Mt. Summerford) through a shrub-dominated upper bajada to a grassland playa on the CDRRC. Resources were homogenized along the treatment transect by fertilizing from the air with ammonium nitrate each year. Field studies quantified plant cover, soil water content, and other indices of ecosystem function along these transects. However, this research, similar to the IBP-funded research of the previous decade, remained focused on individual plants, either shrubs or grasses; the remaining soil in a small spatial extent, such as that provided by a quadrat, was not sampled (Fisher et al. 1988). Relationships developed between dominant shrub species and soil properties along this topographic gradient (e.g., Wondzell et al. 1996) provided paradigms about stability in shrub species-soil patterns that were only recently questioned after more than 30 years of LTER research. Inadvertent fertilization of the control transect in 1986 coupled with poor data management led to a call by NSF for new leadership of the Jornada LTER. This move also provided an opportunity to critically evaluate the catena/watershed model of LTER I, and to forge stronger relationships with the new administrator of the adjoining USDA property.

6.4 Development of Ideas at the CPER/SGS LTER Program

The Early Days (1986–1990) By the mid-1980s, hierarchy theory was emerging as a potential organizing principle for ecological systems (Allen and Starr 1982). This theory states that ecological processes can be distinguished by discrete levels of organization, and by spatial and temporal scales of interaction (O'Neill et al. 1986). The CPER LTER scientists (Lauenroth, Osvaldo Sala, William Parton) benefited from a strong history of simulation modeling and data availability from the IBP that allowed hypotheses to be generated using hierarchy theory as an organizing framework. The second CPER LTER proposal (1986) was the first to develop a nested hierarchy of spatial scales to explain pattern at an LTER site, from individual plants to patches along a catena within the shortgrass steppe in the larger Central Grasslands region (Fig. 6.1). The overall theme revolved around the origin and maintenance of spatial pattern and the rules for transforming information about a particular spatial or temporal scale to observations at the next higher scale. The proposal formed hypotheses about blue grama recovery on disturbances of different sizes and other characteristics using a *shift in thinking away from a square meter sampling approach (i.e., quadrat) to an individual plant-based approach*. This scale-dependent approach based on individual plants (e.g., Coffin and Lauenroth 1988, 1990) was comparable to the individual plant-interspace approach being developed at the Jornada (Schlesinger et al. 1990) and the individual-based simulation models being developed in forests (Shugart 1984). By focusing on individual plants, the research on blue grama could focus on demographic processes needed for this species to recover following disturbance (e.g., seed production, seed germination, seedling establishment and growth), and how these processes are affected by disturbance characteristics (e.g., size, soil type, climatic conditions). This research was also influenced by advances in disturbance ecology and the importance of patch characteristics to recovery dynamics on disturbances of different sizes (Platt 1975; Sousa 1984; Pickett and White 1985).

Major results during this time period reflect the use of simulation modeling to begin to test long-held views about the ecosystem. Blue grama was being managed

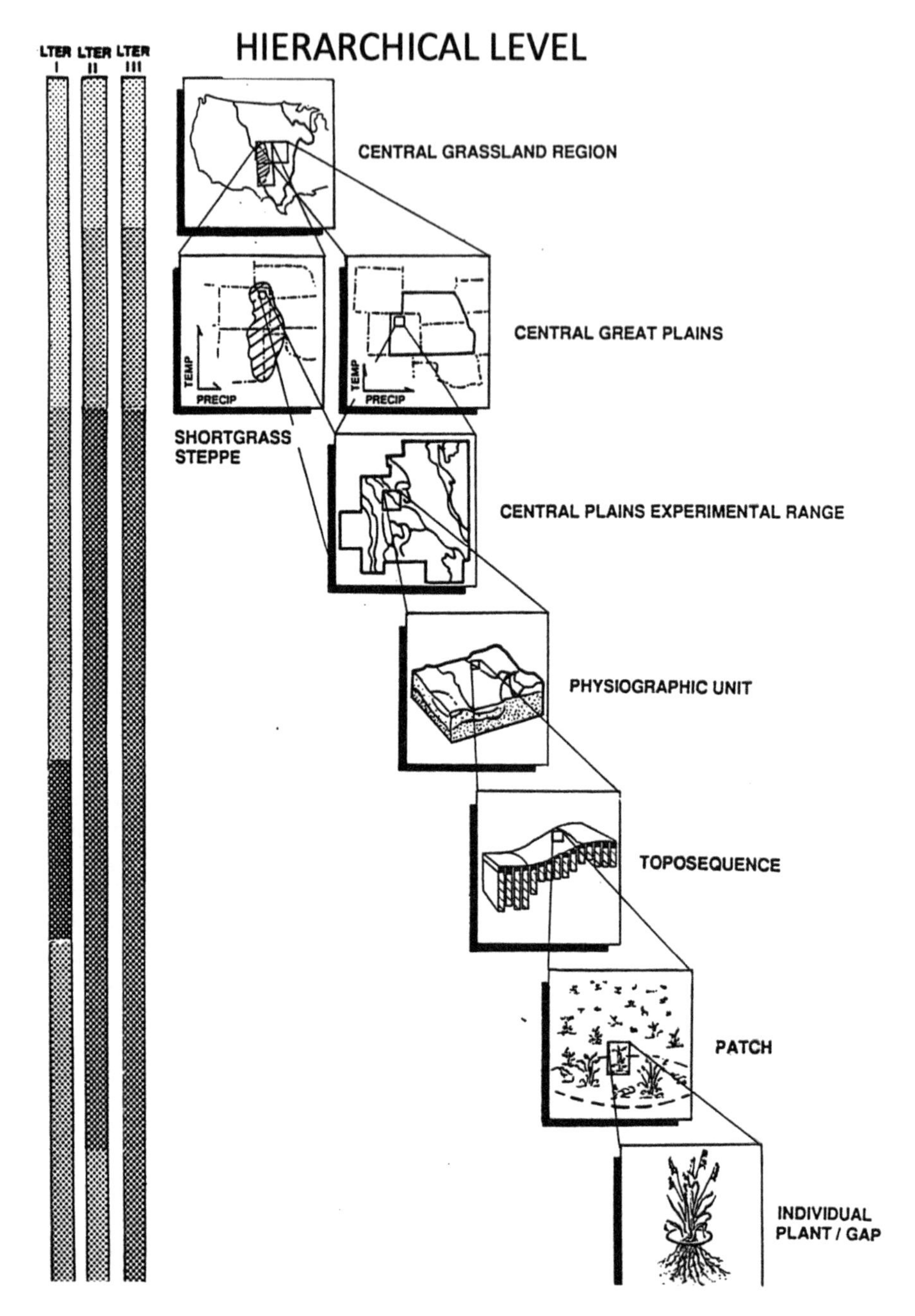

Fig. 6.1 Spatial scales of interest first shown in 1986 (LTER II proposal) for the SGS LTER. Shading at left shows relative importance of each spatial scale for studies in each LTER I-III proposal

as a relict species that could not become established under current climatic conditions (Hyder et al. 1975). LTER researchers using a simulation model (SOILWAT) parameterized with historic data (Briske and Wilson 1977, 1978) ascertained that blue grama seedlings could establish under current climate (Lauenroth et al. 1994), indicating that blue grama is not a relict species. In addition, LTER experimental manipulations in the field and numerical simulations showed that blue grama can recover on disturbances less than 2 m^2, and the rate of recovery depends on soil texture (Coffin and Lauenroth 1988, 1990). These findings suggested a partial rehabilitation to the Clementsian paradigm.

SGS Multi-scale Model (1990–1995) In LTER III, the overall objective for the SGS was to improve understanding of the long-term processes responsible for the origin and sustainability of shortgrass steppe ecosystems. The conceptual framework focused on climate, geomorphology, and land use management as major controls over structure and function of the shortgrass steppe. Each factor controls important spatial and temporal heterogeneity of ecosystem dynamics across a range of scales, and exerts its influence at a particular combination of levels (Fig. 6.1 from LTER II). The research on blue grama recovery used the emerging field of landscape ecology to show that this species can recover through seed dispersal by wind as disturbance size increases to 49 m^2 (Coffin and Lauenroth 1994). Furthermore, objective sampling of a large suite of abandoned fields from the 1930s across the Pawnee National Grasslands showed that blue grama can recover given sufficient time even on fields that are 80 acres or more in size (Coffin et al. 1996). The erroneous belief as to the inability of this species to recover on old fields that prevailed in the 1970s likely had two sources: (1) insufficient time had elapsed for blue grama seeds to disperse across the fields and become established as seedlings and then adults before sampling in the 1970s, and (2) observer bias had restricted observations to only those fields that had visually not recovered, and had overlooked other fields that had recovered (Coffin et al. 1996). The analysis undertaken by Coffin et al. (1996) used 1930s imagery and soils maps to objectively identify and select a large number of fields that had been recently abandoned. This manual overlay of soil maps with imagery in the early 1990s predated computer-aided GIS analyses, yet allowed for the accurate location of fields on the ground more than 50 years later.

SGS Paradigm Shifts Thus, a combination of objective, long-term data combined with simulation modeling led to the conclusion that *blue grama is not a relict species under current climate but, given sufficient time, recovers not only after small to medium-sized naturally-occurring disturbances but also after large, human-made disturbances.* The rehabilitation of the Clementsian paradigm was complete.

Livestock Grazing at the SGS and JRN At the same time that studies of blue grama recovery were ongoing, other SGS researchers were studying the role of livestock grazing to explain differences in vegetation responses between the shortgrass steppe and other grasslands in North America, including the desert grassland at the Jornada. The shortgrass steppe is unique from other semiarid grasslands in North America in that bison played an important role beginning 10,000 years ago until their demise in the late 1800s (Milchunas et al. 2008). Combined with the aridity of the region, plants developed convergent, complementary adaptations to grazing and drought that include a short stature and high proportion of biomass belowground (Milchunas et al. 1988). An extensive literature review showed that evolutionary history of the ecosystem and life history traits of the dominant plant species are important determinants of the effects of livestock grazing (Milchunas and Lauenroth 1993; Milchunas 2006). At the SGS, the resiliency of blue grama (*Bouteloua gracilis*) to livestock grazing results from grazing-adapted plant traits, primarily high proportion of belowground biomass, short stature, and location of growing points within the crown and away from grazers (Milchunas et al. 1988, 2008).

One of the unanswered questions going into this time period at these two LTER sites was: how could a semiarid ecosystem (SGS; 32 cm precipitation/year) be so resistant to livestock grazing and another semiarid to arid ecosystem located 1180 km to the south with only on-average 8 cm fewer precipitation amount per year (JRN; 24 cm/y) be so sensitive to grazing? Large parts of the Southwestern US dominated by the conspecific, black grama (*Bouteloua eriopoda*), had converted to shrublands since 1850. Livestock grazing combined with drought were widely believed to be the drivers, and the consequences of this conversion to reduced air, water, and forage quality were severe (Humphrey 1958). Comparisons between the LTER sites as part of an extensive literature review appeared to provide at least part of the answer: grazing is not tolerated in desert grasslands dominated by black grama because ecosystems dominated by this species have an evolutionary history of light grazing by large herbivores (Milchunas and Lauenroth 1993, Havstad et al. 2006). Black grama is a stoloniferous grass with a large proportion of biomass and growing points aboveground that, when removed by grazers, has a detrimental effect on regrowth (Nelson 1934; Paulsen and Ares 1962). Researchers also recognize that evolutionary history is just one of many factors that need to be considered when attempting to explain complex, multi-scale patterns across Chihuahuan Desert landscapes (Havstad et al. 2006; Milchunas et al. 2008).

SGS Model of Factors Interacting in Space and Time (1996–2012) Starting in 1996, the SGS group developed a framework that persisted until the end of their funding period in 2012. Climate, natural disturbance, physiography, human use, and biotic interactions were hypothesized to be the major determinants of the ecological structure and function of the shortgrass steppe. Specific questions were developed in each funding period within this overarching framework as the project developed and the interests of the group changed. For example, thresholds were studied in the late 1990s, ecosystem resilience was a major part of the proposal in the late 2000s, and climate change and ecosystem services were part of the goals in the last proposal submitted in 2010. The SGS was put on probation for their 2008 proposal that led to 2 years of funding, and was funded for a final 2 years as a result of their 2010 proposal. Major criticisms of the two proposals center around the lack of a vision for the big science questions that would generate new and exciting insights in shortgrass steppe ecosystems by integrating the long-term data and knowledge at this site.

SGS LTER Contributions to Ecology The productivity of the SGS program and contributions to grassland ecology were numerous throughout the time period of the LTER program based on the continued stream of high-impact publications and large number of graduate students trained in ecology. In addition to resolving the 'blue grama recovery problem" described above, and to being critical to the start of the LTER Schoolyard Program by John Moore, a few of the other major contributions are summarized here. More complete descriptions can be found in Lauenroth and Burke (2008).

Livestock grazing studies conducted by SGS researchers were influential in providing both a conceptual and mechanistic understanding of livestock grazing for semiarid and arid grasslands globally (Milchunas et al. 1988, 2008; Milchunas and Lauenroth 1993, 2008; Milchunas 2006).

Regional analyses conducted by SGS researchers were among the first in the LTER network. These analyses focused initially on identifying factors controlling patterns in soil organic matter at the regional scale (Burke et al. 1989, 1990), expanded to vegetation measures (e.g., Epstein et al. 1997, Paruelo et al. 1997), and shifted to encourage LTER researchers to look beyond site boundaries for important influences on within-site dynamics (Burke et al. 1991). Analyses were conducted to evaluate the representativeness of the physical characteristics of an individual LTER site to the rest of the biome (Burke and Lauenroth 1993) that were also repeated by other sites. This seminal work was influential in the NSF Call for Augmentation for Regionalization of LTER sites in 1994 (NSF 94-60).

Development of grassland simulation models has been impacted by the IBP and researchers at the NREL and continuing throughout the LTER program. Numerical models were an integral part of every SGS LTER proposal, and these models have influenced model development at the Jornada and other places globally. Three models, in particular, were developed or modified with LTER funding. The STEPPE individual plant-based model was developed to address the blue grama recovery problem. This model was the first individual plant-based, gap model developed for grasslands, and was based on the belowground resource space occupied by individual plants (Coffin and Lauenroth 1990, 1994). STEPPE later formed the basis for the ECOTONE model of mixed lifeforms at the Jornada (Peters 2002a). The SOILWAT model simulates daily soil water dynamics by depth in the soil profile. This model was developed as part of the ELM model (Parton 1978), and was modified to simulate germination and recruitment events of blue grama (Lauenroth et al. 1994). This model was also coupled to ECOTONE, and modified for arid ecosystems (Peters 2000). Finally, the CENTURY model was developed using LTER data (Parton et al. 1988), and this model has been used extensively by LTER researchers, both as the original CENTURY (e.g., Kelly et al. 1996). and the daily version DAYCENT (e.g., Del Grosso et al. 2005). CENTURY was among the first ecosystem models to extrapolate climate change impacts on carbon cycling at regional to global scales (Schimel et al. 1991; Parton et al. 1994). More recently, detailed process-based nutrient cycling, trace gas flux, and soil organic matter data collected since the 1990s are being used to simulate greenhouse gas emissions (primarily nitrous oxide) and ecosystem dynamics at daily time steps from agricultural soils using DAYCENT. Data from the SGS LTER were instrumental in parameterizing and developing algorithms for the model (Parton et al. 2008). CENTURY, DAYCENT, and their variants have been used globally in grasslands, forests, and savannas.

Patterns in net primary production studies have been an interest of LTER researchers since the IBP. It is generally recognized that more precipitation (PPT) leads to more plant production (Aboveground Net Primary Production or ANPP), in particular at a water-limited site like the SGS. However, there is large variability in the relationship between ANPP and PPT at a site, and researchers at the SGS have been leaders in explaining that variability, both for patterns within the SGS and in comparing the SGS with other sites across environmental gradients (Lauenroth et al. 2008). This research started during the IBP with studies comparing ANPP across grassland sites (Sims and Singh 1978; Lauenroth 1979), and then in recognizing and evaluating errors in estimates of ANPP that could explain these patterns (Singh et al. 1984). During LTER, the focus shifted to explaining long-term observations of ANPP by cross-walking LTER methods with ARS methods to obtain a longer-term record than possible with either method alone (Lauenroth and Sala 1992). A classic experiment started in IBP provided one of the first ball-and-cup examples of thresholds and basins of attraction. Large additions of water and nitrogen had large effects on production and small changes in species composition (Lauenroth et al. 1978), yet long-term observations showed the dynamic nature of this ecosystem with time lags and biotic regulation affecting responses (Milchunas and Lauenroth 1995). In addition to studying landscape-scale controls on patterns in ANPP, SGS researchers conducted a number of regional-scale analyses (e.g., Epstein et al. 1997, Sala et al. 1988). A major finding was that the relationship between ANPP and PPT developed at a site by sampling each year (temporal case) is different than the relationship obtained between mean ANPP and mean PPT for many sites located across a region (spatial case) (Sala et al. 1988). This ANPP space and time relationship that was originally developed for grasslands in the CGP has been repeated for multiple ecosystem types (Huxman et al. 2004).

Soil organic matter and nutrient dynamics have been major themes since the beginning of the LTER, similar to other LTER sites given that these are core areas in the LTER program (Burke et al. 1995). At the SGS, two main threads have driven this research that has focused on belowground processes (Burke et al. 2008). *First,* the fine-scale distribution of carbon and nutrients are determined by the spatial distribution and temporal dynamics of individual plants, primarily the dominant species, blue grama, and the surrounding bare soil interspaces. Stratifying by microsite (plants and bare soil interspaces) at the SGS began with the 1990 proposal, and was the same sampling scheme underway at the Jornada. Numerous studies were conducted evaluating the biotic controls on ecosystem processes using this sampling design (e.g., Hook et al. 1991; Vinton and Burke 1995). *Second*, landscape-scale patterns are controlled by climate, physiography (topography), disturbance, human activities, and biotic interactions. Topography and microsite account for most of the variance in soil organic pools (Burke et al. 1999). Topography was more important for the recalcitrant pools, and microsite was more important for the active pools.

6.5 Development of Ideas at the Jornada LTER Program

The Early Days (1989–2000: Causes and Consequences of Desertification, the Schlesinger Model) In 1991, administration of the program moved to Duke University under the leadership of William Schlesinger. With the change in leadership came changes in group membership. Many of the new members of the team had worked with Schlesinger in the Mojave Desert, and moved their research with him to the Jornada. These changes in leadership and group membership allowed an opportunity for a new conceptual model to be developed to better explain spatial variability as a factor in the causes and consequences of desertification, the major issue associated with drylands globally. The group also recognized that the LTER I transects did not represent a coherent watershed system or a catena. In addition, the retirement of Carlton Herbel from the USDA ARS in the mid-1980s led to Kris Havstad being hired as the new director for the Jornada Experimental Range. The large ARS land base provided the spatial variability needed for the Jornada LTER to test their new ideas, and Havstad was open to collaboration; thus the start of the LTER-USDA team was formed that would be critical to the long-term success of the LTER program. Havstad coupled his training in animal science with the new paradigms of ecosystem ecology being developed by the Jornada LTER to change the way the USDA site was being managed, and was critical in merging USDA research with the Jornada LTER program after its administrative return to NMSU in 1998.

The goal of the second Jornada LTER proposal explicitly addressed the land management problem of the American Southwest, "the causes and consequences of desertification," which has remained a key component of the evolving LTER research to the present. A new conceptual model with two components was developed in 1989 to replace the watershed-based model. The first component focused on how fine-scale *islands of fertility* develop in arid grasslands, eventually causing grasslands to convert to desert shrublands and reinforcing the manifestation and persistence of the fertile islands. The second component, which became known as the *teeter-totter model*, linked this fine-scale change in plant cover to large-scale changes in other key attributes of ecosystem function, including runoff, wind erosion, albedo and nitrogen cycling (Fig. 6.2). This conceptual model of desertification was described in a paper in *Science* (Schlesinger et al. 1990), which became the flagship for subsequent LTER studies at the Jornada until 2003 and remained the foundation for future frameworks (Peters et al. 2006c). The paper was a precursor to modern concepts of ecological thresholds. While the islands of fertility in deserts had been recognized by earlier workers (Noy-Meir 1985), the more than 1500 citations of the *Science* paper indicate that the conceptual model for the development of patchy soil resources has been useful as an index of desertification worldwide.

The general hypothesis tested was that desertification altered a previous, relatively uniform distribution of water and nitrogen by increasing their spatial and temporal heterogeneity, leading to changes in community composition and biogeochemical processes in these ecosystems. A major emphasis was to compare long-term patterns of water and nitrogen availability among ecosystems with different

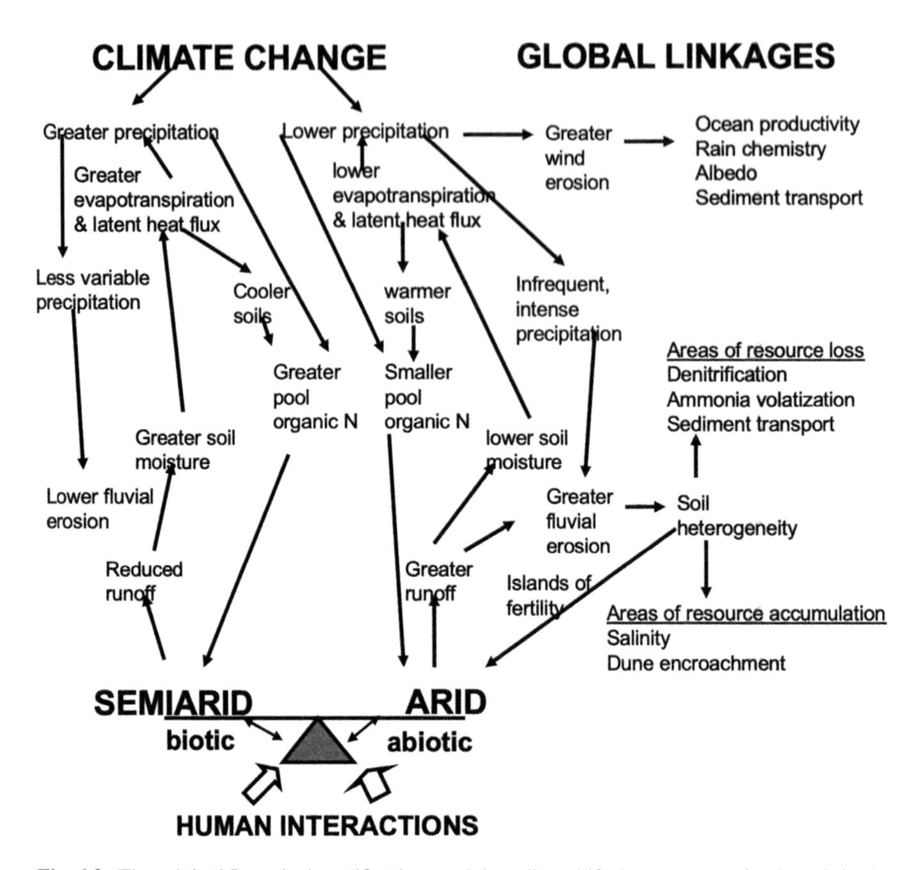

Fig. 6.2 The original Jornada desertification model predicts shifts between grasslands and shrublands resulting from human activities and climate. Islands of fertility between individual patches of plants and bare soil form the finest scale of resolution in the model. Redrawn from Schlesinger et al. (1990)

degrees of desertification that are connected via the movement of water and/or materials. Three levels of study were conducted: (1) comparative process studies in grasslands and shrublands, (2) studies of transport by water and wind between ecosystems, and (3) landscape-level studies involving remote sensing of regional gas fluxes and spectral vegetation indices from satellite imagery. Studies shifted to an area basis to allow an easy comparison of net primary production (NPP) over large areas dominated by different growth forms (Huenneke et al. 2002). Many long-term core studies established during this time period are still maintained by LTER researchers (e.g., primary production, atmospheric deposition, small mammal, and plant phenology studies) and continue to provide new insights (e.g., Peters et al. 2012, 2014a, b).

Nitrogen (N) became a focus of LTER studies, and was shown to be a major limitation to plant production after water was available (Hooper and Johnson 1999; Yahdjian et al. 2011). Measurements of N mineralization and N trace gas losses were distributed randomly in study plots, so that barren areas between shrubs as well as the islands of fertility were sampled in proportion to their coverage of the landscape (Peterjohn and Schlesinger 1991). Not surprisingly, the N cycle under shrubs was markedly greater than in shrub interspaces, and cycling in grasslands was more spatially homogeneous than in shrublands (Schlesinger et al. 1996).

In the renewal proposal submitted in 1995, the group tested their central hypothesis that the distribution of soil resources changes during desertification from spatially homogeneous in semiarid grasslands to heterogeneous in shrublands (Schlesinger et al. 1990). Plant nutrients (nitrogen and phosphorus) were hypothesized to be randomly distributed in grasslands and more concentrated under shrubs than in barren soils between shrubs in shrublands. Isotropic geostatistics showed that the scale of autocorrelation in shrublands is similar to average shrub size (Schlesinger et al. 1996). These data confirmed the conceptual model (Fig. 6.2) and showed that shrubs leave their signature on the biogeochemistry of arid-land soils (Schlesinger and Pilmanis 1998).

Estimates of surface erosion by wind and water showed that the greatest transport occurred from barren soils and the greatest redeposition of mobile materials occurred under shrubs (Schlesinger et al. 1999, 2000). Water and soil nutrients removed from barren soils between shrubs by wind and surface flow of water were deposited in downwind and downslope locations on the landscape (Schlesinger and Jones 1984; Wainwright et al. 2002), and in subsequent LTER studies this landscape connectivity would be shown to reinforce the process of desertification of upland grassland soils (Michaelides et al. 2012; Rachal et al. 2015; Okin et al. 2015).

The Jornada Landscape Linkages Model, (2000–2006: Causes and Consequences of Desertification) In 2000, the LTER group led initially by Laura Huenneke (NMSU) recognized that patterns in desertification across the Jornada landscape required additional components than in the model driven by drought and grazing (Fig. 6.2). A landscape-linkages approach was taken where the plant-interspace scale remained central, but multiple interacting scales were added (Peters and Havstad 2006; Peters et al. 2006a, b, c). The focus shifted to areas consisting of many plants and their associated interspaces within landscape units in order to explain complex patterns that cascade nonlinearly across a landscape; these dynamics mirror those in other systems (Peters et al. 2004). Upon Huenneke's departure in 2003 for the University of Northern Arizona, Debra Peters [previously Debra Coffin from the SGS] of the USDA became lead PI. With the addition of USDA ARS scientists to the team (Peters, Jeffrey Herrick, and later Brandon Bestelmeyer and Al Rango), the LTER team became a close collaboration between academic scientists at New Mexico State (notably H. Curtis Monger) and USDA researchers. Although the causes and consequences of desertification remained the overarching goal of Jornada research, the studies expanded in scope and complexity to include more spatial scales and more drivers of patterns and dynamics, including wind, water, and animals as vectors of redistribution of resources within and among plants and patches (Peters et al. 2006a, b, c). Studies were also expanding to the global extent to include human dimensions more explicitly (Reynolds et al. 2007).

Spatial context became an important feature of many Jornada studies, because distance becomes an important variable as spatial extent increases: e.g., distance to a seed source, either to historic shrublands or to current grass populations, which can either promote expansion or limit recovery (Havstad et al. 1999; Yao et al. 2006). In addition, the conceptual model of cross-scale interactions provides a basis for determining the conditions under which fine-scale processes propagate to have large-scale impacts, such as during shrub expansion, or, alternatively, when broad-scale drivers overwhelm fine-scale heterogeneity, such as during drought.

The Jornada Landscape Linkages Model (2006–2018: Grass Recovery and Alternative States) In 2006, while continuing desertification studies (e.g., Li et al. 2007, 2008), Jornada researchers shifted their focus to include studies on grass recovery following desertification. USDA researchers had been attempting grass remediation since at least the 1930s, but many previous approaches were ineffective because they failed to account for hydrologic, aeolian, and ecological processes that overwhelmed the desired treatment effects (Herrick et al. 2006). In addition, analysis of historic aerial photography starting in the 1930s showed that plot-scale manipulations are not effective at the landscape scale (Rango et al. 2002) and that shrub removals alone are insufficient to promote grass recovery (Rango et al. 2005). Threshold studies were increasingly an important part of Jornada research (Bestelmeyer et al. 2013).

LTER researchers modified the landscape linkages conceptual model to focus on patch structure interacting with transport vectors (wind, water, animals) and environmental drivers (e.g., precipitation, temperature, human activities) to influence cross-scale resource redistribution and ultimately spatial and temporal variation in ecosystem dynamics (Fig. 6.3; Peters et al. 2006a).

These interactions can affect patch structure and dynamics, causing cascading impacts on ecosystem goods and services (Fig. 6.4). Historic management legacies and geomorphic templates are important modifiers of this relationship. LTER researchers used these concepts to attempt grass restoration in a multi-scale manipulative experiment. They first developed *ConMods* (*Con*nectivity *Mod*ifiers) to reduce the flow of material by wind or water to allow grass and forb seedlings to become established. ConMods are small structures placed within bare gaps between shrubs to reduce gap size to less than 50 cm. A pilot study showed that ConMods are successful at trapping soil, litter, and seeds to lead to grass and forb establishment (Peters et al. 2020), a result that was recreated in other arid systems (e.g., Fick et al. 2016). Thus, a long-term, multi-scale experiment focusing on grass recovery was put in place that consisted of: (1) killing shrubs in place (plant scale), (2) adding ConMods to modify connectivity by wind (patch scale), and (3) conducting treatments across a gradient from black grama grassland to mesquite duneland (landscape unit scale).

In addition, long-term observations of NPP, started in 1989 and continued by the current LTER program, showed that perennial grasses (black grama in grasslands, mesa dropseed in mesquite shrublands) can increase in bare interspaces at the plot scale in a sequence of wet years (2004–2008) (Peters et al. 2012). Simulation of

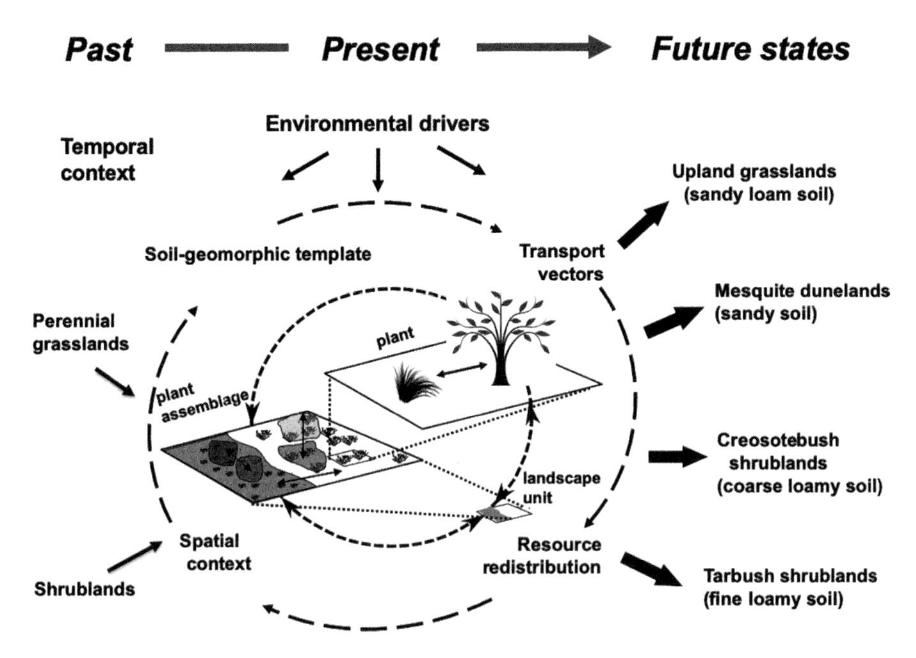

Fig. 6.3 Conceptual framework for Jornada LTER (2006–2018) showing (a) alternative states including shifts from shrublands to perennial grasslands, other shrublands, novel ecosystems, and urban systems, and (b) focus on patch structure and dynamics with consequences for ecosystem services (Peters et al. 2015). Shifts to other shrublands, novel ecosystems, and urban systems were studied by the LTER beginning in 2012. Redrawn from Peters et al. (2006a)

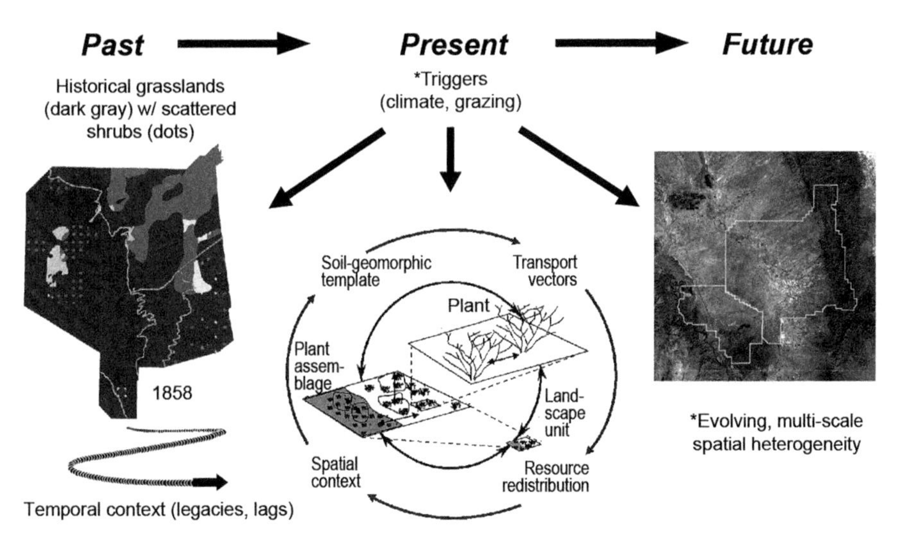

Fig. 6.4 Current conceptual framework of the Jornada LTER (2018–present) that includes triggers, feedbacks, and cross-scale interactions are needed to explain heterogeneity at multiple spatial and temporal scales (Peters et al. 2018). Note the Schlesinger plant-interspace model of resource redistribution (the finest spatial extent shown in the ball in the center) and thresholds (Schlesinger teeter-totter model) are key components of the temporal dynamics. Compare with Fig. 6.2 for development of Jornada LTER research through time

black grama establishment using a numerical model developed at the SGS (SOILWAT: Lauenroth et al. 1994), parameterized for black grama (Minnick and Coffin 1999) and modified for the Jornada, showed that this species can establish in current climatic conditions on certain soil types (Peters et al. 2010). Although black grama has not recovered in shrublands, its increase in neighboring grasslands, its continued presence in degraded shrublands, and the recovery of other perennial grasses in shrublands both naturally and through our manipulations, suggest that *black grama should recover in shrublands given sufficient time* and, in particular, during wet periods (Peters et al. 2006c). It is the sequence of wet years when precipitation is above average that should allow black grama seed production and germination, seedling establishment, and growth created by the gradual accumulation of herbaceous biomass as facilitated by ConMods.

In the 2012 LTER proposal, the focus shifted to understanding and quantifying the mechanisms that generate alternative natural and human-dominated states in dryland ecosystems, and to predicting future states and their consequences for the provisioning of ecosystem services. Here, shifts from grasslands to a desertified state, and a reversal where desertified shrublands shift back to grasslands, were two of the alternative states being studied from the previous funding cycle. The new shifts added were that from dominance of one shrub species to another. The conceptual framework of landscape linkages was modified to account for legacies (Sala et al. 2012; Reichmann et al. 2013), exotic invasive species (e.g., *Eragrostis lehmanniana*), and changes in climate (Bestelmeyer et al. 2011). Plot-scale experimental manipulations of rainfall addition and removal and nitrogen addition in order to simulate climate change, initiated in 2007 (Gherardi and Sala 2013), have been maintained and expanded through time (Gherardi and Sala 2015). Long-term monitoring of small animals provides insights associated with pulses in resources (Schooley et al. 2018).

Results show that grassland to shrubland transitions are a function of: (1) both legacies (Reichmann et al. 2013; Monger et al. 2015) and directional changes in climate (Gherardi and Sala 2015), (2) spatial context and contagion among spatial units (Peters et al. 2010; Schreiner-McGraw and Vivoni 2017; McKenna and Sala 2018; Okin et al. 2018), and (3) the soil-geomorphic template (Weems and Monger 2012). A major outcome was the development of concepts and metrics associated with bare soil gaps (another patch type) that are now used in drylands globally (e.g., Herrick et al. 2017). These gaps control redistribution of soil, water, nutrients, and plant materials that are key to plant-soil feedbacks at multiple scales, and govern erosion-deposition processes. Bare soil gap size distributions are now routinely measured in Jornada experiments and multi-scale pattern analyses. LTER research also influenced the design of the National Ecological Observatory Network being funded by NSF (Peters et al. 2008 and other papers in the special issue; Peters et al. 2014a). A NEON relocatable site was added to the Jornada in 2010.

In the 2018 LTER proposal, the goal shifted to that of developing robust principles governing dryland state changes and applying them across spatially heterogeneous landscapes in order to forecast future states in response to changing climate and land use. In this proposal, the trigger-threshold-feedback process is being quantified during state changes, such as when grasses re-establish in degraded shrublands. Transition triggers, including disturbance (e.g., grazing, drought) or resource pulses (rainfall), push systems across critical thresholds and have effects at the finest spatial scales, for instance the level of individual plants or plant/crust patches (Fig. 6.1). These changes initiate connectivity-mediated feedbacks at expanding spatial scales that involve resource redistribution from bare soil to adjacent plant patches or from one landform to another through hillslope and channel surfaces. Feedbacks create and reinforce patchiness and can lead to pronounced shifts in vegetation composition, soil function, and ecosystem processes resulting in complex patterns in heterogeneity at local to landscape scales with potential consequences for land surface-atmosphere feedbacks. Climate variability and land use affect the rate, type, and magnitude of change in each spatial scale. Changes in patch structure (vegetated and bare) locally modify transport vectors and drivers to influence resource redistribution and connectivity across scales, and attenuate or amplify processes that propagate and culminate in broad-scale state changes. Collectively, these effects control the evolution of spatial heterogeneity in dryland landscapes (Fig. 6.1).

Although the science in this proposal was viewed extremely favorably and scored highly by NSF and the panel, data management was again seen as a sufficient weakness to place this site on probation. Another proposal was submitted in 2020 with similar science goals and a completely revamped information management team.

Jornada Paradigm Shifts The paradigms changed several times for these dynamic landscapes, from a Clementsian model in equilibrium with climate and soils (1900s) to a focus on alternative, desertified states (1980s–2000s) to alternative states where the recovery of black grama is also a possibility (2006–present). Given the longevity of the woody plant species in Chihuahuan Desert landscapes, it is likely that black grama will be a subdominant in shrublands or a component of a savanna in the future as opposed to the return to black grama dominated grasslands envisioned by Clements.

6.6 Relationship Among the SGS, Sevilleta, and Jornada Research

Studies conducted at the Sevilleta LTER by Jornada researchers (primarily Peters on vegetation change; D. Lightfoot on small mammals not discussed here) have illuminated dynamics at all three sites. An LTER site since 1988, the 93,000 ha Sevilleta, located 270 km north of the Jornada, was grazed as a private ranch until the 1970s and has been ungrazed since that time. The SGS, Sevilleta and Jornada occur along a latitudinal gradient that results in differences in annual precipitation and temperature with consequences for the vegetation. Cross-site studies have taken advantage of this multi-dimensional environmental gradient to provide new insight to perennial grass response to climate (Hochstrasser et al. 2002). The Sevilleta represents a currently ungrazed grassland by livestock with dynamic ecotones between grasslands and shrublands that allow comparison with the Jornada, where these ecotones occurred before the early 1900s (Gibbens et al. 2005). Because blue grama dominates at the SGS, and black grama dominates at the Jornada, the multi-scale ecotones between these two species at the Sevilleta are particularly important to understanding the climatic and edaphic controls on their geographic distribution (Gosz 1993). Removal of dominant grasses along ecotones combined with demographic studies of seed production and seedling establishment of blue and black grama have been conducted at the Sevilleta since 1995 (Peters 2000, 2002b; Peters and Yao 2012).

Important insights and information have been shared at the three arid and semiarid LTER sites: an individual plant-based simulation model developed for arid grasslands and shrublands (ECOTONE: Peters 2002a) was based on the grasslands model developed at the SGS (STEPPE: Coffin and Lauenroth 1990), and a multi-scale conceptual model for ecotones at the Sevilleta was based on ideas from the Jornada (Peters et al. 2006b). These studies showed that black grama readily produces viable seed and becomes established at the Sevilleta, which receives 23 cm/y average precipitation, similar to the Jornada; differences in soil properties interacting with higher temperatures (mean 33 °C July Sevilleta; mean 36 °C July Jornada) may explain lower establishment probabilities of this species at the Jornada. Based on these multiple lines of evidence from several LTER sites, *black grama is not currently viewed as a relict species throughout the American Southwest*, in contrast to previous paradigms for Chihuahuan Desert ecosystems (Neilson 1986).

The loss of the SGS from the LTER network in 2014 owing to lack of funding from NSF means that future comparisons will be limited. Although the SGS remains a USDA site with a different mission that includes long-term research as well as management, researchers will probably be unable to conduct long-term experiments comparing these three grasslands with other ecosystem types in the LTER network.

6.7 Conclusions and Lessons Learned

Tremendous knowledge and insights about North American arid and semiarid grasslands and deserts have been gained as a result of the long-term research conducted at the Shortgrass Steppe in north-central Colorado since 1939 and the Jornada in southern New Mexico since 1915. The foresight of the USDA in creating these sites during a period of severe overuse, either from livestock overgrazing (Jornada) or from cultivation and abandonment (primarily SGS), and then maintaining them in collaboration with university researchers, first during the IBP in the 1970s and then with the LTER beginning in 1981 (SGS) and 1982 (Jornada), was critical to the early conceptual insights and paradigm shifts in the 1990s.

Early USDA records (observations, maps, imagery) in the 1930s provided the backdrop for LTER research in the 1990s and 2000s to objectively determine that neither blue grama nor black grama are relict species under current climatic conditions. Results from remediation studies of these grasses in the shortgrass steppe and Chihuahuan Desert are dependent on multiple dimensions of space and time. Understanding the effects of grazing by small and large herbivores has also benefited from long-term studies at both sites. At the Jornada, a multiple, interacting scales approach is needed to understand complex patterns across the landscape that maintains the plant-interspace model of early LTER studies (Schlesinger et al. 1990). This model was expanded to include patches as critical components linking plants to the broader landscape unit (Peters et al. 2006a). Desertification continues to be studied as a state change from grassland to shrubland, but the conceptual framework has been expanded to include alternative states (e.g., recovery of grasses in shrublands, shifts between shrubland types, novel ecosystems). Bare soil gaps are critical in wind and water erosion-deposition to drive state change dynamics.

Both LTER programs have had important and long-lasting impacts on the field of ecology. Each has at least one highly-cited scientific product or publication. The CENTURY biogeochemistry simulation model developed by Parton et al. (1987) and the daily time-step version, DAYCENT for trace gas fluxes (Parton et al. 1998), were developed using data from the SGS, and have been modified and used globally in grasslands, agricultural lands, savannas, and forests. The teeter-totter desertification conceptual model of Schlesinger et al. (1990) developed at the Jornada has been applied in semiarid and arid ecosystems globally. These models have been cited or used in the literature at least 2000 times each.

In addition, these two LTER programs share an exceptionally strong impact on science education. In 1996, John Moore at the SGS received NSF funding to explore the feasibility of establishing small-scale research plots on school grounds. This pioneering project led to the development of the Schoolyard Ecology program (SLTER) across the LTER Network. The Jornada was an early adopter of the SLTER program, and more than 90,000 students, teachers, and other adults were involved in educational outreach programs over a 6y period. The majority of participants were from underserved populations from southern New Mexico and west Texas: ca. 80% were classified as economically disadvantaged, and 75% were Hispanic.

Paradigm shifts through time from a Clementsian model in the 1900s at both sites to alternative states in the 1970s, and then to a more nuanced Clementsian model in Colorado and a more complex, multi-scale model in the Southwest may need to be re-examined again as environmental drivers and landscape properties continue to change. Because the land at both sites is owned by the USDA, ecological research will continue even without LTER funding by NSF. However, the diversity of research projects, the breadth and number of investigators, the collection and accessibility of the long-term data, and, hence, most importantly, inferences across broad spatial extents suffer when a historic long-term research program is lost from ecological science and a national network.

Acknowledgments Funding was provided by the National Science Foundation to New Mexico State University for the Jornada Basin Long Term Ecological Research Program (DEB 12-35828, 18-32194) and by the USDA-ARS CRIS Project at the Jornada Experimental Range (#6235-11210-007). William H. Schlesinger, Justin Derner, Robert Waide, and Sharon Kingsland provided helpful comments on an earlier draft of the chapter.

References

Allen, T.F.H., and T.B. Starr. 1982. *Hierarchical structure: Perspectives for ecological complexity*. Chicago: University of Chicago Press.

Archer, S.A. 1989. Have southern Texas savannas been converted to woodlands in recent history? *American Naturalist* 134: 545–561.

Bement, R.E., R.D. Barmington, A.C. Everson, L.O. Hylton Jr., and E.E. Remmenga. 1965. Seeding of abandoned croplands in the Central Great Plains. *Journal of Rangeland Management* 18: 53–58.

Bestelmeyer, B.T., A.M. Ellison, W.R. Fraser, K.B. Gorman, S.J. Holbrook, C.M. Laney, M.D. Ohman, D.P.C. Peters, F.C. Pillsbury, A. Rassweiler, R.J. Schmitt, and S. Sharma. 2011. Analysis of abrupt transitions in ecological systems. *Ecosphere* 2 (12): 1–26.

Bestelmeyer, B.T., M.C. Duniway, D.K. James, L.M. Burkett, and K.M. Havstad. 2013. A test of critical thresholds and their indicators in a desertification-prone ecosystem: More resilience than we thought. *Ecology Letters* 16: 339–345.

Bormann, F.H., and G.E. Likens. 1967. Nutrient cycling. *Science* 155 (3761): 424–429.

Briske, D.D., and A.M. Wilson. 1977. Temperature effects on adventitious root development in blue grama seedlings. *Journal of Rangeland Management* 30: 276–280.

———. 1978. Moisture and temperature requirements for adventitious root development in blue grama seedlings. *Journal of Rangeland Management* 31: 174–178.

Buffington, L.C., and C.H. Herbel. 1965. Vegetational changes on a semidesert grassland range from 1958 to 1963. *Ecological Monographs* 35: 139–164.

Burke, I.C., and W.K. Lauenroth. 1993. What do LTER results mean? Extrapolating from site to region and decade to century. *Ecological Modelling* 67: 19–35.

Burke, I.C., C.M. Yonker, W.J. Parton, C.V. Cole, K. Flach, and D.S. Schimel. 1989. Texture, climate, and cultivation effects on soil organic matter content in U.S. grassland soils. *Soil Science Society of America Journal* 53: 800–805.

Burke, I.C., D.S. Schimel, W.J. Parton, C.M. Yonker, L.A. Joyce, and W.K. Lauenroth. 1990. Regional modeling of grassland biogeochemistry using GIS. *Landscape Ecology* 4: 45–54.

Burke, I.C., T.G.F. Kittel, W.K. Lauenroth, P. Snook, and C.M. Yonker. 1991. Regional analysis of the Central Great Plains: Sensitivity to climate variability. *Bioscience* 41 (10): 685–692.

Burke, I.C., W.K. Lauenroth, and D.P. Coffin. 1995. Soil organic matter recovery in semiarid grasslands: Implications for the Conservation Reserve Program. *Ecological Applications* 5: 793–801.

Burke, I.C., W.K. Lauenroth, R. Riggle, P. Brannon, B. Madigan, and S. Beard. 1999. Spatial variability of soil properties in the shortgrass steppe: The relative importance of topography, grazing, microsite, and plant species in controlling spatial patterns. *Ecosystems* 2: 422–438.

Burke, I.C., A.R. Mosier, P.B. Hook, D.G. Milchunas, J.E. Barrett, M.A. Vinton, R.L. McCulley, J.P. Kaye, R.A. Gill, H.E. Epstein, R.H. Kelly, W.J. Parton, C.M. Yonker, P. Lowe, and W.K. Lauenroth. 2008. In *Ecology of the shortgrass steppe*, ed. W.K. Lauenroth and I.C. Burke, 306–341. New York: Oxford University Press.

Campbell, R.S., and E.C. Crafts. 1938. *Tentative range utilization standards: Black grama (Bouteloua eriopoda)*. USDA Forest Service, Experimental Statistics Research Note No. 26.

Clements, F.E. 1916. *Plant succession: An analysis of the development of vegetation*. Washington, DC: Carnegie Institute of Washington.

Coffin, D.P., and W.K. Lauenroth. 1988. The effects of disturbance size and frequency on a shortgrass plant community. *Ecology* 69: 1609–1617.

———. 1990. A gap dynamics simulation model of succession in the shortgrass steppe. *Ecological Modelling* 49: 229–266.

———. 1994. Successional dynamics of a semiarid grassland: Effects of soil texture and disturbance size. *Vegetation* 110: 67–82.

Coffin, D.P., W.K. Lauenroth, and I.C. Burke. 1996. Recovery of vegetation in a semiarid grassland 53 years after disturbance. *Ecological Applications* 6: 538–555.

Costello, D.F. 1944. Natural revegetation of abandoned plowed land in the mixed prairie association of northeastern Colorado. *Ecology* 25 (3): 312–326.

Del Grosso, S.J., A.R. Mosier, W.J. Parton, and D.J. Ojima. 2005. DAYCENT model analysis of past and contemporary soil N2O and net greenhouse gas flux for major crops in the USA. *Soil Tillage and Research* 83: 9–24.

Dyksterhuis, E.J. 1949. Condition and trend based on quantitative ecology. *Journal of Rangeland Management* 2: 104–115.

———. 1958. Ecological principles in range evaluation. *Biological Review* 24: 253–272.

Epstein, H.E., W.K. Lauenroth, I.C. Burke, and D.P. Coffin. 1997. Productivity patterns of C3 and C4 functional types in the U.S. Great Plains. *Ecology* 78: 722–731.

Fick, S.E., C. Decker, M.C. Duniway, and M.E. Miller. 2016. Small-scale barriers mitigate desertification processes and enhance plant recruitment in a degraded semi-arid grassland. *Ecosphere* 7 (6): e01354. https://doi.org/10.1002/ecs2.1354.

Fisher, F.M., J.C. Zak, G.L. Cunningham, and W.G. Whitford. 1988. Water and nitrogen effects on growth and allocation patterns of creosote bush in the northern Chihuahuan Desert. *Journal of Rangeland Management* 41: 387–391.

Fredrickson, E., K.M. Havstad, R. Estell, and P. Hyder. 1998. Perspectives on desertification: South-Western United States. *Journal of Arid Environments* 39: 191–207.

Gherardi, L.A., and O.E. Sala. 2013. Automated rainfall manipulation system: A reliable and inexpensive tool for ecologists. *Ecosphere* 4 (2): art18.

———. 2015. Enhanced precipitation variability decreases grass- and increases shrub-productivity. *Proceedings National Academy Sciences* 112: 12735–12740.

Gibbens, R.P., R.P. McNeely, K.M. Havstad, R.F. Beck, and B. Nolen. 2005. Vegetation changes in the Jornada Basin from 1858 to 1998. *Journal of Arid Environments* 61: 651–668.

Gosz, J.R. 1993. Ecotone hierarchies. *Ecological Applications* 3: 369–376.

Havstad, K. 1996. Legacy of Charles Travis Turney: The Jornada Experimental Range. *Archeological Society of New Mexico Annual Volumes* 22: 77–92.

Havstad K, and W.H. Schlesinger. 1996. Reflections on a century of rangeland research in the Jornada Basin of New Mexico. Wildland Shrub Symposium, Proceedings: *Shrubland Ecosystem Dynamics in a Changing Environment.* Gen. Tech. Rep. INT-GTR-338:10–15.

Havstad, K.M., R.P. Gibbens, C.A. Knorr, and L.W. Murray. 1999. Long-term influences of shrub removal and lagomorph exclusion on Chihuahuan Desert vegetation dynamics. *Journal of Vegetation Dynamics* 42: 155–166.

Havstad, K.M., L.F. Huenneke, and W.H. Schlesinger, eds. 2006. *Structure and function of a Chihuahuan Desert ecosystem: The Jornada Basin Long-Term Ecological Research site.* New York: Oxford University Press.

Herbel, C.H. 1983. Principles of intensive range improvements. *Journal of Range Management* 36: 140–144.

Herrick J.E., K.M. Havstad, and A. Rango. 2006. Remediation research in the Jornada Basin: Past and future. In Structure and function of a Chihuahuan Desert ecosystem: The Jornada Basin Long-Term Ecological Research site, ed. Havstad et al., 278–304. New York: Oxford University Press.

Herrick J.E., J.W. Van Zee, S.E. McCord, E.M. Courtright, J.W. Karl, and L.M. Burkett. 2017. *Monitoring Manual for Grassland, Shrubland and Savanna Ecosystems.* Second edition. Volume I: Core methods. Las Cruces, NM: USDA-ARS Jornada Experimental Range.

Hochstrasser, T., Gy. Kröel-Dulay, D.P.C. Peters, and J.R. Gosz. 2002. Vegetation and climate characteristics of arid and semi-arid grasslands in North America and their biome transition zone. *Journal of Arid Environments* 51: 55–78.

Hook, P.B., I.C. Burke, and W.K. Lauenroth. 1991. Heterogeneity of soil and plant N and C associated with individual plants and openings in North American shortgrass steppe. *Plant and Soil* 138: 247–256.

Hooper, D.U., and L. Johnson. 1999. Nitrogen limitation in dryland ecosystems: Responses to geographical and temporal variation in precipitation. *Biogeochemistry* 46: 247–293.

Huenneke, L.F., J.P. Anderson, M. Remmenga, and W.H. Schlesinger. 2002. Desertification alters patterns of aboveground net primary production in Chihuahuan ecosystems. *Global Change Biology* 8: 247–264.

Humphrey, R.R. 1958. The desert grassland: A history of vegetational change and an analysis of causes. *The Botanical Review* 24 (4): 193–252.

Huxman, T.E., M.D. Smith, P.A. Fay, A.K. Knapp, M.R. Shaw, M.E. Loik, S.D. Smith, D.T. Tissue, J.C. Zak, J.F. Weltzin, W.T. Pockman, O.E. Sala, B.M. Haddad, J. Harte, G.W. Koch, S. Schwinning, E.E. Small, and D.G. Williams. 2004. Convergence across biomes to a common rain-use efficiency. *Nature* 429: 651–654.

Hyder, D.N., A.C. Everson, and R.E. Bement. 1971. Seedling morphology and seeding failures with blue grama. *Journal of Rangeland Management* 24: 287–292.

Hyder, D.N., R.E. Bement, E.E. Remmenga, and D.F. Hervey. 1975. *Ecological responses of native plants and guidelines for management of shortgrass range*. U.S. Department of Agriculture Technical Bulletin 1503.

Idso, S.B. 1992. Shrubland expansion in the American southwest. *Climatic Change* 22: 85–86.

Innis, G.S., ed. 1978. *Grassland simulation model*, Ecological studies series. Vol. 26. New York: Springer.

Judd, L.B., and M.L. Jackson. 1939. Natural succession of vegetation on abandoned farmland in the Rosebud soil area of western Nebraska. *Journal of the American Society of Agronomy* 39: 541–547.

Kelly, R.H., I.C. Burke, and W.K. Lauenroth. 1996. Soil organic matter and nutrient availability responses to reduced plant inputs in shortgrass steppe. *Ecology* 77: 2516–2527.

Klipple, G.E., and D.F. Costello. 1960. *Vegetation and cattle responses to different intensities of grazing on shortgrass ranges on the Central Great Plains*. U.S. Department of Agriculture Technical Bulletin No. 1216.

Lauenroth, W.K. 1979. Grassland primary production: North American grasslands in perspective. In *Perspectives in grassland ecology*, ed. N.R. French, 3–24. New York: Springer.

Lauenroth, W.K., and I.C. Burke, eds. 2008. *Ecology of the shortgrass steppe: Perspectives from long-term research*. New York: Oxford University Press.

Lauenroth, W.K., and O.E. Sala. 1992. Long-term forage production of North American shortgrass steppe. *Ecological Applications* 2: 397–403.

Lauenroth, W.K., J.L. Dodd, and P.L. Sims. 1978. The effects of water- and nitrogen-induced stresses on plant community structure in a semiarid grassland. *Oecologia* 36: 211–222.

Lauenroth, W.K., O.E. Sala, D.P. Coffin, and T.B. Kirchner. 1994. The importance of soil water in the recruitment of *Bouteloua gracilis* in the shortgrass steppe. *Ecological Applications* 4: 741–749.

Lauenroth, W.K., D.G. Milchunas, O.E. Sala, I.C. Burke, and J.A. Morgan. 2008. Net primary production in the shortgrass steppe. In *Ecology of the shortgrass steppe*, ed. W.K. Lauenroth and I.C. Burke, 270–305. New York: Oxford University Press.

Laycock, W.A. 1991. Stable steady states and thresholds of range condition on north American rangelands: A viewpoint. *Journal of Range Management* 44: 427–433.

Li, J., G.S. Okin, L. Alvarez, and H. Epstein. 2007. Quantitative effects of vegetation cover on wind erosion and soil nutrient loss in a desert grassland of southern New Mexico, USA. *Biogeochemistry* 85: 317–332.

———. 2008. Effects of wind erosion on the spatial heterogeneity of soil nutrients in two desert grassland communities. *Biogeochemistry* 88: 73–88.

McDonald, J.D. 1981. *North American bison: Their classification and evolution*. Berkeley: University of California Press.

McKenna, O.P., and O.E. Sala. 2018. Playa-wetlands effects on dryland biogeochemistry: Space and time interactions. *Journal of Geophysical Research: Biogeosciences* 123: 1889–1897.

Michaelides, K., D. Lister, J. Wainwright, and A.J. Parsons. 2012. Linking runoff and erosion dynamics to nutrient fluxes in a degrading dryland landscape. *Journal of Geophysical Research* 117. https://doi.org/10.1029/2012JG002071.

Milchunas, D.G. 2006. *Responses of plant communities to grazing in the Southwestern United States*. U.S. Department of Agriculture Forest Service. Rocky Mountain Research Station General Technical Report RMRS-GTR-169: Fort Collins, CO.

Milchunas, D.G., and W.K. Lauenroth. 1993. A quantitative assessment of the effects of grazing on vegetation and soils over a global range of environments. *Ecological Monographs* 63: 327–366.

———. 1995. Inertia in plant community structure: State changes after cessation of nutrient-enrichment stress. *Ecological Applications* 5: 452–458.

———. 2008. Effects of grazing on abundance and distribution of shortgrass steppe consumers. In *Ecology of the shortgrass steppe*, ed. W.K. Lauenroth and I.C. Burke, 459–483. New York: Oxford University Press.

Milchunas, D.G., O.E. Sala, and W.K. Lauenroth. 1988. A generalized model of the effects of grazing by large herbivores on grassland community structure. *American Naturalist* 132: 87–106.

Milchunas, D.G., W.K. Lauenroth, I.C. Burke, and J. Detling. 2008. Effects of grazing on vegetation. In *Ecology of the shortgrass steppe*, ed. W.K. Lauenroth and I.C. Burke, 389–446. New York: Oxford University Press.

Minnick, T.J., and Debra P. Coffin. 1999. Geographic patterns of simulated establishment of two species: Implications for distributions of dominants and ecotones. *Journal of Vegetation Science* 10 (3): 343–356.

Monger, C., O.E. Sala, M.C. Duniway, H. Goldfus, I.A. Meir, R.M. Poch, H.L. Throop, and E.R. Vivoni. 2015. Legacy effects in linked ecological–soil–geomorphic systems of drylands. *Frontiers in Ecology and the Environment* 13: 13–19.

Neilson, R.P. 1986. High-resolution climatic analysis and southwest biogeography. *Science* 232: 27–34.

Nelson, E.W. 1934. *The influence of precipitation and grazing upon black grama grass range*. USDA Technical Bulletin No. 409, Washington, DC.

Noy-Meir, I. 1985. Desert ecosystem structure and function. In *Hot deserts and arid shrublands (Ecosystems of the World, v. 12A)*, ed. M. Evenari, I. Noy-Meir, and D.W. Goodall, 93–103. Amsterdam: Elsevier.

Okin, G.S., M.M. Heras, P.M. Saco, H.L. Throop, E.R. Vivoni, A.J. Parsons, J. Wainwright, and D.P.C. Peters. 2015. Connectivity in dryland landscapes: Shifting concepts of spatial interactions. *Frontiers in Ecology and the Environment* 13: 20–27.

Okin, G.S., O.E. Sala, E.R. Vivoni, and J. Zhang. 2018. The interactive role of wind and water in dryland function: What does the future hold? *Bioscience* 68: 670–677.

O'Neill, R.V., D.L. deAngelis, J.B. Waide, and T.F.H. Allen. 1986. *A hierarchical concept of ecosystems*. Princeton: Princeton University Press.

Parton, W.J. 1978. Abiotic section of ELM. In *Grassland simulation model*, ed. G.S. Innis, 31–53. New York, NY: Springer.

Parton, W.J., D.S. Schimel, C.V. Cole, and D.S. Ojima. 1987. Analysis of factors controlling soil organic matter levels in Great Plains grasslands. *Soil Science Society of America Journal* 51: 1173–1179.

Parton, W.J., J.W.B. Stewart, and C.V. Cole. 1988. Dynamics of C, N, P and S in grassland soils: A model. *Biogeochemistry* 5: 109–131.

Parton, W.J., D.S. Schimel, and D.S. Ojima. 1994. Environmental change in grasslands: Assessment using models. *Climatic Change* 28: 111–141.

Parton, W.J., M. Hartman, D. Ojima, and D. Schimel. 1998. DAYCENT and its land surface submodel: Description and testing. *Global and Planetary Change* 19: 35–48.

Parton, W.J., S.J. Del Grosso, I.C. Burke, and D.S. Ojima. 2008. The shortgrass steppe and ecosystem modeling. In *Ecology of the shortgrass steppe*, ed. W.K. Lauenroth and I.C. Burke, 373–378. New York: Oxford University Press.

Paruelo, J.M., H.E. Epstein, W.K. Lauenroth, and I.C. Burke. 1997. ANPP estimates from NDVI for the Central Grassland region of the United States. *Ecology* 78: 953–958.

Paulsen, H.A., Jr., and F.N. Ares. 1962. *Grazing values and management of black grama and tobosa grasslands and associated shrub ranges of the Southwest.* U.S. Department of Agriculture Technical Bulletin No. 1270.

Peterjohn, W.T., and W.H. Schlesinger. 1991. Factors controlling denitrification in a Chihuahuan Desert ecosystem. *Soil Science Society of America Journal* 55: 1694–1701.

Peters, D.P.C. 2000. Climatic variation and simulated patterns in seedling establishment of two dominant grasses at a semiarid-arid grassland ecotone. *Journal of Vegetation Science* 11: 493–504.

———. 2002a. Plant species dominance at a grassland-shrubland ecotone: An individual-based gap dynamics model of herbaceous and woody species. *Ecological Modelling* 152 (1): 5–32.

———. 2002b. Recruitment potential of two perennial grasses with different growth forms at a semiarid-arid ecotone. *American Journal of Botany* 89: 1616–1623.

Peters, D.P.C., and K.M. Havstad. 2006. Nonlinear dynamics in arid and semiarid systems: Interactions among drivers and processes across scales. *Journal of Arid Environments* 66: 196–206.

Peters, D.P.C., and J. Yao. 2012. Long-term experimental loss of foundation species: Consequences for dynamics at ecotones across heterogeneous landscapes. *Ecosphere* 3. https://doi.org/10.1890/ES11-00273.1.

Peters, D.P.C., R.A. Pielke Sr., B.T. Bestelmeyer, C.D. Allen, S. Munson-McGee, and K.M. Havstad. 2004. Cross-scale interactions, nonlinearities, and forecasting catastrophic events. *Proceedings of the National Academy of Sciences* 101: 15130–15135.

Peters, D.P.C., B.T. Bestelmeyer, J.E. Herrick, H.C. Monger, E. Fredrickson, and K.M. Havstad. 2006a. Disentangling complex landscapes: New insights to forecasting arid and semiarid system dynamics. *Bioscience* 56: 491–501.

Peters, D.P.C., J.R. Gosz, W.T. Pockman, E.E. Small, R.R. Parmenter, S.L. Collins, and E. Muldavin. 2006b. Integrating patch and boundary dynamics to understand and predict biotic transitions at multiple scales. *Landscape Ecology* 21: 19–33.

Peters, D.P.C., I. Mariotto, K.M. Havstad, and L.W. Murray. 2006c. Spatial variation in remnant grasses after a grassland to shrubland state change: Implications for restoration. *Rangeland Ecology and Management* 59: 343–350.

Peters, D.P.C., P.M. Groffman, K.J. Nadelhoffer, N.B. Grimm, S.L. Collins, W.K. Michener, and M.A. Huston. 2008. Living in an increasingly connected world: A framework for continental-scale environmental science. *Frontiers in Ecology and the Environment* 5: 229–237.

Peters, D.P.C., J.E. Herrick, H.C. Monger, and H. Huang. 2010. Soil-vegetation-climate interactions in arid landscapes: Effects of the north American monsoon on grass recruitment. *Journal of Arid Environments* 74 (5): 618–623.

Peters, D.P.C., J. Yao, O.E. Sala, and J. Anderson. 2012. Directional climate change and potential reversal of desertification in arid and semiarid ecosystems. *Global Change Biology* 18: 151–163.

Peters, D.P.C., H.W. Loescher, M.D. SanClements, and K.M. Havstad. 2014a. Taking the pulse of a continent: Building on site-based research infrastructure for regional to continental scale ecology. *Ecosphere* 5. https://doi.org/10.1890/ES13-00295.1.

Peters, D.P.C., J. Yao, D.B. Browning, and A. Rango. 2014b. Mechanisms of grass response in grasslands and shrublands during dry or wet periods. *Oecologia* 174: 1323–1334.

Peters, D.P.C., K.M. Havstad, S.R. Archer, and O.E. Sala. 2015. Beyond desertification: New paradigms for dryland landscapes. *Frontiers in Ecology and the Environment* 1: 4–12.

Peters, D.P.C., N.D. Burrus, L.L. Rodriguez, D.S. McVey, E.H. Elias, A.M. Pelzel-McCluskey, J.D. Derner, T.S. Schrader, J. Yao, S.J. Pauszek, J. Lombard, S.R. Archer, B.T. Bestelmeyer, D.M. Browning, C.W. Brungard, J.L. Hatfield, N.P. Hanan, J.E. Herrick, G.S. Okin, O.E. Sala, H. Savoy, and E.R. Vivoni. 2018. An integrated view of complex landscapes: A big data-model integration approach to trans-disciplinary science. *Bioscience* 68: 658–669.

Peters, D.P.C., G.S. Okin, J.E. Herrick, H. Savoy, J.P. Anderson, S.L. Scroggs, and J. Zhang. 2020. Modifying connectivity to promote state change reversal in drylands. *Ecology*. https://doi.org/10.1002/ecy.3069.

Pickett, S.T.A., and P.S. White, eds. 1985. *The ecology of natural disturbance and patch dynamics*. Orlando: Academic.

Platt, W.J. 1975. The colonization and formation of equilibrium plant species associations on badger disturbances in a tall-grass prairie. *Ecological Monographs* 45: 285–305.

Pound, R., and F.E. Clements. 1898. A method of determining the abundance of secondary species. *Minnesota Botanical Studies* 2: 19–24.

Rachal, D.M., G.S. Okin, C. Alexander, J.E. Herrick, and D.C. Peters. 2015. Modifying landscape connectivity by reducing wind driven sediment redistribution, northern Chihuahuan desert, USA. *Aeolian Research* 17: 129–137.

Rango, A., S. Goslee, J.E. Herrick, M. Chopping, K.M. Havstad, L.F. Huenneke, R.P. Gibbens, R. Beck, and R.P. McNeely. 2002. Remote sensing documentation of historic rangeland remediation treatments in southern New Mexico. *Journal of Arid Environments* 50: 549–572.

Rango, A., L. Huenneke, M. Buonopane, J.E. Herrick, and K.M. Havstad. 2005. Using historic data to assess effectiveness of shrub removal in southern New Mexico. *Journal of Arid Environments* 62: 75–91.

Reichhardt, K.L. 1982. Succession of abandoned fields on the shortgrass prairie, northeastern Colo. *The Southwestern Naturalist* 27: 299–304.

Reichmann, L.G., O.E. Sala, and D.P.C. Peters. 2013. Precipitation legacies in desert grassland primary production occur through previous-year tiller density. *Ecology* 94: 435–443.

Reynolds, J.F., D.M. Stafford Smith, E.F. Lambin, B.L. Turner II, M. Mortimore, and S.P.J. Batterbury. 2007. Global desertification: Building a science for dryland development. *Science* 316: 847–851. https://doi.org/10.1126/science.1131634.

Riegel, A. 1941. Life history habits of blue grama. *Kansas Academy of Sciences Transactions* 44: 76–83.

Sala, O.E., W.J. Parton, L.A. Joyce, and W.K. Lauenroth. 1988. Primary production of the central grassland region of the United States: Spatial pattern and major controls. *Ecology* 69: 40–45.

Samuel, M.J. 1985. Growth parameter differences between populations of blue grama. *Journal of Range Management* 38: 339–342.

Sala, O.E., L.A. Gherardi, L.G. Reichmann, E. Jobbagy, and D.P.C. Peters. 2012. Legacies of precipitation fluctuations on primary production: Theory and data synthesis. *Philosophical Transactions of the Royal Society B* 367: 3135–3144.

Savage D.A. 1937. *Drought survival of native grass species in the central and southern Great Plains, 1935*. U.S. Department of Agriculture Technical Bulletin No. 549, Washington, DC.

Schimel, D.S., T.G.F. Kittel, and W.J. Parton. 1991. Terrestrial biogeochemical cycles: Global interactions with the atmosphere and hydrology. *Tellus* 43AB: 188–203.

Schimel, D.S., M.A. Stillwell, and R.G. Woodmansee. 1985. Biogeochemistry of C, N, and P in a soil catena of the shortgrass steppe. *Ecology* 66: 276–282.

Schlesinger, W.H., and C.S. Jones. 1984. The comparative importance of overland runoff and mean annual rainfall to shrub communities of the Mojave Desert. *Botanical Gazette* 145: 116–124.

Schlesinger, W.H., J.F. Reynolds, G.L. Cunningham, L. Huenneke, W.M. Jarrell, R.A. Virginia, and W.G. Whitford. 1990. Biological feedbacks in global desertification. *Science* 247: 1043–1048.

Schlesinger, W.H., J.A. Raikes, A.E. Hartley, and A.F. Cross. 1996. On the spatial pattern of soil nutrients in desert ecosystems. *Ecology* 77: 364–374.

Schlesinger, W.H., and A.M. Pilmanis. 1998. Plant-soil interactions in deserts. *Biogeochemistry* 42: 169–187.

Schlesinger, W.H., A.D. Abrahams, A.J. Parsons, and J.C. Wainwright. 1999. Nutrient losses in runoff from grassland and shrubland habitats in southern New Mexico: I. Rainfall simulation experiments. *Biogeochemistry* 45: 21–34.

Schlesinger, W.H., T.J. Ward, and J. Anderson. 2000. Nutrient losses in runoff from grassland and shrubland habitats in southern New Mexico: II: Field plots. *Biogeochemistry* 49: 69–86.

Schooley, R.L., B.T. Bestelmeyer, and A. Campenella. 2018. Shrub encroachment, productivity pulses, and core-transient dynamics of Chihuahuan Desert rodents. *Ecosphere* 9: e02330. https://doi.org/10.1002/ecs2.2330.

Schreiner-McGraw, A.P., and E.R. Vivoni. 2017. Percolation observations in an arid piedmont watershed and linkages to historical conditions in the Chihuahuan Desert. *Ecosphere* 8 (11): e02000.

Senft, R.L., L.R. Rittenhouse, and R.G. Woodmansee. 1985. Factors influencing patterns of cattle grazing behavior on shortgrass steppe. *Journal of Rangeland Management* 38 (1): 82–87.

Senft, R.L., M.B. Coughenour, D.W. Bailey, L.R. Rittenhouse, O.E. Sala, and D.M. Swift. 1987. Large herbivore foraging and ecological hierarchies. *Bioscience* 37: 789–799.

Shoop, M., M. Kanode, and M. Calvert. 1989. Central Plains experimental range: 50 years of research. *Rangelands* 11: 112–117.

Shreve, F. 1917. A map of the vegetation of the United States. *Geographical Review* 3 (2): 119–125.

Shugart, H.H. 1984. *A theory of forest dynamics: The ecological implications of forest succession models*. New York: Springer.

Sims, P.L., and J.S. Singh. 1978. The structure and function of ten western North American grasslands. III. Net primary production, turnover and efficiencies of energy capture and water use. *Journal of Ecology* 66: 573–597.

Sims, P.L., J.S. Singh, and W.K. Lauenroth. 1978. The structure and function of ten western north American grasslands: I. Abiotic and vegetational characteristics. *Journal of Ecology* 66: 251–285.

Singh, J.S., W.K. Lauenroth, H.W. Hunt, and D.M. Swift. 1984. Bias and random errors in estimators of net root production: A simulation approach. *Ecology* 65: 1760–1764.

Sousa, W.P. 1984. Intertidal mosaics: Patch size, propagule availability, and spatially variable patterns of succession. *Ecology* 65: 1918–1935.

Swanson, F.J., T.R. Kratz, N. Caine, and R.G. Woodmansee. 1988. Land-form effects on ecosystem patterns and processes. *Bioscience* 38 (2): 92–98.

Van Auken, O.W. 2000. Shrub invasions of North American semiarid grasslands. *Annual Review of Ecology and Systematics* 31: 197–215.

Van Devender, T.R. 1995. Desert grassland history: Changing climates, evolution, biogeography, and community dynamics. In *The desert grassland*, ed. M.P. McClaren and T.R. Van Devender, 68–99. Tucson: University of Arizona Press.

Vinton, M.A., and I.C. Burke. 1995. Interactions between individual plant species and soil nutrient status in shortgrass steppe. *Ecology* 76: 1116–1133.

Wainwright, J., A.J. Parsons, W.H. Schlesinger, and A.D. Abrahams. 2002. Hydrology–vegetation interactions in areas of discontinuous flow on a semi-arid bajada, southern New Mexico. *Journal of Arid Environments* 51 (3): 319–338.

Weems, S.L., and H.C. Monger. 2012. Banded vegetation-dune development during the medieval warm period and 20th century, Chihuahuan Desert, New Mexico, USA. *Ecosphere* 3 (3): art 21.

Westoby, M., B. Walker, and I. Noy-Meir. 1979. Opportunistic management for rangelands not at equilibrium. *Journal of Range Management* 42 (4): 266–274.

Wondzell, S.M., G.L. Cunningham, and D. Bachelet. 1996. Relationships between landforms, geomorphic processes, and plant communities on a watershed in the northern Chihuahuan Desert. *Landscape Ecology* 11 (6): 351–362.

Wooton, E.O. 1908. *The range problem in New Mexico*. New Mexico College of Agriculture and Mechanic Arts Agricultural Experiment Station Bulletin No. 66. Albuquerque: Albuquerque Morning Journal.

Yahdjian, L., L. Gherardi, and O.E. Sala. 2011. Nitrogen limitation in arid-subhumid ecosystems: A meta-analysis of fertilization studies. *Journal of Arid Environments* 75: 675–680.

Yao, J., D.P.C. Peters, K.M. Havstad, R.P. Gibbens, and J.E. Herrick. 2006. Multi-scale factors and long-term responses of Chihuahuan Desert grasses to drought. *Landscape Ecology* 21: 1217–1231.

Yonker, C.M., D.S. Schimel, E. Paroussis, and R.D. Heil. 1988. Patterns of organic carbon accumulation in a semiarid shortgrass steppe, Colorado. *Soil Science Society of America Journal* 52: 478–483.

Chapter 7
Cold War Origins of Long-Term Ecological Research in Alaska

Sharon E. Kingsland

Abstract This chapter examines the history of ecology in Alaska leading up to the creation of two Long Term Ecological Research (LTER) projects in Alaska in 1987, the Arctic LTER at Toolik Lake, and the Bonanza Creek LTER in Alaska's interior boreal forest. Starting with the founding of the Navy's Arctic Research Laboratory at Point Barrow, Alaska, in 1947, the chapter explores how Cold War imperatives, as well as programs such as the International Biological Program (IBP), and concerns about the environmental impact of energy development, all helped to support basic and applied ecological research in the Arctic. The evolving multidisciplinary research program at Barrow embraced human ecology, as well as physiological ecology, population ecology, and ecosystem ecology. When the Alaskan projects were added to the LTER network in 1987, many of the ecologists involved had substantial experience at Barrow, both in connection with the IBP and with research that preceded the IBP. Using a long-term historical perspective demonstrates how LTER projects built upon a solid foundation of earlier ecological work. I examine in particular the place and role of human ecology in such enterprises.

Keywords LTER program · Long-term ecological research · Naval Arctic Research Laboratory · Cold war science · Human ecology · International Biological Program · Point Barrow

---

S. E. Kingsland (✉)
Department of History of Science and Technology, Johns Hopkins University, Baltimore, MD, USA
e-mail: sharon@jhu.edu

R. B. Waide, S. E. Kingsland (eds.), *The Challenges of Long Term Ecological Research: A Historical Analysis*, Archimedes 59,
https://doi.org/10.1007/978-3-030-66933-1_7

7.1 Introduction

Tom Callahan, the program officer who was instrumental in promoting the new Long Term Ecological Research Program (LTER) within the National Science Foundation (NSF), published a report in 1984 in which he drew attention to the value of long-term ecological research and lauded NSF's program, created in 1980, as a pioneering effort to support such enterprises. He pointed out that the usual short-term funding cycles were inadequate for the support of long-term research, and stressed that the need for long-term research had already been "proven at all levels of government and in most sectors of the private economy" (Callahan 1984). He identified several precedents showing that support for long-term research had been steadily growing from the 1960s on. These precedents were: national parks, wildlife refuges and preserves, experimental forests and ranges, research parks run by the national laboratories under the Department of Energy, and the cataloging of federal lands available for research by the Federal Committee on Research Natural Areas, and by its successor, the Federal Committee on Ecological Reserves.

He noted too the importance of the International Biological Program's biome studies (1968–1974), and the designation of an international system of biosphere reserves as part of UNESCO's Man and the Biosphere Program. The passage of the National Environmental Policy Act, which went into effect in January 1970, had also created new demands for ecological research. In the lead-up to the creation of NSF's program, a report commissioned by NSF and published in 1977 recommended creating a network of experimental ecological reserves across the United States and its territories (The Institute of Ecology 1977). That report built on earlier discussions of the need for protected natural ecosystems as sites for baseline studies, monitoring, and experimental programs (Franklin et al. 1972).

Callahan was aware that the LTER program was benefiting from momentum that had been building for many years within ecology. It was clear that there had been growing support for the idea of long-term research, and that there were several sites and projects already underway that could be identified as precedents of long-term ecological research. It is worth inquiring further into these precedents, in order to understand how ecologists successfully built support for long-term ecological research in the era before the LTER program.

One of the goals of this chapter is to explore how Cold War imperatives helped to stimulate ecological research from the 1940s on, and how these developments created a foundation for long-term ecological research. My case study will focus mainly on Arctic ecology and on the Arctic Long Term Ecological Research project in Alaska, which joined the LTER network of sites in 1987. However, I will devote attention also to the Bonanza Creek LTER created the same year, and located farther south in the boreal forest of Alaska's interior. To establish the broader historical context for these two LTER projects, we first need to explore how ecological programs expanded during the early Cold War years, giving rise to a robust program of ecological research in northern Alaska.

7.2 The Gestation of LTER: The Cold War Context

In 1978 an NSF-sponsored workshop preceding the formation of the LTER program identified several existing sites that appeared promising for long-term research and monitoring (National Science Foundation 1978). Potential sites included the Naval Arctic Research Laboratory in Alaska, as well as several research parks connected to the national laboratories, such as Los Alamos Environmental Research Park, Nevada Test site, Oak Ridge National Environmental Research Park, Argonne Experimental Forest, Savannah River Environmental Research Park, Brookhaven National Laboratory, and Luquillo Experimental Forest. The appearance in this report of several national laboratories as potential long-term research sites acknowledges the key role that the national laboratories played in the growth of postwar ecology. These laboratories also served U.S. national interests during the Cold War, being centers for both military and peaceful development of atomic energy. The link between the growth of ecology and Cold War concerns is particularly clear in these locations.

Stephen Bocking (1995), has explored in depth the way ecological programs grew within the context of the national laboratories, which were controlled by the Atomic Energy Commission (AEC) during the postwar decades. Bocking analyzed the growth and evolution of ecology at the Oak Ridge National Laboratory in Tennessee, paying attention to how the relationship between ecology and environmental protection was shaped by changes in the social and political contexts from the 1950s to the 1970s. Oak Ridge was a former Manhattan Project site for separating uranium isotopes and producing plutonium for the first atomic bombs. After the war its work focused on nuclear science, including production of radioactive isotopes for research in physics, chemistry, and medicine. Interest in ecology developed initially within the field known as "health physics", which studied the effects of exposure to radiation on health. Ecological research at Oak Ridge in the 1950s was closely related to health physics and to problems of radioactive waste contamination. Bocking's point was that ecology gained a foothold at Oak Ridge, not because the AEC recognized the value of ecology for its work, but because health physicists at Oak Ridge had identified ecological research as a component of health physics.

But once ecology had gained a foothold under Stanley Auerbach's direction at Oak Ridge, it started to expand, gradually taking on greater independence and reflecting the goals, ideas, and approaches of ecologists. Auerbach achieved a degree of autonomy for ecology by focusing on the new concept of the "ecosystem" and pursuing research that took advantage of the fact that radioisotopes were a novel and useful tool for studying ecosystem processes. Research on radionuclides served the needs of the AEC, but ecologists developed a degree of independence that enabled them to determine their own research directions. They broadened their analysis to general ecological problems that they found of interest, including the development of ecological theory and modeling.

As Bocking points out, this balancing act succeeded brilliantly, and by 1968 the Oak Ridge Radiation Ecology Section was "one of the largest ecological research groups in the United States" (Bocking 1995, p. 17). Bocking suggests that ecologists were able to serve the needs of the AEC, while also developing basic ecological science in keeping with their ideas about what problems, methods, and approaches would advance the science. The ecosystem approach proved to be well suited both for applied and fundamental research. By the time that the LTER program was being contemplated in the late 1970s, the Oak Ridge Laboratory appeared to be a promising site for an LTER project because it already had a robust ecological program under way, with particular strengths in ecosystem analysis.

This chapter's goal is to examine another candidate for long-term research on the list of the same workshop report: the Naval Arctic Research Laboratory (NARL) at Point Barrow, Alaska. The Laboratory, created in August 1947, was first known as the Arctic Research Laboratory (ARL) and was renamed the Naval Arctic Research Laboratory (NARL) in 1967. By the late 1960s it had become the world's largest facility for the support of Arctic research across a range of disciplines. Here too we find an exceptionally robust ecological research tradition that had been developing for several decades prior to the LTER program. Hence when two Alaskan LTER projects were added to the list of sites in 1987, their scientists were able to draw on an ecological research legacy going back four decades.

The Arctic Research Laboratory bears some similarity to national laboratories like Oak Ridge in that it was created immediately after the war, in this case with funding from the Navy rather than the AEC. The Navy was convinced of the need for a scientific laboratory in Alaska in part because of the possibility that the U.S. might be attacked by the Soviet Union from the north, requiring a strong U.S. military presence there, and therefore better knowledge of how to support a military force under arctic conditions. One surprising aspect of this history is the way the sales pitch to the Navy focused on the importance of developing what was called "human ecology," based on the argument that supporting a military presence in the north depended on understanding human adaptation to this extreme environment. From the outset, this laboratory's raison d'être was linked to the idea of establishing a base for wide-ranging scientific research, research that would include humans as well as other animals and plants. This goal was different from NSF's later mission in the LTER program, but it helped to create a vibrant ecological research community with valuable experience that would support the LTER programs established in Alaska decades later.

ARL's research agenda was not wholly dominated by the Navy's military mission, however, because its funding came from the newly created Office of Naval Research (ONR), which was established in 1946 to support civilian scientists working at universities and colleges. The ONR also served as a model for the National Science Foundation, created in 1950; NSF's first director, Alan T. Waterman, had been the ONR's Chief Scientist. The Arctic Research Laboratory, one of the first projects undertaken by the ONR, received enthusiastic endorsement from the civilian sector because of its potential to open up basic scientific research in many disciplines, ecology being one. In the 1950s and early 1960s, much of the biological research at ARL dealt with physiological and ecological problems, with an emphasis on physiological ecology and population ecology.

Ecosystem ecology was also developing quickly, especially from the mid-1960s. Ecologists began promoting the concept of the ecosystem as an organizing idea for ecology in the 1950s, and ecosystem ecology received a boost when the International Biological Program (IBP) got underway in the 1960s (Hagen 1992). The IBP started in imitation of the International Geophysical Year (1957–58), which had included some biological projects, although its focus was on geological, geophysical, oceanographic, meteorological, and atmospheric studies. Soon after the end of IGY, ideas were proposed for an International Biological Year, but this quickly evolved into a plan for a multi-year program that was less centralized than the IGY had been (Aronova et al. 2010).

Early themes identified in 1961 were human genetics, conservation, and improvement in the use of natural resources. By the late 1960s the ecological analysis of large biomes emerged as one of the central foci for North American research, and the National Science Foundation funded five major biome studies as part of IBP from 1968 to 1974 (Coleman 2010; Golley 1993). A "biome" was a region controlled by climate, in which certain species of plants or animals dominated the ecosystems. Initially four biome types were chosen: grassland, coniferous forest, deciduous forest, and desert, and the arctic tundra biome was later added as a fifth. Because of the strong ecological research tradition at Point Barrow, it was selected as the site for the Arctic Tundra Biome study, which ran from 1970 to 1974.

As Gus Shaver noted in a review of Arctic tundra research in 1996, the biome study was "the first multi-investigator ecological research project in northern Alaska to be designed as an integrated, ecosystem-level program" (Shaver 1996a, p. 21). Because the ecological research tradition in the Arctic was so strong, and because its existence prompted growth in arctic biology at other institutions, its history helps us to understand how the confluence of many goals and enterprises, unfolding over four decades, created a foundation for the later LTER projects in Alaska. Shaver observed that it was at Point Barrow's laboratory where groups of ecologists first "began working on interrelated projects, each focusing on different aspects of the same ecosystem type and often working on the same sites or experimental plots" (Shaver 1996a, p. 19). Moreover, Shaver continued, the stable presence of this laboratory "allowed the kind of long-term field studies that are important in separating signal from noise in ecological processes and patterns" (Shaver 1996a, p. 19).

Shaver's research on arctic ecology started in the early 1970s and he was involved in the Arctic LTER from its inception. He has written an overview of this early history for the Arctic LTER's website (Shaver 1996b). This chapter expands upon that history in order to illustrate the degree of entrepreneurial ingenuity that it took for biologists to start and maintain these early research programs, as they took advantage of opportunities created by the Cold War to further their own research interests. The next section probes in more detail the origins of ARL and the struggle to define a scientific research program that would promote the laboratory's longevity, which was partly a consequence of continuing Cold War imperatives, and partly of dogged determination by various scientific leaders.

We can think of this period as the "gestational period" for LTER, even if LTER was not foreseen at the outset. We then will consider in more detail some of the ecological programs that developed at ARL, leading to the biome studies of the IBP, further studies related to energy development in the Arctic, and then to the Arctic LTER. The conclusion uses the comparison between the earlier research program at ARL and the later LTER program to ask general questions about how ecological science has been and could be defined.

7.3 The Navy's Arctic Research Laboratory: A Strategy for Survival

The story of the Arctic LTER begins with the creation of the Office of Naval Research (ONR) in August 1946. The ONR was the only government agency of any significant size to fund civilian research broadly after the war, and along with the Atomic Energy Commission, it was an important source of support for ecology in the 1950s. (NSF was created in 1950, its first director coming from the ONR, but it had a very small budget until the launch of the Soviet Sputnik satellite in 1957 spurred its growth). The Office of Naval Research was an idea hatched by a group of young naval reserve officers who happened to be scientists, and who thought it would be a great benefit to science if the Navy would undertake the support of civilian scientific research after the war (Sapolsky 1990). They made a persuasive case for the creation of the Office, which was placed under the command of Vice-Admiral Harold Bowen. As Harvey Sapolsky explains in his history of the ONR, it proved to be a flexible and accommodating funding agency: "It gave scientists control over the direction of its research program, believed in open research, and even liked graduate students. Not surprisingly, many scientists soon came to prefer ONR's contracts to any other agency's grants" (Sapolsky 1990, p. 45).

One particular project emerged as an early favorite in the discussions of what the ONR should support: the creation of a laboratory dedicated to Arctic research. We do not know who first proposed this idea, but a physiologist named Moses C. Shelesnyak strongly promoted the idea within ONR and may have initiated it (Reed and Ronhovde 1971, p. 32). He was a Lieutenant Commander in the Office of Naval Research and in 1946 was Head of ONR's Environmental Physiology and Ecology Branch in its Medical Sciences Division. His scientific interests were in stress physiology, heat regulation, human ecology and polar research in general. It fell to him to review the plans for Arctic research in relation to the Navy's needs, and to develop a concrete plan that would suit both the government's needs and non-government research interests (Shelesnyak 1947a; Reed 1969).

As Reed and Ronhovde (1971) later commented in their history of the first quarter century of ARL, Shelesnyak masterfully linked his personal interests in physiology and human ecology to the general and specific needs of the Navy and the principles that guided the ONR. These principles included the idea that in funding basic research, the ONR was entering into a collaboration with civilian scientists that would mutually benefit both sides. The scientist was offered freedom to initiate and carry out basic research under contract with the Navy, with no restrictions on publishing or teaching, while the Navy would benefit from any new knowledge that might lead to the strengthening of Naval power (Shelesnyak 1948b, p. 100).

Shelesnyak's strategy for enlisting the ONR's support for arctic research built on his experience as a U. S. naval observer during Operation *Muskox*, a Canadian military exercise that took place from February to May, 1946, and included British and American observers (Lajeunesse and Lackenbauer 2017). The motive for the exercise was to learn more about how to conduct military maneuvers in the Arctic, given the possibility of a Soviet attack from the north. While a large-scale Soviet ground invasion seemed unlikely, it was possible that the Soviets might set up northern bases for airborne attacks on North America. Since the Soviets already had extensive experience fighting in northern conditions, both Canadians and Americans felt they needed to catch up by gaining direct experience in living, working, and travelling in the Arctic. *Muskox* involved moving a mechanized force of forty people over 3200 miles across Canada's north, but the expedition's members also collected scientific data as they travelled. Shelesnyak also wanted to compare mechanized travel with traditional "Eskimo style" travel, so he and a companion made an independent excursion by dog-sled from Coppermine, Northwest Territories, to Cambridge Bay, Victoria Island, in Canada's northern archipelago (Shelesnyak 1947c).

That experience, immediately preceding the creation of ONR, clearly gave him a strong sense of what ONR's mission should include: an arctic research laboratory in Alaska. The Arctic was strategically important for national defense, and Americans, like their Canadian counterparts, had to figure out how to support a military force there. Shelesnyak translated this strategic defense need into a pitch for the development of what he called "human ecology," by which he meant the study of how humans responded to, or could adapt to, the extreme arctic climate. Thus the interdisciplinary subject of human ecology, and the need for its development, was from the very start part of the justification for the arctic laboratory.

This strategy was cleverly designed to serve naval needs, but at the same time appeal to wider civilian interests in the biological sciences. Shelesnyak envisioned human ecology as a broad field of study concerning human responses to the environment. In linking human ecology to national security, he argued that for military work it was vital to select people who would function well in the Arctic, or adapt well to their work environment (Shelesnyak 1947a). Therefore studies had to be made of how different races, living in different geographical locations, responded to their environments. Examples of successful adaptation would provide models of what type of person to select for a given job. Broader studies of natural history would be needed to investigate how plants, animals, and micro-organisms affected humans. Human ecology also included studies of public health and disease. Studies of hypertension and cardiovascular disease, growth, development, longevity, and nutrition of the indigenous and immigrant populations would all be of interest to researchers. Observing that northern cultural groups treated children "with a real acceptance of the child as a person," he also recommended study of psychophysiological development. Finally, he included within human ecology various engineering problems involving such things as sanitation, food, transportation and community planning.

Shelesnyak envisioned a multi-disciplinary enterprise where the human ecologist would be working with engineers, geographers, and physiologists. "Human ecology" in Shelesnyak's appeal was not the same concept that is captured by the term "social-ecological systems" in the modern LTER program. It was much broader, was not tied to any particular ecological concept (such as the ecosystem concept, which had not yet been widely adopted), and could include branches of science that focused mainly on humans, including medical sciences and any kind of study involving human biology, human health, or psychology. That breadth was meant to serve the Navy's interest in understanding human adaptation to extreme environments.

But Shelesnyak also intended "human ecology" to involve general biology, so that while he was pitching the human ecology angle to the Navy, he was touting the advantages of the Arctic for general biology to the academic community. He was clearly imagining a research operation that would continue for several years as a multidisciplinary effort. As he pointed out, "it has been apparent for years that long-term studies are essential to the understanding of Arctic phenomena" (Shelesnyak 1948a). Study of arctic regions, he insisted, demanded "large, highly organized, fully supported, research teams and expeditions" (Shelesnyak 1947c, p. 477). By supporting research at this scientific frontier, the Navy was keeping up the heroic traditions of polar exploration for which it was justly celebrated.

Although the Navy was interested in the idea, it preferred to have a civilian group coordinate the research, and for this purpose turned to the Arctic Institute of North America, a bi-national institute with headquarters in Canada and the U.S. The Arctic Institute was itself a new entity created in 1945. Its director from 1945 to 1950 was Lincoln Washburn, who had been an intelligence officer during the war in the Arctic, Desert, Tropic Information Center (ADTIC) of the U.S. Army Air Forces (Benson 2007). He and Laurence Gould, chief of the Arctic section of the ADTIC, realized it would be valuable to create a bi-national organization that would bring together the interests of Canada and the U.S. in the Arctic, and this became the Arctic Institute of North America (today it continues its mission of research and education and since 1976 has been part of the University of Calgary in Canada).

In 1946 the Institute outlined a broad agenda for research that covered a range of scientific fields: mapping and description, meteorology and climatology, oceanography, geology, atmospheric studies and magnetism, biology, anthropology, ethnology and archaeology, and soils and agriculture (Arctic Institute of North America 1946). Washburn initially endorsed the idea of the ARL, and the Arctic Institute provided grants for research which would later be conducted out of Point Barrow. In the biology section of the Arctic Institute's research plan, studies of animal populations, mammalogy, invertebrate biology, marine biology and fisheries dominated. Shelesnyak explained that human biology and medicine had been omitted from that plan by design, but he made sure that human ecology was at center stage in the plan presented to the Navy (Shelesnyak 1947b).

The obvious location for an arctic research laboratory was Point Barrow, the most northerly U.S. settlement in North America, on the seacoast between the Chukchi Sea to the west and the Beaufort Sea to the east, and about four miles from the native town of Barrow. Point Barrow was then the base of supply for oil-exploring parties in Alaska's interior. In 1946, when the idea of building a scientific laboratory at the base came up, Vice-Admiral Bowen, who was directing naval research, invited comments from within the Naval Department as well as from Arctic scientists about the possibilities of creating a laboratory. The Arctic experts consulted were Vilhjalmur Stefansson (arctic explorer), Sir Hubert Wilkins (famed Australian polar explorer), Laurence Gould (President of Carleton College and geologist to the Byrd Antarctic Expedition), Harald U. Sverdrup (Director of Scripps Institution of Oceanography), Paul Siple (geographer, U.S. War Department General Staff), H. B. Collins (ethnographer, Smithsonian Institution), and Lincoln Washburn (Director of the Arctic Institute of North America). They offered enthusiastic support (Shelesnyak 1948b).

In February 1947 Shelesnyak made a reconnaissance trip to Point Barrow for the Chief of Naval Research and deemed the location suitable for a research laboratory. The petroleum personnel already at the site were interested in research, and the logistical support provided by the Bureau of Yards and Docks was a big advantage, although, as it turned out, the initial cost estimates were woefully underestimated (Reed and Ronhovde 1971, p. 40). The new Arctic Research Laboratory was approved later that year.

Shelesnyak immediately approached Laurence Irving at Swarthmore College, enlisting him to begin the scientific program at Barrow. Irving, a comparative physiologist, was already interested in arctic biology. He was part of a group of highly talented comparative physiologists who rose to prominence in the 1950s and 1960s and became leaders in the study of environmental physiology. At the Navy's request Swarthmore proposed research on human physiology, but the group intended to pursue more general animal studies of how temperature affected the rate of metabolism, as well as to encourage research in oceanography, ecology, and botany, which would involve contracts with other colleges and universities (Schindler 2001, p. 30). In August 1947 Irving arrived at Point Barrow with four other biologists from Swarthmore College, and in September two more biologists from Cornell University joined them (Shelesnyak 1948b; Reed and Ronhovde 1971). At the same time Shelesnyak created and became head of a Human Ecology Branch at ONR in Washington, D. C., to provide support for this initiative within ONR.

What Shelesnyak apparently failed to tell Irving was that no laboratory facilities existed yet at Point Barrow. When Irving and his group arrived, they found a landscape of tundra dotted with lakes of various sizes, stretching as far as the eye could see toward the Brooks Range to the south. Around them was "the noise and hustle of an oil-exploration camp," with tractors hauling equipment over the soft sand, and "weasels," or small tracked vehicles, scooting in all directions (Reed 1969, p. 177). Scattered across the landscape were fuel drums, "that ubiquitous trade-mark of the American developer in out-of-the-way places all over the world." On the beach where they landed, power barges stood ready to bring freight to shore. But there was no laboratory in sight. One of the first things the scientists had to do was remodel a small one-story Quonset hut, 20 by 40 feet, to serve as a physiological laboratory, and for the first season it served the small group well enough.

The Swarthmore group worked on oxygen consumption of animals at different temperatures, metabolic rates of arctic animals, and adaptations for heat conservation. Accompanying Irving was Per Scholander, a Swedish-born biologist who had been educated in Norway and who had gone to Swarthmore just before the war broke out. He had extremely broad interests ranging from botany to animal physiology. He was interested in the physiology of diving mammals, among other problems.

Scholander and Irving were central to the development of Point Barrow during its first 2 years, and over subsequent years became leaders in the field of physiological ecology, or the comparative study of physiological responses to the environment (Dawson 2007; Scholander 1990). In September two biologists, Donald Griffin and Raymond Hock, arrived from Cornell University to study the metabolism and heat economy of arctic birds. Griffin stayed only 3 months, but Hock remained longer and collaborated with Irving's group (Scholander et al. 1950). Griffin later became well known for his studies of bat echolocation and bird navigation and migration. The first group of biologists at ARL were men of great creativity, with wide-ranging interests at the intersection of ecology and physiology. Robert Elsner later commented that Irving and Scholander approached their research as a team effort, and David Norton and Gunter Weller also suggested that this style of collaborative teamwork was a "hallmark that dated from the laboratory's founding in 1947" (Elsner 2001; Norton and Weller 2001, p. 236).

In February 1948 Irving was appointed as the first Scientific Director of ARL, and he left Barrow that month to make a tour of the United States and build support for ARL. That winter he travelled widely, giving talks, speaking to colleagues across the country, and in general publicizing the opportunity that ARL offered, while a larger laboratory was being erected to accommodate what Irving hoped would be a growing community of scientists (Irving 1948; Reed and Ronhovde 1971, pp. 46–47). By 1948 a larger Quonset hut laboratory, 40 by 100 feet and two stories high, was available to researchers (Shelesnyak 1948b). It had laboratories and workshop on the ground floor, and a library, office, storeroom and seminar room on the second floor. The Swarthmore group continued to use the original small hut. By summer of 1948 the projects had expanded from two to nine, and in addition to physiology, natural history, and ecological studies, they included research on the health of the Barrow Eskimos. At Irving's suggestion the ONR established a scientific advisory board in 1948 to review and recommend proposals for research, and the laboratory grew to a total staff of about thirty scientists and technicians.

Irving was delighted to discover that the native residents were also exceptionally able collaborators in the scientific work, being "skillful hunters, reliable observers, able mechanics and interesting and most agreeable associates" (Irving 1950, p. 1047). The basic natural history that the biologists had expected to do, had largely already been done by those who inhabited the land. An important long-term collaborator was Simon Paneak, a Nunamiut native who had learned to read and write English and was highly knowledgeable about natural history, geology, and archeology. Paneak worked with Irving for 20 years, especially in research on Alaskan birds. He kept accurate journals of the seasonal cycles of birds and other animals and continued to collaborate with academic and government scientists until his death in 1975 (Brewster 1997; Huryn and Hobbie 2012, p. 253).

Shelesnyak, in his role at ONR in Washington, and Irving, in his role as Scientific Director, had to build support for their arctic enterprise not only by keeping ONR interested, but also by convincing the scientific community that arctic biology was worth pursuing. In an article in *Science* published in March 1948, Irving appealed to scientists to consider the advantages of arctic biology with its rich possibilities for studies in limnology, ecology, physiology, oceanography, climatology and geology (Irving 1948). For ecologists, an advantage was the small number of species and the relative simplicity of ecosystems compared to other regions. Studies of human acclimatization to cold benefited from having a range of human populations present, from natives to newcomers.

However, Irving's travels across the U.S. and talks with colleagues also made him realize that it might be difficult to persuade people to come to the Arctic for any length of time. As he wrote to Shelesnyak in January 1948, scientists were not ready to commit themselves to "prolonged programs of exclusive arctic research". There was no program in the U.S. that offered lasting opportunities, programs, or terms of employment for arctic research (Reed and Ronhovde 1971, p. 95).

He therefore devised an approach markedly different from Shelesnyak's vision, in order to appeal to the wider interests of biologists. His idea was to promote what he called "expeditionary physiology", which meant physiological studies of adaptation that were conducted in the field, rather than in the artificial environment of the laboratory. The term "expeditionary physiology" had been adopted by a small group of comparative physiologists who were associated with Irving, and who had connections to the eminent Danish comparative physiologist and Nobel laureate, August Krogh. Irving himself knew Krogh, and had invited him in 1939 to give a lecture series at Swarthmore on the comparative physiology of respiratory systems (Krogh 1941). Krogh knew Per Scholander and with Irving's help arranged for a Rockefeller Foundation fellowship so that Scholander could work with Irving at Swarthmore (Schmidt-Nielsen 1987).

The term "expeditionary physiology" likely came from Scholander, who used the term to describe the research excursions that he led during the war years, when he was working at Harvard's Fatigue Laboratory (Folk 2010). Later his enthusiasm for expeditionary physiology led him to help design a new type of research vessel, the *Alpha Helix*, launched in 1965 and supported by the National Science Foundation and Scripps Institution of Oceanography, to serve as a national facility carrying scientists to remote locations worldwide (Schmidt-Nielsen 1969). Another "expeditionary physiologist" was Knut Schmidt-Nielsen, a Norwegian biologist whom Scholander introduced to Irving after the war, and who was similarly interested in adaptations to extreme environments, particularly to desert conditions (Vogel 2008). Schmidt-Nielsen had been Krogh's student and had married Krogh's daughter, Bodil, who was a physiologist. After the war Irving offered Knut and Bodil year-long research positions at Swarthmore, and then arranged for them to spend a year at Stanford University. In 1952 the Schmidt-Nielsens settled permanently at Duke University.

"Expeditionary physiology," as understood by this group, meant comparative physiology done in the field, with the goal of gaining a full understanding of adaptations to environmental stressors by studying animals in their natural habitats (Hagen 2015). We would today identify this subject as physiological ecology. Studying animals in extreme environments was seen as a good way to discover basic physiological principles, because the adaptations were more obvious when animals were pushed to extremes (Wood 2007). Irving's collaborations with Scholander and Schmidt-Nielsen must have strengthened his conviction that comparative "expeditionary physiology" was a growing field attracting scientific interest, and that military funding could effectively be steered in this direction.

In 1947 Irving chaired a panel on Expeditionary Physiology for the Research and Development Board, an advisory board to the military on scientific matters related to national security and military interests and goals (Anonymous 1947). Before the creation of the National Science Foundation in 1950, this board had great power to decide the direction of scientific research and development in the U.S. Irving noted that the National Research Council was also taking an interest in current discussions of how geographic and climatic stresses affected life. These discussions suggested to him that "expeditionary physiology" was ripe for development (Irving 1950). Irving emphasized how the work done at ARL contributed to this initiative, but the key question was whether to focus exclusively on arctic research, or to use ONR's support to develop a broader comparative research program that would include other extreme environments.

Irving's concept of "expeditionary physiology" took the broader approach and involved proposing research in arctic, tropical, and desert climates. For tropical and desert research respectively, he recommended Barro Colorado in the Panama Canal Zone and the Santa Rita Experimental Range just south of Tucson, Arizona, which was where the Schmidt-Nielsens were conducting studies of rodent adaptations to the desert (Schmidt-Nielsen and Schmidt-Nielsen 1950; Reed and Ronhovde 1971, p. 60).

Irving's plan to use ONR funds to expand the field of expeditionary physiology brought him into direct conflict with Shelesnyak by the end of 1948. Shelesnyak did not think he could justify this broad plan to the Navy, and he insisted that the focus must be on developing the Arctic program and proving its merits, especially since additional funds were not available for research elsewhere (Reed and Ronhovde 1971, pp. 60–71). Shelesnyak prevailed in that argument. Irving was not willing to continue as Director of ARL given Shelesnyak's lack of support for his program, and left the directorship when his term ended in June 1949. But he remained active in arctic biology, leaving Swarthmore the same year to become chief of the Physiology Section of the Arctic Health Research Center, a division of the U.S. Public Health Service in Anchorage. In 1962 he moved to the University of Alaska, Fairbanks, where he helped to establish the Institute of Arctic Biology, serving as its director until 1966. That Institute, two decades later, would provide management support for the two Alaskan LTER projects.

Despite the initial growing pains and conflicts over research focus, the strong ecological foundation that began at ARL in 1947 continued through the appointment of the next two scientific directors. The biological interests of the first three directors ensured that a wide range of ecological subjects was supported through ARL. Shelesnyak persuaded the President of Johns Hopkins University, Detlev Bronk, to take over the management of the research program at ARL from Swarthmore when its contract ended in 1949. Shelesnyak himself left the ONR in September 1949 to head the Washington-Baltimore office of the Arctic Institute of North America, which had relocated to Johns Hopkins University. Although Shelesnyak soon departed from Hopkins in 1950 to take up a research post at the Weizman Institute in Israel, the university continued to coordinate the research program until March 1954 (Catling 1953).

George MacGinitie, a marine invertebrate ecologist from the California Institute of Technology, became the Scientific Director of ARL after Irving's departure. MacGinitie strengthened the research program and expanded the laboratory by acquiring and improving ARL buildings during his relatively short 14-month tenure, which ended early due to ill health. During his tenure the laboratory's research expanded into studies of life histories of organisms and ecology, especially of marine invertebrates. His wife Nettie accompanied him and participated in the taxonomic work. One of MacGinitie's students, Howard Feder, who arrived at ARL in June 1949, noted that MacGinitie had hoped that his work on benthic species in the environments next to Barrow would lead to long-term investigations, including field and laboratory studies. However, because these long-term studies were not done, some of the questions that MacGinitie had raised in the 1950s remained unresolved a half century later (Feder 2001, p. 53).

Ira Wiggins, a botanist from Stanford University, became Director in August 1950 and remained through January 1954. During his tenure botanical field work greatly expanded, along with biological surveys of flora and fauna of the coastal plain, freshwater lakes and lagoons, and coastal marine waters of northern Alaska (Wiggins 1952). Wiggins also broadened research on crystallography of ice, magnetic storms, the aurora, permafrost, paleontology, oceanography, and microclimatology (Reed and Ronhovde 1971, p. 169).

Wiggins pointedly reminded ARL's Advisory Board in 1952 that the Russians were still ahead in the game: there were 76 stations for arctic research on Russian territory, and only one on American territory (Reed and Ronhovde 1971, p. 167). There was, however, a clear spike in interest in the military and economic significance of the North at this time, due, as Lincoln Washburn noted, to the "proximity of Communist Russia, only 2.1 miles away" (Washburn 1952). New research centers were being created in Alaska and Canada for studies of the atmosphere, geophysics, geology, physical oceanography, and weather. Arctic biology, as Washburn noted, was also benefiting from the ONR's support of the Arctic Research Laboratory. Anthropological research was an active area of study, with the University of Alaska in the forefront of archeological investigation. Washburn estimated that "more than fifty universities and private research groups in the United States and Canada are actively engaged in some phase of northern research," although many areas were still scientifically unknown (Washburn 1952).

Despite this attention, in 1953 the Navy's oil exploration program stopped and it was assumed that ARL would also close. The scientists were very reluctant to leave, and fortunately for them the Cold War provided a reason to keep it open. This was right when discussions were underway about building the Distant Early Warning, or DEW Line, a set of radar stations along the 69th parallel across Alaska and Canada, which could provide early warning of a Soviet air invasion (Farish 2006). The research done at the lab turned out to be extremely valuable for the construction of the DEW Line between 1954 and 1957. The DEW Line was under control of the U.S. Air Force, and it agreed to support the Arctic Research Laboratory while using the Navy's base (one valuable form of support was access to the DEW Line's dining hall, which not only provided three large meals for hundreds of workers and scientists, but also a fourth meal in the middle of the night).[1] In 1954 the Laboratory became associated with the University of Alaska at Fairbanks, and the university took over its operation, providing its director and staff.

In the mid-1950s the Laboratory acquired two staff who remained in place for many years, providing stability. In 1955 Maxwell Britton, a botanist from Northwestern University who was interested in the ecology of the Arctic tundra, assumed responsibility for the lab from the Washington end, where he reportedly battled for the laboratory at every turn (Reed 1969, p. 180). He remained with ONR's arctic program until 1971. In late 1956 the University of Alaska arranged for the appointment of a permanent director, Max Brewer, a geophysicist involved in the Arctic Lab's permafrost project; he also remained for many years. When John Hobbie arrived at the laboratory to do ecological research later in the 1960s, he was informed that Britton and Brewer, along with ecologist Frank Pitelka, served as a scientific advisory group for the laboratory.[2]

By the summer of 1957 the International Geophysical Year (IGY) (which lasted 18 months) was underway, focusing attention on the Arctic and bringing in more researchers from the physical and earth sciences. With strong leadership both at ONR and at Barrow, the lab entered a period of stability and growth by the late 1950s. In its first 25 years of operation the ARL supported over 700 projects in the physical, biological and social sciences. Of the new projects undertaken during that time, 140 were in biological sciences (Reed 1969). Ecological research extended well beyond the lab at Point Barrow, as ARL also had a small air force of several light planes as well as a large R4D two-engine freight plane (a military version of the DC-3 plane). In the late-1950's these supported John Hobbie's thesis research on the year-round biology of Lake Peters, about 300 miles to the east, and the thesis research of Jerry Brown on soils in the present-day Arctic National Wildlife Reserve about 330 miles to the east.[3] After September 1960, the airplanes also supported

[1] John E. Hobbie, personal communication, 31 August 2018.

[2] John E. Hobbie, personal communication, 31 August 2018.

[3] John E. Hobbie, personal communication, 31 August 2018.

research at a laboratory set up on an ice floe in the Arctic Ocean, hundreds of miles from Barrow.

By the mid-1960s the Arctic Research Lab had a permanent staff of nearly fifty, several buildings at the base camp, and a network of field stations, including four drifting research stations on ice islands, an idea that Britton helped initiate (Britton 1964b; Schindler 2001). In the summer of 1963, biological projects constituted 28 of the 60 research programs at ARL. Some were under direct contract with ONR, others were supported under subcontracts between ONR and the Arctic Institute of North America, and the remainder were supported by other government agencies. In 1964 plans were underway to construct a modern laboratory complex to replace temporary and outmoded structures. The laboratory was renamed the Naval Arctic Research Laboratory (NARL) in 1967, and the new laboratory building was dedicated in 1969.

The International Biological Program, which in certain respects was modeled on IGY, strengthened support for ecological studies by the early 1970s (Britton 2001). The IBP had another important impact on the scientific community at NARL by creating research opportunities for more women scientists. Until this time the scientific community at NARL was predominantly male. A few wives, such as Nettie MacGinitie, accompanied their husbands to the laboratory and often assisted in the research, but there were no single young women scientists, for there were no living quarters for them. The laboratory had some of the features of a military base: scientists wore Navy-issued clothing, and if they wanted to leave the laboratory for recreation, they had to request permission from the director.[4] The Arctic biome project of the IBP opened new opportunities for women graduate students, assistants, and senior investigators at Barrow (Brown 2001).

During the 1960s the ONR's support of research shifted away from physiological studies of adaptation in humans and other animals, which had dominated in the early years, and focused more on field programs. Studies of animal physiology and cold adaptation were still considered important, although funding for the Animal Research Facility was on an ad hoc basis until 1974, when it became a line item in NARL's budget (Philo et al. 2001). Only then was it possible to make the facility upgrades needed to accommodate work on larger mammals, although the facilities had been adequate for work on smaller mammals and birds.

The links between ecology and the interests of the Navy were spelled out in a report by Helen Hayes on the ONR's support of arctic biology in 1964 (Hayes 1964). Biogeography and ecology were well supported "to obtain a well-rounded picture of the interrelationships which produce the character of arctic environments" (Hayes 1964, p. 29). Since the Navy operated from shore-based installations, Hayes explained, "every aspect of arctic environment ashore must be understood if operations are to proceed on a rational basis with some guarantee of success" (Hayes 1964, p. 29). That included the soil and substrate composition, the animal and plant

---

[4] Interview of Jerry Brown by Karen Brewster and David Norton, 2008. In Elmer E. Rasmuson Library, University of Alaska, Fairbanks.

populations it supported (and which might be sources of food for humans), microbial populations in the soil, and studies of permafrost.

The Navy in fact had considerable interest in ecology, studying not just the tundra ecosystems but also the ecology of the arctic seas and the planktonic organisms that were of interest "because of their all-too-often-exhibited potential for interfering with Navy activities"' (Hayes 1964, p. 31). Adaptations to cold and to extreme variations in the length of day were also central problems in research on animals and on humans, one question being the relationship between disruption of biological rhythms and stress. The study of daily, seasonal, and multi-annual cycles of animals were important parts of ONR's biological program. Although much of this research was basic ecology and physiology, it was seen as relevant to the Navy's needs, providing "background information upon which the Navy can base its operational plans" (Hayes 1964, p. 32).

But in addition to these practical needs, a lot of the ecological work was basic curiosity-driven research that served the interests of academic scientists who were free to propose their own research plans. These research projects helped to develop ecological theory and served as a good foundation for studies of the effects of disturbance on arctic ecosystems. A broad spectrum of ecological work was supported, some of which evolved into long-term research projects. The next section considers in more detail some of these general studies, and their contributions to ecological theory.

7.4 An Expanding Program of Arctic Ecology in Alaska

It would be impossible to do justice here to the range of ecological studies that came from ARL in its first two decades. However, a few general points can be developed, drawing on Shaver's 1996 review of arctic tundra research, and on reviews of research at ARL and NARL by John E. Hobbie and Jerry Brown, to support the idea that long-term ecological research was well represented within ecological science prior to 1970. Hobbie pointed out that studies of arctic freshwaters really went back to early arctic expeditions starting in the 1880s. After ARL was founded, studies of lakes and their flora and fauna were a prominent area of research, but this research, even in the supposedly simpler Arctic systems, required years to answer key questions. As Hobbie remarked in 1973, although there had been a remarkable increase of knowledge over the prior two decades, it would still require many years of research to answer some basic questions "such as control of chemical cycles or the factors controlling distribution of certain species" (Hobbie 1973, p. 163). He projected in 1973 that progress would require "adequate arctic laboratories and a group of investigators from every possible discipline" (Hobbie 1973, p. 163).

Study of life cycles of the lakes on the coastal plain was one area where research was advancing well in the 1950s. The coastal plain has many lakes and ponds that are shallow thaw basins in the permafrost; these are elongated and oriented in the same direction. The prevailing east-west winds appeared to be responsible for their shape and orientation; the long axes were perpendicular to the prevailing wind direction because erosion and thawing of permafrost were greatest on the sides. Max Britton in 1957 proposed a thaw-lake cycle hypothesis that proved helpful for later assessments of human impact on the arctic environment (Britton 1967). The shorelines of these lakes would erode in the summer during the thaw period. As the lakes grew larger, they would drain if they merged with a stream, another lake, or the seacoast.

Britton assumed that succession occurring in the drained lake bottoms would explain the diversity of soil and vegetation in the area. He proposed that frost mediates succession at the bottoms of these lakes, especially through "polygonization," or polygonal patterns that appeared in the bottom of drained lake basins. The patterns were caused by the formation of ice wedges in cracks that formed when the permafrost contracted as it cooled in winter. Repeated cracking and growth of ice wedges over many years can produce wedges several feet across at the top and extending more than thirty feet into the ground. These cracks formed in long lines which crossed each other, eventually forming rectangles or many-sided polygons. The polygonal landforms so produced could coalesce into small thaw ponds, and the ponds would grow by marginal erosion into larger thaw lakes.

Britton's studies of these thaw lake cycles revealed that the vegetation of ice-wedge polygons appeared "chaotic" in distribution, and that over a large region there were mosaics of communities that repeated themselves. The "apparent chaos" of the surface features, he noted, was "duplicated by equal chaos below ground where bewildering mixtures of organic and inorganic components occur" (Britton 1967, p. 111). He concluded that tundra landscapes were highly diverse, and that moreover there was nothing to indicate "progression of vegetation toward equilibrium states of great uniformity" (Britton 1967, p. 126), which was a challenge to prevailing ecological theories of succession that were based on the idea that the endpoint of a successional process was a state of relative equilibrium. Britton was circumspect in stating his challenge, noting that on the Coastal Plain his "tentative" conclusion was that "concepts of vegetation equilibrium must incorporate marked diversity as an end product". This work was important because it showed the role of natural change and disturbance in the arctic landscape over a relatively long time period. Such research on natural disturbance was valuable for evaluating how human impact from resource development would affect the environment, and how long it might take for the environment to recover from such impact (Shaver 1996a, pp. 20–21).

As an example of how Britton's work contributed to later studies of disturbance, W. Dwight Billings and K. M. Peterson re-evaluated Britton's hypothesis in the 1970s by returning to the locations Britton had studied (Billings and Peterson 1980). Britton had investigated succession in four lakes that had been artificially drained in 1950 in order to prevent flooding of nearby gas wells. Billings and Peterson returned to the same basins two decades later with three questions in mind. First, were the successional pathways following geomorphic or climatic changes predictable? The answer was "yes"; successional pathways were predictable following natural changes. Second, was vegetation change (or succession) linear, cyclic, or a combination of both? The answer was a combination of both, but they also found that the vegetation cycles observed at Barrow did not fit modern successional models. The reason was that most models assumed that physical changes in the environment were insignificant, whereas at Barrow, physical factors operating above and below ground were dominant and the physical environment determined which species could invade and when.

Third, when humans caused disturbances, did the ecosystem respond the way it did to natural disturbances? The answer here depended on the scale of the human disturbance, but they concluded that modern vehicular traffic had high potential for producing permanent damage, and that if large numbers of lakes were drained artificially, the "natural tundra ecosystem with its cyclic diversity would cease to exist" (Billings and Peterson 1980, p. 430). They further observed that disturbances elsewhere in the world could have major impact on arctic systems. Accumulation of carbon dioxide in the atmosphere, already recognized as a problem, along with projections of temperature rises in the Arctic, could lead to highly destructive effects. Decomposition rates would increase (releasing carbon dioxide into the atmosphere from peat) and there would be greater depth of thaw in the permafrost, which combined with a possible rise in sea level "could eliminate much of the coastal tundra." Thus Britton's work in the 1950s, revisited and revised 20 years later, culminated in a warning that climate change would cause the coastal tundra ecosystem to disappear. The study of disturbance would become one of the five core research areas of the LTER program that was just getting underway, and at the arctic LTER sites especially the study of the impact of climate change would become a dominant theme.

Another hypothesis, developed by Frank Pitelka and Arnold M. Schultz, provided predictions about the role of natural change and disturbance at short time scales. Pitelka first came to ARL in 1951, after Ira Wiggins invited him to study the breeding birds on the coastal tundra (Batzli 2006). Pitelka had just published an article on interspecific competition in hummingbirds, which in 1953 would earn him the prestigious Mercer Award from the Ecological Society of America. At ARL he began a research program on bird populations that continued for 30 years. As it happened, 1953 was a lemming population peak, and he also got interested in the role of small mammals in the arctic tundra ecosystem. Studies of lemming and vole populations, and their ecological relationship to bird and mammal predators, began in 1955 and lasted 20 years, but Pitelka was also able to draw on lemming studies that had started earlier, in 1949. As he remarked in 1957, only at Barrow was it possible "to make a proper beginning in ecological studies of arctic microtines requiring long-term records, because of the opportunities afforded by the Arctic Research Laboratory for continuing research" (Pitelka 1967, p. 156).

Pitelka recruited Schultz, an agronomist and plant physiologist who was then working at the California Agricultural Experiment Station in Berkeley, to study lemming-plant-soil relations. David Lack, an ecologist at Oxford University, had argued that microtine population levels were linked to food supply: when food supplies deteriorated, the populations crashed and did not recover until enough nutrients were available. Pitelka and Schultz developed this idea. Their "nutrient recovery hypothesis", published in 1964, was meant to explain the regular, dramatic changes in the abundance of lemmings every 3–4 years (Pitelka 1964; Schultz 1964). A lemming peak could only occur when a sufficient amount of vegetation had accumulated to support the lemmings' growth and reproduction. In consuming the vegetation, the nutrients sequestered in the plants would shift rapidly into the lemmings. As the populations peaked, the nutrients would return to the environment through the lemmings' urine and feces, and from the excretions of the predators eating the lemmings.

Pitelka and Schultz's hypothesis linked the cyclical population changes of lemmings and their predators to shifts in the availability of key nutrients. They argued that to explain these herbivore cycles it was necessary to understand other fluctuations connected to the quantity and quality of the food supply, for instance changes in plant growth, nutritional quality of the forage, decomposition rates, and changes in soil properties (Schultz 1964). Most importantly, their work showed that alternative hypotheses, which linked population cycles to intraspecific factors (such as the effects of crowding on the physiology, behavior, and genetics of the lemmings) were not likely to be promising, because they ignored the larger environmental context in which these cycles occurred.

As Shaver explained, Pitelka and Shultz's hypothesis "helped change our focus from individual organism-environment interactions to an integrated approach encompassing interactions among trophic levels as well as ecosystem-level element budgets" (Shaver 1996a, p. 20). It stimulated research on plant mineral nutrition and decomposition. Though later found to be over-simplified and wrong in some details, it "provided a starting point and was of heuristic value." In 1992, when Pitelka received the Eminent Ecologist Award from the Ecological Society of America, the citation noted that ecologists were still mining the data accumulated by Pitelka's research group at Barrow decades earlier.

Where the Navy's support of ARL "exemplified a successful, long-term approach to arctic research," in Britton's words, another arctic project funded by the Atomic Energy Commission, and also involving scientists at ARL, illustrated what could be achieved by a "rather massive, relatively short-term program" (Britton 1964a). This was the ecological assessment conducted to evaluate the possible impact of using atomic bombs to excavate a harbor on Alaska's west coast. This idea, which seems fantastical today and was never tried, was called Project Chariot, part of the broader Project Plowshare program that began under the Eisenhower administration to explore peaceful uses of atomic energy. Project Chariot was first proposed in 1957 and had the backing of Edward Teller, known as the father of the hydrogen bomb (O'Neill 1994). The project was to detonate several nuclear devices in Alaska to move 30 million cubic yards of material, carving out a harbor on the coast near Cape Thompson in northwestern Alaska. The project would not really be of great economic advantage to Alaska, for a harbor at that location would be ice-locked for 9 months of the year, but one purpose was to generate data for the next big project, which was to use nuclear weapons to carve out a sea-level canal in Panama (and which was abandoned after much debate).

John Wolfe, a plant ecologist from Ohio State University, was at the time the Chief of the Atomic Energy Commission's Environmental Sciences Division, which had been created in 1958. After Project Chariot was proposed, Wolfe pushed for major ecological studies to be done both before and after the nuclear detonations, in order to assess their impact. The first thing Wolfe did was approach Britton, whom he had known for 20 years, to get his advice about whom to select for the Committee on Environmental Studies for Project Chariot. Britton himself became a member of the committee. Some members of the committee were funded by the AEC, and the committee did not have the power to cancel the project, so there was uncertainty about what the role of this environmental assessment would really be.

Daniel O'Neill's historical analysis of this episode revealed that some of the scientists involved in this study were deeply unhappy with how Wolfe and the AEC accepted that the blasts could be done safely, and did not seem to evaluate the scientific evidence objectively (O'Neill 1994). A group of four dissenting scientists, who later referred to themselves as the "Gang of Four", included Leslie Viereck, who was assistant professor at the University of Alaska at Fairbanks. Viereck felt obliged to resign from the AEC contract because of his opposition to how the project was being run, and he quickly discovered that the University of Alaska's president no longer wanted to employ him. He returned to the university in 1963 by joining the Forest Service's Institute of Northern Forestry, located on the university campus; his research at the nearby Bonanza Creek Experimental Forest would provide an important foundation for the LTER established in 1987.

In the end, Project Chariot succumbed to determined opposition from scientists, environmentalists, and the local community, which led the AEC to suspend the project and end the bioenvironmental studies in 1962 (O'Neill 1994). But since the preliminary ecological studies had been done from 1959 to 1961, Wolfe pressed to keep the committee active so that the results could be published (Wolfe 1964). The studies were published in 1966 as *The Environment of the Cape Thompson Region, Alaska* (Wilimovsky and Wolfe 1966). This ecological project was a landmark in multi-disciplinary research: almost 100 scientists participated in the environmental studies, and the final volume had 71 authors. The Environmental Committee hedged in its assessment of the use of nuclear bombs for this particular project, believing that the ecological and human risks in this case appeared to be "exceedingly remote". However, the report did note that "massive nuclear undertakings involve far more than radiation, blast, seismic shock and heat – they involve also the total biota and physical environment and impinge abundantly on man himself" (Wilimovsky and Wolfe 1966, pp. vii-viii). One outcome of the study was the finding that some of the native communities already had high body burdens of radioactive cesium-137 from nuclear fallout from bomb tests, because they consumed reindeer and caribou that had fed on contaminated lichens (O'Neill 1994; Hanson 2001).

In 1969, at the dedication symposium for the newly named Naval Arctic Research Laboratory, Pitelka argued for the importance of tundra research, much as Irving and others had argued earlier in favor of "expeditionary physiology". Just as comparative physiologists thought that examining organisms in extreme environments was the ideal way to yield insights into basic biological processes and adaptations, so Pitelka argued that ecosystems in extreme environments, such as arctic tundra, were "particularly suited for comparative and analytic work about how ecosystems are organized and how they function. Indeed, tundra is a low-temperature extreme among ecosystem-types on the land areas of the earth and hence it assumes a special importance to the theory of ecosystems" (Pitelka 1969, p. 335).

That importance, he went on, justified pursuing faunistic and floristic work (taxonomy) on a tighter, local scale, but this kind of work had to be closely related to the questions that ecologists and physiologists were asking. He argued that to advance the field, "organized teamwork of ecologists with physiologists, geomorphologists, soil scientists, climatologists and others is essential" (Pitelka 1969, p. 337). Moreover, the recent discovery of oil in northern Alaska created a "Texas-size threat to a land doubtfully able to take it," which made it crucial to understand the ecology of normal tundra as the specter of dealing with the ecology of damaged tundra loomed. The discovery of oil in 1968 prompted renewed interest in arctic science within the federal government, which steered more funds into arctic ecosystem ecology.

7.5 The Impact of Oil Discoveries: IBP and Its Aftermath

Oil discoveries at Prudhoe Bay to the east of Point Barrow on Alaska's North Slope created the rationale for a new biome program in the Arctic as part of the IBP (Brown 2001). Pitelka was instrumental in organizing and getting support for the Tundra Biome Program, which was centered at Barrow because of its long heritage of ecological research, but included comparative studies at Prudhoe Bay, Eagle Summit in Alaska, and Niwot Ridge in Colorado (alpine tundra). Given the legacy of ecological research at Barrow, a small scale one-year pilot program was started there in 1969 and ran for a year, leading to NSF's approval of additional funding for research in 1971–1973. The biome research areas were on the coastal plain near Barrow and at Prudhoe Bay.

The early to mid-1970s saw general strengthening of the NSF's role in funding arctic research. In 1969, in response to the discovery of oil, Vice-President Spiro Agnew, in his capacity as chairman of the National Council on Marine Resources and Engineering Development, designated the NSF as the lead agency for the extension of arctic research. In April 1970, in response to pressure from the public and from industry, President Nixon designated arctic environmental research as a national priority. In July 1970, the NSF expanded its Office of Antarctic Programs to include arctic programs under the aegis of an Office of Polar Programs, which by the mid-1970s grew into the Division of Polar Programs. This meant that NSF established arctic research as a budget line-item beginning in fiscal year 1971. These funds supported a few large multi-disciplinary programs, including the Tundra Biome project.

Jerry Brown became the director of the Tundra Biome Study and although he was not then based in Barrow, he had many years of experience working there.

Brown had first gone to Barrow as an undergraduate from Rutgers University in 1957, returning to complete his Ph.D. thesis research on the study of soils and soil classification. In 1961 he took a position at the Cold Regions Research and Engineering Lab of the U.S. Army Corps of Engineers in Hanover, New Hampshire. With his extensive experience at Barrow it made sense to return there in the 1960s for studies of permafrost. During the IBP, the Barrow summer programs were directed on-site by Larry Tieszen, a plant ecologist and physiologist with an interest in cold adaptation, while John Hobbie supervised the aquatic sites. Hobbie's PhD research on mountain lakes was supported by ARL and he had also worked on lakes in Greenland, in the Antarctic, and in the Swedish far north. This long experience gave strength to the IBP program.

Scientists spent the summer of 1974 at the Mountain Research Station of the University of Colorado preparing the synthesis volumes, which were published in 1978 and 1980 (Brown et al. 1980; Hobbie 1980; Tieszen 1978). Since a portion of the biome project funds came from the petroleum industry, which provided unrestricted grants through the University of Alaska, those funds were used to enable scientists' families to stay at the research station for that summer. The project, involving hundreds of scientists and dozens of institutions and agencies, received crucial logistical support from the Office of Naval Research through the Naval Arctic Research Laboratory (Brown et al. 1980, p. xi).

Hobbie's IBP synthesis volume, *Limnology of Tundra Ponds: Barrow, Alaska*, published in 1980, was praised for being one of the "finest efforts" of the IBP, which the reviewer found ironic given that it treated "little more than three puddles on the tundra landscape," in sharp contrast to the "grand scale and sweeping prognostications" of the IBP (Kerfoot 1982). What was important about this study, the reviewer noted, was its experimental manipulations and attention to state-of-the art techniques. In the mid-1970s, as part of the IBP study, Hobbie broke new ground in the study of bacterial roles in nutrient cycling by applying techniques for making direct bacterial counts using fluorescent stains (Daley and Hobbie 1975; Hobbie et al. 1977). Bacterial decomposition controlled the cycling of nutrients in streams and lakes, but it was difficult using standard counting techniques to assess their numbers and level of activity, because they were hard to observe. By staining the bacteria with acridine orange, which binds with DNA or RNA and emits a fluorescent color, they were made visible and could be counted with an epifluorescence microscope. Using these new techniques transformed microbiological studies of both freshwater and marine systems.

Coinciding roughly with the end of the IBP, the energy policies of the U. S. government were being reviewed under the Nixon administration, especially in response to the Arab embargo on oil exports that extended from October 1973 to March 1974. The Atomic Energy Commission, headed by Dixy Lee Ray, was one casualty of this reorganization. The AEC had for years been involved in both the development and regulation of nuclear power, but it came under fire because it was responsible for regulating the same industry that it had helped to create. It was dissolved in October 1974 under the Ford administration, and its responsibilities were assumed by the Energy Research and Development Administration (ERDA) and the Nuclear

Regulatory Commission. ERDA in turn because part of the new Department of Energy, which was created in 1977 under the Carter administration. The Department of Energy became the leading government agency supporting ecological research in Alaska; this research was driven by concerns about the environmental impacts of energy-related development.

The oil discoveries at Prudhoe Bay prompted the construction of the trans-Alaskan pipeline running from Prudhoe Bay to Valdez in the south. Construction of the pipeline required building a gravel access road in 1974–1976 (the "Haul Road", later named the Dalton Highway), which ran from the north about half the length of the roughly 800 mile pipeline. That road opened access to the previously inaccessible interior regions of the North Slope (Reynolds and Tenhunen 1996). By the end of the IBP, ecologists were already looking for new sites on the North Slope, and the building of the Dalton Highway "suddenly created access to a magnificent environmental transect across the heart of northern Alaska" (Hobbie and Kling 2014, p. 5). Taking advantage of pipeline-related construction, ecologists quickly moved inland to exploit new research opportunities.

They were able to do this with support from NSF as well as from the Environmental Protection Agency and ERDA. Together, these agencies began funding a new program in June 1975 called Research on Arctic Tundra Environments (or RATE), which involved teams of scientists from fifteen universities. Research focused on manmade stresses on the tundra and deep lakes of Alaska's North Slope. At this time, NSF was one of a dozen federal agencies supporting arctic research, but its projects represented only 12 percent of the federal budget going toward arctic studies. Within NSF, five divisions supported arctic research, and of these the Division of Polar Programs was the most important for the North Slope ecological studies.

In 1976 the Polar Research Board of the National Research Council (part of the National Academy of Sciences) was asked to evaluate all arctic programs supported by NSF. The chair of the committee that made the evaluation was John Cantlon, an ecologist. Two people who had earlier advocated the creation of NARL, Lincoln Washburn and Laurence Gould, participated in the study, and Washburn in particular was actively promoting polar research (Washburn 1980). The Board's report, issued in 1977, recommended that NSF support a vigorous basic arctic research program, that the Division of Polar Programs continue to serve NSF for arctic research, and that NSF funding for arctic research be increased substantially (Polar Research Board 1977).

It is interesting that the committee's report, which recommended support for basic arctic research of all kinds, also included social science research and continued study of human adaptability. While no longer using the term "human ecology", the report noted that Alaska was facing urgent problems of resource development, land use, environmental protection, and labor-management relations. As a result of these rapid changes, Alaska was "emerging as an excellent, modestly scaled 'laboratory', reasonably well separated from the rest of the United States, for studying theories and techniques of modern social science. It offers an unparalleled opportunity to develop these data from which to follow, in real time, the simultaneous social, economic, cultural, and political changes" (Polar Research Board 1977, p. 92).

In addition, the arctic region was home to native populations that were examples of extraordinarily successful human adaptability to a harsh environment going back thousands of years. That success justified continued support of anthropological, archeological, and biomedical research on these communities. Moreover, the Alaska Native Claims Settlement Act of 1971 had settled aboriginal claims by compensating Alaska's natives for land taken by the federal government. Native groups now owned more than a tenth of the state's land, and many of those areas were in key regions of scientific interest. Those native communities were concerned about the stresses caused by accelerating development on their culture, their value systems, and their health and well-being, and they were interested in working with scientists to understand how to manage these changes. They valued scientific research for its role in helping with their land claims, and saw themselves as working in partnership with scientists.

In effect, this report reiterated the argument for wide-ranging study of human ecology, similar to the one used to promote the creation of the Arctic Research Laboratory in the 1940s, but without specific reference to the Navy's military interests. However, the report also noted that the disciplinary divisions of NSF had failed to maintain long-term multidisciplinary research involving humans, and therefore concluded that the principle responsibility for this research should reside with the Division of Polar Programs. The problem of how and where to incorporate the human dimension into ecology would be revisited in the later LTER program.

7.6 Research on Arctic Tundra Environments (RATE)

The RATE program ran two large integrated projects, one at Toolik Lake (led by John Hobbie) and another terrestrial site at Meade River (now Atqasuk) about 60 miles south of Barrow. These funds helped to extend the research done under the IBP. At Toolik Lake a research camp was established on a former airstrip and near the 400-person Toolik Lake Camp pipeline construction facility. The lake was chosen because it was deep, contained fish, and offered a contrast to the shallow ponds (which had no fish) that had been studied during the IBP (Hobbie and Kling 2014, p. 5). Two of the shallow pond IBP scientists, Vera Alexander and Robert Barsdate from the University of Alaska and their students, studied Toolik Lake year-round. Connected to these studies of the lake were studies of the nearby Kuparuk River, begun in the late-1970s by Bruce Peterson, who did pioneering stream experiments in his later Arctic LTER research. In 1984, when the Institute of Arctic Biology and the University of Alaska, Fairbanks, took charge of logistics at the camp, it moved to a disused gravel pit on the south shore of the lake and several new trailers were constructed.

Jerry Brown led the terrestrial study, and many of the scientists involved at both sites would become key advocates of long-term ecological research, not just in relation to the LTER program but also on a wider international scale. For instance, Patrick Webber, who was principal investigator with the terrestrial group both in the tundra biome project of the IBP and in the RATE program, promoted the importance of long-term studies along several fronts. He was the founding principal investigator of the Niwot Ridge LTER, an alpine tundra and sub-alpine forest site in Colorado that was among the first group of LTER sites created in 1980. He also helped develop the International Tundra Experiment (ITEX), a Pan-Arctic experiment that started in the late-1990s to assess the effects of long-term warming on plant growth.

Terrestrial ecologists also started using the Toolik site for studies of nutritional controls over plant growth. Two of the first terrestrial ecologists to work at Toolik were Gus Shaver and F. Stuart (Terry) Chapin III. Shaver had been a graduate student in the Tundra Biome project, receiving his Ph.D. from Duke University in 1976, for work on the ecology of roots and rhizomes in graminoid plants in the Alaskan coastal tundra. Chapin received his Ph.D. from Stanford University in 1973, with a dissertation that explored how cold-adapted plants in the arctic took up phosphorus, an important nutrient and often a limiting factor on plant growth. His advisor was Harold (Hal) Mooney, but since his fieldwork was in Alaska, he benefited from informal mentoring from Keith Van Cleve, who let Chapin use his forest-soils laboratory at the University of Alaska. The Institute of Arctic Biology in Fairbanks provided facilities for Chapin's experimental work, and he did fieldwork at Barrow with logistical support from NARL as part of the Tundra Biome project of the IBP (Chapin 1974).

In the late 1970s and early 1980s Chapin continued to work with Van Cleve, along with Ted Dyrness and Leslie Viereck, on the effects of soil temperatures on ecosystems, as well as studying forest succession at the sites that Viereck had established at Bonanza Creek in the 1960s, when he was working for the Forest Service. This research laid the intellectual foundation for the Bonanza Creek LTER proposal, and Van Cleve served as principal investigator for the LTER project from its start in 1987 until the mid-1990s. Both Shaver and Chapin also became principal investigators for the Arctic and Bonanza Creek LTER projects, and here too, their work built on previous studies. As Shaver noted, "In 1978 … Terry and I began the long term work on plant growth and its controls that led to the present LTER terrestrial experiments. Many of the current LTER experimental plots were set up in 1981, 6 years before the Arctic LTER was funded."[5] These comments underscore the importance of having a long, continuous research program in place to anchor the later LTER projects. As Tom Callahan noted in his 1984 article about the new LTER program, there was already ample recognition within the scientific community, as well as within government agencies, of the value of long-term research.

In the early 1980s, a committee of the National Research Council, chaired by Harold Mooney of Stanford University, evaluated the Department of Energy's terrestrial environmental research programs, recommending in 1982 that the Department of Energy support long-term intensive studies of environmental effects of energy-related activities in the Arctic. This recommendation led to the

---

[5] Gus Shaver, personal communication, 29 August 2018.

establishment in 1983–1984 of a program known as "R4D", where the letters stood for "response, resistance, resilience, and recovery" in arctic ecosystems that have been subject to "disturbance." (R4D was also the name of the military version of the DC-3 airplane that was used at Barrow.) The R4D program's site was the Imnavait Creek watershed, near Toolik Lake in the southern part of the northern foothills of the Brooks Range, and about 130 miles south of Prudhoe Bay. Preliminary conclusions from the R4D project, which ran for 8 years, were summarized in the 1996 volume edited by James F. Reynolds and John D. Tenhunen, *Landscape Function and Disturbance in Arctic Tundra*, in which they noted that ecological research in the Alaskan tundra had been "directly and indirectly linked to social and economic interests since World War II" (Reynolds and Tenhunen 1996, p. 419).

In 1980, just as the IPB synthesis volumes were being published and as the LTER program was coming into being, the Navy left NARL. The local community at Barrow was interested in running the facility, partly to serve community needs, and in 1989 NARL was formally taken over by the Ukpeagvik Iñupiat Corporation (UIC) (Kelley and Brower 2001). When the Arctic tundra site joined the LTER network in 1987, the Arctic Research Laboratory at Barrow was at a low point. However, as Shaver noted, its legacy remained important for the LTER site, for many of the ecologists who worked on the LTER were trained while working on the Tundra Biome Study at Barrow years earlier (Shaver 1996a). The LTER research base was at the Toolik Lake Field Station, which by then was managed by the Institute of Arctic Biology that Laurence Irving had founded at the University of Fairbanks in the 1960s. After joining the LTER program, support from NSF enabled the Toolik station facilities to be built up substantially, starting in 1994.

7.7 The Arctic LTER: Continuity with the Past, New Directions

These traditions, as Shaver has noted, provided the training ground for the scientists who became principal investigators for the Arctic LTER, and they joined the LTER with many years of experience in arctic ecology under their belts. The institutional base for many of them was The Ecosystems Center located at the Marine Biological Laboratory at Woods Hole, Massachusetts, and this became the Arctic LTER project's base as well. Shaver had been on the scientific staff at the Center since 1979 and had been engaged in arctic research for several years. In 1982 John Hobbie became director of The Ecosystems Center. Seven of the fourteen investigators supported by the NSF grant for the LTER were on the Center's staff: in addition to Hobbie and Shaver they were Brian Fry, Anne Giblin, Knute Nadelhoffer, Bruce Peterson, and Ed Rastetter. As Hobbie has remarked, the shift to the LTER did not involve personal changes for him. He had been doing "large-project long-term research for 20 years in estuaries and in the Arctic", and had been "testing ways to operate the ARC [Arctic LTER] site for many years before its establishment as part of the LTER program in 1987" (Hobbie 2016, p. 95).

The Center, as it turns out, also had roots in Cold War ecological science. It was founded in 1975 by George Woodwell, who came from the Brookhaven National Laboratory on Long Island, New York, where he had been conducting ecological research since 1961. At Brookhaven, Woodwell had organized long-term studies of the effect of ionizing radiation on plant and animal communities, and those studies made its oak-pine forest one of the best studied ecological systems in the world (Woodwell 1967). Here again we see how Cold War activities, in this case nuclear weapons testing, motivated research programs at the national laboratories and gave rise to long-term ecological studies. Woodwell's research would take a direction that would later make it highly relevant to arctic studies, for he was one of the earliest scientists to draw attention to the dangers of climate change as a result of human activity.

At Brookhaven, Woodwell was looking for possible changes in the metabolism of the forest (its photosynthesis and respiration) and to this end he designed and built equipment that could measure carbon dioxide concentrations in the atmosphere (Woodwell 2016). The Brookhaven group started measuring CO_2 concentrations both locally and generally, and kept in touch with Charles David Keeling, who had been monitoring carbon dioxide levels in the atmosphere since 1957. Keeling's long-term monitoring, showing ever-rising levels of atmospheric carbon dioxide, would become one of the clearest signals of impending climate change, and Woodwell recognized those signs by the 1970s. When Billings and Peterson drew attention to the impact of human activity on the arctic, it was Woodwell's *Scientific American* article on the subject that they cited (Woodwell 1978).

Woodwell was unequivocal in stating the danger in 1978: "There is almost no aspect of national and international policy that can remain unaffected by the prospect of global climatic change. Carbon dioxide, until now an apparently innocuous trace gas in the atmosphere, may be moving rapidly toward a central role as a major threat to the present world order" (Woodwell 1978, p. 43). (While he and other scientists had the ear of the Carter administration in the 1970s, the Reagan administration ignored these warnings in the 1980s and undid many of the initiatives undertaken by Carter.) Woodwell brought these concerns to The Ecosystems Center, where he built a scientific and educational program that he hoped would respond to a set of interlocking challenges caused by human impact on the biosphere. These included global warming and sea level rise, changes in the circulation of nitrogen and sulfur, increase in extinction of species, and acid rain (Ecosystems Center 1979).

How would the world's ecosystems respond to such profound changes? These same questions carried over into the Arctic LTER project, which found a natural home at The Ecosystems Center, where there was an active program on arctic Alaska already underway in the 1970s. The Arctic would experience significant warming over the next three decades, and the effects of climate change would be a prominent question. Prior to the LTER in the 1980s, Hobbie and his colleagues at the Center were working on the analysis of the sources of carbon dioxide in the atmosphere, a problem that had generated controversy (Hobbie et al. 1984).

Geochemists and ecologists had different views of why carbon dioxide was building up in the atmosphere. Fossil fuel use was implicated as an important source, and was the focus of geochemical calculations, but the release of carbon dioxide from deforestation and changes in land use was not well understood, and that was where ecological studies were needed. The problem Hobbie and his colleagues were working on was how to identify what kinds of studies and measurements were needed to reconcile the calculations of these two groups. These kinds of big questions, involving cross-disciplinary communications and reconciliation of data coming from different scientific communities, were very much at the core of the Center's work.

But the Arctic LTER site was also suited to studying many traditional ecological problems, working toward linking different fields within ecology itself. Animals, plants, and soil were all within the purview of the terrestrial LTER group, and the whole LTER was organized around four interest groups focused on terrestrial studies, lakes, streams, and land-water interactions. The location of the Arctic LTER in the Toolik Lake region offered a contrast to the other Low Arctic regions studied at Barrow during the IBP. The site, being farther south, was in warmer foothills, where tundra, deep lakes, and rivers could be compared to the better-known cooler coastal regions. As Hobbie and Kling (2014, p. 5) noted, "later other advantages of the site became obvious: small lakes and headwater streams were nearby, and a complex glacial history provided a variety of soils and vegetations." While the Toolik site was warm compared to other arctic vegetation zones, and therefore had higher productivity and diversity, the ecological processes and relationships observed there were common to all systems throughout the Arctic. Research conducted at Toolik was thus relevant to understanding all arctic ecosystems.

The site was also unusual, compared to Canadian and European arctic locations, in that there was relatively little airborne pollution. European sites received pollution from industries to their south, and some of the polluted air moved across the pole to Canadian sites. The effects of pollution from industrial processes elsewhere (such as lead and mercury contamination) were measured, but the LTER site represented a nearly pristine system, physically isolated, with the nearest permanent native Alaskan settlement, Anaktuvik Pass, 68 miles away and only accessible by air.

The recent disturbances stemming from road and pipeline building did, however, become subjects of the LTER project, since the study of disturbance is one of the five core research areas common to the LTER network. The gravel mining required to construct the highway caused permafrost to melt, exposing the soil to weathering and affecting the chemistry of nearby streams even 30 years later. Fire, which is relatively rare in the tundra, has not been totally absent. A dry summer in 2007 combined with lightning strikes on the North Slope created a large fire northwest of the lake, resulting in the release of large quantities of carbon to the atmosphere. John Hobbie and George Kling, writing in the Arctic LTER synthesis volume of 2014, noted that "disturbance due to wildfire will likely become another major disturbance of the tundra landscape in the same way as wildfire dominates large-scale disturbance in the boreal forest" (Hobbie and Kling 2014, p. 308).

As the Arctic LTER synthesis volume makes clear, the evolving story leading to LTER represents no discontinuity with the past in terms of research. It was the outcome of a decades-long process of development originating in the early Cold War, which enabled ecologists interested in Arctic environments to build a robust research tradition over three decades. Funding streams, however, remained short to mid-term in length. In 1980, NSF took the next logical step by creating a program that would allow for funding at time scales that might extend to decades, although at the beginning of the LTER program no one knew how long the program or individual projects would last. As Shaver noted, doing long-term research became much easier compared to the days when "projects with long-term goals had to be renewed and refreshed on a 3-year cycle" (Shaver 2016, p. 102).

From the perspective of Arctic LTER scientists such as Shaver, the changing research agenda that I have described has unfolded historically toward a certain endpoint, that endpoint being the ecosystem approach of the LTER program. The long-term shift in scientific perspective went from organism-centered to ecosystem-centered ecology, occurred over three decades, and gave rise to a different way of understanding the ecosystem. The synthesis volume on the Arctic LTER characterizes this evolutionary process as a gradual change in conceptual approach, driven by shifts in research emphasis as ecosystem ecology emerged and grew dominant (Shaver et al. 2014).

In the 1960s and earlier, research focused on the problem of how species were adapted to cold climate and to a short growing season. In the cold-dominated environment of the Arctic, the key question was how species survived in such an extreme landscape. An example of this approach are the studies by H. A. Mooney and W. Dwight Billings (1961) on arctic and alpine plants that have wide distribution, such as alpine sorrel (*Oxyria digyna*). Here the central question was how species were able to grow in diverse environments, that is, what physiological characteristics were of adaptive significance. These questions reflected research problems that were being explored in evolutionary ecology and physiological ecology in the 1950s and 1960s, and were centered on species and populations, not ecosystems.

Given that earlier emphasis on physiological ecology, two other problems received less attention: how species interacted through food webs, and whole-system biogeochemical cycles. A shift in perspective toward food webs and nutrient cycles was evident in Pitelka's work on population regulation in the 1950s, but the study of biogeochemical cycles got a boost from the International Biological Program in the 1970s. As the experimental focus shifted to the study of nutrient cycles, ecologists found that slow inputs and turnover of nitrogen and phosphorus were important limiting factors in tundra ecosystems. At the same time, species interactions were seen to be more important in determining the composition of ecological communities, while the composition of these communities in turn affected biogeochemistry as well as vegetation, soil, and air temperature.

The historical trajectory described here thus went from a focus on how species adapted to extreme environments, to the study of how species affected ecosystem properties. One conclusion of these studies was that the species effects and the slow change in element cycles made "the tundra ecosystem much more resistant and resilient to climate warming than was expected when the research at Toolik Lake began" (Shaver et al. 2014, p. 133). Ecologists described this shift as a "major change in how ecologists think about tundra ecosystems", and one that opened the way "for a rich new era of whole-system experimentation, observation, and understanding" (Shaver et al. 2014, p. 134). This is a story of progress and greater sophistication as ecologists have come to understand patterns and processes in more detail, over longer time periods and at larger scales.

7.8 The Bonanza Creek LTER: Human Adaptability Revisited

Scientists who worked at the Bonanza Creek LTER, which joined the network at the same time as the Arctic LTER, also benefited from the long research experiences at Barrow, especially during the IBP. As the IBP ended, scientists from both the Coniferous Forest and the Tundra biome projects combined forces to develop a project on the taiga (forest) biome at Washington Creek, just outside of Fairbanks. The leaders who designed the intellectual framework for the Bonanza Creek LTER, in Chapin's view, were Leslie Viereck, Keith Van Cleve, and Ted Dyrness.[6] Here, partnership with the U. S. Forest Service was extremely important, and remains so today (Geier 1998).

In 1963 the Forest Service established the Bonanza Creek Experimental Forest on state-owned land leased to the Forest Service, along with a new laboratory on the campus of the university in Fairbanks. This was where Viereck settled after his unhappy experience with Project Chariot. Viereck, who continued to work at the LTER site after his retirement from the Forest Service, drew on his understanding of forest succession around Fairbanks, as well as on the interdisciplinary studies at Cape Thompson that had been part of Project Chariot. Van Cleve had interests in basic ecosystem theory, and took up the "state factor theory" developed in the mid-twentieth century by Hans Jenny, a leading Swiss-American soil scientist at the University of California-Berkeley (Jenny 1980). Jenny conceived of soil ecosystems as a function of five interdependent "state factors" (time, climate, parent material, topography, and biota). Dyrness had studied forestry in the Pacific Northwest, where he had worked with Jerry Franklin, who had a leadership role both in the Coniferous Forest Biome project of the IBP and the LTER program. Before coming to Fairbanks in 1974, Dyrness had extensive experience at the H. J. Andrews Experimental Forest in Oregon. These diverse multidisciplinary perspectives on forest ecosystems combined to provide the intellectual framework for the Bonanza Creek LTER. The two main research sites are at the Bonanza Creek Experimental Forest (about 20 miles southwest of Fairbanks) and the Caribou-Poker Creeks Experimental Watershed (about 30 miles north of Fairbanks). Research questions focused on understanding the process of ecological succession, especially

[6] Terry Chapin, personal communication, 17 February 2019.

succession in the floodplain of the Tanana River, and succession in upland forests following wildfires.

The career of Terry Chapin, now distinguished professor emeritus at the University of Alaska, Fairbanks, provides a good example of how initial interest in problems of physiological ecology gradually opened up broader questions and led to a wider, ecosystem-level perspective. Eventually that wider perspective also raised the question of how social and natural systems interact. At the Bonanza Creek LTER, we can see an interesting return to the question of human adaptability, the problem that motivated the founding of the Arctic Research Laboratory back in the 1940s, but now reimagined in the era of climate change.

Chapin joined the faculty of the Institute of Arctic Biology at the University of Alaska immediately following his Ph.D. and remained there through the years when the LTER program started and when the two Alaskan sites were added in 1987. In 1989 he left Fairbanks to take a position as Professor of Integrative Biology at the University of California, Berkeley, then rejoined the University of Alaska as Professor in 1996, when he also became principal investigator for the Bonanza Creek LTER, a position he held until 2010. Throughout this period he maintained his research interests in arctic plants and physiological problems, but his perspective gradually broadened.

In a profile written shortly after he was elected to the National Academy of Sciences in 2004 (the first Alaskan elected to the Academy), Chapin described his intellectual trajectory over about three decades (Downey 2006). Having studied plant nutritional adaptations to low temperature during the 1970s, he realized during a sabbatical in 1979–1980 that he did not understand how plants adapted to low nutrient availability. Pursuing this problem led him to conclude that plants conserved nutrients or prevented loss of nutrients, and this in turn caused him to think about how plants interacted with and affected their environments, and therefore to think about ecosystem processes. During 1979 and 1980 he and Gus Shaver designed field experiments in the arctic tundra, in the area close to Toolik Lake, to learn what environmental factors limited the species in an entire community (Chapin and Shaver 1985). The study showed that no single factor limited growth of all species within these communities. The experiments involved using small plastic greenhouses to increase the air temperature, and Chapin drew a connection to the effects of climate warming on growth (Downey 2006).

The two Alaskan LTER projects joined the LTER network at the time that global climate change, which would likely occur more rapidly at high latitudes, was being recognized as a serious problem. Ecologists, alarmed by the intensifying impact of human population growth on the Earth's systems, became more vocal in their advocacy of basic research, which they viewed as a priority if we were to prevent collapse of entire ecological systems. The Ecological Society of America started an ambitious strategic planning process in 1988, culminating in "The Sustainable Biosphere Initiative," launched in 1991, which focused on three research priorities: global climate change, biological diversity and its effects on ecological systems, and understanding how to protect both natural and managed ecological systems (Lubchenco et al. 1991; Risser et al. 1991).

Chapin turned his attention to these problems as well. During the 1990s he chaired the National Academy of Sciences global change working group on terrestrial ecosystems, and served on the Science Steering Committee of an international committee on Global Change in Terrestrial Ecosystems. By the mid-1990s Chapin was rethinking the problem of ecosystem sustainability, especially in light of global environmental changes, including climate change (Chapin et al. 1996). As part of a special issue in *Science* published in July 1997 on the human domination of Earth's ecosystems, Chapin and his colleagues called attention to how changes in abundance of species were affecting the structure and function of ecosystems (Chapin et al. 1997; Vitousek et al. 1997).

These kinds of problems led Chapin, along with a growing community of ecological thinkers in the 1990s, to the realization that the problem of sustainability—understood as preventing the collapse of ecological systems—required analysis of the interdependence of ecological and social systems. As Chapin and coauthor Gail Whiteman noted in 1998: "It is increasingly clear that products of the natural environment strongly shape the functioning of social systems, and that human activities influence the activity of all natural ecosystems. These interactions between the social and ecological systems must influence the long-term sustainability of each system" (Chapin and Whiteman 1998).

Alaska's boreal forest, they noted, was a perfect example of the need to examine the relationship between ecological and social systems, and the Bonanza Creek LTER became one of the first sites to consider the forest as a coupled social and ecological system. While Alaska's interior was mostly sparsely populated outside of Fairbanks (with a population of just over 30,000), changes in the boreal forest would have far-reaching effects on the people living there, many of whom still depended on the land for food and survival. These changes included "modification to the biota, soil stability, hydrologic regime, forest productivity, fire regime, insect outbreaks, recreational opportunities, and many ecosystem goods and services" (Chapin et al. 2006a, p. 4). Developing tools for effective forest management also entailed explicitly recognizing that humans were part of the ecosystem, making it essential to engage with broader communities (Sparrow et al. 2006).

As Chapin assumed the leadership of the Bonanza Creek LTER in 1996, he and his colleagues also took up the challenge of developing a new kind of human ecology. In 1997 the LTER held a workshop on the ecological role of forest harvest, bringing together state and federal land managers, native groups, commercial interests and ecologists.[7] That workshop led to a new conceptual model of human activity as an integral component of the boreal forest. In the inaugural article in the *Proceedings of the National Academy of Sciences* following Chapin's election in 2004, he and his colleagues placed the problem of human adaptability at center stage. They were responding to the realization that social and ecological systems were changing in a directional way, and sought to identify a suite of policy strategies

[7] 1998–2000 NSF Final Report, website of Bonanza Creek Long Term Ecological Research, Reports and Proposals, at: http://www.lter.uaf.edu/publications/reports. Accessed 14 January 2019.

designed to enhance the sustainability of these systems (Chapin et al. 2006b). Along with other leading ecologists, these ideas became the basis for promoting "Earth Stewardship," defined as the "active shaping of the trajectories of change in coupled social-ecological systems at local-to-global scales to enhance ecosystem resilience and promote human well-being" (Chapin et al. 2011). ESA's President, Mary Power, first articulated the goal of promoting Earth Stewardship during her term in 2009–2010. Chapin, who followed her as President in 2010–2011, followed suit, as did Steward T. A. Pickett, president in 2011–2012.

In numerous publications in the early 2000s, Chapin and colleagues built on earlier initiatives, including the results of long-term studies in Alaska's boreal forest, to develop a comprehensive vision of sustainability in the era of climate change (Chapin et al. 2009). Central to that vision was the need to adopt an integrated social-ecological conceptual framework. It included efforts to integrate Western science with traditional ecological knowledge, and gaining insight from the way that Indigenous Peoples have always viewed humans as part of ecological systems (Sparrow et al. 2006). Thus the Bonanza Creek LTER came full circle, returning to the problem of human adaptability—the question that had preoccupied the original investigators at the Navy's arctic laboratory, but now made urgent not by the threat of war with the Soviets, but the threat of climate change.

7.9 Conclusions: Reflections on the Nature of Ecology

There was considerable momentum building toward long-term ecological research in the Arctic for several decades prior to the formal establishment of the LTER program. This momentum owed much to the fact that research in the Arctic directly addressed national needs, including military concerns about defending against a Soviet attack, as well as meeting the energy needs of the country. Scientists used these opportunities to pursue an extraordinarily diverse set of research projects, drawing also on the local knowledge of the indigenous community. The result, as Irving noted, was that Arctic Alaska went from being "a blank in knowledge to one of the well known regions of the world" by the late 1960s (Irving 1969, p. 331). The origins of LTER go back in this case to the early years of the Cold War, and reflect the early understanding that support of science, including curiosity-driven science, would benefit the nation's military and economic goals. These military and economic needs created opportunities that ecologists could creatively exploit. Individual leadership and vision were very important: people fought for their visions of science and did their best to promote long-term studies when opportunities arose.

Some of the ecological projects undertaken at ARL became the objects of long-term study, such as Pitelka's work on birds and rodents, while others, such as MacGinitie's studies of benthic fauna, did not generate long-term study in the way he had initially hoped. The aspiration of doing long-term research was there from the start, even though it was not always possible to find support for long-term studies. But the Navy's long-term support of science at ARL helped to strengthen the discipline of ecology in important ways.

The research that preceded the LTER program was wide-ranging and opportunistic, in the sense that research questions reflected the various opportunities afforded by the arctic environment and species living there. At ARL, foci of research shifted with the changing priorities of the laboratory's directors, but the ecological approaches also evolved as the discipline of ecology evolved. Physiological ecology dominated in the early years, while population ecology and ecosystem ecology developed later as these fields within ecology grew and achieved coherence. With the early emphasis on physiological ecology as well as anthropology, human ecology—broadly conceived as human adaptation to extreme environments—could find a place at ARL alongside the ecological study of other animals, plants, and soil. The fact that the laboratory supported ecological research in the broadest sense meant that scientists working there were contributing to the growth of the discipline of ecology on many fronts. The Arctic Laboratory thus served as an engine for the growth and expansion of the discipline of ecology broadly speaking. Its history illustrates a form of evolutionary divergence or branching over time, as new ecological fields developed and new problems arose. During the long history of arctic ecology leading up to the LTER projects, ecology advanced in many directions, and several sub-fields within the discipline grew stronger.

This expansive process was important in giving arctic biology greater authority and visibility, helping to link ecology with other disciplines, and establishing its presence in new institutions. As Irving noted in 1969, research at the Navy's Arctic Research Laboratory "stimulated the rapid growth of biological research at the University of Alaska, and in the Arctic Health Research Center" (Irving 1969, p. 331). A key reason, he argued, was the support of relatively long-term research, for the Laboratory "provided means whereby many scientists, be they young or old, could count upon the continuity of their studies through enough years to serve as important parts of their scientific careers" (Irving 1969, p. 331). This process benefited ecology too, and what mattered was not that all ecologists adopted the same perspective or addressed the same core areas, but that a research community could be built and continue operating in the same region for decades, providing opportunities for ecological research across many fields of study. I suggest that this form of disciplinary strengthening, achieved through the maintenance of this evolving research enterprise in the Arctic, was necessary for the success of the later LTER program.

Human ecology, or the problem of human adaptability, was seen as an important component of the Arctic Research Laboratory's mission, in part because it could appeal to the Navy's interest in supporting a military force in the north, but in part because biologists and anthropologists were genuinely interested in how humans (like other organisms) adapted to their environments. By the time we reach 1980 and the origins of the LTER program, human ecology was not an integral part of the program, although humans were present in an indirect way, because the study of disturbance could involve the study of human impact on the structure and function of ecosystems. There are good reasons to study natural systems, for example to produce good baseline data to assess human impact on the environment. Arguably human ecology belongs to the disciplines of anthropology and sociology, but not to ecology, and therefore should not be funded through NSF's biological divisions. These are all valid arguments, and they reflect the difficulty of overcoming modern disciplinary barriers to achieve a truly holistic interpretation of an ecological system.

However, it is interesting to see how the question of human adaptability, now seen as a problem in the context of climate change, and reimagined as Earth Stewardship, resurfaced in the Bonanza Creek LTER. There, ecological leaders have incorporated the problem of human adaptability into a new vision of what ecology is about as a science, and what its mission to society should be. This vision insists that social and ecological systems must be understood as interdependent, and values the integration of traditional ecological knowledge with modern science. Today, discussion of the need for socioecological research is important for many LTERs (Hobbie et al. 2003). However, it is not important enough to be designated a core research area, which suggests it is still marginal to how ecologists conceive of or define their discipline. At ARL, in contrast, human ecology was very much a central interest and a justification for support from the outset. As ecology evolved and moved toward an ecosystem-level analysis, it relegated the problem of human adaptability to disciplines outside of ecology, such as anthropology or sociology or the public health disciplines. Now the need to develop an integrated or "human" ecology has resurfaced as an urgent challenge.

One cannot rewind the tape of history, but one can still ask whether, in reflecting on what has happened, there is any need to reimagine what ecology should be about, or how one might define it or identify its boundaries in relation to other disciplines. Does that moment of exclusion, which ecologists are now trying to remedy by thinking of ways to link natural and social systems, represent a problem? If we start to look at humans in a different, more direct way, would we still consider that study to be part of ecology? Do we care about human adaptability, our greatest ecological challenge in the coming era of climate change? The concept of "Earth Stewardship," with its multidimensional and holistic approach to understanding human-environment relationships, certainly suggests that ecologists today are acutely aware of the need to take a very broad view of ecology in order to achieve the kind of transformations needed to sustain ecological systems.

These are partly questions about how to define ecology as a science, and where it draws its limits and how it links to other disciplines. These questions will be raised in other chapters of this book. What is interesting about the LTER program, apart from its contributions to the advancement of ecosystem ecology, is that long-term place-based studies of this kind raise questions about the definition of the science, and force one to think about whether there is a need for fundamental rethinking of goals. This book explores some of those deeper questions.

Acknowledgments I thank Gus Shaver, John Hobbie, and Terry Chapin for sharing reminiscences, clarifying scientific points, and fleshing out biographical details, and Robert B. Waide for several editorial suggestions that have improved the chapter's arguments.

References

Anonymous. 1947. Board to plan research. *Science News-Letter* 52: 119.

Arctic Institute of North America. 1946. *A program of desirable scientific investigations in Arctic North America*. Montreal: Arctic Institute of North America.

Aronova, E., K.S. Baker, and N. Oreskes. 2010. Big science and big data in biology: From the international geophysical year through the international biological program to the Long Term Ecological Research (LTER) Network, 1957-present. *Historical Studies in the Natural Sciences* 40: 183–224.

Batzli, George O. 2006. Frank A. Pitelka, 1916-2003. *Arctic* 59: 91–93.

Benson, C.S. 2007. Albert Lincoln Washburn, 1911-2007. *Arctic* 60: 212–214.

Billings, W.D., and K.M. Peterson. 1980. Vegetation change and ice-wedge polygons through the thaw-lake cycle in arctic Alaska. *Arctic and Alpine Research* 12: 413–432.

Bocking, Stephen. 1995. Ecosystems, ecologists, and the atom: Environmental research at Oak Ridge National Laboratory. *Journal of the History of Biology* 28: 1–47.

Brewster, Karen. 1997. Native contributions to Arctic science at Barrow, Alaska. *Arctic* 50: 277–288.

Britton, Max. 1964a. Arctic biology. *Bioscience* 14: 11–13.

———. 1964b. ONR: Arctic research laboratory. *Bioscience* 14: 44–48.

———. 1967. Vegetation of the Arctic tundra. In *Arctic biology: Ten papers presented at the 1957 and one presented at the 1965 biology colloquium at Oregon State University*, ed. H.P. Henson, 67–130. Corvallis: Oregon State University Press.

———. 2001. The role of the Office of Naval Research and the International Geophysical Year (1957-1958) in the growth of the Naval Arctic Research Laboratory. In *Fifty more years below zero: Tributes and meditations for the Naval Arctic Research Laboratory's first half century at Barrow, Alaska*, ed. D. Norton, 65–70. Fairbanks: University of Alaska Press.

Brown, Jerry. 2001. NARL-based terrestrial bioenvironmental research including the U.S. IBP Tundra Biome Program, 1957 to 1997. In *Fifty more years below zero: Tributes and meditations for the Naval Arctic Research Laboratory's first half century at Barrow, Alaska*, ed. D. Norton, 191–200. Fairbanks: University of Alaska Press.

Brown, Jerry, P.C. Miller, L.L. Tieszen, and F. Bunnell, eds. 1980. *An arctic ecosystem: The coastal tundra at Barrow, Alaska*, U.S. IBP Synthesis Series, 12. Stroudsberg: Dowden, Hutchinson & Ross.

Callahan, J.T. 1984. Long-term ecological research. *Bioscience* 34: 363–367.

Catling, P.S. 1953. *Hopkins men at top of world for Navy*. Baltimore Sun, March 1.

Chapin, F.S., III. 1974. Morphological and physiological mechanisms of temperature compensation in phosphate absorption along a latitudinal gradient. *Ecology* 55: 1180–1198.

Chapin, F.S., III, and G. Shaver. 1985. Individualistic growth response of tundra plant species to environmental manipulations in the field. *Ecology* 66: 564–576.

Chapin, F.S., III, and G. Whiteman. 1998. Sustainable development of the boreal forest: interaction of ecological, social, and business feedbacks. *Ecology and Society* 2 (2): article 12.

Chapin, F.S., III, Margaret S. Torn, and Masaki Tateno. 1996. Principles of ecosystem sustainability. *American Naturalist* 148: 1016–1037.

Chapin, F.S., III, B.H. Walker, R.J. Hobbs, D.U. Hooper, J.H. Lawton, O.E. Sala, and D. Tilman. 1997. Biotic control over the functioning of ecosystems. *Science* 277: 500–504.

Chapin, F.S., III, J. Yarie, K. Van Cleve, and L.A. Viereck. 2006a. The conceptual basis of LTER studies in the Alaskan boreal forest. In *Alaska's changing boreal forest*, ed. F.S. Chapin III, M.W. Oswood, K. Van Cleve, L.A. Viereck, and D.L. Verbyla, 3–11. New York: Oxford University Press.

Chapin, F.S., III, A.L. Lovecraft, E.S. Zavaleta, J. Nelson, M.D. Robards, G.P. Kofinas, S.F. Trainor, G.D. Peterson, H.P. Huntington, and R.L. Naylor. 2006b. Policy strategies to address sustainability of Alaskan boreal forests in response to a directionally changing climate. *Proceedings of the National Academy of Sciences of the United States of America* 103: 16637–16643.

Chapin, F.S., III, S.T.A. Pickett, M.E. Power, R.B. Jackson, D.M. Carter, and C. Duke. 2011. Earth stewardship: A strategy for social-ecological transformation to reverse planetary degradation. *Journal of Environmental Studies and Sciences* 1: 44–53.

Chapin, F.S., G.P. Kofinas, and Carl Folke, eds. 2009. *Principles of ecosystem stewardship: Resilience-based natural resource management in a changing world*. New York: Springer.

Coleman, David C. 2010. *Big ecology: The emergence of ecosystem science*. Berkeley: University of California Press.

Daley, R.J., and J.E. Hobbie. 1975. Counts of aquatic bacteria by a modified epifluorescence technique. *Limnology and Oceanography* 20 (5): 875–882.

Dawson, William R. 2007. Laurence Irving, an appreciation. *Physiological and Biochemical Zoology* 80: 9–24.

Downey, P. 2006. Profile of F. Stuart Chapin III. *Proceedings of the National Academy of Sciences of the United States of America* 103: 16634–16636.

Ecosystems Center. 1979. *Annual report*. Woods Hole: Marine Biological Laboratory.

Elsner, Robert. 2001. Cold adaptations and fossil atmospheres: Polar legacies of Irving and Scholander. In *Fifty more tears below zero: Tributes and meditations for the Naval Arctic Research Laboratory's first half century at Barrow, Alaska*, ed. D. Norton, 77–80. Fairbanks: University of Alaska Press.

Farish, M. 2006. Frontier engineering: From the globe to the body in the Cold War Arctic. *Canadian Geographer* 50: 177–196.

Feder, Howard M. 2001. A year at NARL: Experiences of a young biologist in the Laboratory's early days. In *Fifty more years below zero: Tributes and meditations for the Naval Arctic Research Laboratory's first half century at Barrow, Alaska*, ed. D. Norton, 33–60. Fairbanks: University of Alaska Press.

Folk, G.E. 2010. The Harvard Fatigue Laboratory: contributions to World War II. *Advances in Physiology Education* 34: 119–127.

Franklin, J.F., R.E. Jenkins, and R.M. Romancier. 1972. Research natural areas: Contributors to environmental quality programs. *Journal of Environmental Quality* 1: 133–139.

Geier, Max G. 1998. *Forest science research and scientific communities in Alaska: A history of the origins and evolution of USDA Forest Service research in Juneau, Fairbanks, and Anchorage*. Portland: US Department of Agriculture, Forest Service.

Golley, Frank B. 1993. *History of the ecosystem concept in ecology: More than the sum of the parts*. New Haven: Yale University Press.

Hagen, Joel B. 1992. *An entangled bank: The origins of ecosystem ecology*. New Brunswick: Rutgers University Press.

———. 2015. Camels, cormorants, and kangaroo rats: Integration and synthesis in organismal biology after World War II. *Journal of the History of Biology* 48: 169–199.

Hanson, Wayne C. 2001. Monitoring of worldwide fallout-related radiation in North Slope people and animals. In *Fifty more years below zero: Tributes and meditations for the Naval Arctic Research Laboratory's first half century at Barrow, Alaska*, ed. D. Norton, 71–74. Fairbanks: University of Alaska Press.

Hayes, Helen. 1964. Office of Naval Research. *Bioscience* 14: 29–32.

Hobbie, John E. 1973. Limnology. In *Alaskan arctic tundra*, Technical paper no. 25, ed. Max Britton, 127–168. Washington, DC: Arctic Institute of North America.

———., ed. 1980. *Limnology of tundra ponds, Barrow, Alaska*, U.S. IBP synthesis series, 13. Stroudsburg: Dowden, Hutchinson & Ross.

———. 2016. Long-term ecological research in the Arctic: Where science never sleeps. In *Long-term ecological research: Changing the nature of scientists*, ed. M.R. Willig and L.R. Walker, 91–97. New York: Oxford University Press.

Hobbie, John E., and G.W. Kling, eds. 2014. *Alaska's changing arctic: Ecological consequences for tundra, streams, and lakes*. New York: Oxford University Press.

Hobbie, John E., R.J. Daley, and J. Jasper. 1977. Use of nuclepore filters for counting bacteria by fluorescence microscopy. *Applied and Environmental Microbiology* 33: 1225–1228.

Hobbie, John E., J. Cole, J. Dungan, R.A. Houghton, and B. Peterson. 1984. Role of biota in global CO_2 balance: The controversy. *Bioscience* 34: 492–498.

Hobbie, John E., S.R. Carpenter, N.B. Grimm, J.R. Gosz, and T.E. Seastedt. 2003. The US long-term ecological research program. *Bioscience* 53: 21–32.

Huryn, A.D., and J.E. Hobbie. 2012. *Land of extremes: A natural history of the North Slope of arctic Alaska*. Fairbanks: University of Alaska Press.

Irving, Laurence. 1948. Arctic research at Point Barrow, Alaska. *Science* 107: 284–286.

———. 1950. Measurement of some physiological reactions to Arctic conditions. *Annals New York Academy of Sciences* 51: 1045–1050.

———. 1969. Progress of research in zoology through the Naval Arctic Research Laboratory. *Arctic* 22: 327–332.

Jenny, Hans. 1980. *The soil resource: Origin and behavior*. New York: Springer.

Kelley, John J., and Arnold Brower Sr. 2001. The NARL and its transition to the local community. In *Fifty more years below zero: Tributes and meditations for the Naval Arctic Research Laboratory's first half century at Barrow, Alaska*, ed. D. Norton, 259–264. Fairbanks: University of Alaska Press.

Kerfoot, W.C. 1982. Review of *Limnology of tundra ponds: Barrow, Alaska*, by John E. Hobbie, ed. *Ecology* 63: 598–599.

Krogh, August. 1941. *The comparative physiology of respiratory mechanisms*. Philadelphia: University of Pennsylvania Press.

Lajeunesse, Adam, and P. Whitney Lackenbauer, eds. 2017. *Canadian Arctic operations, 1941–2015: Lessons learned, lost, and relearned*. Frederickton: Gregg Centre for the Study of War and Society.

Lubchenco, J., A.M. Olson, L.B. Brubaker, S.R. Carpenter, M.M. Holland, S.P. Hubbell, S.A. Levin, J.A. MacMahon, P.A. Matson, J.M. Melillo, H.A. Mooney, C.H. Peterson, H.R. Pulliam, L.A. Real, P.J. Regal, and P.G. Risser. 1991. The sustainable biosphere initiative: An ecological research agenda. *Ecology* 72: 371–412.

Mooney, H.A., and W.D. Billings. 1961. Comparative physiological ecology of Arctic and Alpine populations of Oxyria digyna. *Ecological Monographs* 31: 1–29.

National Science Foundation. 1978. *Pilot program for long-term observation and study of ecosystems in the United States*. Report of a conference, Woods Hole, Mass. https://lternet.edu/wp-content/uploads/2010/12/78workshop.pdf. Accessed 4 March 2019.

Norton, David, and G. Weller. 2001. NARL's scientific legacies through outer continental shelf environmental studies. In *Fifty more years below zero: Tributes and meditations for the naval Arctic research Laboratory's first half century at Barrow, Alaska*, ed. D. Norton, 233–236. Fairbanks: University of Alaska Press.

O'Neill, Daniel T. 1994. *The firecracker boys*. New York: St. Martin's Press.

Philo, Mike, L. Underwood, and A. Callahan. 2001. History of the NARL Animal Research Facility (ARF) and its contribution to the development of a rabies control program for the North Slope. In *Fifty more years below zero: Tributes and meditations for the Naval Arctic Research Laboratory's first half century at Barrow, Alaska*, ed. D. Norton, 255–258. Fairbanks: University of Alaska Press.

Pitelka, Frank A. 1964. The nutrient-recovery hypothesis for Arctic microtine cycles. I. Introduction. In *Grazing in marine and terrestrial environments*, ed. D.J. Crisp, 55–56. Oxford: Blackwell Scientific.

———. 1967. Some characteristics of microtine cycles in the Arctic. In *Arctic biology*, ed. H.P. Hansen, 153–184. Corvallis: Oregon State University.

———. 1969. Ecological studies on the Alaskan arctic slope. *Arctic* 22: 333–340.

Polar Research Board. 1977. *An evaluation of arctic programs supported by the National Science Foundation*. Washington, DC: National Academy of Sciences.

Reed, John C. 1969. The story of the Naval Arctic Research Laboratory. *Arctic* 22: 177–183.

Reed, John C., and A.G. Ronhovde. 1971. *Arctic laboratory: A history (1946–1966) of the Naval Arctic Research Laboratory at Point Barrow, Alaska*. Washington, DC: Arctic Institute of North America.

Reynolds, J.F., and J.D. Tenhunen. 1996. Ecosystem response, resistance, resilience, and recovery in Arctic landscapes: Introduction. In *Landscape function and disturbance in Arctic tundra*, ed. J.F. Reynolds and J.D. Tenhunen, 3–18. Berlin: Springer.

Risser, Paul G., Jane Lubchenco, and Simon A. Levin. 1991. Biological research priorities: A sustainable biosphere. *Bioscience* 41: 625–627.

Sapolsky, H.M. 1990. *Science and the Navy: The history of the Office of Naval Research*. Princeton: Princeton University Press.

Schindler, John F. 2001. Naval Petroleum Reserve No. 4 and the beginnings of the Arctic Research Laboratory (ARL). In *Fifty more years below zero: Tributes and meditations for the Naval Arctic Research Laboratory's first half century at Barrow, Alaska*, ed. D. Norton, 29–32. Fairbanks: University of Alaska Press.

Schmidt-Nielsen, Knut. 1969. The Alpha-Helix, a research opportunity. *Bioscience* 19: 59.

———. 1987. Per Scholander, 1905-1980. *Biographical Memoirs of the National Academy of Sciences United States of America* 56: 387–412.

Schmidt-Nielsen, B., and K. Schmidt-Nielsen. 1950. Evaporative water loss in desert rodents in their natural habitat. *Ecology* 31: 75–85.

Scholander, Per F. 1990. *Enjoying a life in science: The autobiography of P.F. Scholander*. Fairbanks: University of Alaska Press.

Scholander, P.F., R. Hock, V. Walters, and L. Irving. 1950. Adaptation to cold in Arctic and tropical mammals and birds in relation to body temperature, insulation, and basal metabolic rate. *Biological Bulletin* 99: 259–271.

Schultz, A.M. 1964. The nutrient recovery hypothesis for Arctic microtine cycles. II. Ecosystem variables in relation to Arctic microtine cycles. In *Grazing in marine and terrestrial environments*, ed. D.J. Crisp, 57–68. Oxford: Blackwell Scientific.

Shaver, G.R. 1996a. Integrated ecosystem research in Northern Alaska, 1947-1994. In *Landscape function and disturbance in arctic tundra*, ed. J.F. Reynolds and J.D. Tenhunen, 19–33. Berlin: Springer.

———. 1996b. *History of research at Toolik*. Arctic long term ecological research. http://arc-lter.ecosystems.mbl.edu/history-research-toolik. Accessed 21 Aug 2018.

———. 2016. Forty arctic summers. In *Long-term ecological research: Changing the nature of scientists*, ed. M.R. Willig and L.R. Walker, 99–105. New York: Oxford University Press.

Shaver, G.R., J.A. Laundre, M.S. Bret-Harte, F.S. Chapin III, J.A. Mercado-Díaz, A.E. Giblin, L. Gough, S.E. Hobbie, G.W. Kling, M.C. Mack, J.C. Moore, K.J. Nadelhoffer, E.B. Rastetter, and J.P. Schimel. 2014. Terrestrial ecosystems at Toolik Lake, Alaska. In *Alaska's changing arctic: Ecological consequences for tundra, streams, and lakes*, ed. J.E. Hobbie and G.W. Kling, 90–142. New York: Oxford University Press.

Shelesnyak, M.C. 1947a. *Across the top of the world: A discussion of the Arctic*. Washington, DC: U.S. Office of Naval Research.

———. 1947b. Some problems of human ecology in polar regions. *Science* 106: 405–409.

———. 1947c. The Navy explores its northern frontiers. *Naval Engineers Journal* 59: 471–485.

———. 1948a. Arctic Research Laboratory, Office of Naval Research, Point Barrow, Alaska. *Science* 107: 283.

———. 1948b. The history of the Arctic Research Laboratory, Point Barrow, Alaska. *Arctic* 1: 97–106.

Sparrow, Elena B., Janice C. Dawe, and F. Stuart Chapin III. 2006. Communication of Alaskan boreal science with broader communities. In *Alaska's changing boreal forest*, ed. F.S. Chapin, M.W. Oswood, K. Van Cleve, L.A. Viereck, and D.L. Verbyla, 323–331. New York: Oxford University Press.

The Institute of Ecology. 1977. *Experimental ecological reserves: A proposed national network*. Washington, DC: U.S. Government Printing Office.

Tieszen, Larry L., ed. 1978. *Vegetation and production ecology of an Alaska Arctic Tundra*. New York: Springer.

Vitousek, Peter M., Harold A. Mooney, Jane Lubchenco, and Jerry M. Melillo. 1997. Human domination of earth's ecosystems. *Science* 277: 494–499.

Vogel, Steven. 2008. Knut Schmidt-Nielsen. 24 September 1915 — 25 January 2007. *Biographical Memoirs Fellows of the Royal Society* 54: 319–331.

Washburn, A.L. 1952. Arctic research in North America. *Science* 115: 3.

———. 1980. Focus on polar research. *Science* 209: 643–652.

Wiggins, Ira L. 1952. Arctic botanical research. *Science* 115: 3a.

Wilimovsky, N.J., and J.N. Wolfe, eds. 1966. *Environment of the Cape Thompson Region, Alaska*. Oak Ridge: U.S. Atomic Energy Commission.

Wolfe, J.N. 1964. Atomic energy commission. *Bioscience* 14: 22–25.

Wood, Chris. 2007. Knut Schmidt-Nielsen plenary lecture: In praise of expeditionary physiology. *Comparative Biochemistry and Physiology, Part A* 148: S1.

Woodwell, G.M. 1967. Radiation and the patterns of nature. *Science* 156: 461–470.

———. 1978. The carbon dioxide question. *Scientific American* 238 (1): 34–43.

———. 2016. Dr. George Woodwell, climate science pioneer, Woods Hole Research Center. *The Climate Times*, April 23. https://www.theclimatetimesus.org/essays/dr-george-woodwell-climate-science-pioneer-woods-hole-research-center. Accessed 22 Aug 2018.

Part III
Experiments in Broadening the Social Significance of LTER

Chapter 8
How LTER Site Communities Can Address Major Environmental Challenges

Frederick J. Swanson, David R. Foster, Charles T. Driscoll, Jonathan R. Thompson, and Lindsey E. Rustad

Abstract Long-term, place-based research programs in the National Science Foundation-supported Long Term Ecological Research (LTER) Network have had profound effects on public policies and practices in land use, conservation, and the environment. While less well known than their contributions to fundamental ecological science, LTER programs' commitment to serving broad public interests has been key to helping achieve their mission to advance basic science that supports society's need to address major environmental challenges. Several attributes of all LTER programs are critical to these accomplishments: highly credible science, strong site-level leadership, long-term environmental measurements of ecosystem attributes that are relevant to the public and to resource managers, and effective and accessible information that supports sound management practices. Less recognized attributes of three case study LTER sites (Andrews Forest, Harvard Forest, Hubbard Brook) which have contributed to major impacts include strong interdisciplinary research communities with cultures of openness, dispersed leadership within those communities, a commitment to carry science perspectives to society through multiple governance processes, strong public-private partnerships, and communications programs that facilitate the exchange of information and perspectives among sci-

F. J. Swanson (✉)
Pacific Northwest Research Station, US Forest Service, Corvallis, OR, USA
e-mail: fred.swanson@oregonstate.edu

D. R. Foster · J. R. Thompson
Harvard Forest, Harvard University, Petersham, MA, USA
e-mail: drfoster@fas.harvard.edu; jthomps@fas.harvard.edu

C. T. Driscoll
Department of Civil and Environmental Engineering, Syracuse University, Syracuse, NY, USA
e-mail: ctdrisco@syr.edu

L. E. Rustad
Northern Research Station, U.S. Forest Service, Durham, NH, USA
e-mail: lrustad@fs.fed.us

R. B. Waide, S. E. Kingsland (eds.), *The Challenges of Long Term Ecological Research: A Historical Analysis*, Archimedes 59,
https://doi.org/10.1007/978-3-030-66933-1_8

ence communities, policy-makers, land managers, and the public. Taken together, these attributes of sites drive on-the-ground outcomes. These case studies reveal a virtue of the long-term nature of LTER not anticipated when the program began: that the decades-long engagement of a place-based, science community can have a major impact on environmental policies and practices. These activities, and the cultivation of science communities that can accomplish them, go beyond the initial directives and review criteria for LTER site proposals and programs.

Keywords LTER program · Long-term ecological research · Acid rain · Ecosystem experiments · Environmental legislation · Environmental policy · Forest ecology · Forest management · Interdisciplinary research

8.1 Introduction

The US Long Term Ecological Research (LTER) Program has grown and evolved dramatically since its inception in 1980. In an era of short-term, single-investigator projects, the National Science Foundation (NSF) initiated a large-scale, pioneering experiment to understand key ecological processes and their interactions that unfold over decadal scales. The NSF launched this novel approach to environmental science by doubling the duration of single grants (from about 2 to 3 years to 5 to 6 years with potential for renewal), encouraging longer-term research planning with expectation of greater interdisciplinary scope and collaboration, promoting intersite work, assuring continuity of leadership, and demanding a high level of attention to data management and sharing (see Jones and Nelson, Chap. 3, this volume). The LTER program has been fulfilling its initial objectives of establishing long-term experiments, collecting ongoing measurements, developing and applying models, interpreting long-term observations and results, and conducting synthesis across five core research themes (disturbance patterns, primary productivity, mineral cycling, organic matter cycling, and population studies). Over the past several decades, the network of LTER sites has grown in number, disciplinary scope, types of ecosystems studied, and support across NSF divisions (Jones and Nelson, Chap. 3, this volume). At the core of LTER are the individual site-based programs run by communities of scientists, students, support staff, and collaborators such as public and private land manager and land trust partners, foundations, and even artists. Cross-site meetings, research and synthesis, organized and coordinated through initiatives from groups of sites and through a Network Office, have facilitated network cohesion and identity. LTER has had sufficient success to stimulate government funders, science communities, and institutions in about 40 other countries to establish their own LTER-like programs and participate in a network referred to as International LTER (Vanderbilt and Gaiser 2017). These accomplishments have made LTER NSF's longest running program other than graduate research fellowships.

Over its 40-year history, the US LTER program has also exerted a profound impact on formal and informal education and on public outreach, as documented in the histories of individual programs (e.g., site synthesis volumes in the Oxford Press series) and in several multi-site syntheses (e.g., Colman 2010; Driscoll et al. 2012). In particular, LTER has had notable success in delivering science information to decision-makers and a broad public audience, consistent with the requirement for "broader impacts" in the NSF evaluation criteria. The sustained long-term nature of the LTER site research communities allows for greater impacts than programs funded by a kaleidoscope of short-term grants (Hughes et al. 2017). LTER communities provide (i) conceptual frameworks for exploring the coupled nature-human system in an increasingly human-dominated world, (ii) improved environmental literacy of the engaged public, (iii) information about plausible futures for ecosystems and the environment through modeling and scenario analysis, (iv) well-managed, readily-accessible, long-term environmental data for future use in addressing issues challenging society, and (v) insights into landscape vulnerability and resilience to global change (Robertson et al. 2012). These tasks require continuity and strong, lasting partnerships; the mission and sustained funding of the LTER program support those qualities at a site and network level.

All LTER sites must accomplish substantive broader impacts to maintain their program funding, but several sites have had distinctive, major, direct impacts on environmental issues at regional and national scales. Three examples of major impact are reviewed here: Hubbard Brook's work on the effects of air pollution on terrestrial and freshwater ecosystems, Harvard Forest's conservation program for the New England landscape, and H. J. Andrews Experimental Forest's role in regional forest conservation planning. In each case, we acknowledge the existing site strengths and research programs that made these impacts possible and the pre-LTER roots of the site that established the basis for taking on the issues, but we also explore and highlight several critical but less known features. We pay particular attention to the confluence of attributes of these LTER sites that made the major impacts possible: attributes of the LTER program community itself; its science program; the major environmental issues it addressed; the "governance" systems in which the impacts played out; formation of partnerships that facilitate connections with society; and communications systems directed toward policy-makers and the public. In this context, the term *governance* refers to formal and informal processes through which public and/or private individuals and institutions can guide policies and actions concerning the environment. For example, governance includes public policy channels through which science can inform land-use planning and the management or regulation of environmental quality. This overview is undertaken in part to highlight some notable accomplishments by these particular sites and the whole LTER program. The larger purpose of this chapter is to identify the distinctive qualities of these individual research programs in the hope that this will aid other institutions, research communities, the LTER network, and NSF leadership in their administrative and management decisions moving forward.

8.2 Case Examples

The commonalities and differences among these cases are instructive. In order to highlight these and facilitate comparison, the case studies follow a parallel structure. Each begins with a brief review of the pre-LTER context of the site and program, the significance of each site's LTER programs in terms relevant to the major environmental issues addressed, and key features of the science emerging from the long-term research that influenced policy, planning, and execution of a path forward. This is followed by discussion of the governance context that was critical in connecting the site's scientists and their research findings with policy and management. Finally, we address special features of partnerships and communications programs that emerged in dealing with the big issues and have then persisted as channels for conversation between science communities and society on other topics.

We consider these cases with two important caveats: first, more complete description of the science involved and its connection with society are presented in greater detail elsewhere, and, second, in each case the issues are so vast that LTER science and research communities are only two of many factors in the process of dealing with big issues.

8.2.1 Hubbard Brook Ecosystem Study, LTER Program, and Experimental Forest

Perhaps the prime example of a long-term research site having a major impact on public policy is the story of the Hubbard Brook Ecosystem Study (HBES) and its role in the discovery and remediation of atmospheric deposition, or "acid rain" (Likens and Bormann 1977; Driscoll et al. 2001; Bocking 2016; Holmes and Likens 2016). This research is based on long-term studies conducted at the Hubbard Brook Experimental Forest in the White Mountains of New Hampshire. The Hubbard Brook is a 3519-ha experimental forest, established by the US Forest Service in 1955 as a center for forest hydrology research in New England. In keeping with the research methods of nearly a dozen other Forest Service sites, paired-watershed experiments were established for the study of forest hydrology (streamflow gauging at Hubbard Brook started in 1956). That watershed research approach prompted several professors at Dartmouth College, led by F. Herbert Bormann and Gene Likens, to realize that whole-ecosystem biogeochemistry research—the inputs, cycling, and outputs of chemical elements—could be piggy-backed on the small watershed hydrology study in what became known as the Hubbard Brook Ecosystem Study (HBES), commencing in 1963. The study advanced through a partnership of the academic scientists with Forest Service researchers led by Robert Pierce.

Although the term "acid rain" was first used in the mid-nineteenth-century in Britain, the first observations in North America were reported from Hubbard Brook, based on early measurements from the Hubbard Brook Ecosystem Study (Likens et al. 1972). (The simple yet evocative phrase "acid rain" conveys the notion that the rain that nourishes us can be fouled by acidity.) Thus, began a long and multi-faceted engagement of the Hubbard Brook community of scientists with issues related to air pollution and biogeochemistry. The initial identification of acid rain at Hubbard Brook did not require long-term research, but the long-term studies, including those funded by LTER beginning in 1988, made it ultimately possible to track the cascade of effects of atmospheric pollutants through forest, soil, stream, and lake ecosystems and observe subsequent ecosystem recovery from declining pollution as policies regulating emissions were implemented. Hubbard Brook scientists have gone on to address other air pollutants, such as nitrogen, mercury, and carbon dioxide (Driscoll et al. 2001, 2003, 2007, 2012, 2015, 2016).

The Hubbard Brook Ecosystem Study and Hubbard Brook LTER had several important synergies; indeed, accomplishments of the HBES were used as support for NSF's decision to begin the LTER program in 1980. In his seminal paper introducing the LTER concept to the science community, for example, Callahan (1984, p. 363) cites Bormann and Likens' (1979) argument that long-term studies are essential for the study of "effects of atmospheric pollution, forest harvesting practices, and forest development cycles" on forest productivity. These early results emerged over the initial 20-year history of the HBES through a series of individual, short-term grants. At that initial stage of LTER development, and still four years before Hubbard Brook joined LTER, the benefits of the new LTER program were yet to materialize. As the HBES grew in thematic scope, facilitated substantially by LTER funding beginning in 1988, so too did the size and the disciplinary and institutional diversity of its community of researchers and educators. From its roots in hydrology and then biogeochemistry, the program grew to include studies of forest bird populations, tree community development and dynamics, organic matter budgets of streams, limnology, and many other topics. The LTER program has been an important means to both diversify and integrate the Hubbard Brook scientific enterprise, and integration has been particularly challenging with its researchers spread across many institutions and states and minimal senior academic science staff in residence at the site.

The core contribution of Hubbard Brook science in characterizing and understanding effects of air pollution was to reveal broad yet nuanced interpretations of complex effects throughout the ecosystem and over time. The supporting science included tracing the inputs, transport and fate of atmospheric contaminants through the forest canopy, soils, streams, and into lakes, as well as examining their cascading effects on plant, soil and microbial processes. Environmental monitoring was a foundational component of the work, which was complemented by process studies, the development and application of models (Gbondo-Tugbawa et al. 2001), innovative whole-watershed experiments (Peters et al. 2004), and other approaches. A critical dimension was to view ecosystem effects of air pollution in the context of other drivers of environmental change, such as climate variability and loss of species to species-specific pests and pathogens.

The national and regional discourse on air pollution and its effects provided the opportunity for Hubbard Brook science to impact federal policy. Well before the seminal 1970 amendments to the Clean Air Act, many elements of federal and state government involved in air and water pollution issues were informed by research from Hubbard Brook. Over the decades Hubbard Brook researchers and science have interacted with all three branches of the Federal government: the legislative process included amendments to the Clean Air Act, administrative rules were promulgated to reduce air pollution, and the judiciary enforced compliance (Driscoll et al. 2011). As intense political disputes about amending the Clean Air Act unfolded, the credibility of "high-quality, long-term data on precipitation and stream water chemistry helped ward off aggressive attacks from various science deniers and vested interests" (Holmes and Likens 2016, p. 216).

In 2012, Hubbard Brook joined with three other LTER sites (Harvard Forest, Plum Island Sound, Baltimore Ecosystem Study) and five other institutions to form the Science Policy Exchange (SPE) to promote the use and synthesis of long-term observations and ecosystem science in science translation and to inform initiatives on energy, land and water policy. The mission of the SPE is to promote the use of long-term observations and science in environmental policy decisions consistent with their mission "to harness the power of science to generate environmental solutions for people and nature." Another critical, if informal, aspect of the SPE has been contacts between scientists with members of the media and with individuals in state and federal agencies and non-governmental organizations (NGOs). Some of these relationships have been sustained over many years, and they remain vital in the ever-changing social and political environment.

Science communications from the HBES began with the prolific publication of articles in scientific journals and the synthesis of findings in landmark books on biogeochemistry and forest dynamics (Likens and Bormann 1977; Bormann and Likens 1979). But as the magnitude and increasing threat of air pollution in the form of acid precipitation became more apparent, an effort was made to synthesize and translate the science on effects at Hubbard Brook and regionally (Driscoll et al. 2001) and to use models to project ecosystem response and recovery in various emission reduction scenarios (Chen and Driscoll 2005). As public attention to the issue grew, venues for communication reached the highest levels of government, including a briefing of President Reagan and numerous Congressional hearings, contributing substantially to the 1990 Amendments to the Clean Air Act of 1970 (Holmes and Likens 2016, p. 216) and to other air quality management policies such as the Cross-State Air Pollution Rule, the Mercury and Air Toxics Standard, and the Affordable Clean Energy Rule. The initial effort and success in engaging in science communication prompted the HBES to institutionalize their outreach program through a series of reports collectively called *Science Links*, produced through the Hubbard Brook Research Foundation since 1998 (Driscoll et al. 2012). The Science Links reports are developed by interdisciplinary teams of scientists and policy advisers who frame policy-relevant questions and analyze alternatives in text and illustrations suitable for a readership of policy-makers and journalists. These seminal publications were accompanied by outreach to the public via op-ed pieces in major national print media, extensive media coverage, and presentations to civic and professional groups in order to more effectively affect policy shifts. Many lines of evidence attest to the success of this communications program, including numbers of citations in the scientific literature, quantity and quality of media coverage, and reference to Science Links reports in drafts of legislation (Driscoll et al. 2011).

8.2.2 Harvard Forest and its LTER Program

Harvard University established the Harvard Forest in north-central Massachusetts in 1907 as a center for forest and forestry research, education, and demonstration (Foster and Aber 2004). A central theme for the research, education, and outreach programs at the Harvard Forest was the trajectory and consequences of the four-century history of human land use involving European settlement, deforestation, and agricultural development (ca. 1650–1850) that dramatically reduced the nearly-complete regional forest cover, followed by the progressive expansion of forest cover in the wake of farmland abandonment, industrialization, and urbanization. Incredibly detailed dioramas in the Fisher Museum at the Harvard Forest depict this history of landscape dynamics and the resiliency of the forests in the face of sustained environmental degradation. These dioramas, as well as stonewalls, the former farmland fence lines that run through the region's forests, remind all who visit or work at the Forest that the modern landscape is strongly conditioned by its past and that ongoing recovery from that history will strongly control every ecosystem's future. During the pre-LTER period, much of the research by Harvard Forest staff centered on silviculture, forestry and forest ecology, and related studies of soil properties and processes and wildlife, on the Forest's 1200 hectares and more broadly across southern and central New England.

Upon entering the LTER network in 1988, Harvard Forest greatly expanded its research portfolio of major processes shaping forest ecosystems by establishing large, long-term, ecosystem experiments concerning hurricane damage to forests, climate change and soil warming, and nitrogen deposition from air pollution (Foster et al. 2014). The establishment of one of the first forest-based eddy flux towers initiated what has become the world's longest running record of exchanges between the atmosphere and a forest ecosystem. Research on environmental history and prehistory intensified, using archival, paleoecological, archaeological (of both indigenous and European peoples), and even literary sources (Foster 1999). Despite the differences in cultures between experimentalist and historical researchers, synergies emerged early in the twenty-first century when it became clear that land-use history provided an indispensable foundation for the interpretation of the modern landscape and results emerging from experimental ecosystem studies. The legacies from the history of land use and natural disturbances shaped the landscape with enduring consequences for ecosystem structure and function (Foster and Aber 2004).

The understanding of long-term landscape change also helped to galvanize a vision for the future of the region's forests. Among the many components of the Harvard Forest portfolio of research and outreach activities, the Wildlands and Woodlands regional conservation strategy stands out as exerting a growing influence on regional policy and management. This program, grounded in LTER science and environmental history, was a response to residential and urban sprawl into rural forest and farm lands that by the late twentieth century had reversed the century-long expansion of forest cover across all New England states (Foster et al. 2014). This conservation strategy was first applied to Massachusetts, with the goal of permanently protecting 50% of the state in forest cover (Foster et al. 2014).

The Wildlands and Woodlands vision comprises two major components: "wildlands", large forest reserves covering about 10% of the conserved forest area, which sustain landscape-scale ecological process in the absence of active management, and expansive "woodlands" across the remaining 90%, in which sustainable forest management is encouraged for a diversity of private and public objectives. Harvard Forest scientists built their case on the dynamic and resilient properties of New England's forests and the "illusion of preservation," the argument that strongly protectionist policies in populated regions with intensive resource use and resilient ecosystems, like New England, can displace the environmental impacts of resource production to more pristine and vulnerable ecosystems elsewhere in the country or globe (Berlik et al. 2002; Foster et al. 2014).

The Wildlands and Woodlands vision gained traction following its initial application to Massachusetts in 2005 (Foster et al. 2005) and was widely endorsed at a state level. With growing support by conservation groups, active involvement of leading scientists across the region (including many LTER collaborators and the two principal investigators of the Hubbard Brook LTER), and major collaboration from the Highstead Foundation and New England Forestry Foundation, the Wildlands and Woodlands program was expanded to the forests within the six New England states (Foster et al. 2010), and then expanded further to include the entire landscape, including farmlands and associated communities (Foster et al. 2017).

Wildlands and Woodlands is founded in the recognition that, although New England is one of the nation's most densely populated (> 15 million people) and economically thriving regions, it is also the country's most heavily forested area (81% forest cover; 7% farmland, 10% developed area). It therefore has the potential to support much more strategically focused development and land conservation activity in ways that will conserve the bulk of its forest and farmland to support both nature and society. The vision's regional goal for 2060 is to conserve approximately 80% of the region: 70% of it as intact forest (63% woodlands and 7% wildlands), 7% as agricultural land, and the rest as other semi-wild lands (wetland, water, etc.). Since 2010, protected forest has increased from 22 to 26%, but forest conversion by development continues to proceed at a rate of 9700 ha a year and wildland reserves remain less than 1% of the landscape (Foster et al. 2017). The regional vision has spawned an integrated network of regional conservation partnerships, expanded capacity focused on conservation finance, and recognized the role of well-managed forests and farmlands in providing conservation infrastructure that supports human health and well-being. These advances also represent a more strategic linkage between academic research and policy and management needs.

Wildlands and Woodlands, and its strong private and public partnerships and linkage of basic research to conservation applications, has become an increasingly important element of the *broader impacts* in the Harvard Forest LTER program over the past three funding cycles. LTER science has helped leverage significant funding from other NSF grants and programs (e.g., Research Coordination Network and Coupled Natural and Human Systems) and private foundations, thus providing the scientific input needed for extensive stakeholder engagement and outreach to policy and decision makers. Harvard Forest Director and LTER Principal Investigator David Foster has been the central catalyst and leader of both efforts, but the breadth of topics and lists of co-authors make clear that a large, diverse, collaborative group of colleagues—forest and landscape ecologists, policy specialists, environmental historians, biogeochemists—from many institutions has figured prominently throughout. In the formative steps of Wildlands and Woodlands, this Harvard Forest community identified a critical regional issue, laid the conceptual framework for addressing it, mapped a conservation strategy for the region, and began to pursue a solution, including working with state government and land trusts. A distinctive contribution of the LTER program has been co-designing future scenarios of land change with hundreds of diverse stakeholders from throughout New England, and then evaluating their consequences for people and nature using ecosystem models developed using LTER science. By using a participatory process for scenario creation, the scientists improve the relevance and maximize the uptake of the results (Thompson et al. 2012, 2014, 2016; McBride et al. 2017, 2019).

The governance of regional conservation strategies set in a predominantly private lands context, such as Wildlands and Woodlands, involves navigating a challenging blend of governmental (mainly municipal and state level) and private entities (e.g., land trusts and other conservation enterprises) (Foster et al. 2014). The New England setting puts a premium on development of institutional partnerships and social networking to carry findings from long-term ecological research into the public sector. A critical step has been partnering with the Highstead Foundation, which supports Regional Conservation Partnerships comprised of 42 groups of partnerships and land trusts covering 60% of the New England region (Labich et al. 2013; Foster et al. 2017). Broadly, Wildlands and Woodlands seeks to advance the success of partner organizations as they advance conservation either through the direct purchase of land or, increasingly, by securing conservation easements on private land parcels, often in collaboration with state and federal agencies, private foundations, and individual philanthropic support. Harvard Forest's partnership activities extend beyond Highstead Foundation to include teaming up with Hubbard Brook and other institutions such as the New England Forest Foundation and as a founding partner of the New England Science Policy Exchange to promote the use of long-term observations and science in land policy decisions (Templer et al. 2015; Lambert et al. 2018).

Through much of the twentieth century, the Harvard Forest community had a rather traditional communications program that included scientific publications and various forms of reports in many cases aimed at local audiences of the general public, small woodlot owners, and foresters. And, of course, the Fisher Museum collection of dioramas and other displays concerning forest and land history gave visitors distinctive learning opportunities. With the development of research topics of broad significance and interest during its LTER era, Harvard Forest's communication efforts diversified greatly to include editorials and various forms of reporting in major regional and national outlets, the special communications of the Wildlands and Woodlands reports, and arts/humanities programs, including installation art exhibits in the forest and on campus (Leigh et al., Chap. 11, this volume). A collection of beautifully illustrated and engagingly written books has found a wide readership. More recently, the Schoolyard LTER program has incorporated a module on forest and landscape dynamics to integrate some of the historical, ecological, and conservation perspectives of Wildlands and Woodlands into the classroom for some 6000 students across Massachusetts and adjoining states.

8.2.3 H.J. Andrews Experimental Forest LTER Program

The H.J. Andrews Experimental Forest is well known for its science-rooted role in the major forest policy shift in the 1990s to stop logging of old-growth forests on the extensive Federal lands of the Pacific Northwest (Johnson and Swanson 2009; Spies and Duncan 2009; Colman 2010; Robbins 2020). Ironically, this Forest Service experimental forest (originally named Blue River Experimental Forest) had been established in 1948 for use in applied studies to support conversion of the native forest, notably old growth, to intensively-managed tree plantations. With the inception of the NSF funding of the International Biological Program (IBP) at Andrews Forest in 1969, academic ecosystem scientists joined Forest Service researchers in intensive, basic, multi-disciplinary investigations of forests, streams, and whole watersheds. Forest Service scientist Jerry Franklin and Oregon State University professor Dick Waring teamed up to lead the group during the IBP era, and Franklin continued to lead into the LTER period beginning in 1980. Blending applied forestry and watershed research with basic ecosystem science, since 1970 the program has been managed jointly by the US Forest Service's Pacific Northwest Research Station, Willamette National Forest, and Oregon State University working in tight partnership at the research-management interface. Franklin's leadership in the site's participation in IBP and then LTER, and his role in the inception of LTER as a whole, including a stint as a program officer at NSF in the early 1970s and later as coordinator of the LTER Network while he was a professor at University of Washington, all proved vital in advancing long-term ecological research globally.

The trajectory of Federal forestry issues in the Pacific Northwest both influenced and was influenced by the Andrews Forest program from the start of the Timber Era at the end of World War II through its transition to the present era emphasizing biodiversity on Federal land forests of the Northwest. In accord with the founding charge to help guide development of the Federal forestry program, logging and road construction in the experimental forest proceeded during the 1950s and 1960s. Applied studies by US Forest Service scientists addressed effects of forest operations on plants, animals, soil, and watershed processes, especially streamflow and water quality. This set the stage for IBP ecosystem research of the 1970s to inadvertently reveal salient features of old-growth forests and for a solitary Oregon State University M.S. student, Eric Forsman, to begin his career-long investigations of the preference of northern spotted owl (*Strix occidentalis caruina*) for that habitat and the progressive decline of their numbers. The IBP directive was to study the native forest ecosystem; it happened to be old growth in the Andrews Forest, because of the mid-1940s decision to locate the research property where there was extensive old growth in order to study its liquidation. A published synthesis of 1980-vintage knowledge of old growth (Franklin et al. 1981) and the emerging understanding of the spotted owl prompted environmentalists to work over the 1980s to achieve an injunction in 1990 to stop logging in the 10 million ha of Federal lands in the range of the spotted owl from San Francisco to the Canadian border. The injunction was not lifted until a team of scientists, including many from the Andrews Forest program, crafted the foundation for the Northwest Forest Plan (NWFP) (Duncan and Thompson 2006; Robbins 2020).

As with Hubbard Brook and Harvard Forest, participation in LTER proved critical in sustaining and expanding a diverse, interdisciplinary community of academic and federal science personnel at the Andrews Forest. Existing long-term studies of vegetation dynamics and watershed processes could be extended with LTER support and new topics added. Entirely new experiments were undertaken, such as Mark Harmon's monumental 200-year log decomposition experiment, which became a stage for public discussion of future management of dead wood on land and in streams in terms of habitat, carbon dynamics and sequestration, soil fertility, fuels for wildfire, and other topics. Regional networks of forest plots, remote sensing studies, and landscape modeling have been used to assess past and alternative future forest change in response to land use, wildfire, and forest growth (Thompson et al. 2012).

The governance context of change in Federal forest policy in which Andrews Forest scientists participated involved all three branches of government and a variety of formal and informal venues over time (Swanson 2004; Johnson and Swanson 2009). For example, findings from IBP and LTER science of forest-stream interaction research found their way into management policy and practice at the scale of local collaborations between researchers and Ranger District staff concerning timber sales, at the scale of the 700,000-ha Willamette National Forest as it developed its management plan of 1990, and across the entire Pacific Northwest though the Northwest Forest Plan. The original basic-science work on old-growth forests and streams influenced the arc from a Federal policy of old-growth liquidation through to its protection in the Northwest Forest Plan in the mid-1990s with many governance instruments along the way. The intensity of conflict over old growth drew global attention, stimulating conservation efforts and calling for consultations in places (Tasmania, Taiwan, Scandinavia) where very different governance contexts prevailed.

A long-standing partnership between scientists and land managers and a sustained culture of close cooperation between academic and agency scientists have been central to the capacity of the Andrews Forest program to participate in these processes (Swanson et al. 2010). These relationships are institutionalized in part in a funded Research Liaison position and in regular monthly meetings. The intensity of public issues waxes and wanes, and the topics shift with time, but the partnership has persisted for decades.

The communications portfolio of Andrews Forest scientists and science along this trajectory represents its diversity of roles in public decision-making concerning forests and watersheds. Scientists offered suggestions about ways to mitigate impacts of forestry operations, beginning with the first researchers stationed at the Andrews Forest (e.g., Timberman 1957). In the 1970s, basic science revealed the incredible richness and complexity of native forests, especially old growth, at a time when it was derisively referred to as a "biological desert," "decadent," and "over-mature" (Johnson and Swanson 2009). During the 1980s, as the "Old-growth War" unfolded, and more recently, scientists and their findings were featured in the *New York Times*, coffee table books with evocative writing and photography, books for a general readership, and many other venues for public outreach, including the arts and humanities (e.g., Kelly and Braasch 1988; Luoma 2006; Brodie et al. 2016). Applied studies of impacts of forestry operations on watersheds and ecosystems, and the resulting publications, were used extensively in challenges to continued logging.

Scientists presented interpretations of the past, present, and possible future states of the environment to leaders in the legislative and executive branches of government, including President Clinton in his Forest Summit in April, 1993. Central features of the IBP and LTER eras of ecosystem science--forest-stream interactions, roles of dead wood in terrestrial and aquatic ecosystems, forest succession following disturbance by fire and logging, watershed processes, biodiversity of the major components of the ecosystem, landscape dynamics in response to fire and flood—all found a place in the regional conservation strategy, the Northwest Forest Plan (e.g., Harmon et al. 1986; Gregory et al. 1991). Andrews Forest scientists had many forums to communicate: Congressional hearings, NAS-NRC committee reports, National Forest planning processes, as well as the one-of-a-kind NWFP processes. Science discoveries have been delivered to a wide readership, including Congressional staffers, through the Pacific Northwest Research Station's *Science Findings* and *Science Update* print and digital communications. Some LTER experiments, such as Mark Harmon's 200-year log decomposition experiment, have come to symbolize the commitments of scientists and land manager colleagues to long-term learning and adoption of new information. Messages about the science and these commitments to learning have been conveyed during the hundreds of field tours that serve as forums for discussion of the future of the forest.

8.3 Discussion

These case studies reveal features of all LTER site programs that create the potential for research and activities at LTER sites to have important impacts on major, societally-relevant environmental issues. Several up-front, intrinsic properties stipulated in requests for proposals of all LTER sites facilitate their capacity to contribute (Jones and Nelson, Chap. 3, this volume). First, the five core areas of LTER research direct sites to address key components of the environment, which are commonly central to big environmental challenges. Second, this research charge from NSF requires LTER site communities to have specialists in a wide range of disciplines to address environmental topics of interest and concern to the public. Third, NSF requires these science communities to operate a conceptually integrated research program which is reviewed by NSF-designated panels every three years to affirm this integration (i.e., at times of grant renewal and mid-term review). Consequently, over the past 40 years, these required features of LTER programs have produced strong science communities that are actively engaged in societally-relevant environmental issues to varying degrees.

Another feature common to many of the early sites to enter the LTER network is a deep, highly relevant pre-LTER history, which was important in the site's selection to join the LTER network. For example, the framework of long-term monitoring of precipitation and surface water hydrology and chemistry, along with the penchant of the Forest Service for establishing experimental watershed studies long before LTER was established prepared research communities in the culture and practices that are critical to successful long-term ecosystem research, including data management. Therefore, it is not surprising that several Forest Service sites--H.J. Andrews, Bonanza Creek, Coweeta, Hubbard Brook Experimental Forest, and Luquillo Experimental Forest—are or were part of the LTER network. In some cases, participation in the International Biological Program during the late 1960s and 1970s prepared site communities to successfully compete in calls to join the LTER network by giving them a decade head start in long-term, interdisciplinary ecosystem research.

As LTER sites have matured, they have acquired key characteristics that positioned them to make the big step to addressing major environmental issues. A strong background of highly credible science and records from many types of environmental monitoring that are relevant to emerging issues lends credibility and brings value-neutral information resources to public deliberations. For a site to effectively contribute on important issues over the long term, strong site leadership at the top has to be complemented by strong distributed leadership among the disciplines and institutions within the communities. Members of site communities must be able to articulate a strong sense of the history of past environmental conditions and prospective future conditions across their home bioregions. Key leaders in the site communities must be willing to participate in governance processes, even though they sacrifice time and resources that would otherwise go to traditional science activities that are rewarded in their home institutions (Lach et al. 2003). As big issues emerge and evolve, it is important for a site community to be ready, receptive, and adaptable to take on new, emergent science questions, such as assessing outcomes of alternative futures or the actual responses of an ecosystem to efforts to deal with the issue. Adaptive science is an integral part of adaptive management processes that are commonly an implicit part of addressing big environmental issues.

Emergence and persistence of an issue and the relevance of that issue to an LTER program occurred in similar ways across the three cases. In each case, a lingering regional problem festered, site science was highly relevant, and a science community with strong leadership stepped forward with willingness to participate in the often-messy governance processes. As a first step, site scientists clearly and with authority called out the issues: Hubbard Brook science showed that elevated air pollution was affecting forests and waters; Harvard Forest research revealed that urban and rural sprawl had reversed a 150-year trend of forest expansion and recovery throughout the New England landscape, and Andrews Forest science characterized the complexity and diversity of old-growth forests. Once a site community engaged with an emerging issue, the scope of relevant science expanded as did the scope and demands of the governance processes, such as providing science input to policy-makers.

These cases provide a small but diverse sample of possible governance contexts in which LTER sites exist. The Hubbard Brook case of air pollution played out in the national legislative, judicial and executive arenas, resulting in modification of a highly influential federal law (the Clean Air Act) directly affecting the entire country. The Andrews Forest case of contributing to the development of an ecosystem conservation strategy was regional in its biological and policy scope; but it rose to the national political stage with a federal judge's injunction stopping logging on National Forest and Bureau of Land Management lands throughout the vast Pacific Northwest region, and then the executive branch stepped in to develop the Northwest Forest Plan when Congress would not enact a solution. This case stimulated interest and actions to protect old native forests in other parts of the world. Harvard Forest's Wildlands and Woodlands regional forest conservation strategy takes place in a dominantly private landownership context, so "governance" must be more a bottom-up than a Federal top-down approach and must be carried out through individual landowner decisions, land trusts, and local or state-level governance mechanisms of policy and practice.

The full suite of LTER sites operate in a wide array of governance settings, raising interesting issues about the potential for LTER science impacts on major issues. Urban sites, for example, are in densely populated human landscapes with complicated governance contexts. Some of the polar sites, on the other hand, are in areas nearly devoid of human residents, yet their subject matter has profound significance to global change and governance. Despite this great difference in proximity to population centers, the potential for big impacts may be more a matter of the alignment of research themes, societal-relevant issues, and systems of governance that can connect the science with society.

A critical feature of engagement in governance processes is that it is a two-way street of communication; generally, scientists are learning throughout the process, and do not serve simply as providers of data and information (Lach et al. 2003). Conversations with land managers, state and federal agency personnel, and NGOs not only inform decisions and actions in land and resource management but also provide perspective for new site science to address important management questions.

These case studies reveal a value of the long-term nature of LTER not anticipated when the program began—decades-long engagement of a place-based science community with environmental issues can have major impact. Each of these efforts played out over several decades and continues today, despite the illusion of finality in the form of a culminating or landmark accomplishment, such as a piece of federal legislation or regulation or the publication of a regional conservation plan. Persistence of engagement has been very important, and LTER provided a necessary base of continuity. The extended tenure of LTER-based communities facilitates development of social networks with the public and policy-makers through a wide variety of formal and informal channels. The long intergenerational tenure common among scientists and other members of research communities at LTER sites facilitates long-term institutional memory, in contrast to the characteristic short-term memory of the political realm.

Partnerships with non-science entities are extremely important in connecting the science communities with society. For example, in the case of regional conservation strategies, partnerships with land management agencies (e.g., federal and state) and non-governmental land stewardship organizations (e.g., land trusts) can channel science information and perspectives to land management decision-making, and understanding of research needs back to the science community. Partner relationships forged during periods of social conflict persisted in some cases, because the science communities experienced their effectiveness and wished to channel other science findings into policy and practice. This is evident, for example, in the New England Science Policy Exchange involving Harvard Forest and Hubbard Brook. In the Andrews Forest case, the research-management partnership of the science community with the Willamette National Forest was staffed by a Research Liaison position charged with flow of information between the two communities and the two cultures and to society at large.

In all three case studies, the LTER site communities undertook new forms of communications essential to serving society more directly than traditional science communications. In terms of print media, all three sites issue attractive, color booklets aimed at public and policy-maker readers. The Hubbard Brook group, for example, launched *Science Links*, developed collaboratively by scientists and advisers to policy-makers. Harvard Forest has been reporting to a general readership on the Wildlands and Woodlands project at five-year intervals, and Andrews Forest has shared information via the *Science Findings* and *Science Updates* communiques of the Forest Service. A variety of other outreach engagements have been deployed, including field tours, workshops, opinion pieces in traditional and social news media, and sustained, one-on-one, informal relations with key contacts in policy, news media, and land management arenas.

In summary, important impacts occur when there is a confluence of science community capacity and an environmental issue for which that community and its LTER science has been relevant. These confluences have occurred in part because LTER scientists strive to communicate their science to broader communities of policy-makers, land managers, educators, students, and the general public. These self-selected participants have a commitment to the wellbeing of their home bioregion.

LTER site research and the careers of participants have now spanned a professional lifetime, so the accumulated wealth of knowledge and social networking are mature on topics of broad social relevance – air, water, climate, vegetation, animals, and use of natural resources. The well-known examples of major impacts outlined above show that big accomplishments are possible, but they require patience, flexibility, and willingness to sacrifice work on science objectives in order to participate in governance processes (Lach et al. 2003). Hopefully, these successes will encourage others to take on the challenge.

8.4 Looking Forward

Speculation about the future development of LTER-site communities in this regard is quite challenging, given their continuing evolution, the dynamism of big environmental problems, funding limitations for such efforts at the interface of science and society, and, above all, the funding future for long-term, place-based ecological science. Maintenance of adventuresome communities is not a review criterion in an NSF proposal; it is outside the scope of this NSF program. However, it is important for all involved—the communities themselves, administrators in their home institutions, NSF program officers, and reviewers of proposals and mid-term site reviews—to recognize the importance of community culture as LTER moves forward with generational change in leadership and participation within the LTER ranks. Engagement with the arts and humanities, an emerging feature of many LTER programs, may influence site community culture and the ability to reach a wider public (Swanson 2015; Leigh et al., Chap. 11, this volume). Clearly, this topic of LTER community connection with society deserves careful scholarship; the case study research sites discussed here have extensive archives that contain relevant resources.

Acknowledgements We especially thank R. Waide and S. Kingsland for creating this book project, bringing the participants together to reflect on the history of LTER, and for many helpful comments. We also thank J. Jones for helpful discussions of various phases of this work and H. Gosnell for perspectives on governance. This chapter is a contribution from the NSF-sponsored LTER programs of Andrews Forest (NSF 1440409), Harvard Forest (NSF grants:1237491 and 1832210), and Hubbard Brook (NSF grant: DEB- 1637685). The Harvard Forest contribution is also supported by a Coupled Natural and Human Systems grant (NSF grant: 1617075) and long-term funding from the Highstead Foundation.

References

Berlik, Mary M., David B. Kittredge, and David R. Foster. 2002. The illusion of preservation: A global environmental argument for the local production of natural resources. *Journal of Biogeography* 29: 1557–1568.

Bocking, Stephen. 2016. Forest experiment and practice at Hubbard Brook. In *Ecologists and environmental politics: A history of contemporary ecology*, ed. Stephen Bocking, 116–147. Morgantown: West Virginia University Press.

Bormann, F. Herbert, and Gene Likens. 1979. *Pattern and process in a forested ecosystem*. New York: Springer.

Brodie, Nathaniel, Charles Goodrich, and Frederick J. Swanson, eds. 2016. *Forest under story: Creative inquiry in an old-growth forest*. Seattle: University of Washington Press.

Callahan, James T. 1984. Long-term ecological research. *Bioscience* 34 (6): 363–367.

Chen, L., and C.T. Driscoll. 2005. Regional application of an integrated biogeochemical model to northern New England and Maine. *Ecological Applications* 15: 1783–1797.

Colman, David C. 2010. *Big ecology: The emergence of ecosystem science*. Berkeley: University of California Press.

Driscoll, Charles T., Gregory B. Lawrence, Arthur J. Bulger, Thomas J. Butler, Christopher S. Cronan, Christopher Eagar, Kathleen F. Lambert, Gene E. Likens, John L. Stoddard, and Kathleen C. Weathers. 2001. Acidic deposition in the northeastern United States: Sources and inputs, ecosystem effects, and management strategies. *BioScience* 51: 180–198.

Driscoll, Charles T., D. Whitall, J. Aber, E. Boyer, M. Castro, C. Cronan, C.L. Goodale, P. Groffman, C. Hopkinson, K. Lambert, G. Lawrence, and S. Ollinger. 2003. Nitrogen pollution in the northeastern United States: Sources, effects and management options. *BioScience* 53: 357–374.

Driscoll, Charles T., Y.-J. Han, C.Y. Chen, D.C. Evers, K.F. Lambert, T.M. Holsen, N.C. Kamman, and R.K. Munson. 2007. Mercury contamination in forest and freshwater ecosystems in the northeastern United States. *BioScience* 57: 17–28.

Driscoll, Charles T., Kathy F. Lambert, and Kathleen C. Weathers. 2011. Integrating science and policy: A case study of the Hubbard Brook Research Foundation Science Links program. *BioScience* 61: 791–801.

Driscoll, Charles T., Kathleen F. Lambert, F. Stuart Chapin III, David J. Nowak, Thomas A. Spies, Frederick J. Swanson, David B. Kittredge, and Clarisse M. Hart. 2012. Science and society: The role of long-term studies in environmental stewardship. *BioScience* 62: 354–366.

Driscoll, Charles T., K.F. Lambert, D. Burtraw, J.J. Buonocore, S.B. Reid, and H. Fakhraei. 2015. US power plant carbon standards and clean air and health co-benefits. *Nature Climate Change* 5: 535–540. https://doi.org/10.1038/nclimate2598.

Driscoll, Charles T., Kimberley M. Driscoll, Habibollah Fakhraei, and Kevin Civerolo. 2016. Long-term temporal trends and spatial patterns in the acid-base chemistry of lakes in the Adirondack region of New York in response to decreases in acidic deposition. *Atmospheric Environment* 146: 5–14.

Duncan, S.L., and J.R. Thompson. 2006. Forest plans and ad hoc scientist groups in the 1990s. *Forest Policy and Economics* 9: 32–41.

Franklin, Jerry F., Kermit Cromack, Jr., William Denison, Arthur McKee, Chris Maser, James Sedell, Fred Swanson, and Glen Juday. 1981. Ecological characteristics of old-growth Douglas-fir forests. US Forest Service, Pacific Northwest Forest and Range Experiment Station, Gen. Tech. Report PNW-118.

Foster, David R. 1999. *Thoreau's country: Journey through a transformed landscape*. Cambridge: Harvard University Press.

Foster, David R., and John D. Aber. 2004. *Forests in time: The environmental consequences of 1,000 years of change in New England*. New Haven: Yale University Press.

Foster, David R., D.B. Kittredge, B. Donahue, G. Motzkin, D. Orwig, A. Ellison, B. Hall, E.A. Colburn, and A. D'Amato. 2005. *Wildlands and woodlands: A vision for the forest of Massachusetts*. Harvard forest paper no. 27. Petersham, MA: Harvard University Press.

Foster, David R., B.M. Donahue, D.B. Kittredge, K.F. Lambert, M.L. Hunter, B.R. Hall, L.C. Irland, R.J. Lilieholm, D.A. Orwig, A.W. D'Amato, E.A. Colburn, J.R. Thompson, J.N. Levitt, A.M. Ellison, W.S. Keeton, J.D. Aber, C.V. Cogbill, C.T. Driscoll, T.J. Fahey, and C.M. Hart.

2010. *Wildlands and woodlands: A vision for the New England landscape*. Cambridge, MA: Harvard University Press.
Foster, David R., David Kittredge, Brian Donahue, Kathy Fallon Lambert, Clarisse Hart, and James N. Levitt. 2014. The Wildlands and Woodlands Initiative of the Harvard Forest, Harvard University. In *Conservation catalysts: The Academy as nature's agent*, ed. James N. Levitt, 3–30. Cambridge, MA: Lincoln Institute of Land Policy.
Foster, David R., K.F. Lambert, D.B. Kittredge, B.M. Donahue, C.M. Hart, W.G. Labich, S. Meyer, J. Thompson, M. Buchanan, J.N. Levitt, R. Pershel, K. Ross, G. Elkins, C. Daigle, B. Hall, E.K. Faison, A.W. D'Amato, R.T.T. Forman, P. Del Tredici, L.C. Irland, B.A. Colburn, D.A., Orwig, J.D. Aber, A. Berger, C.T. Driscoll, W.S. Keeton, R.J. Lilieholm, N. Pederson, A.M. Ellison, M.L. Hunter, and T.J. Fahey, T.J. 2017. *Wildlands and woodlands, farmlands and communities: Broadening the vision for New England*. Harvard Forest, Petersham, MA: Harvard University Press.
Gondo-Tugbawa, Solomon S., Charles T. Driscoll, John D. Aber, and Gene E. Likens. 2001. Evaluation of an integrated biogeochemical model (PnET-BGC) at a northern hardwood forest ecosystem. *Water Resources Research* 37: 1057–1070.
Gregory, Stanley V., Frederick J. Swanson, W. Arthur McKee, and Kenneth W. Cummins. 1991. An ecosystem perspective of riparian zones: Focus on links between land and water. *BioScience* 41: 540–551.
Harmon, M.E., J.F. Franklin, F.J. Swanson, P. Sollins, S.V. Gregory, J.D. Lattin, N.H. Anderson, S.P. Cline, N.G. Aumen, J.R. Sedell, G.W. Lienkaemper, K. Cromack, and K.W. Cummins. 1986. Ecology of coarse woody debris in temperate ecosystems. In *Advances in ecological research*, ed. A. MacFadyen and E.D. Ford, vol. 15, 133–302. Orlando, FL: Academic Press.
Holmes, Richard T., and Gene E. Likens. 2016. *Hubbard Brook: The story of a forested ecosystem*. New Haven: Yale University Press.
Hughes, Brent B., Rodrigo Beas-Luna, Allison K. Barner, et al. 2017. Long-term studies contribute disproportionately to ecology and policy. *BioScience* 67: 271–281.
Johnson, K. Norman, and Frederick J. Swanson. 2009. Historical context of old-growth forests in the Pacific Northwest--Policy, practices, and competing worldviews. In *Old growth in a new world: A Pacific Northwest icon reexamined*, ed. Thomas A. Spies and Sally L. Duncan, 12–28. Covelo, CA: Island Press.
Kelly, David, and Gary Braasch. 1988. *Secrets of the old-growth forest*. Salt Lake City, Utah: Gibbs-Smith Publisher.
Labich, W.G., E.M. Hamin, and S. Record. 2013. Regional conservation partnerships in New England. *Journal of Forestry* 111: 326–334.
Lach, Denise, Peter List, Brent Steel, and Bruce Shindler. 2003. Advocacy and credibility of ecological scientists in resource decision making: A regional study. *BioScience* 53: 171–179.
Lambert, K.F., M.F. McBride, M. Weiss, J.R. Thompson, K.A. Theoharides, and P. Field. 2018. *Voices from the land: Listening to New Englanders' views of the future*. 24 p. Harvard Forest, Harvard University and the Science Policy Exchange.
Likens, Gene E., F. Herbert Bormann, and Noye M. Johnson. 1972. Acid rain. *Environment: Science and Policy for Sustainable Development* 14 (2): 33–40.
Likens, Gene, and F. Herbert Bormann. 1977. *Biogeochemistry of a forested ecosystem*. New York: Springer.
Luoma, Jon R. 2006. *The hidden forest: The biography of a forest ecosystem*. Corvallis: Oregon State University Press.
McBride, Marissa F., Kathleen F. Lambert, Emily S. Huff, Kathleen A. Theoharides, Patrick Field, and Jonathan R. Thompson. 2017. Increasing the effectiveness of participatory scenario development through codesign. *Ecology and Society* 22 (3): 16. https://www.ecologyandsociety.org/vol22/iss3/art16.
McBride, Marissa F., Matthew J. Duveneck, Kathleen F. Lambert, Kathleen A. Theoharides, and Jonathan R. Thompson. 2019. Perspectives of resource management professionals on the

future of New England's landscape: Challenges, barriers, and opportunities. *Landscape and Urban Planning* 188: 30–42.

Peters, S.C., Joel D. Blum, Charles T. Driscoll, and Gene E. Likens. 2004. Dissolution of wollastonite during the experimental manipulation of Hubbard Brook Watershed 1. *Biogeochemistry* 67: 309–329.

Robbins, William G. 2020. *A place for inquiry, a place for wonder: The Andrews Forest*. Corvallis: Oregon State University Press.

Robertson, G. Philip, Scott L. Collins, David R. Foster, and 10 additional authors. 2012. Long-term ecological research in a human-dominated world. *BioScience* 62:342-353.

Spies, Thomas A., and Sally L. Duncan, eds. 2009. *Old growth in a new world: A Pacific Northwest icon reexamined*. Covelo, CA: Island Press.

Swanson, Frederick J. 2004. Roles of scientists in forestry policy and management: views from the Pacific Northwest. In *Forest futures: Science, politics, and policy for the next century*, ed. Karen Arabas and Joe Bowersox, 112–126. Lanham, MD: Rowman & Littlefield Publishers.

———. 2015. Confluence of arts, humanities, and science in sites of long-term ecological inquiry. *Ecosphere* 6 (8): 1–23.

Swanson, Frederick J., Steve Eubanks, Mary Beth Adams, John C. Brissette, and Carol DeMuth. 2010. *Guide to effective research-management collaboration at long-term environmental research sites*. Gen. Tech. Rep. PNW-GTR-821. Portland, OR: U.S. Department of Agriculture, Forest Service, Pacific Northwest Research Station.

Templer, P.H., K.F. Lambert, M. Weiss, J.S. Baron, C.T. Driscoll, and D.R. Foster. 2015. Using science-policy integration to improve ecosystem science and inform decision-making: lessons from U.S. LTERs. Proceedings of Special Session at 100th Ecological Society of America Meeting. Baltimore, Maryland.

The Timberman. 1957. Can there be orderly harvest of old growth? *The Timberman* 58: 48–52.

Thompson, Jonathan R., Arim Wiek, Frederick J. Swanson, Stephen R. Carpenter, Nancy Fresco, Theresa Hollingsworth, Thomas A. Spies, and David R. Foster. 2012. Scenario studies as a synthetic and integrative research activity for long-term ecological research. *BioScience* 62: 367–376.

Thompson, Jonathan R., K. Fallon-Lambert, D.R. Foster, M. Blumstein, E.N. Broadbent, and A.M. Almeyda Zambrano. 2014. *Changes to the land: Four scenarios for the future of the Massachusetts landscape*. Petersham, MA: Harvard Forest, Harvard University.

Thompson, Jonathan R., K.F. Lambert, D.R. Foster, E.N. Broadbent, M. Blumstein, A.M.A. Zambrano, and Y. Fan. 2016. Four land-use scenarios and their consequences for forest ecosystems and services they provide. *Ecosphere* 7: 1–22.

Vanderbilt, Kristin, and Evelyn Gaiser. 2017. The International Long Term Ecological Research Network: A platform for collaboration. *Ecosphere* 8 (2): e01697.

Chapter 9
Seeing the Invisible Present and Place: From Years to Centuries with Lake Ice from Wisconsin to the Northern Hemisphere

John J. Magnuson

Abstract Ecologists involved in long-term research created the metaphors of the "Invisible Present" and the "Invisible Place" to communicate the purposes of Long-Term Ecological Research to the National Science Foundation, other scientists, and the wider public. This chapter discusses using lake ice cover to tackle the joint problems of the Invisible Present and the Invisible Place, by expanding the time and space scales not only at the North Temperate Lakes Long-Term Ecological Research site, but across Wisconsin, North America, and the Northern Hemisphere. LTER formed an international Lake Ice Analysis Group in the mid-1990s, to create and make publicly available a common lake ice phenology database. By 2000, analyses of ice cover provided strong evidence for warming winters in the Northern Hemisphere, and for the impact of the Second Industrial Revolution on climate. Importantly, for ice freeze time series, 20 and 50 years of data were insufficient to detect climate change while longer records did; large-scale climate drivers like El Niño, the North Atlantic Oscillation, as well as local weather influenced inter-year dynamics of lake ice seasonality; inter-year dynamics between lakes were coherent locally and persisted regionally at lower values; extreme ice dates changed in the direction of a warming climate; and changes in lake ice already have affected people negatively. Research on inland water ice and winter limnology continues today and has depended on successes in continuing leadership and participation within and beyond the LTER program.

Keywords LTER program · Long-term ecological research · Limnology · Climate change · Global ecology · International networks · Time and space scales · Long-term monitoring · Lake ice phenology

---

J. J. Magnuson (✉)
Center for Limnology, University of Wisconsin-Madison, Madison, WI, USA
e-mail: john.magnuson@wisc.edu

R. B. Waide, S. E. Kingsland (eds.), *The Challenges of Long Term Ecological Research: A Historical Analysis*, Archimedes 59,
https://doi.org/10.1007/978-3-030-66933-1_9

9.1 Introduction

> Time is sort of a river of passing events, and strong is its current; no sooner is a thing brought to sight than it is swept by and another takes its place, and this too will be swept away.
>
> Emperor Marcus Aurelius (121–180).

> It takes a leap of the imagination to … accelerate … a process of change in the environment enough to see it in a more familiar frame and thus discern its meaning.
>
> Al Gore (1992).

> The field cannot be well seen from within the field.
>
> Ralph Waldo Emerson (1909).

> A fundamental characteristic of complex human systems [is that] 'cause' and 'effect' are not close in time and space [yet] most of us assume, most of the time, that cause and effect are close in time and space.
>
> Peter M. Senge (1990)[1]

The metaphors of the "Invisible Present" and the "Invisible Place" were published in a trilogy of papers in *BioScience* describing the benefits of a new research program, the Long Term Ecological Research (LTER) Program, initiated in 1980 by the National Science Foundation (Franklin et al. 1990; Magnuson 1990; Swanson and Sparks 1990). We introduced these metaphors to emphasize the values of setting observations on ecological systems in a broader temporal and spatial context and to quickly give them meaning to more general audiences. We first used the metaphor of the "Invisible Present" in 1983, only a couple of years after our LTER site, North Temperate Lakes, located in Wisconsin, was funded (Magnuson et al. 1983); the purpose was to inform our university about the new LTER Program on lake ecosystems.

What did those metaphors signify? The "Invisible Present" expressed the idea that a single observation may reveal little or nothing of the processes that produced the observed datum, but considering that datum in the historical context established by a series of observations may reveal those underlying processes and so make it possible to imagine the future. Magnuson (1990) posited that even scientists had underestimated the slow changes occurring around us. These occur over decades, and thus multiple-year lags between cause and effect are easily missed. Such changes and processes were hidden in the "Invisible Present" (Magnuson et al. 1983).

Similarly, by the "Invisible Place" Swanson and Sparks (1990) suggested that our understanding of the processes at work in a particular place may be greatly enhanced by considering it in a broader spatial context comprising multiple spatial scales ranging from the plot/patch to watershed, landscape, region, and continent. To do otherwise separated that place from the important spatial context, the influences and dynamics at that place. For example, without knowing the position

[1] While Senge was writing about complex human systems, I think the same could be said of complex ecological systems.

of a research site with respect to regional variations of disturbance regimes or temperature-moisture conditions, the significance of results cannot be understood; that particular plot or patch resides in an "invisible place."

Both Magnuson (1990) and Swanson and Sparks (1990) argued that, without the temporal and spatial context provided by long-term ecological research, attempts to understand and predict changes in the world around us may be flawed and our attempts to manage our environment, or adapt to environmental change and variability, weakened. The metaphors were intended to help persuade the National Science Foundation, our peer reviewers, the scientific community at large, and as we later found, the general public of the value of the LTER. The metaphors provided a quick way to communicate the importance of the LTER program in "elevator conversations" or in a few minutes with a policy maker.

As the epigraphs that begin this introduction illustrate, other writers have reflected on the difficulty of understanding events in their broad contexts of space and time. These and other writers have helped me understand the importance of these concepts. The "Invisible Present" and "Invisible Place" characterized the situation we were in, or at least partially in, in the 1970s. The community of ecological scientists and the National Science Foundation organized in the late 1970s to respond to the need for long-term, sustained research on ecosystems across longer time and broader space scales. My purpose here is to recount one long-term effort to move beyond that situation since 1981. I will discuss the use of analyses of lake ice cover to expand the time and space scales not only at the North Temperate Lakes LTER Site, but across Wisconsin, North America, and the Northern Hemisphere.

9.2 Historical Background

When we were writing our first LTER proposal we pointed out the legacies of sustained research that were already in place, as did others. For us, the legacies were an active interdisciplinary faculty, lake-side research facilities, a diversity of lake types, and a long history of limnological science. In North America, limnology began at the University of Wisconsin-Madison in the late 1800s with Edward A. Birge and Chancey Juday (Beckel 1987; Magnuson 2002b). Birge initially studied the natural history and taxonomy of a major group of zooplankton. Birge brought Juday to Wisconsin and by 1906 they had begun to ask questions about water circulation, dissolved oxygen, thermal structure, and color, mostly in Lake Mendota. Birge used the metaphor of a "house," analogous to Stephen Forbes's description of "The Lake as a Microcosm" (Forbes 1887). Birge referred to the physical, chemical and biological processes in a lake as "housekeeping"; he drew the analogy between different rooms in a house and different zones of a lake (Sellery 1956). Forbes' "microcosm" and Birge's "house" were early precursors of the "ecosystem" concept used today.

Perhaps more relevant to our first LTER proposal was that Birge and Juday's activities moved to the Northern Highlands Lake District in Wisconsin. At the Trout Lake Station, Juday led a comparative study on the physics, chemistry, and biology of a diverse set of more than 500 lakes from 1925 to 1941 (Beckel 1987; Magnuson 2002b). A quote explains their approach: "If enough data are gathered, the data will speak for themselves." In 1936, five years before the research in the Northern Highlands ended, Birge wrote his essay "A House Half Built."[2] Advancing age and lack of funding brought their work to an end before they could finish the synthesis they appeared to have intended. The next generation of limnologists at UW–Madison led by Arthur Davis Hasler (Beckel 1987; Magnuson 2002b) went in new directions with experimental limnology, eutrophication, and fish orientation at the core.

The present faculty at the Center for Limnology have adopted the approaches of both of these early schools and added more as the science of limnology evolved and advanced (Magnuson 2002b). One major research activity is the North Temperate Lakes LTER Program.[3]

It would be fair to ask how many years of long-term data are needed to see results from an LTER program. What can we do with 1 year, 3 years, or N years? We did have 17 years of data (1925–1941) from many lakes in the Northern Highlands from the Birge and Juday era and a survey of many of the same lakes from the 1960s conducted by the Wisconsin Department of Natural Resources. Perhaps we could do something with that. We eventually did, but first we had to find the data in dusty boxes in the University Archives and develop a digital database for researchers to use. Our first use of the old data analyzed the temporal and spatial variability in the long-term zooplankton data (Kratz et al. 1987). Over the years, many uses have been made of this legacy database.

9.3 Jumping into Long-Term Ecological Research

Our first group of graduate students and their faculty advisors began the core measurements in 1980 and 1981, on various parameters in their disciplinary specialties, such as lake physics and chemistry, groundwater hydrology, phytoplankton, zooplankton, fish ecology, benthos, and macrophytes. After their first summer, we had 1 year of data. Not surprisingly, the resulting student theses and subsequent publications were process-oriented and descriptive research in their disciplinary specialties. Some of our faculty believed that would be the flavor of all graduate student research in a long-term study; that early judgement proved to be incorrect. However, I wanted to quickly move out of the invisible present in a major way. So what could we do in the early LTER years to analyze much longer-term data than was available from the slowly emerging LTER time series on Wisconsin lakes?

---

[2] Birge' essay, "A House Half Built", was an address before the Madison Literary Club, October 12, 1936. In Edward A. Birge Papers, Wisconsin Historical Society, Madison, Wisconsin.

[3] The first LTER grant to the Center for Limnology from NSF's Division of Environmental Biology was DEB-8012313 (1980–1986).

Fig. 9.1 Dale Robertson sampling Mary Lake in the Northern Highlands of Wisconsin in the mid-1980s. (Photo: P. Jacobsen)

One Ph.D. student, Dale Robertson (Fig. 9.1), took a different tack than other students who were analyzing the relatively short data sets from our own measurements. Dale had finished his MS degree studying the thermal structure of Trout Lake, one of our LTER primary lakes (Robertson and Ragotzkie 1990). This research was consistent with the short time and space scales being researched by the other graduate students. However, I told him that if he wished to continue his LTER funding, his Ph.D. research must make use of long-term limnological data. I suggested he analyze the long records of ice freeze and breakup dates on Lake Mendota in south central Wisconsin. We knew of the Mendota ice records: one of our principal investigators, Robert Ragotzkie, previously advised Jon Scott's Ph.D. research on the ice and winter heat budgets of lakes in Wisconsin including Lake Mendota (Scott 1964). Serendipitous long-term ice records made prior to the LTER years gave us an opportunity to quick-start analyses of longer time series.

Robertson decided to study Lake Mendota's unusually long ice record, which began in the winter of 1855–1856 and continued annually to 1986–87, or for 132 years at that point in our history.[4] Dale and I both remember his Ph.D. thesis defense in 1989, at which he concluded that the changing ice dates provided

[4]The record likely originated because the local citizenry used the ice for their kitchen ice-boxes, and also for easier travel than the overland trails and wagon routes over land over the wetland-rich region.

evidence of climate warming and part of the variability was related to El Niño events documented in the Pacific Ocean. These conclusions were challenged by the paleoclimatologist on his committee, who stated that climate did not change that fast and suggested that the phrase "climate change" be removed from the thesis. He also stated that El Niño was a Central Pacific event and that the entire chapter on El Niño should be deleted. Fortunately, the climate scientist signed the thesis: Dale did not remove the challenged conclusions and phrases; he had learned to respect his data and his analyses, and ultimately published the conclusions that they revealed (Robertson 1988; Robertson 1989; Robertson et al. 1992; Robertson et al. 1994; Anderson et al. 1996). Also, in a conclusion ahead of its time, he forecast the ice seasonality into the future with air temperature scenarios from one of the climate change models. The resulting ice cover scenarios suggested that Mendota might on average be ice free by about 2050. Unfortunately, while these graphics are in his thesis (Robertson 1989) he omitted them from his publication (Robertson et al. 1992).

The initial pushback from a climate scientist was somewhat surprising; Robertson had cited several recent papers in his thesis that supported the idea of climate change (Liss and Crane 1983; Hansen et al. 1984; Hansen and Lebedeff 1987). Dale's conclusions were modest, suggesting that lake ice cover could be a good indicator of climate change or were consistent with global warming associated with the "Greenhouse Effect." He also cited papers on the role of El Niño causing warmer winter temperatures in some parts of Canada and the United States. His thesis work on short ice covers on lakes in the year following a strong El Niño were the first for lake ice (Robertson 1989; Robertson et al. 1994; Anderson et al. 1996). That the paleoclimatologist, after reflection, signed Robertson's thesis is not surprising.

That our work on lake ice was not cited in the 1990 Assessment Report of the International Panel on Climate change (Houghton et al. 1990) is not surprising; Robertson's thesis did not appear until 1989 and our published papers appeared in 1990 and later. The climate science community was just beginning to make its case for climate change based on research from the 1980s and earlier. At Wisconsin, Reid Bryson (Bryson and Murray 1979) had recently argued that sulfate aerosols primarily from burning fossil fuels were causing the earth's atmosphere to cool; eventually carbon dioxide, a greenhouse gas, won the argument as the world warmed.

The three lead papers presented in 1987 at the first LTER briefing of the National Science Foundation were early versions of our 1990 *BioScience* publications (Franklin et al. 1990; Magnuson 1990; Swanson and Sparks 1990). I had been looking for a long annual time series to help persuade reviewers of the LTER Program that results would reveal or unveil new dynamics not apparent from observations over a few years. My idea was to sequentially open up a time series from a single year to longer and longer windows of history. I had spent part of a sabbatical at the Windermere Laboratory in the English Lake District in 1983–84. This laboratory was known for long-term research carried out by specialists within the sub-disciplines of limnology and aquatic ecology. I was thinking of using their northern pike database, which had a strong 4-year oscillatory dynamic. I ended up choosing the changes in the Lake Mendota ice time series because it was much longer than the northern pike time series, was on a Wisconsin Lake, seemed interestingly complex, and, perhaps most importantly, Robertson had conducted quality control and thorough analyses of the time series.[5] For the 1987 NSF briefing, I persuaded my colleagues to use "Invisible Present" and "Invisible Place" to clarify and simplify the science issues in our communication with a wide audience.

In 1991, an additional effort to advance long-term ecological research was made with an international perspective, and was discussed in a volume edited by Paul G. Risser (Risser 1991).[6] Risser chaired the Scientific Advisory Committee of five other scientists from the USA, Australia, Germany, Scotland, and Venezuela to plan this contribution. The diversity of papers reinforced its international scope, with contributions from the United States, Europe, Africa, and Australia. As an LTER lead Investigator at the North Temperate Lakes Site, I found Tom Callahan's descriptive chapter on the first 10 years of LTER from the National Science Foundation's perspective most interesting (Callahan 1991) (Fig. 9.2). At the outset he reiterated that the LTER was aimed at ecosystems and was a research program rather than a monitoring program. He also pointed out that an ecosystem property "might be assessed at the level of a tree, the stand, or the forest or for minutes, days, or years." This suggested to me that he was thinking in the context of the thrust of my chapter and its metaphors. Two reviews of the book (Kolasa 1992; White 1993)

Fig. 9.2 James T. Callahan at an LTER meeting in winter 1985, in Albuquerque, New Mexico. (Photo: J. Magnuson; Magnuson et al. 2006, on p. vi)

[5] At the time I was unaware of an equally striking, but shorter, time series in the freezing and breakup of Lake Kallavesi in Finland (Simojoki 1961) from winters 1883–84 to 1957–58. Warming trends after about 1880 and oscillatory dynamics were apparent.

[6] Published on behalf of the Scientific Committee on Problems of the Environment (SCOPE) of the International Council of Scientific Unions (ICSU).

were positive and also raised questions about what science could best be addressed by long-term ecological research, and how it should be conducted.

Our chapter in Risser's volume (Magnuson et al. 1991) focused on expanding the temporal and spatial scales of ecological research and comparing divergent ecosystems. We again presented the idea of the "Invisible Present" and provided approaches to conducting research in the "Invisible Place" by comparing dissimilar ecosystems. Peter White (1993), in his review of the volume, noted a contrast between the US sites and non-US sites. The chapters dealing with the US LTER program, he observed, appeared to adopt what he called a "bottom-up" approach in that they were organized around investigators and specific research hypotheses, but paid little or no attention to large-scale monitoring over wide regions. This observation conformed to Callahan's comment that LTER was a research and not a monitoring program. White noted that the non-US sites, on the other hand, appeared to exhibit a more "top-down" approach, in that more attention was given to spatial extensiveness and monitoring, in addition to intensive study at particular sites. As an illustration of the dual nature of the non-US sites, White referred to O. W. Heal's discussion of British long-term research, which had posed the question of "how a network of sites throughout Great Britain should combine intensive measurements and experimentation at some sites and less intensive measurements at other sites in order to achieve an extensive spatial context" (White 1993, p. 637).

In his review, White reflected on how the NSF LTER program might also combine both of these functions and thereby take on a larger role, and he used our chapter on North Temperate Lakes to illustrate how this expanded function might operate. I found White's comments about our chapter encouraging and appreciated his belief that we were going in a good direction (White 1993):

> That the NSF LTER sites can play a larger role is beautifully illustrated by the work at the North Temperate Lakes LTER as described in the chapter by John J. Magnuson and collaborators. This extremely valuable contribution supplies the book's best image: the role of long-term research in making visible the "invisible present" (that is supplying temporal context to current observations) and the role of spatial extensiveness in making visible the "invisible place" (supplying spatial and environmental context to observations at one point). The researchers at the North Temperate Lakes LTER have complemented NSF funding with many other sources and created a regional context for their observations and experiments.

I take from these kind words that the metaphors were useful in making the case for the LTER program and that implementing long-term research in space and time was underway. But White was also identifying some key problems that he thought needed to be addressed: "How should we integrate observation with experimentation, extensive networks with intensive research sites, and monitoring with research? How should we support institutions for the long-term and retain a focus on the creativity and scientific merit of individual proposals?" The present chapter discusses how these questions about integrating different goals could be addressed successfully by building on our experience in the North Temperate Lakes LTER program and adopting an ever-widening temporal and spatial scale.

9.4 A Closer Look at the Invisible Present

Let's take a closer look at the "Invisible Present" as it melts away in the face of long-term data. When we look at a single year in the ice-cover duration time series, the winter of 1982–83 (Fig. 9.3 top), we only see that the lake was ice covered for 52 days. We were in the "Invisible Present" with no context to make that observation more meaningful. When we opened the window to 10 years, we then saw that the winter of 1982–83 was an unusual year, with less than half the duration of ice cover observed in the other years. When we opened the window to 50 years, the influence of repeating El Niño events was apparent. Ice cover in 1982–83 and other strong-to-moderate El Niño years was shorter than the other years. Thus, ice duration was being influenced by a large-scale climatic teleconnection from the Central Pacific Ocean. This was instructive, as we were often asked whether ice cover change and variation were primarily being determined by local events. When we opened up the window to 132 years, the period of record at that time, a long-term trend toward shorter ice cover was apparent with a significance level of $p = 0.05$. All three of these conclusions were apparent in Robertson's 1989 Ph.D. thesis.

The final view (Fig. 9.3 bottom) of the time series removed essentially half of the invisible-present problem by putting the present in the context of its numerical history, or at least 132 years of it. The other half, the future, had not been revealed because the series, of course, only extended to the period of record. But future years would be revealed with long-term research underway. Today, in winter 2019–2020, the period of record for Lake Mendota reached 164 winters, with 32 more years added to the now recorded future. New features are unveiled as the time series is extended. However, as Emperor Marcus Aurelius, quoted above, laments: "no sooner is a thing brought to sight than it is swept by and another takes its place, and this too will be swept away." Clearly, the expanding time series reveals what was once the unknown future, but the task never is really completed. This makes a strong case for long-term ecological research; if the time series measurements were to stop, then any future 'presents' will be invisible without the temporal context of the past.

Our first effort with a long-term flavor analyzed the coherent dynamics between our northern lakes. Temporal coherence between two lakes is a measure of the degree to which the two lakes have coincident or synchronous changes from year to year versus whether the two lakes change independently from year to year. For example, do ice breakup dates of the two lakes over the years track each other or not? We calculate the correlation between them and often express it as the shared variance (variability) between the two. A shared variance of 1.0 means they track each other perfectly; a coherence of 0.5 means that half of their variability is shared over the time series and the other half is independent. If they are completely independent, then the shared variance is 0.0.

The ice breakup times series for the northern and southern Wisconsin lakes illustrates the meaning of coherence (Magnuson et al. 2004) (Fig. 9.4). The seven northern Wisconsin lakes were highly coherent; this is apparent in the shared

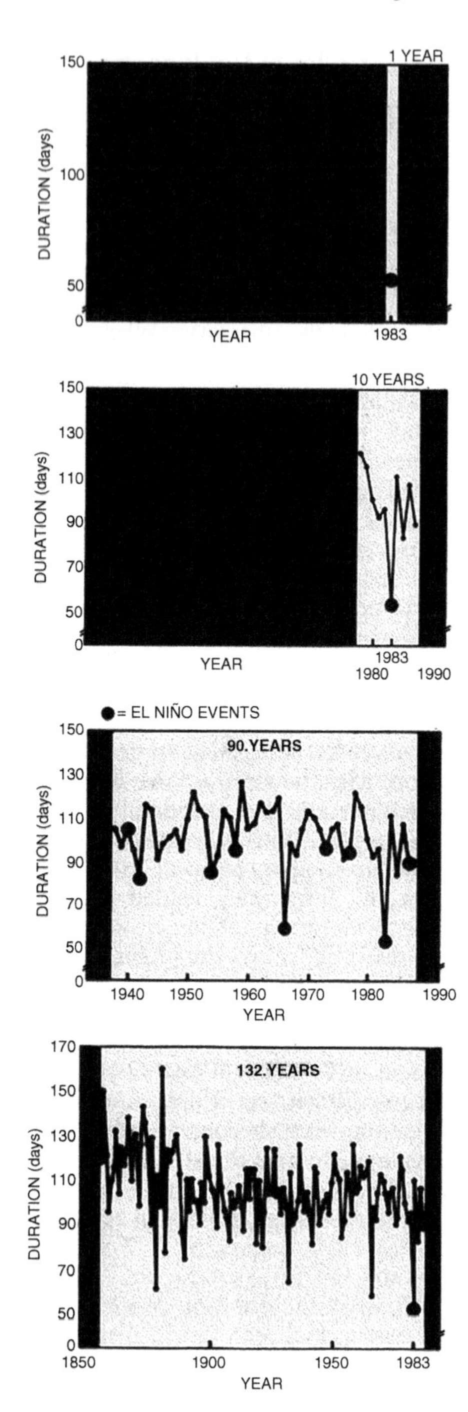

Fig. 9.3 When a time series is opened up, new phenomena become apparent and the present is put into a context that makes it more understandable and more interesting. The duration of ice cover on Lake Mendota, Wisconsin, at Madison, has been recorded, originally by interested citizens beginning in 1855 and now by the state's climatologist, providing the longest record in the state. The record was analyzed by Dale M Robertson (1989). (Magnuson 1990, on p. 496)

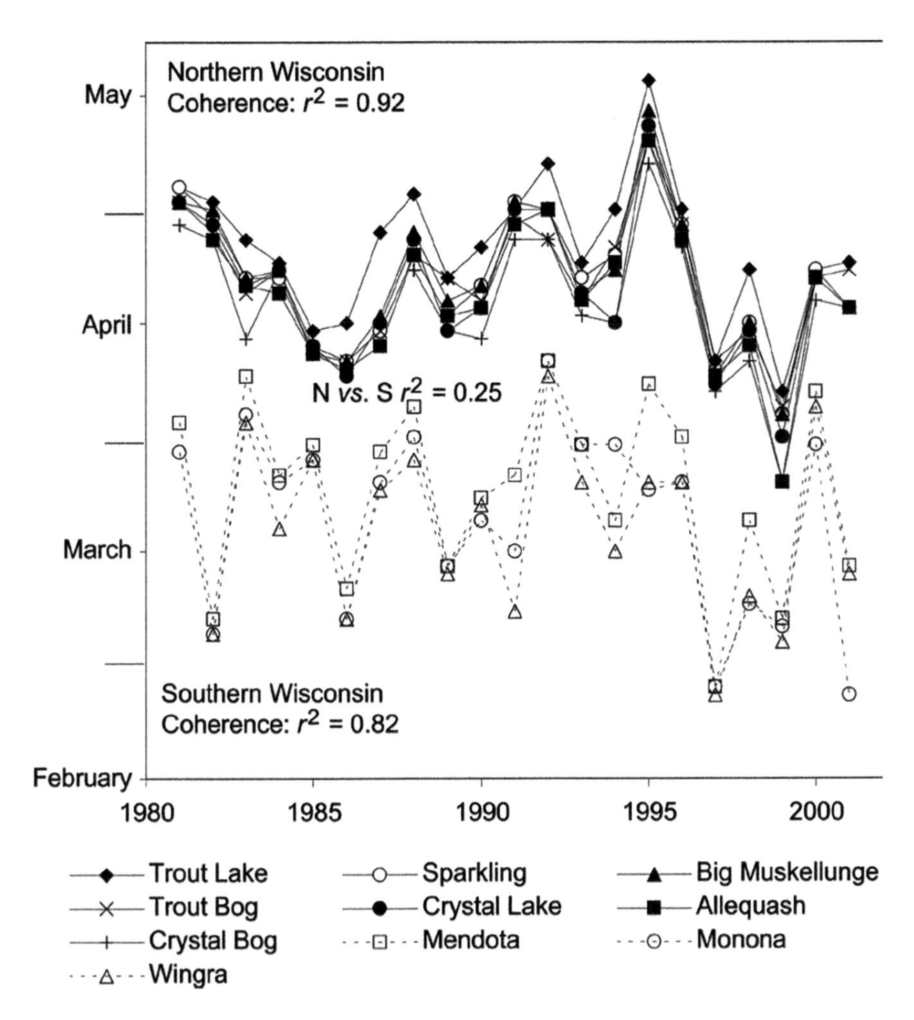

Fig. 9.4 Coherence in ice breakup time series for LTER lakes from 1981–2001 LTER lakes. Coherences are measured as r^2 or shared variance between pairs of 7 northern, 3 southern, and between northern and southern LTER lakes (Magnuson et al. 2004 on page 361)

variance of 0.92 or by just viewing the data and noting how well the lakes track each other. The southern lakes were similar but a bit less coherent with a shared variance of 0.82. However, the coherence between a pairing of each southern with each northern lake was much lower (0.25). Again, you can see by inspection that the northern lakes did not share all the dips and rises with the southern lakes. The north-south distance between the northern and southern group of lakes is about 200 km; average annual temperatures are 8.1 °C in the south and 4.5 °C in the north (Magnuson et al. 2006). Even these moderate differences in location and annual temperature influenced the differences in the year-to-year dynamics of the lakes.

Among all the variables we measure, breakup date provides the cleanest measure of the climate (Magnuson et al. 2004). Breakup dates, regardless of lake depth and area, are influenced by interactions with climate processes at the lake's surface on a per unit area basis. On the other hand, freeze dates are markedly less coherent with a shared variance of 62% because they are influenced by the different areas and volumes of lakes that influence wind mixing and heat transfer. Other physical limnological properties of lakes differ in coherence as well. For example, lake surface temperatures are highly coherent between the lakes (86%), but only 6% for summer bottom temperature).

To return to our first long-term paper on temporal dynamics using our still short time series of LTER data, we calculated coherence between pairs of the 7 lakes in Northern Wisconsin when we had only 7–8 years of data (Magnuson et al. 1990). We estimated coherence of 15 physical, chemical, and biological variables of annual values. Surprisingly, the conclusions with these short time series were supported in a second paper (Kratz et al. 1998) with our then 13-year time series from 1982 to 1994 using 61 physical, chemical, and biological variables. Physical variables were more coherent between lakes, on average, than chemical variables, which in turn were more coherent than biological variables (Fig. 9.5). We concluded from our northern lake site that physical variables were more closely linked to a common climate signal; climate signals of chemical variables were dampened and complicated by within-lake differences in hydrology and biology. Climate signals of biological data were even further dampened and complicated owing to within lake interactions in biology and chemistry. We were encouraged by these findings that began to move us out of the invisible present (1 year) and place (one lake) with our ongoing LTER measurements.

What were the implications of these results from these short time series? First, that some dynamic features of lake ecosystems were apparent even in time series only 8 years long. Second, temporal coherence between lake pairs was not always well estimated in short time series but changed with longer time series for some variables (Magnuson et al. 2004; Magnuson et al. 2006). For example, the coherence

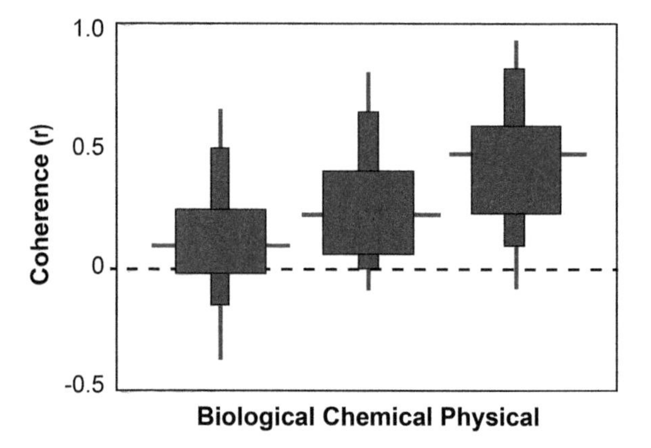

Fig. 9.5 The correlation (r) between each biological, chemical, and physical parameter between all pairings of seven northern Wisconsin LTER lakes. The median, 50 percentiles, 90 percentiles, and range are shown. (Modified from Magnuson and Kratz 2000, on p. 79, and from Kratz et al. 1998, on p. 282)

of acid neutralizing capacity of the lakes' waters increased from a shared variance of 0.25 to 0.83 as the time series lengthened. Third, we also learned that many questions other than coherent dynamics are important if we are to understand lake ecology and consider potential changes in the future.

9.5 Taking a Wider View of the Invisible Place

Our first LTER proposal (1981–85) indicated that our original goals were to expand both the temporal and the spatial scale of ecological analyses on lake ecosystems (Magnuson et al. 2006). We proposed to do long-term research on seven lakes in Northern Wisconsin which differed greatly from each other, were near our Trout Lake field station, and had been sampled earlier by Birge and Juday. Our first scientific review panel, also in 1981, helped us choose the suite of lakes from high to low in the landscape in the same groundwater flow system. The few long-term studies of lakes prior to our LTER focused on single important lake ecosystems; examples are Mirror Lake, New Hampshire (Likens 1985), Lake Mendota, Wisconsin (Brock 1985), Lake Washington, Washington (Edmondson 1991), and Lake Tahoe, California (Goldman 2000).

We had first used Robertson's analysis of Lake Mendota ice seasonality to make the case for our metaphor, the "Invisible Present." We realized that one lake in one place did not suffice. Records and analyses from a single place/lake should not be blindly extrapolated to a lake district, state, country, continent, hemisphere, or certainly to the globe. For example, the reduction in ice cover on Lake Mendota did not mean that the trend in winter warming was global. Our next obvious goal was to add lakes with long time series and that were more widely distributed around the Northern Hemisphere.

The seven lakes we chose for LTER primary lakes were individual lake ecosystems that differed greatly from each other. Comparing them already initiated an attack on the "Invisible Place." But we wished to address that question more broadly and bring the northern Wisconsin group of seven LTER lakes and the Mendota ice records into a larger spatial context (Swanson and Sparks 1990). Fortunately, an augmentation proposal to NSF (1994–96) was funded to add four southern Wisconsin lakes into the North Temperate Lakes LTER program.[7] Just as importantly, this proposal, as far as the challenge to face the invisible place was concerned, added more distant regions in North America and around the Northern Hemisphere.

To address the invisible place issue, we formed a Lake Ice Analysis Group (LIAG). The group's title signified that we intended to create and analyze a global

---

[7]The augmentation from NSF Division of Environmental Biology was added to Grant DEB-9011660 to the Center for Limnology.

Fig. 9.6 Participants in the Lake Ice Analysis Group (LIAG) workshop at the Trout Lake Station, Center for Limnology, UW–Madison in October 1996 (photo received from D. Robertson). Left to right
Front Row: Kenton M. Stewart (State University of New York–Buffalo), Peter Biro (Balaton Limnological Research Institute, Hungary), Dale M. Robertson (U.S Geological Survey, Madison WI), Lynn Herche (Great Lakes Ecosystem Research Laboratory, Michigan), Rita Adrian (Institute of Water Ecology and Inland Fisheries Muggelsee, Berlin, Germany), Pam Montz (Center for Limnology, UW–Madison), David Jewson (University of Ulster Traad Point, Northern Ireland), John J. Magnuson (Center for Limnology, UW–Madison)
Middle Row: John D. Lenters (Center for Limnology, UW–Madison), Aija-Rittaa Elo (U. Helsinki, Finland), William Chang (National Science Foundation), Sarah E. Walsh (Climate, People, and Environment Program, UW–Madison), Barbara J. Benson (Center for Limnology, UW–Madison), Raymond Assel (Great Lakes Ecosystem Research Laboratory, Michigan), Bruce P. Hayden (Virginia Coast Reserve LTER), Nicolas Grannin (Limnological Institute, Irkutsk, Russia)
Back Row: Esko Kuusisto (Finnish Environmental Institute, Helsinki), Timothy K. Kratz (Center for Limnology, UW–Madison), Valery Vuglinski (Hydrological Institut, St. Petersburg, Russia), Steven J. Vavrus (Center for Climatic Research, UW–Madison), Roger Barry (World Data Center for Glaciology, U. Colorado), Randy Wynne (Virginia Tech,-Blacksburg), David W. Bolgrien (Center for Limnology, UW–Madison)

ice seasonality database emphasizing networking and analysis. We knew of scientists in other states and countries with time series of lake ice dates. We all met at our Trout Lake Station in October 1996 (Fig. 9.6) with the requirement that all data be shared in a common database for all contributors to use; we made it publicly available in 1999 at the National Snow & Ice Data Center, National Oceanic and

Atmospheric Agency.[8] We also began analyses on 16 specific topical ice papers to be presented at a symposium of a meeting of the Society of International Limnology held in Dublin, Ireland in 1998 (Magnuson et al. 2000b), and followed up with two synthesis papers (Walsh et al. 1998; Magnuson et al. 2000a). Many papers emerged from the LIAG scientists with access to data from a large number of lakes for many years.

We soon discovered that successive records as long as 20 or 50 years in a 100-year time series produced alternating trends of increasing and decreasing ice breakup date (Wynne 2000; Magnuson 2002a). This persuaded us that, if we wished to determine whether the climate was warming or cooling, neither 20 nor 50 years of observation were long enough. This was important because letters to the editor, and other such sources, often confused weather with climate, arguing that climate was not undergoing long-term warming from only a few recent years or a few decades of data.

Consequently, we next focused our ice cover research on records of 100 years or longer. Our first Northern Hemisphere synthesis using the LIAG database was published in *Science*: Magnuson et al. (2000a). Many of the LIAG participants were co-authors. We reported the ice dates on lakes and rivers from 1846 to 1995 (Fig. 9.7). Freeze dates of the lakes averaged 5.7 days later per 100 years, breakup dates averaged 6.3 days earlier per 100 years. For freeze dates, 14 of 15 sites were occurring later and for breakup 23 of 24 were occurring earlier. Rates of change at most, but not all, sites were statistically significant. For Lake Mendota significance levels were $p = 0.008$ for freeze and 0.001 for breakup.[9]

National Public Radio picked up this paper and interviewed me by phone; many other reporters contacted me as well. A common question was "How do you know these early records are accurate?" I told them we could, of course, no longer talk to the people who wrote the numbers down and their criteria were not always documented. However, I believed that the data were credible because I could not imagine that it was purely coincidental that records from observers around the Northern Hemisphere revealed the same major features such as decreasing ice cover and concurrent peaks and valleys. We concluded that ice cover changes provided strong evidence for climate change around the Northern Hemisphere in the direction of warmer winters and that lakes (in this case their ice on and off dates) were responding to the warming climate. Changes in lake ice were, on one hand, an indicator of climate change and, on the other hand, evidence that lake ecosystems

---

[8] "Global Lake and River Ice Phenology Database," https://nsidc.org/data/g01377. Updated in 2012. "The Global Lake and River Ice Phenology Database contains freeze and thaw/breakup dates as well as other descriptive ice cover data for 865 lakes and rivers in the Northern Hemisphere. Of the 542 water bodies that have records longer than 19 years, 370 are in North America and 172 are in Eurasia. 249 lakes and rivers have records longer than 50 years, and 66 have records longer than 100 years. A few water bodies have data available prior to 1845." B. J. Benson led the management of the LIAG database.

[9] This compares with the $p = 0.05$ in Robertson's research about 10 years earlier on lake ice duration (Robertson 1989).

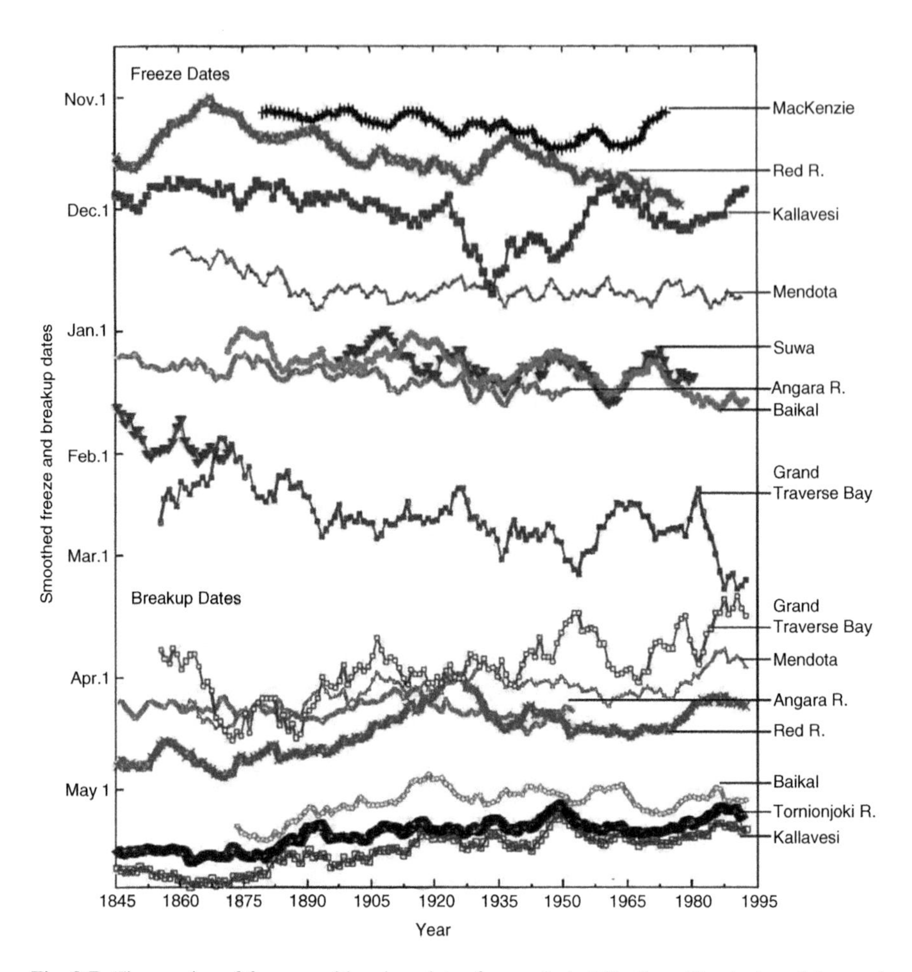

Fig. 9.7 Time series of freeze and breakup dates from selected Northern Hemisphere lakes and rivers (1846–1995). Data were smoothed with a 10-year moving average. (Magnuson et al. 2000a, on p. 1744)

were being affected by climate change. This article had significant public impact and is widely cited.[10]

In April 2008, we held a second, smaller meeting of a subset of LIAG researchers at the Lake Erken Laboratory of Uppsala University, Sweden (Fig. 9.8). The meeting focused on lakes in northern Europe and the Laurentian Great Lakes Region. We believed that the data quantity and quality in these two areas were suitable for researchers to address more in-depth questions. Some of these results are presented below.

[10] Google scholar had recorded 1082 citations by June 2019.

Fig. 9.8 The second LIAG Workshop at Lake Erken Laboratory, Uppsala University, Sweden in April 2008 (Photo: J. Magnuson). Left to right: Markus Meili (Uppsala University), David A. Livingstone (Swiss Federal Institute of Aquatic Science and Technology, Dübendorf), Lauri Arvola (University of Helsinki, Finland), Gesa A. Weyhenmeyer (Uppsala University), Barbara J. Benson (Center for Limnology, UW–Madison), Johanna Korhonen (Finnish Environment Institute, Freshwater Centre, Helsinki), James A. Rusak (Center for Limnology, Wisconsin), John J. Magnuson (Center for Limnology, Wisconsin), Olaf P. Jensen (Center for Limnology, Wisconsin), Thorsten Blenckner (Uppsala University)

We updated the LIAG database and published a more detailed and comprehensive hemispheric paper with more lakes and three different time periods (Benson et al. 2012). The major conclusions were: (1) trends in ice cover in the most recent 30-year period were steeper than those in the most recent 100- and 150-year periods, and trends in the 150-year period were steeper than in the 100-year period; (2) during the 1900s the frequencies of winters for lakes without complete ice cover increased; and (3) winters with extremely long ice covers became less frequent while those with extremely short ice covers became more frequent (Fig. 9.9). This change from cold to warm extremes was best explained by the overall trend toward shorter ice durations than by changes in inter-year variability. The relations between the extreme ice dates and large-scale climate drivers like El Niño and the North Atlantic Oscillation were discussed in Benson et al. 2012 and are a major subject treated below in sect. 9.8.

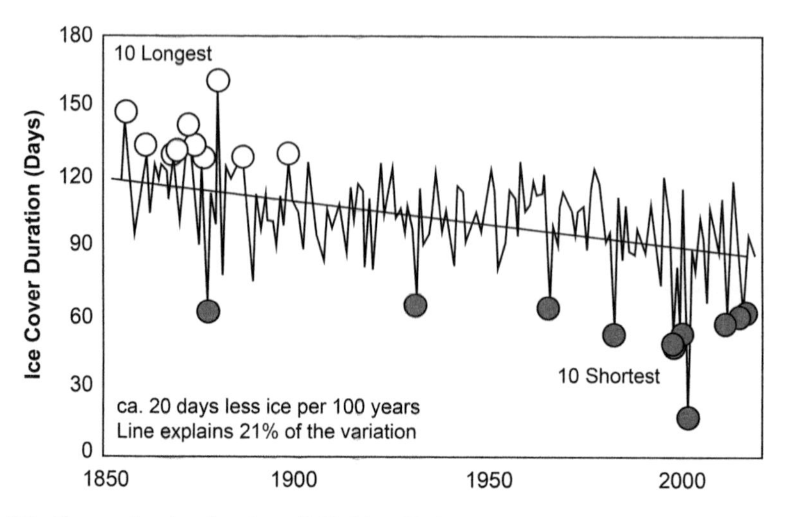

Fig. 9.9 Changes ice duration from 1855–56 to 2018–19 for Lake Mendota, Wisconsin and the shift from cold-related extremes to warm-related extremes in ice duration. Plotted from the NTL-LTER database https://portal.edirepository.org/nis/mapbrowse?packageid=edi.267.2

9.6 How Far Does Coherence Between Lakes Persist?

The ice data from our two workshops made it possible to determine how far coherence in the inter-year dynamics of lake ice dates extended. We present findings on the Laurentian Great Lakes Region from Minnesota to New York (Magnuson et al. 2005) and from Wisconsin, to New York and Maine, to Finland, and perhaps to Lake Baikal in Siberia (Magnuson et al. 2004). One would expect that lakes that were farther apart would have less similar dynamics. That is the case, but differences in the north/south direction have a greater rate of decline in coherence than those in east/west direction (Magnuson et al. 2005). For example, lake pairs at the same longitude had a coherence of about 0.8, those separated by 400 km had a coherence of about 0.6, and those separated by 900 km had a coherence of 0.2. In contrast lakes that differed in the north/south direction by only 400 km already had a coherence of only 0.2. The greater persistence in the east/west direction makes sense. Dominant weather usually moves from west to east in this region, and in winter energy from sunlight declines from south to north, but not from west to east.

Temporal coherence between lakes for ice dates declined exponentially with increasing distances between lakes (Fig. 9.10 top). Coherent dynamics between Wisconsin lakes and New York or Maine lakes were considerably lower than those between lakes in the same lake district. Coherence continued to decline with intercontinental comparisons between the lakes in Finland or eastern Russia. Our conceptual view of what contributed to both the decline of and the spatial persistence of coherent dynamics is diagrammed (Fig. 9.10 bottom). Highest coherences are within lake districts where lakes are exposed to the same weather. Even at this scale, in-lake processes play out a little differently for each lake as modulated by lake specific features. At greater distances between lakes, local weathers, climates, and

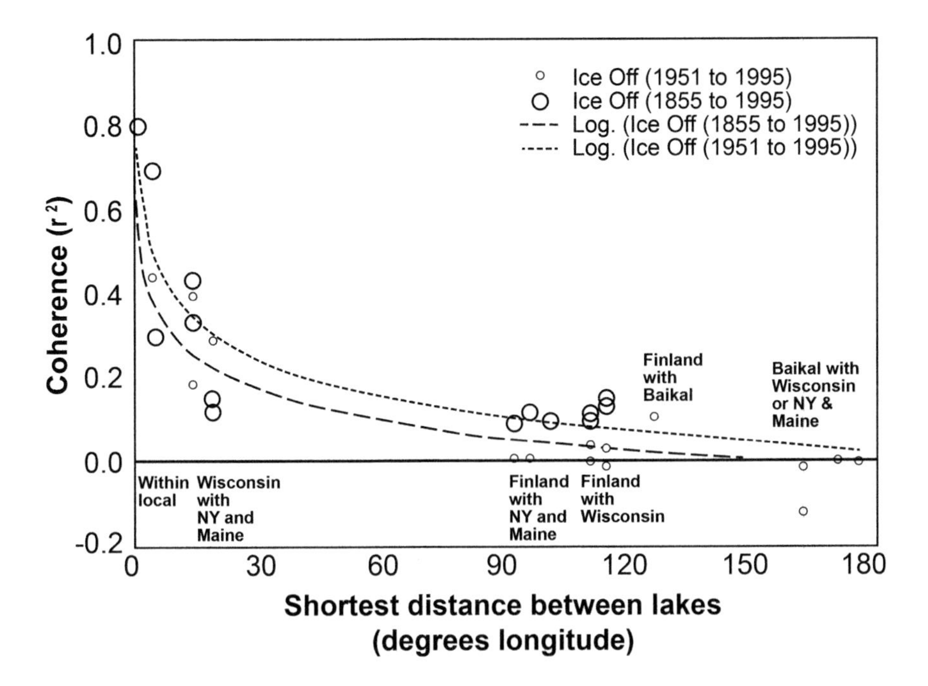

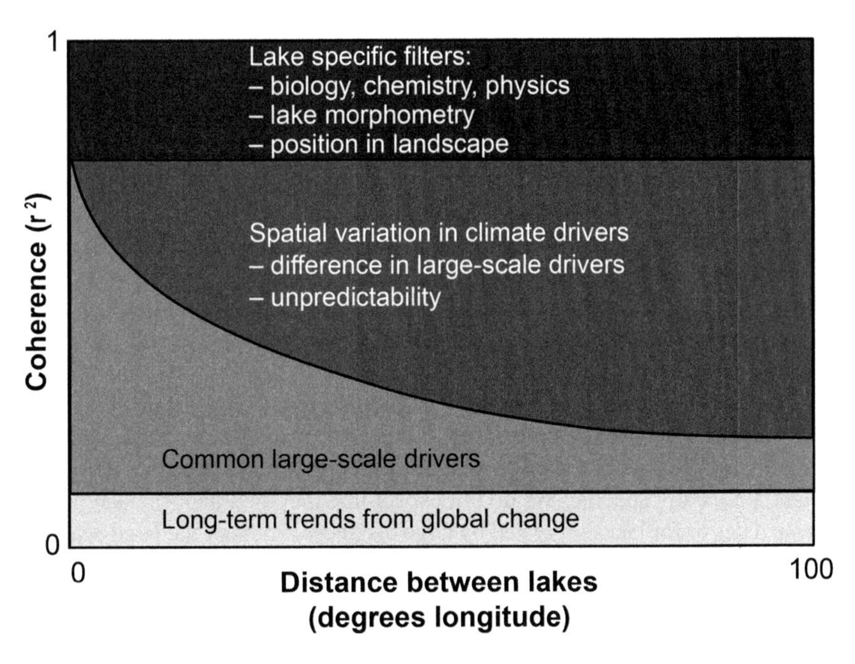

Fig. 9.10 Temporal coherence of ice breakup between near and distant lake pairs (Magnuson et al. 2004); Lakes Mendota and Monona from Wisconsin, Lake Otsego in New York and Moosehead in Maine, Lakes Kallavesi and Näsijärvi in Finland, and Lake Baikal in eastern Russia. Coherence was computed with a 45-year and a 151-year time series. Baikal did not have enough years for use for the 151-year coherences. Values of r^2 are plotted as negative if they were calculated from a negative correlation coefficient. (Modified from Magnuson et al. 2004, on pp. 365–366)

large-scale climate drivers differ. But 10% of the variability is still shared between North America and Europe because the warming trends of earlier breakup occur at all sites.

We also observed a coherent pattern in the northward movement of ice dates for 65 lakes across the Laurentian Great Lakes Region. During the warming period from 1975–2004, average breakup dates observed on April 15 moved northward 3.8 km/year (Jensen et al. 2007). The northward movement occurred from Minnesota on the west to New York on the east. This rate of northward movement was similar to that of temperature isotherms in the Northern Hemisphere that moved northward at about 4 km per year from 1975 to 2000 (Hansen et al. 2006).

9.7 Why Are Ice Dates So Variable from Year to Year?

Trends for later freeze dates, earlier breakup dates, and shorter ice cover durations are related closely to the warming climate (Benson et al. 2012). What also caught our attention from our earliest analyses was that long-term trends toward warmer conditions only explained about 20% of the variation in the total ice record (Fig. 9.9). Many researchers have found relations between ice dates and air temperature (Palecki and Barry 1986; Assel and Robertson 1995; Ruosteenoja 1986; Vavrus et al. 1996; Gao and Stefan 1999; Adrian and Hintze 2000; Livingstone 1997, 1999; Williams et al. 2004; Holopainen et al. 2009; Livingstone and Adrian 2009).

When I give public talks about lake ice that support the global climate warming hypothesis, I usually get a question about whether the changes are all just normal cyclic oscillations in climate. We wondered whether some of the remaining 80% around the linear trendline could result from cyclic dynamics. Many researchers have analyzed the role of large-scale climatic drivers in oscillatory dynamics for lake ice time series (Assel and Rodionov 1998; Livingstone 2000; Benson et al. 2000; Robertson et al. 2000; Magnuson 2002a; Magnuson et al. 2004; Bonsal et al. 2006; Bai et al. 2012; Mudelsee 2012). In most studies the North Atlantic Oscillation (NAO) and the Southern Ocean Index (SOI) were found to be statistically significant, but still left most of the variation unexplained.

We have attempted to determine the contributions of large-scale climate drivers, the 11-year solar oscillation, and day-to-day local weather to the variation around trendlines. The date of the passing of a stormfront, a rain event, or a particular cold night are locally specific. We do not expect that they will happen each year on the same dates or for lakes separated by large distances. Some of the large-scale climate drivers, such as the El Niño, referred to by climatologists as the Southern Oscillation Index (SOI), and perhaps the North Atlantic Oscillation (NAO), are better known generally. Other large-scale drivers include the Quasi Biennial Oscillation (QBO), the North Pacific Oscillation Index (NP). So what variation can we account for with local weather and large-scale climate drivers? These analyses are much more challenging than calculating trend lines; methodologies are complex, and we make no attempt here to explain how we did the analyses.

We present results from three papers that used time series approaches to help understand the influence of the various drivers of variation around the trendlines of climate change (Ghanbari et al. 2009 on Lake Mendota, Sharma et al. 2013 on Lakes Mendota and Monona, and Sharma and Magnuson 2014 on 13 lakes scattered around the Northern Hemisphere). Each analyzed a suite of climate drivers interacting with each other rather than considering each driver as a separate and independent actor as had the earlier researchers cited above.

Ghanbari et al. (2009) studied interactions in a cascade of dynamics between the large-scale climate drivers, to local weather, and to ice cover; they painted a complex picture for Lake Mendota (Fig. 9.11). Statistically significant coherencies are shown for frequencies with periods less than 10 years (interannual, top panel) and more than 10 years (interdecadal, bottom panel). These coherencies reveal a cascade of influences between the large-scale climate drivers and local weather and then in turn between local weather and lake ice seasonality. I point out only a few of the stronger interactions. Air temperature is the weather variable that most significantly

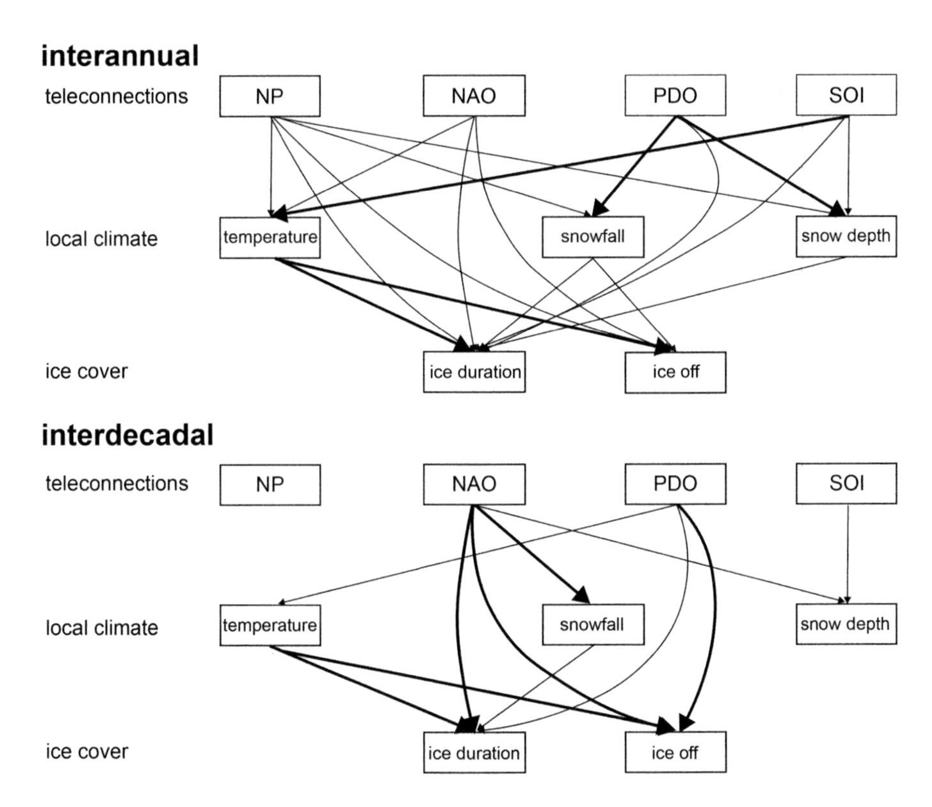

Fig. 9.11 Diagram of the complexity of inter-annual (top) and inter-decadal relations (bottom) of coherencies between a cascade of influences between time series of teleconnections, local weather, and lake ice (Ghanbari et al. 2009). Thick lines are significant at the 99% confidence level and thin lines at the 95% confidence level. The arrow points indicate the direction of influence. (Redrawn from Ghanbari et al. 2009, on p. 291)

influences ice duration and ice-off date both at inter-annual and inter-decadal frequencies; snow variables were less influential. Both the Pacific Decadal Oscillation (PDO) to the west and the

North Atlantic Oscillation (NAO) to the east were related to ice-cover duration and ice-off date, particularly at interdecadal frequencies. At interannual frequencies, the PDO was the most significant variable describing changes in the amount of snow, while the Southern Oscillation Index (SOI) was the most significant variable describing changes in air temperature. At interdecadal frequencies the NAO, rather than the PDO was most strongly related to changes in snowfall. This set of results was, to say the least, complex.

I think the complexity of the interactions and the questions raised by Reza Namdar Ghanbari and his colleagues are perhaps even more important than the specific results. Ghanbari did this research, his Ph.D. thesis, with his mentor, Hector Bravo, at the University of Wisconsin–Milwaukee. The rest of us helped provide data and discussion. For example, what is implied when an interaction arrow goes directly from a teleconnection to a lake ice variable without going through at least one of the three weather variables? It might have been good to have additional weather variables, such as cloud cover or solar radiation. Also, even though the diagram of significant interactions already is complex, these four large-scale drivers do not constitute the entire suite of drivers that might have been used. This really interesting analysis suggests many lines of additional research.

Sapna Sharma and I (Fig. 9.12) have continued to challenge ourselves to determine the amount of variability in lake ice time series that can be explained by the linear trends and large scale climate drivers. We first met in 2007, when I served as

Fig. 9.12 Sapna Sharma, York University, and John J. Magnuson in Toronto when I visited in March 2014 to work with her and chat with the students she advised. (Photo: J. Magnuson)

the outside examiner on her Ph.D. committee at the University of Toronto. She later came to the University of Wisconsin–Madison as a postdoc at the Center for Limnology. She and I began to talk about the lake ice time series. I thought time series analyses would be useful and that she had the talents to do it.[11]

Our first analysis (Sharma et al. 2013) evaluated large-scale climate drivers (SOI, and NAO), the 11-year solar cycle, and daily weather. Most lake sites could not be used for the analyses because they did not have long-term daily weather data; Madison, Wisconsin did for 100 years. The bottom line of this analysis was that climate change trendlines, local weather, and indices of sunspots and large-scale climatic drivers statistically explained 59% of the total variation. Of this, about half was associated with local weather. This still left 41% of the variation unexplained by our model. Perhaps future analyses and models will explain more, but for now explaining 59% seems important, as does the significant influence of local weather.

Our second analysis (Sharma and Magnuson 2014) expanded the spatial and temporal scales of analysis by using 13 lakes from around the Northern Hemisphere and 150-year time series (1854 and 2004). Linear trends and oscillatory dynamics explained 41% of the total variation in ice breakup time series. Linear trends accounted for 15% of the variation and the oscillations 26%. This is less than the model for Lakes Mendota and Monona because we did not include daily weather data.[12] Among the significant periods of oscillation were those 2–3 years long, or the same period as the Quasi Biennial Oscillation (QBO); these accounted for 9.4% of the variance. This oscillation is what makes the records look like a sharply-jagged crosscut saw blade (Figs. 9.9 and 9.13 top). Other significant periods were 3–6 years coincident with the periods of El Niño (the Southern Oscillation Index, SOI); these accounted for 8.2% of the variance. Periods of the sunspot cycle of 10–12 years, explained only 1.6% of variation around the trend lines. Longer decadal oscillations explained little, likely because the 150-year time was too short and did not include enough repetitions of longer multidecadal oscillations.

For the 13 lakes around the Northern Hemisphere, the oscillations do not mask the linear trends (Sharma and Magnuson 2014) even though they collectively explain more of the variance than did the trends (Fig. 9.13). The trend is quite visible both before and after sequentially removing the oscillatory dynamics. After removing all oscillations up through 12-year periods, the remaining signal in the time series does appear to include longer interdecadal oscillations.

Overall, large-scale climate drivers reveal a complex set of dynamics that differ somewhat depending on the lakes, the number of years in the time series, and the analytical methods used. This complexity is part of the reality of the inter-year variation that occurs in the lake ice data. In the above publications, 40–60% of the variability remains unexplained even after accounting for trends, large-scale drivers

[11] Sapna is now at York University in Toronto where she has become a leader in lake-ice research; I settled happily into the roles of advisor, historian, advocate, critic, and coauthor. I retired in 2000 but continued to be an active researcher at the Center for Limnology, UW–Madison, as an Emeritus Professor.

[12] Daily weather data did not exist for these lakes for 150 years.

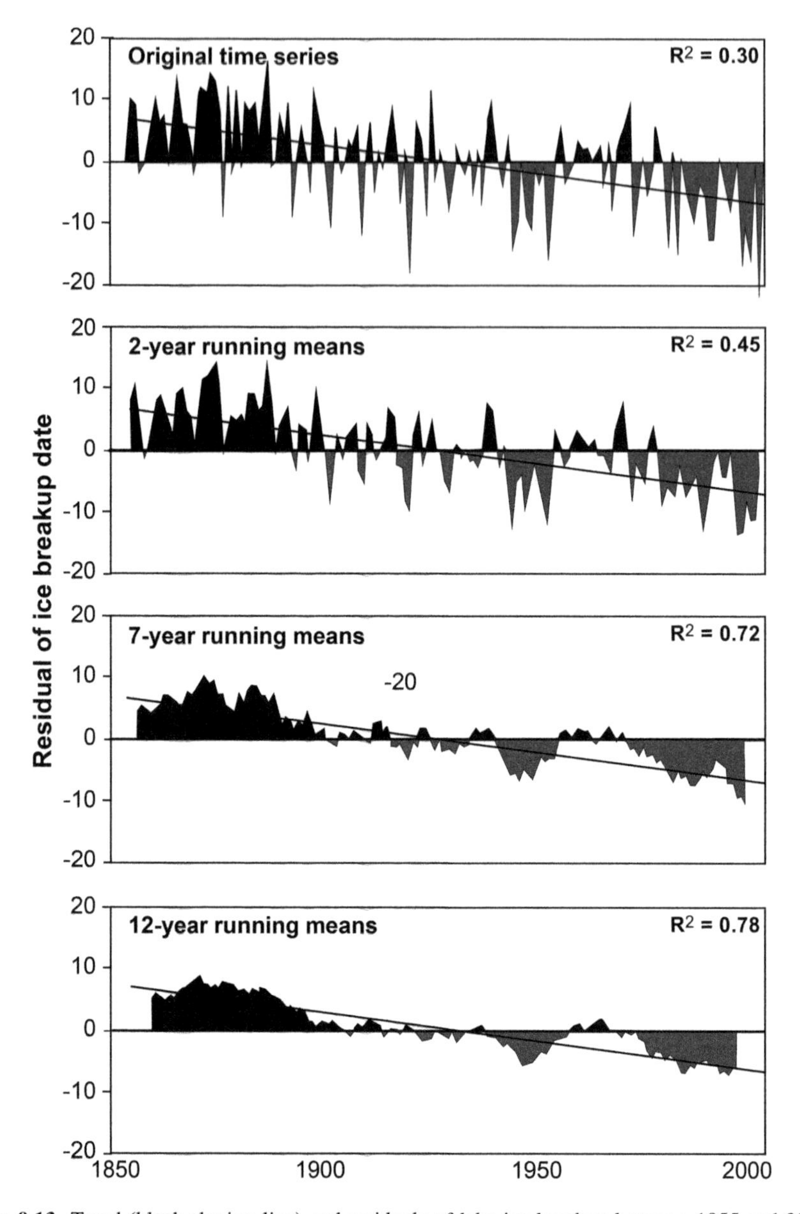

Fig. 9.13 Trend (black sloping line) and residuals of lake-ice breakup between 1855 and 2004 compared as inter-year variability components are successively removed using running means. Plotted in sequence from top to bottom: as original data, 2-year running mean, 7-year running mean, and 12-year running mean. Shared variances, r^2, are given between the trend and the actual values. (Modified from Sharma and Magnuson 2014, on p. 844)

and, in some cases, local weather. No wonder that lake and river ice freeze and breakup dates are difficult to forecast with any accuracy, a fact that makes them ideal subjects for local ice event lotteries. The high inter-year variability also makes forecasting relative safety threats to people on the ice the next winter difficult or impossible at this time. A positive aspect of the high interannual variability in lake ice seasonality is that limnological researchers do not have to wait for future climate change to occur to analyze the effects of longer or shorter ice cover on lake ecosystems. Long-term ecological research provides the response of lake ecosystems for a wide variety of winter conditions that may represent average conditions 50 or more years ago or to be encountered 50 or more years in the future.

9.8 The Longest Annual Ice Time Series

The longest inland water ice records directly observed by humans extend over multiple centuries (Sharma et al. 2016). The ice breakup time series from the Torne River in Finland/Sweden runs for 320 years, and the ice freeze time series from Lake Suwa in Japan for 581 years. Importantly, both ice records began many years prior to the beginning of the so-called Second Industrial Revolution, the intensification of fossil fuel-driven industrialization globally that began around 1870. Thus, we were able to ask whether a rate of change in ice dates coincided with the Second Industrial Revolution and the burning of fossil fuels. All the 150-year ice records we had analyzed earlier began only a few years before the start of the Second Industrial Revolution.

To access and become familiar with these two long time series, we had invited Finnish (Kuusisto, Elo, and Korhonen). and Japanese (Arai) scientists to the LIAG workshop in 1996. Arai was unable to attend but did present a paper about Lake Suwa at our subsequent 1998 symposium (Arai 2000).

Publication on these long records (Sharma et al. 2016) eluded us for 20 years owing to challenges that can arise when using long historic time series. In our first publication showing Lake Suwa data (Magnuson et al. 2000a), the freeze date data looked peculiar in the years prior to 1900; for example, note the gap in the Suwa data during which a 30-year step change occurred (Fig. 9.7). I posited that we should impose a 30-day correction to the original time series. Tadashi Arai told me that calendar change from the Japanese moon calendar to the Gregorian calendar in 1872 already had been applied to the time series by Arakawa (1954). Tadashi Arai quality controlled and translated the Arakawa tables for us. Arai and the earlier papers (Tanaka and Yoshino 1982) suggested that the data from the winters of 1682–83 to 1922–23, while interesting, should not be used for quantitative analyses. We left those years out for Fig. 9.14 and did not include those years in our analyses.

In November 2005, Tadashi Arai took me to the Shinto Shrine on Lake Suwa where we met and talked with Mr. Kiyoshi Miyasaki, the Shinto Priest who kept the

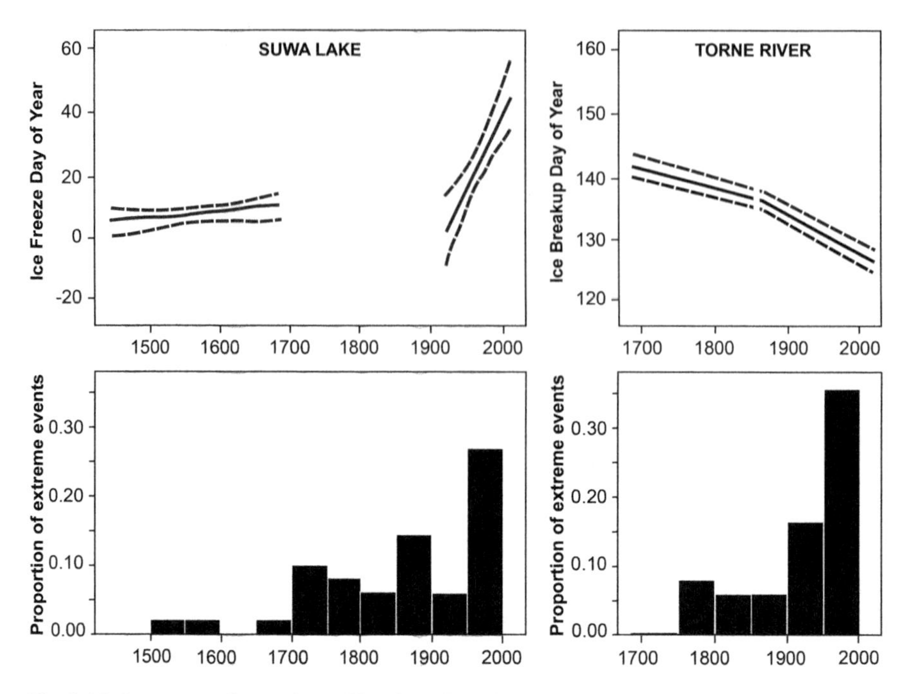

Fig. 9.14 Long term observations of ice dates (best fit mean lines and confidence are shown) and the proportion of years that are extreme events by 50-year intervals for Lake Suwa, Japan, and River Torne, Finland. (Modified from Sharma et al. 2016, Fig. 1 and 2)

data and participated in the ice-based Omwatari ceremony on the lake.[13] He was most gracious and very helpful. As the years went by, we had to contact him several times to update the database. Challenges of language and the absence of email at the shrine complicated the effort.[14]

Sapna and I pulled together a group of scientists to finally finish the paper about Lake Suwa and River Torne. Unfortunately, Arai decided not to join the effort as he was retired and no longer had access to the original logs and diaries pertaining to Suwa. We needed scientists from Japan (Yasuyuki Aono, who had published on the Kyoto Cherry Blossom time series) and Finland (Johanna Korhonen with whom we had worked before) to advise us on local knowledge.

The results from these longest records were striking. For both waters the rates of change after the start of the Industrial Revolution were more rapid than those before the Second Industrial Revolution (Fig. 9.14 top). From the Torne data we estimated when the changing trend reflected the start of the Industrial Revolution; the year was 1867. We also analyzed the changing occurrence of extreme warm events; these

[13] https://blog.nationalgeographic.org/2016/04/26/lake-suwas-shinto-legend

[14] A University of Wisconsin-Madison graduate student (Masami Nii Glines) and a news reporter in Suwa (Asuka Momose) worked us through those challenges through a series of phone calls and emails.

increased at both sites (Fig. 9.14 bottom). The extreme years for Suwa were for years when ice did not completely cover the lake. In regression analyses of the long time series, CO_2 was a significant predictor of freeze dates but sunspots were not.

9.9 Impacts of Ice Loss on People

The network of researchers who study lake and river ice are turning their attention to the importance of the ice to people. Ice loss decreases many valued uses (Magnuson and Lathrop 2014; Sharma et al. 2019; Knoll et al. 2019). Losses include many cultural values such as recreation, education and research, artistic, and ceremonial as well as subsistence for fishing and hunting.

Examples and analyses are emerging. Magnuson and Lathrop (2014) provided examples of beauty, winter festivals, ice skating, ice boating, skating and ice hockey, ice fishing, education, research, and minor shoreline damage mostly on Lake Mendota, Wisconsin. Knoll et al. (2019) provided a schematic for considering all cultural values and quantitative analyses of losses associated with warming for ice roads in Canada, ice skating in Sweden, ice fishing in Minnesota, and religious ceremonies in Japan and Europe.

In northern regions where ice cover occurs in winter, some lakes already are not freezing over completely every winter; the number of such lakes increased from winter 1905–6 to 2004–5 (Benson et al. 2012). Sharma et al. (2019) estimated that 14,800 lakes with intermittent ice exist presently in the Northern Hemisphere. The number of such lakes could increase by 35,300 with a temperature increase of 2 °C, by 57,000 with an increase of 3.2 °C, and by 90,200 with an increase of 4.5 °C. Presently 30 countries contain such lakes; the number increases to 47 countries with a 4.5 °C increase in global temperature. Sharma et al. (2019) estimated that Lake Mendota could start to experience intermittent ice cover in the 2020s based on annual temperature. Recall that Robertson (1989) forecast that the lake might be ice free on average by about 2050.

9.10 Concluding Thoughts and the Future

I made a pragmatic choice in the 1980s to use lake-ice time series to demonstrate the value of Long-Term Ecological Research through the metaphors of the "Invisible Present" and "Invisible Place." We have come a long way using lake ice seasonality to expand the temporal and spatial scales of ecological research. For example, the research has increased the temporal extent of the lake ice analyses from a few years to 570 years and the spatial extent from Wisconsin to the Northern Hemisphere. This occurred even though we have been collecting data at the North Temperate Lakes LTER site on only 10 Wisconsin lakes. Networking with scientists from around the Northern Hemisphere enabled the longer and broader analyses.

I believe that no definitive answer exists for "how long is long enough" for Long-Term Ecological Research. The question is: long enough for what? In this chapter I presented examples of what we could do that was new with time series less than a decade long. Yet a decade of measurements cannot detect climate change or even the influence of relatively short climate oscillators like the Quasi Biennial Oscillation (1–2 years) or El Niño (4–7 years). And multidecadal low-frequency climate oscillators require numerous repetitions to identify with assurance. For example, we could not even determine whether long-term climate trends of warming or cooling were occurring with only 20 or 50 years of data on lake ice seasonality; large-scale multidecadal climate drivers resulted in the slope switching back and forth. Longer-term changes important to understand climate change and variability became apparent as the duration of record increased to 150 years and beyond. For example, only the two longest, multi-century ice records that we analyzed could be used to address whether direct human observations revealed a change in climate that began at the start of the Industrial Revolution in the 1860s or '70s. These direct observations supported what had been previously quantified using paleoclimate data in ice cores, tree rings, and the like.

My belief is that the utility of sustained, long-term ecosystem research would continue to increase as the extent of the record increases. My chapter supports this belief. Ecosystem sites funded by the Long-Term Ecological Research Program at NSF will continue to reveal dynamics that have been missed in the Invisible Present and, for that matter, in the Invisible Place. A search for "how long" is less helpful than is the knowledge that sustained research on ecosystems over years, decades, and centuries will reveal surprises and new dynamics and processes in what used to be the Invisible Present.

We recognized that long-term studies and analyses were important, not only for variables tightly related to climate, but also to address a plethora of previously unknown influences and impacts of climate change and variability on lake ecosystems. For Lake Mendota these included: that the increasing frequency of extreme rain events was a significant cause for increases in excess algae blooms; that the cisco, a coldwater fish, was becoming rare and was forecast to become locally extinct; and that 2/3rds of the inter-year variability in water clarity of lake Mendota was related to processes related to climate change (Lathrop and Carpenter 2013). The approach of the LTER program, with its diversity of ecosystem measurements and analyses, was able to penetrate the wide-ranging complexity of climate-related dynamics in lake ecosystems.

An LTER program on an ecosystem type, in this case lakes, can be especially important in the study of climate change and variability. The 1995 Intergovernmental Panel on Climate Change (IPCC) Assessment from Working Group II on impacts of climate change almost went to press without a section on inland lakes and streams.[15] Prepublication reviewers of the IPCC document, especially Diane McKnight at the University of Colorado-Boulder, advocated that this deficiency be eliminated. I was

[15] Only the Great Lakes had been included.

asked by the IPCC to lead the effort; I put together a small group of knowledgeable leaders in lake and stream aquatic ecology. The results of our LTER-funded ice papers appeared in that and subsequent IPCC Assessments (Arnell et al. 1996; Foland et al. 2001; Gitay et al. 2001; Lemke et al. 2007). I suspect that one reason I was invited to lead the addition of inland waters was that we were known as the LTER site focused on lakes. We also used the ice data in the publication "Freshwater ecosystems and climate change in North America" published in *Advances in Hydrological Processes* (see Magnuson et al. 1997), "Confronting Climate Change in the Great Lakes Region," published by the Union of Concerned Scientists and the Ecological Society of America (Kling et al. 2003), and in the 2011 publication of the Wisconsin Initiative on Climate Change Impacts (www.wicci.wisc.edu).

Research on inland water ice and winter limnology is continuing and expanding. More papers are emerging that consider the impact of loss of ice cover on the wellbeing of people and the cultural uses of ice-covered lakes (Magnuson and Lathrop 2014; Sharma et al. 2019; Knoll et al. 2019). The most recent limnological faculty hire at UW–Madison, Hilary Dugan at the Center for Limnology, has begun to use the LTER time series to understand winter limnology, an under-studied area of limnological science. Ongoing studies continue with Sharma, her students at York University in Toronto, new colleagues, and sometimes me (Hewitt et al. 2018; Lopez et al. 2019). Sharma is updating the LIAG database with more recent ice dates and additional lakes around the globe. Ice-oriented research groups can be found in Canada, central Europe, Finland, Mongolia, Poland, Russia, Sweden, USA, and other countries.

Networking on lake ice research continues nationally and internationally. Now most networking occurs over the Web, where groups can meet to discuss, gather new data, assign analyses, and work on manuscripts without traveling halfway around the globe as we did in the 1990s. One of the new networks is a subset of researchers networking through the Global Lake Ecological Observatory Network (GLEON),[16] formed initially out of NSF support to researchers in the USA. A recent surge in larger-scale analyses is underway on many topics for lake ecosystems, such as surface temperature (O'Reilly et al. 2015), chloride concentrations (Dugan et al. 2017), phytoplankton and nutrients (Oliver et al. 2017). LTER data from the Wisconsin lakes is usually included in such efforts. Being able to gather researchers and data through networking tools that are now available is exciting and opens up many ways to expand the temporal and spatial scales of ecological research. Extension of time series from available records from before the North Temperate Lakes LTER began is a common practice: see, for example, our analyses of the nutrient dynamics and water quality of the Madison area lakes, especially Lake Mendota (Lathrop et al. 1999; Lathrop and Carpenter 2013).

I use the metaphors of Invisible Present and Invisible Place, and the example of opening up a time series (Fig. 9.3) and the spatial scale of analyses (Fig. 9.7), in

[16] http://gleon.org; https://lter.limnology.wisc.edu/project/global-lakes-ecological-observatory-network-gleon

public and class presentations. I presented 10–40 such public presentations each year from 2003 to 2014. My sequence of slides enabled audiences to discover what was happening to climate from the experience of "seeing it for themselves" along with apparently simple graphics of lake time series. Some in the audience would come up afterwards and say: "I get it."

The last times I used the words "Invisible Present" and "Invisible Place" in our peer reviewed science publications were in our LTER Oxford Synthesis Book (Magnuson et al. 2006) and in Waller and Rooney's book "The Vanishing Present" (Magnuson 2008). Both Bob Waide and I used the ideas in Willig and Walker (2016), which asked scientists whether participation in the LTER program had changed the way they thought about and did science. I think the metaphors have been catalysts (see Hobbie et al. 2003), and that most LTER scientists now work naturally across an increasing range of temporal and spatial scales in their search to understand ecological systems. I also think that the two metaphors will continue to inform new audiences and NSF staff as they and the scientific community consider and reconsider the future of the LTER Program and ecosystem sites.

Acknowledgements I thank the National Science Foundation for supporting the North Temperate Lakes LTER program and special competitions for inter-site projects and workshops from the early 1980s to 2019. The NTL LTER program supported final figure preparation. I recognize the investments of time and talents and data from participants in the Lake Ice Analysis Group and the many other colleagues and students at Wisconsin and around the Northern Hemisphere. I appreciate reviews of the manuscript and suggestions received from Sharon Kingsland, Robert Waide, Sapna Sharma, Emily Stanley, Hilary Dugan, and Curt Meine.

References

Adrian, R., and T. Hintze. 2000. Effects of winter air temperature on the ice phenology of the Müggelsee (Berlin, Germany). *Verhandlungen des Internationalen Verein Limnologie* 27: 2808–2811.

Anderson, W., D.M. Robertson, and J.J. Magnuson. 1996. Evidence of recent warming and ENSO variation in ice breakup of Wisconsin lakes. *Limnology and Oceanography* 41: 815–821.

Arai, Tadashi. 2000. The hydro–climatological significance of long-term ice records of Lake Suwa, Japan. *Verhandlungen des Internationalen Verein Limnologie* 27: 2757–2760.

Arakawa, H. 1954. On five centuries of freezing dates of Lake Suwa. *Journal of Geography (Tokyo)* 63: 41–48.

Arnell, N., B. Bates, H. Land, J.J. Magnuson, P.I. Mullholland, S. Fischer, C. Lui, D. McKnight, O. Starosolov, and M. Taylor. 1996. Hydrology and freshwater ecology. In *Climate change 1995: Impacts, adaptations and mitigations of climate change*, ed. R.T. Watson, M.C. Zinyowera, and R.H. Moss, 325–363. Cambridge: Cambridge University Press.

Assel, R.A., and D.M. Robertson. 1995. Changes in winter air temperature near Lake Michigan, 1851–1993, as determined from regional lake-ice records. *Limnology and Oceanography* 40: 165–176.

Assel, R., and S. Rodionov. 1998. Atmospheric teleconnections for annual maximum ice cover on the Laurentian Great Lakes. *International Journal of Climatology* 18: 425–442.

Bai, X., J. Wang, C. Sellinger, A. Clites, and R. Assel. 2012. Interannual variability of ice cover and its relationship to NAO and ENSO. *JGR Oceans* 117: CO3002.

Beckel, Annamarie L. 1987. *Breaking new waters: A century of limnology at the University of Wisconsin*. Madison: Wisconsin Academy of Sciences, Arts, and Letters.

Benson, B.J., J.J. Magnuson, R.L. Jacob, and S.L. Fuenger. 2000. Response of lake ice breakup in the Northern Hemisphere to the 1976 interdecadal shift in the North Pacific. *Verhandlungen des Internationalen Verein Limnologie* 27: 2770–2774.

Benson, Barbara J., John J. Magnuson, Olaf P. Jensen, Virginia M. Card, Glenn Hodgkins, Johanna Korhonen, David M. Livingstone, Kenton M. Stewart, Gesa A. Weyhenmeyer, and Nick G. Granin. 2012. Extreme events, trends, and variability in Northern Hemisphere lake-ice phenology (1855-2005). *Climatic Change* 112: 299–323.

Bonsal, B.R., T.D. Prowse, C.R. Duguay, and M.P. Lacroix. 2006. Impacts of large-scale teleconnections on freshwater-ice break/freeze-up dates over Canada. *Journal of Hydrology* 330: 340–353.

Brock, T.D. 1985. *A eutrophic lake: Lake Mendota, Wisconsin*. New York: Springer.

Bryson, Reid A., and Thomas J. Murray. 1979. *Climates of hunger: Mankind and the world's changing weather*. Madison: University of Wisconsin Press.

Callahan, J. T. 1991. Long-term ecological research in the United States: A federal perspective. In *Long-term ecological research, an international perspective*, ed. Paul G. Risser, 9–22. New York: Wiley.

Dugan, H.A., S.L. Bartlett, S.M. Burke, J.P. Doubek, F.E. Krivak-Tetley, N.K. Skaff, J.C. Summers, K.J. Farrell, I.M. McCullough, A.M. Morales-Williams, et al. 2017. Salting our freshwater lakes. *Proceedings of the National Academy of Sciences* 114 (17): 4453–4458.

Edmondson, W.T. 1991. *The uses of ecology: Lake Washington and beyond. Seattle*. University of Washington Press.

Emerson, Ralph Waldo. 1909. *Essays and English traits*. Vol. V. New York: P.F. Collier & Son.

Foland, C.K., T.R. Karl, J.R. Christy, R.A. Clarke, G.V. Gruza, J. Jouzel, M.E. Mann, J. Oerlemanns, M.J. Salinger, and S.-W. Wang. 2001. Observed climate variability and change. In *Climate change 2001: The scientific basis. Contribution of working group I to the third assessment report of the Intergovernmental Panel on Climate Change*, ed. J.T. Houghton, Y. Ding, D.J. Griggs, M. Noguer, P.J. van der Linder, X. Dai, K. Maskell, and C.A. Johnson, 99–182. Cambridge: Cambridge University Press.

Franklin, J.F., C.S. Bledsoe, and J.T. Callahan. 1990. Contributions of the long-term ecological research program. *Bioscience* 40: 509–523.

Gao, Shaobai, and Heinz G. Stefan. 1999. Multiple linear regression for lake ice and lake temperature characteristics. *Journal of Cold Regions Engineering* 13 (2): 59–77.

Ghanbari, Reza Namdar, Hector R. Bravo, John J. Magnuson, William G. Hyzer, and Barbara J. Benson. 2009. Coherence between lake ice cover, local climate and teleconnections (Lake Mendota, Wisconsin). *Journal of Hydrology* 374: 282–293.

Gitay, H., S. Brown, W. Easterling, B. Jallow, J. Antle, M. Apps, R. Beamish, T. Chapin, W. Cramer, J. Frangi, J. Laine, L. Erda, J. Magnuson, I. Noble, J. Price, T. Prowse, T. Root, E. Schulze, O. Sirotenko, B. Sohngen, and J. Soussana. 2001. Ecosystems and their goods and services. In *Climate change 2001: Impacts, adaptation, and vulnerability*, ed. J.J. McCarthy, O.F. Canziani, N.A. Leary, D.J. Dokken, and K.S. White, 237–342. Cambridge: Cambridge University Press.

Goldman, C.R. 2000. Four decades of change in two subalpine lakes. *Verhandlungen des Internationalen Verein Limnologie* 27: 7–26.

Gore, A. 1992. *Earth in the balance: Ecology and the human spirit*. Boston: Houghton Mifflin.

Hansen, J., and S. Lebedeff. 1987. Global trends of measured surface air temperature. *Journal of Geophysical Research* 92: 13345–13372.

Hansen, J.I., A. Lacis, D. Rind, G. Russel, P. Stone, R. Ruedy, and J. Lerner. 1984. Climate sensitivity: Analysis of feedback mechanisms. In *Climate processes and climate sensitivity. Geophysical monograph series*, ed. J. Hansen and T. Takahashi, vol. 29, 130–163. Washington, DC: American Geophysical Union.

Hansen, J., M. Sato, R. Ruedy, K. Lo, D.W. Lea, and M. Medina-Elizade. 2006. Global temperature change. *Proceedings of the National Academy of Sciences* 103: 14288–14293.

Hewitt, B.A., L.S. Lopez, K.M. Gaibisels, A. Murdoch, S.N. Higgins, J.J. Magnuson, A.M. Paterson, J.A. Rusak, H. Yao, and S. Sharma. 2018. Historical trends, drivers, and future projections of ice phenology in small north temperate lakes in the Laurentian Great Lakes watershed. *Water* 10. https://doi.org/10.3390/w10010070.

Hobbie, John E., Stephen R. Carpenter, Nancy B. Grimm, James R. Gosz, and Timothy R. Seastedt. 2003. The US long term ecological research program. *Bioscience* 53: 21–32.

Holopainen, J., S. Helama, J.M. Kajand, J. Korhonen, J. Launiainen, H. Nevanlinna, A. Reissell, and V.-P. Salonen. 2009. A multiproxy reconstruction of spring temperatures in south-west Finland since 1750. *Climatic Change* 92: 213–233.

Houghton, J.T., G.J. Jenkins, and J.J. Ephraums, eds. 1990. *Climate change: The IPCC scientific assessment*. Cambridge: Cambridge University Press.

Jensen, O.P., B.J. Benson, J.J. Magnuson, V.M. Card, M.N. Futter, P.A. Soranno, and K.M. Stewart. 2007. Spatial analysis of ice phenology trends across the Laurentian Great Lakes region during a recent warming period. *Limnology and Oceanography* 52: 2013–2026.

Kling, G.W., K. Hayhoe, L.B. Johnson, J.J. Magnuson, S. Polasky, S.K. Robinson, B.J. Shuter, M.M. Wander, D.J. Wuebbles, D.R. Zak, R.L. Lindroth, S.C. Moser, and M.L. Wilson. 2003. Confronting climate change in the Great Lakes region: Impacts on our communities and ecosystems. Report by the Union of Concerned Scientists and the Ecological Society of America. https://www.ucsusa.org/resources/confronting-climate-change-great-lakes. Accessed 28 June 2020.

Knoll, L.B., S. Sharma, B.A. Denfeld, G. Flaim, Y. Hori, J.J. Magnuson, D. Straile, and G.A. Weyhenmeyer. 2019. Consequences of lake and river ice loss on cultural ecosystem services. *Limnology and Oceanography Letters*. https://doi.org/10.1002/lol2.10116.

Kolasa, Jurek. 1992. Getting to grips with ecosystems. Review of Risser (1991). *Trends in Ecology and Evolution* 7: 281–282.

Kratz, T.K., T.M. Frost, and J.J. Magnuson. 1987. Inferences from spatial and temporal variability in ecosystems: Long-term zooplankton data from lakes. *American Naturalist* 129: 830–846.

Kratz, T.K., P.A. Soranno, S.B. Baines, B.J. Benson, J.J. Magnuson, T.M. Frost, and R.C. Lathrop. 1998. Interannual synchronous dynamics in north temperate lakes in Wisconsin, USA. In *Management of lakes and reservoirs during global climate change*, ed. D.G. George, J.G. Jones, C.S. Puncochar, C.S. Reynolds, and D.W. Sutcliffe, 273–287. Dordrecht: Kluwer Academic Publishers.

Lathrop, R.C., and S.R. Carpenter. 2013. Water quality implications from three decades of phosphorus loads and trophic dynamics in the Yahara chain of lakes. *Inland Waters* 4: 1–14.

Lathrop, R.C., S.R. Carpenter, and D.M. Robertson. 1999. Summer water clarity responses to phosphorus, Daphnia grazing, and internal mixing in Lake Mendota. *Limnology and Oceanography* 44: 137–146.

Lemke, P., J. Ren, R.B. Alley, I. Allison, J. Carrasco, G. Flato, Y. Fujii, G. Kaser, P. Mote, R.H. Thomas, and T. Zhang. 2007. Observations: Changes in snow, ice and frozen ground. In *Climate change 2007: The physical science basis. Contribution of working group I to the fourth assessment report of the Intergovernmental Panel on Climate Change*, ed. S. Solomon, D. Qin, M. Manning, Z. Chen, M. Marquis, K.B. Averyt, M. Tignor, and H.L. Miller, 339–383. Cambridge: Cambridge University Press.

Likens, G.E., ed. 1985. *An ecosystem approach to aquatic ecology: Mirror Lake and its environment*. New York: Springer.

Liss, R.S., and A.J. Crane. 1983. *Man-made carbon dioxide and climatic change: A review of scientific problems*. Norwich: GeoBooks.

Livingstone, D.M. 1997. Break-up dates of alpine lakes as proxy data for local and regional mean surface air temperatures. *Climatic Change* 37 (2): 407–439.

———. 1999. Ice break-up on southern Lake Baikal and its relationship to local and regional air temperatures in Siberia and to the North Atlantic Oscillation. *Limnology and Oceanography* 44 (6): 1486–1497.

———. 2000. Large-scale climatic forcing detected in historical observations of lake ice breakup. *Verhandlungen des Internationalen Verein Limnologie* 27: 2775–2783.

Livingstone, D.M., and R. Adrian. 2009. Modeling the duration of intermittent ice cover on a lake for climate-change studies. *Limnology and Oceanography* 54 (5): 1709–1722.

Lopez, L.S., B.A. Hewitt, and S. Sharma. 2019. Reaching a breaking point: How is climate change influencing the timing of ice breakup in lakes across the northern hemisphere. *Limnology and Oceanography* 64 (6): 2621–2631.

Magnuson, J.J. 1990. Long-term ecological research and the invisible present. *Bioscience* 40: 495–501.

———. 2002a. Signals from ice cover trends and variability. In *Fisheries in a changing climate*, ed. N.A. McGinn, 3–14. Bethesda: American Fisheries Society.

———. 2002b. Three generations of limnology at the University of Wisconsin. *Verhandlungen des Internationalen Verein Limnologie* 28: 856–860.

———. 2008. Challenge of unveiling the invisible present. In *The vanishing present: Wisconsin's Changing lands, waters, and wildlife*, ed. Donald Waller and Thomas Rooney, 31–40. Chicago: University of Chicago Press.

Magnuson, J.J., and T.K. Kratz. 2000. Lakes in the landscape: Approaches to regional limnology. *Verhandlungen des Internationalen Verein Limnologie* 27: 74–87.

Magnuson, J.J., and R.C. Lathrop. 2014. Lake ice: Winter, beauty, value, changes, and a threatened future. *Lakeline* 34: 18–27.

Magnuson, J.J., C.J. Bowser, and A.L. Beckel. 1983. The invisible present: Long term ecological research on lakes. *Letters & Science Magazine, University of Wisconsin-Madison* 1: 3–6.

Magnuson, J.J., B.J. Benson, and T.K. Kratz. 1990. Temporal coherence in the limnology of a suite of lakes in Wisconsin, U.S.A. *Freshwater Biology* 23: 145–149.

Magnuson, J.J., T.K. Kratz, T.M. Frost, B.J. Benson, R. Nero, and C.J. Bowser. 1991. Expanding the temporal and spatial scales of ecological research and comparison of divergent ecosystems: Roles for the LTER in the United States. In *Long term ecological research*, ed. P.G. Risser, 45–70. Chichester: Wiley.

Magnuson, J.J., K.E. Webster, R.A. Assel, C.J. Bowser, P.J. Dillon, J.G. Eaton, H.E. Evans, D.J. Fee, R.I. Hall, L.R. Mortsch, D.W. Schindler, and F.H. Quinn. 1997. Potential effects of climate change on aquatic systems: Laurentian Great Lakes and Precambrian Shield region. In *Freshwater ecosystems and climate change in North America: A regional assessment*, ed. C.E. Cushing, 7–53. Chichester: Wiley.

Magnuson, J.J., D.M. Robertson, B.J. Benson, R.H. Wynne, D.M. Livingstone, T. Arai, R.A. Assel, R. G Barry, V. Card, E. Kuusisto, N.G. Granin, T.D. Prowse, K.M. Stewart, and V.S. Vuglinski. 2000a. Historical trends in lake and river ice cover in the Northern Hemisphere. *Science* 289: 1743–1746 and Errata 2001, *Science* 291: 254.

Magnuson, J.J., R.W. Wynne, B.J. Benson, and D.M. Robertson. 2000b. Lake and river ice as a powerful indicator of past and present climates. *Verhandlungen des Internationalen Verein Limnologie* 27: 2749–2756.

Magnuson, John J., Barbara J. Benson, and Timothy K. Kratz. 2004. Patterns of coherent dynamics within and between lake districts at local to intercontinental scales. *Boreal Environment Research* 9: 359–369.

Magnuson, John J., Barbara J. Benson, Olaf P. Jensen, Taryn B. Clark, Virginia Card, Martin N. Futter, Patricia A. Soranno, and Kenton M. Stewart. 2005. Persistence of coherence of ice-off dates for inland lakes across the Laurentian Great Lakes region. *Verhandlungen des Internationalen Verein Limnologie* 29: 521–527.

Magnuson, John J., T.K. Kratz, and Barbara J. Benson, eds. 2006. *Long-term dynamics of lakes in the landscape: Long-term ecological research on North Temperate Lakes*. Oxford: Oxford University Press.

Mudelsee, M. 2012. A proxy record of winter temperatures since 1836 from ice freeze-up/breakup in Lake Nasijarvi, Finland. *Climate Dynamics* 38: 1413–1420.

O'Reilly, C.M., S. Sharma, D.K Gray, S.E. Hampton (+ 53 other authors). 2015. Rapid and highly variable warming of lake surface waters around the globe. *Geophysical Research Letters* 42, 10,773-10,781. https://doi.org/10.1002/2015GL066235.

Oliver, Samantha K., Sarah M. Collins, Patricia A. Soranno, Tyler Wagner, Emily H. Stanley, John R. Jones, Craig A. Stow, and Noah R. Lottig. 2017. Unexpected stasis in a changing world: Lake nutrient and chlorophyll trends since 1990. *Global Change Biology* 23 (12): 5455–5467.

Palecki, M.A., and R.G. Barry. 1986. Freeze-up of lakes as an index of temperature changes during the transition seasons: a case study for Finland. *American Meteorological Society* 25: 893–902.

Risser, Paul G., ed. 1991. *Long-term ecological research: An international perspective*. Chichester: John Wiley & Sons.

Robertson, D.M. 1988. Lakes as indicators of and responders to climate change in climate variability and ecosystem response. In *Proceedings of a long-term ecological research workshop*, ed. D. Greenland and S.W. Swift Jr., 38–46. Boulder: University of Colorado.

———. 1989. *The use of water temperature and ice cover as climate indicators*. Ph.D. dissertation, University of Wisconsin, Madison.

Robertson, D.M., and R.A. Ragotzkie. 1990. Thermal structure of a multibasin lake: Influence of morphometry, interbasin exchange, and groundwater. *Canadian Journal of Fisheries and Aquatic Sciences* 47: 1206–1212.

Robertson, D.M., R.A. Ragotzkie, and J.J. Magnuson. 1992. Lake ice records used to detect historical and future climatic changes. *Climatic Change* 21: 407–427.

Robertson, D.M., W. Anderson, and J.J. Magnuson. 1994. Relations between El Niño/Southern Oscillation events and the climate and ice cover of lakes in Wisconsin. In *El Niño and long-term ecological research (LTER) sites*, ed. D. Greenland, 48–57. Seattle: LTER Network Office, University of Washington.

Robertson, D.M., Randolph M. Wynne, and William Y.B. Chang. 2000. Influence of El Niño on lake and river ice cover in the Northern Hemisphere from 1900 to 1995. *Verhandlungen des Internationalen Verein Limnologie* 27: 2784–2788.

Ruosteenoja, K. 1986. *The date of break-up of lake ice as a climatic index*. Helsinki: Department of Meteorology, University of Helsinki.

Scott, J.T. 1964. *A comparison of the heat balance of lakes in winter*. Ph.D. dissertation, University of Wisconsin–Madison.

Sellery, G.C. 1956. *E. A. Birge, a memoir*. Madison: University of Wisconsin Press.

Senge, P.M. 1990. *The fifth discipline: The art & practice of the learning organization*. New York: Doubleday/Currency.

Sharma, Sapna, and John J. Magnuson. 2014. Oscillatory dynamics do not mask linear trends in the timing of ice breakup for Northern Hemisphere lakes from 1855 to 2004. *Climatic Change* 124: 835–847.

Sharma, Sapna, John J. Magnuson, Gricelda Mendoza, and Stephen R. Carpenter. 2013. Influences of local weather, large-scale climatic drivers, and the ca. 11-year solar cycle on lake ice breakup dates; 1905-2004. *Climatic Change* 118: 857–870.

Sharma, Sapna, John J. Magnuson, Ryan D. Batt, Luke A. Winslow, Johanna Korhonen, and Yasuyuki Aono. 2016. Direct observations of ice seasonality reveal changes in climate over the past 320-570 years. *Scientific Reports* 6: 2506.

Sharma, S., K. Blagrave, J.J. Magnuson, C. O'Reilly, S. Oliver, R.D. Batt, M. Magee, D. Straile, G. Weyhenmeyer, L. Winslow, and R.I. Woolway. 2019. Widespread loss of lake ice around the Northern Hemisphere in a warming world. *Nature Climate Change* 9: 227–231.

Simojoki, H. 1961. *Climatic change and long series of ice observations at Lake Kallavesi*. International Association of Scientific Hydrology, General Assembly of Helsinki, 1960, Publication no. 54, pp. 20–24.

Swanson, F.J., and R.E. Sparks. 1990. Long-term ecological research and the invisible place. *Bioscience* 40: 502–508.

Tanaka, M., and M.M. Yoshino. 1982. Re-examination of the climatic changes in central Japan based on freezing dates of Lake Suwa. *Weather* 37: 252–259.

Vavrus, S.J., R.H. Wynne, and J.A. Foley. 1996. Measuring the sensitivity of southern Wisconsin lake ice to climate variations and lake depth. *Limnology and Oceanography* 4 (1): 822–831.

Walsh, Sarah E., Stephen J. Vavrus, Jonathan A. Foley, Veronica A. Fisher, Randolph W. Wynne, and John D. Lenters. 1998. Global patterns of lake ice phenology and climate: Model simulations and observations. *Journal of Geophysical Research* 103: 28825–28837.

White, Peter S. 1993. Will we be able to detect, understand, and predict ecosystem change? Review of Paul G. Risser, ed. *Long-term ecological research: An international perspective. Ecology* 74: 636–637.

Williams, G., K.L. Layman, and H.G. Stefan. 2004. Dependence of lake ice covers on climatic, geographic and bathymetric variables. *Cold Region Science and Technology* 40: 145–164.

Willig, Michael R., and Lawrence R. Walker. 2016. *Long-term ecological research: Changing the nature of scientists*. New York: Oxford University Press.

Wynne, R.H. 2000. Statistical modeling of lake ice phenology: Issues and implications. *Verhandlungen des Internationalen Verein Limnologie* 27: 2820–2825.

Chapter 10
Evolution of Social-Ecological Research in the LTER Network and the Baltimore Ecosystem Study

J. Morgan Grove and Steward T. A. Pickett

Abstract The addition of two urban sites, based in Baltimore, Maryland, and Phoenix, Arizona, to the Long Term Ecological Research (LTER) program in 1997, posed challenges for creating a truly integrated social-ecological framework. Proposals to include social "core areas" to sites with social science agendas were developed in concert with national environmental priorities being set within the scientific community. Although the National Science Foundation rejected these proposals for LTER as a whole, researchers at the urban sites pursued their goals of developing more sophisticated multi-disciplinary frameworks for the study of urban patterns and processes. The Baltimore Ecosystem Study (BES) illustrates how the conceptual basis of urban ecology evolved over 20 years, developing new strategies to manage cross-disciplinary interactions and relationships. Researchers drew especially on the concept of "boundary objects" as articulated by S. L. Star and J. Griesemer, and on General Stanley A. McChrystal's conception of the "team of teams" as a way to engage with complex problems. Looking forward, BES is positioned now to expand to a long-term, transdisciplinary science platform which we call the Baltimore Ecosystem Alliance. Its goals include developing an authentic urban ecology on local and regional levels, while also serving as a national and international leader in urban ecology.

Keywords LTER program · Long-term ecological research · Urban ecology · Socio-ecological research · Boundary objects · Team science · Transdisciplinary science

J. M. Grove (✉)
USDA Forest Service, Northern Research Station, Baltimore, MD, USA
e-mail: morgan.grove@usda.gov

S. T. A. Pickett
Cary Institute of Ecosystem Studies, Millbrook, NY, USA
e-mail: picketts@caryinstitute.org

R. B. Waide, S. E. Kingsland (eds.), *The Challenges of Long Term Ecological Research: A Historical Analysis*, Archimedes 59,
https://doi.org/10.1007/978-3-030-66933-1_10

10.1 Two Trajectories for Resolving Ecology's People Problem

Prologue This chapter traces two intertwining trajectories that embody the struggle of ecological science to accommodate people in its thinking and research. The steps taken along these two trajectories to resolve the problematic issue of people as components of ecological systems point to challenges and to opportunities for future progress. One trajectory is the programmatic role of the Division of Environmental Biology (DEB) at the National Science Foundation (NSF). The other trajectory is exemplified by the Baltimore Ecosystem Study, a Long-Term Ecological Research (LTER) project that has operated from 1997 through the present. The Baltimore Ecosystem Study (BES) represents a radical solution to ecology's people problem, a problem we describe just below. BES developed research strategies that joined biological, physical, and social sciences, as well as pursued dialog and cooperation with public agencies, community groups, and non-governmental organizations to embed ecological science and insights into the civic decisions around environmental quality and social equity in Baltimore. The chapter is the first to identify two of the larger strategies that BES developed for research synthesis: "boundary-spanning" tools, and structuring the effort as a "team of teams." The chapter explores the increasingly embedded nature of BES in the Baltimore social-ecological system, leading us to suggest that the next generation of effort might be as a social-ecological-practical platform for generating shared knowledge and action within and about Baltimore. The chapter explores the question of the potential divergence between the programmatic trajectory of the LTER program in DEB and the trajectory of long-term research and practice as embedded in its urban home.

Trajectory 1: The NSF LTER Program Ecology has struggled with how, or even whether, to include humans, human agency, and human institutions within its scope since the very founding of the discipline (Kingsland 2005, 2019). The struggle is exemplified in how the Long-Term Ecological Research (LTER) program has approached humans as ecological entities. The LTER program was an extraordinarily bold move at its founding in 1980, because it extended ecological research projects to time spans long enough to understand slow processes, long-lasting environmental phenomena, rare events, and slowly emerging indirect effects (Likens 1989). The usual ecological grants at the time were blind to these longer-term phenomena.

This boldness was repeated several times as LTER sought to integrate humans into the thinking about ecosystems. However, the LTER program can equally be said to vacillate in its exploration of the human frontier. Interdisciplinary recommendations from the LTER community concerning the importance and framing of social-ecological research were ignored or rejected by NSF during this time. This history suggests that ecology's discomfort about including people in its scientific scope has not been entirely vanquished. Indeed, this history has shown repeated retreat from consistently and firmly embedding human motivations, behaviors, institutions, and reciprocal environmental interactions into its support of ecological system science.

Trajectory 2: Baltimore Ecosystem Study as an LTER Project The most forward looking and transformative step that LTER took at the suture between the natural and the human was to establish two urban sites in 1997, charging them to include human actions and structures in their guiding frameworks and long-term research projects. This step, manifested in the founding of the Baltimore Ecosystem Study (BES) and the Central Arizona-Phoenix project (CAP), signaled to ecological science and to those who sought to apply ecological knowledge to the pressing problems at the cusp of the twenty-first century, that the research community was ready to accept humans and their actions and artifacts as ecological processes on par with the biotic, organismal, and physical features usually recognized as components of all ecological systems.

This radical step, though only taken in two urban regions, put ecological science at the nexus of two of the Earth's major transitions: global climate change, and the exponential growth of urban populations associated with urban spatial expansion and regional-to-global connectivities. The intersection of these two complex and multifaceted trajectories embeds ecology squarely in the realm of societies' most complex, multifaceted, and contested problems. Long-term research is well-suited to disentangle the mechanisms, feedbacks, and practical implications of such problems.

BES, along with its sibling urban site, CAP, employed and vetted ecologically unfamiliar research approaches. This chapter traces this radical history using the authors' experience with BES, complemented by interactions with CAP and the growing cadre of interdisciplinary urban researchers around the world. We identify three important strategies that have facilitated the establishment, growth, and adaptation of transdisciplinary social-ecological research and application.

Chapter Roadmap The chapter begins with an overview of Trajectory 1, based on available reports and our perception of the history of how the LTER program in DEB approached urban ecology. Our own research and understanding of the literatures of urban history, design, planning, and sociology convince us that the institutional trajectory of LTER requires a social-ecological approach. As the chapter moves to Trajectory 2, we discuss the shift in urban research approaches from an ecology embedded *in* the city, to an ecology *of* the city that investigates the biological, physical, and social aspects of the city together, and ending with a socially interactive ecology *for* or *with* the city, which is co-produced by researchers, residents, social, and governmental institutions. Section 10.4 discusses tools that BES developed to allow researchers from different disciplines to learn with each other and to develop trust. Examples include information repositories, ideal schematic charts to communicate conceptual models, standardized forms and practices such as shared mapping and field trips, and conscious learning about disciplines' different perspectives on the same geographies. Closely coupled with the use of boundary objects is an adaptive use of a "team-of-teams"approach to organize and enhance the interactions required by researchers and stakeholders to conduct transdisciplinary research and application.

Finally, we discuss a strategy for adapting to changing urban realities, for facilitating interdisciplinary work, and for matching the increasing complexity of cities with the complexity of their dynamic regional and global contexts. We end by discussing the promise of moving from urban social-ecological research projects to transdisciplinary platforms in which researchers, policy makers, managers, and urban residents interact to jointly envision and improve their shared urban places. Although the LTER program has played a pivotal role in bringing ecology to this threshold, it is unclear whether, as a formal funding program, it is poised to carry the transformation to fruition.

10.2 Trajectory 1: Toward Social-Ecological Research in the LTER Program

The seeds for social-ecological research in LTER were sown by the first decadal review of the program, covering 1980 to 1990. Led by Paul Risser and Jane Lubchenco, the *Ten-year Review of the National Science Foundation's Long-term Ecological Research Program* made several important recommendations (Risser et al. 1993). Among those that ultimately matured into social-ecological projects, the report suggested that the NSF LTER program consider expanding the spatial scale of sites; increasing flexibility and perhaps adding to its core areas; moving from "pristine" sites in forested, grassland, desert, coastal, or alpine ecosystems to other types of ecosystems, particularly human-dominated ones; including social and economic sciences; developing education programs; and expanding the number of sites to as many of 40 through partnerships with other federal agencies. Based upon this report, the NSF LTER program requested proposals to "regionalize" two existing LTER sites in 1993 by increasing their spatial extent and including a human dimension. In 1994, two sites successfully competed for this regionalization: the North Temperate Lakes LTER (NTL), based at the University of Wisconsin-Madison and Coweeta LTER (CWT), based at the University of Georgia. In both of these cases, social and economic sciences were added to existing, long-term ecological research (Gragson and Grove 2006).[1]

[1]During this time, a 1991 Cary Conference on humans as components of ecosystems was held to examine the state of the science and the needs and opportunities for examining the role of humans in ecosystems. This conference was funded largely by NSF and led to a 1993 book by the same name (McDonnell and Pickett 1993). NSF program officers from the Social, Behavioral, and Economic Sciences Directorate and the Division of Environmental Biology attended the conference.

In 1997, the NSF LTER program requested proposals for up to two urban LTERs, perhaps the most obvious example of human-dominated systems. In addition to the existing core areas, two new research areas and one programmatic feature were to be added. NSF specified in its Urban LTER request for proposals (RFP):

> LTER research should be developed around a site-specific conceptual framework that generates questions requiring experiments and observations over long time frames and broad spatial scales. The conceptual frameworks of the existing LTER sites are broadly focused around five core areas: 1) Pattern and control of primary production; 2) Spatial and temporal distribution of populations selected to represent trophic structure; 3) Pattern and control of organic matter accumulation in surface layers and sediments; 4) Patterns of inorganic inputs and movements of nutrients through soils, groundwater, and surface waters; and 5) Patterns and frequency of disturbance to the research site.
>
> The five core areas help to focus and integrate LTER research within and across sites. These core areas are broadly defined and must be incorporated into the research to be conducted in an urban ecosystem. In addition to the traditional LTER core areas, an Urban LTER will: 1) Examine the human impact on land use and land-cover change in urban systems and relate these effects to ecosystem dynamics; 2) Monitor the effects of human-environmental interactions in urban systems, develop appropriate tools (such as GIS) for data collection and analysis of socio-economic and ecosystem data, and develop integrated approaches to linking human and natural systems in an urban ecosystem environment; and 3) Integrate research with local K-12 educational systems.

The NSF LTER program received 23 proposals to their call (Collins 2019). Two sites were funded: the Baltimore Ecosystem Study LTER, led by Cary Institute of Ecosystem Studies with numerous partners, and the Central Arizona-Phoenix LTER, led by Arizona State University. Social and economic scientists were involved from the initial development of the two urban LTERs.

In 2004, another human-dominated site, representing an agricultural system, was added with the Kellogg Biological Station (KBS), led by Michigan State University. However, social and economic scientists were not involved in the initial planning of this project nor was there core funding for social and economic science.

The two regionalized sites (CWT and NTL) and the agricultural site (KBS) began and operated for some time as "disciplinary, but co-located" or "multi-disciplinary" sites, requiring few interactions and little interdependence among ecological, economic, and social science theories, methods, or data. In contrast, the two urban sites began as interdisciplinary sites, with social-ecological questions and interdependent social and ecological theories and data (Collins et al. 2000; Pickett et al. 2019, 2020).

Importantly, the two additional research areas and school system interactions required for the urban LTERs were not official core areas of equal standing to the original five ecological core areas of the LTER program. That is, although the urban LTERs were required to add human impacts of land use and land cover change on ecosystem dynamics to their research portfolio, the RFP did not explicitly establish those as core areas. In essence, there were no "constitutional amendments" ensuring that future LTER projects in human-settled ecosystems would be required to conduct research in these two additional topics. The two new topics were criteria that might or might not persist, either for these two sites or for new sites in the future.[2]

An Expansion among LTER Sites and a Network Rejected Over the decade after the establishment of the two urban LTERs in 1997, the LTER network progressed intellectually and organizationally from thinking prospectively about what it meant to do long-term social-ecological science to reflecting on the experience at a range of initial LTER sites, to considering how to expand social-ecological science at all the LTER sites and beyond, and finally to using LTER social-ecological science as a model for other NSF programs and observatories. One of the first steps on this journey was the LTER Science Council meeting in 1998, in Madison, Wisconsin, where a social science committee was established to guide the development of social science and advance long-term, interdisciplinary science in the LTER network.

To correct the fact that the urban LTER RFP from 1997 did not rely on or establish new social or economic core areas, Tom Baerwald, Director of NSF's Social, Behavioral, and Economic (SBE) Sciences Directorate, held a workshop with this new LTER social science committee to consider social and economic core areas for the LTER network. In January 2000, LTER scientists and colleagues from other large interdisciplinary projects[3] funded by NSF gathered in Tempe, Arizona, to craft a network-wide approach to integrate social and economic science into long-term ecological research.

Members of the workshop developed an integrative conceptual framework, recommended core social and economic science areas, and suggested practical strategies for interdisciplinary science (Redman et al. 2004). As a corrective to the urban LTER Request for Proposals, the group recommended moving from a human impact framework (Fig. 10.1), which separates "natural" and human systems, to an integrated, social-ecological system framework (Fig. 10.2). This shift is significant because ecology in general has long regarded humans as separate from natural

[2]As of April 2020, the LTER Network website notes: "Two additional themes emerged with the addition of urban LTER sites, but it has become clear that they are also relevant for the rest of the Network." https://lternet.edu/core-research-areas/

[3]Charles Redman (CAP) and Morgan Grove (BES) led the workshop. Many of the workshop participants were recommended by Tom Baerwald and were internationally-recognized leaders in their fields. Workshop participants and their affiliations included Charles L. Redman, Nancy Grimm, Ann Kinzig, Lauren H. Kuby, and Ed Hackett (Central Arizona–Phoenix LTER); J. Morgan Grove, Bill Burch, and Steward Pickett (Baltimore Ecosystem Study); Steve Carpenter and Peter Nowak (North Temperate Lakes LTER); F. Stuart Chapin (Bonanza Creek Experimental Forest); Ted Gragson (Coweeta Hydrologic Laboratory); Craig Harris (Kellogg Biological Station); Bob Waide (LTER Network); Tom Baerwald (NSF/Division of Behavioral and Cognitive Sciences); Anthony de Souza (National Research Council); Grant Heiken (Los Alamos National Laboratories); Peter Kareiva (National Oceanic and Atmospheric Administration); Emilio Moran and Elinor Ostrom (Indiana University); Sander van der Leeuw (Sorbonne); Tom Wilbanks (Oak Ridge National Laboratory); and Brent Yarnal (Penn State).

systems and has narrowly focused on the impacts of human actions on ecosystem structure and function. Furthermore, these impacts are most often assumed to be negative. This assumption weakened in the 1980s as ecologists began to acknowledge a shift toward a research approach, also called a "non-equilibrium paradigm," that considers systems to be open to outside influence, and to have open-ended dynamics. Furthermore, as historical and contemporary data identified how humans and their actions were embedded in many ecosystems that had been thought to be pristine, the external, negative impact model was also questioned (Simberloff 1980; Spirn 1984; Southgate 2019). The contrary assumption that humans were components of ecosystems (e.g., McDonnell and Pickett 1993) helped fuel the work of the interdisciplinary group that generated the framework in Fig. 10.2.

The workshop used the integrated social-ecological framework (Fig. 10.2) and built on six focus areas had been identified as high priorities in a National Research Council (1999) report on understanding the human dimensions of global environmental change. The workshop proposed six social core areas of patterns and processes that were intended to be "constitutional" requirements for LTER sites having social science agendas and to complement the five biophysical core areas that were already required for all LTER sites. The core areas' components included, but were not limited to:

1. Demography: growth, size, composition, distribution, and movement of human populations;
2. Technological change: accumulated cultural knowledge about how to adapt to, use, and act on the biophysical environment and its material resources to satisfy human needs and wants;
3. Economic growth: institutional arrangements through which goods and services are produced and distributed;
4. Political and social institutions: enduring sets of ideas about how to accomplish goals recognized as important in a society. For instance, most societies have some form of family, religious, economic, educational, health, and political institutions that characterize their way of life;
5. Culture: attitudes, beliefs, and values that purport to characterize aspects of collective reality, sentiments, and preferences of various groups at different scales, times, and places; and

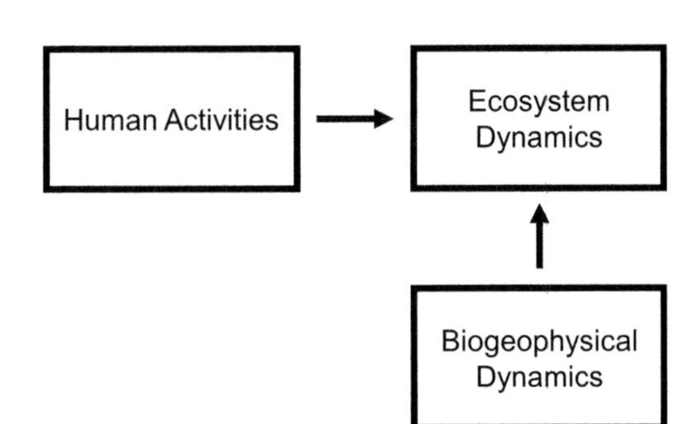

Fig. 10.1 Human Impact Framework, focusing on how human activities affect ecosystem dynamics. Note the absence of interactions and feedbacks among human activities, biogeophysical drivers, and ecosystem dynamics

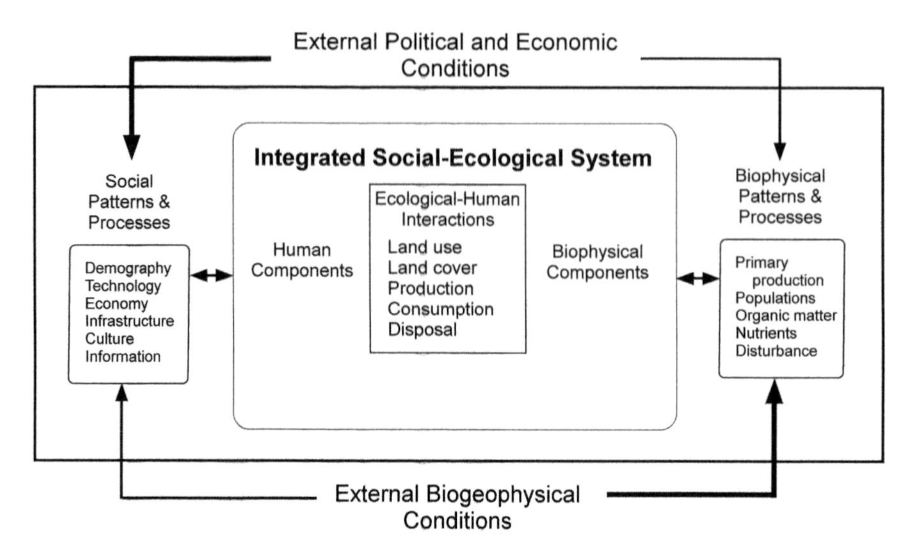

Fig. 10.2 The integrated social-ecological framework identifies external conditions that may be political, economic, or biogeophysical conditions; and social and ecological patterns and processes that interact with an integrated social-ecological system. Importantly, components within the integrated social-ecological system are considered both human and ecological simultaneously

6. Knowledge and information exchange: genetic and cultural communication of instructions, data, ideas, and so on.

The workshop also identified practices for collaboration and integration to operationalize the integrated social-ecological framework and the now eleven core areas. These included the use of geographic information systems (GIS) and mapping of physical, biological, and socioeconomic data; the use of historical analyses of places to understand the "how" and "why" of social-ecological change; and comparative, cross-site analyses to identify common themes and generalizable knowledge.

The report and recommendations were provided to both NSF's Social, Behavioral, and Economic Sciences Directorate and the Division of Environmental Biology (DEB), which was the NSF lead for the LTER Program. BES and CAP led methodological and conceptual advances suggested by these reports for the field of urban research. LTER could not sign on in real time, but BES and CAP managed to sustain these advances and infuse them into the field as a whole, as will be described in Sect. 10.4. Neither the conceptual framework nor the six social core areas were adopted by the LTER program, however.

The LTER Network's social science committee continued to conduct workshops and sessions at the triennial LTER All Scientists Meetings during the early 2000s (Gragson and Grove 2006; Zimmerman et al. 2009). In these meetings, the social science committee focused on how to create and support a functional social science community and improve its capacity and effectiveness in the LTER Network. Reports from these workshops observed that the conceptual framework and core areas proposed by Redman and others (2004) were necessary but insufficient. Also needed was support for an information system of demographic, social, and economic data that would compile standardized data for all sites. This would facilitate research at individual sites and cross-site comparisons. Some of these data would come from existing sources such as the U.S. Census and the Bureau of Labor Statistics, while other data would be produced through practices standardized among LTER social scientists. A second major concern was the capacity for long-term research support. Similar to the needs of LTER biophysical scientists, the social scientists emphasized the need for long-term support because it was extremely difficult and risky to cobble together long-term research through a series of three-year grants.

The reports and recommendations for capacity building were provided to NSF's Division of Environmental Biology and Directorate for Social, Behavioral, and Economic Sciences offices. Neither the multi-site, socio-demographic information system nor long-term support for social science were adopted by the LTER program.

Despite this chilly reception, interest and support for long-term interdisciplinary research grew over time within the LTER Network and the larger science community (e.g., Gordon et al. 2019). A new conceptual framework was produced (Collins et al. 2007; Collins et al. 2011), which incorporated new thinking about ecosystem services and press and pulse phenomena of change (Fig. 10.3). Press changes are those that persist over time, whereas pulse changes are episodic and usually of brief duration.

This framework was presented in 1998 as part of a report to NSF and a proposal for a new strategic research initiative, "Integrative Science for Society and Environment" (ISSE).[4] This proposal was intended to support social-ecological research for all LTER sites as well as to inform support for social-ecological research for all of NSF, other government agencies, and the larger scientific community (Grove et al. 2015a, b). As SBE considered integrating social science into NSF environmental observatories, both specific LTER sites and the ISSE report were considered exemplary models for how to proceed (Vajjhala et al. 2007). As before, the reports and recommendations were provided to DEB and SBE, but were not adopted into division policy.

From 1998 to 2008, numerous efforts were made and a detailed rationale provided to expand the scope and capacity for social-ecological research in the LTER Network. Over this period, several other national scientific intuitions were also exploring the nature and importance of addressing joint biological, social, physical, and engineering frontiers. In particular, reports from the National Research Council (2000), and "Grand Challenges" from both the LTER Network and NSF advanced the scientific community's thinking and priorities for socio-environmental science. While these efforts were unsuccessful for shaping the LTER Network as whole, they provided important insights into strategies for collaboration to foster social-ecological research. BES is an example of how these strategies were realized at an LTER site. We switch to that trajectory in the next section.

---

[4] https://lternet.edu/wp-content/themes/ndic/library/pdf/reports/ISSE_complete_30April.pdf

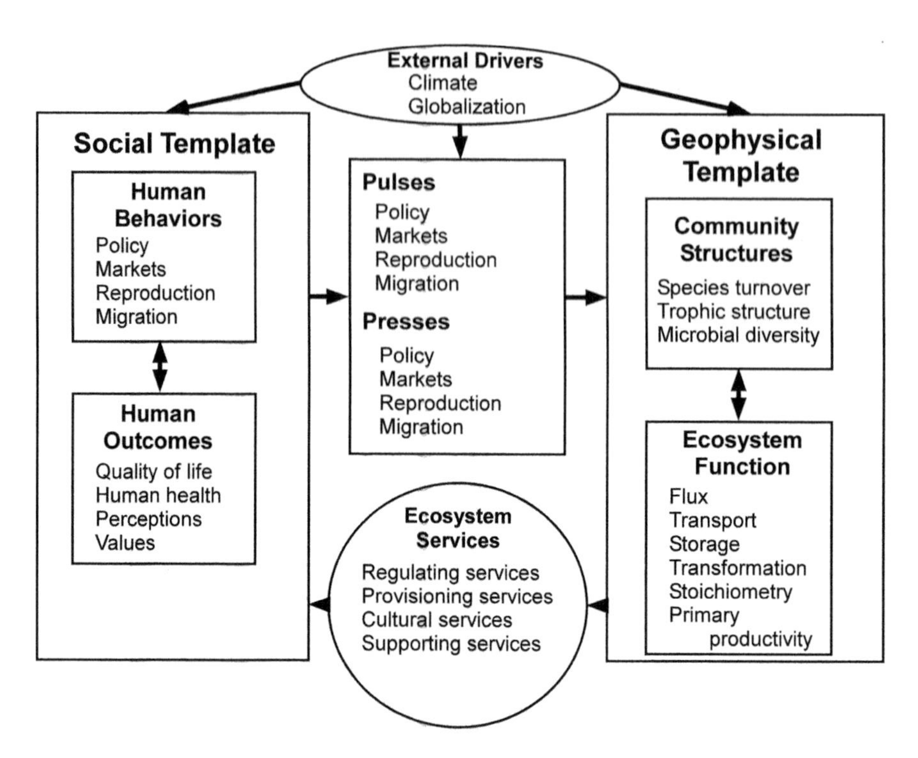

Fig. 10.3 The Integrative Science for Society and Environment (ISSE) framework was devised by an interdisciplinary group of social and biophysical researchers and policy makers. The framework is intended to give equal weight to social and biophysical structures and processes, to clearly identify services of relevance to humans, to show that biophysical and social processes and structures are reciprocally linked, that changes in the drivers of the system can occur as short term, or as chronic. The framework includes key kinds of indicators and phenomena that must be measured in each step of the feedback between human and natural phenomena. Integrative research hypotheses appear in the original and are indicated by the labels on the interaction arrows. (From Collins et al. 2007)

10.3 Trajectory 2: Two Decades of Social-Ecological Research in BES

Two processes characterize the pioneering era of social-ecological research from 1998 to 2018. One was conceptual and the other methodological. The first, conceptual process was the shift in ecology's fundamental background assumptions and research approaches, that is, its paradigm (Simberloff 1980; Pickett et al. 2007). The second, methodological process was the adoption of strategies and tactics to advance interdisciplinary research and application, including "boundary objects," a "team of teams" approach, and the use of organizations that focus on coordinating the effort.

Conceptual Shifts in Ecological Science The efforts to regionalize two sites in the LTER Network – Coweeta and North Temperate Lakes – were built primarily upon existing, traditional disciplines and university programs. In contrast, the new urban LTER sites were more radical and demonstratively transformative because they were built upon new intellectual infrastructure and forged new scientific partnerships (Grimm et al. 2000; Cressey 2015; Grove et al. 2015b; Pickett et al. 2020). The vision of BES was to improve the understanding of metropolitan Baltimore as a social-ecological system and to inform residents and private, governmental, and civic decision makers in their efforts to improve the quality of their environments and their lives. The initial BES team drew on and grew out of the expertise and experiences of two programs that had already been working in urban areas. The Urban-Rural Gradient Ecology (URGE) program had begun in New York City in 1986 as a partnership between the New York Botanical Garden and Cary Institute of Ecosystem Studies (McDonnell and Hahs 2008). The URGE team included soil scientists, biogeochemical scientists, plant ecologists, and landscape ecologists. The Urban Resources Initiative (URI) began in Baltimore in 1989 as a collaboration among the Yale School of Forestry & Environmental Studies, the City of Baltimore, and Parks & People Foundation (Grove and Carrera 2019). The URI team included natural resource sociologists, community foresters, and community organizers.

These two teams provided the foundation for BES, which was constituted to conduct research in an integrated social-ecological system; pursue educational activities with the local school system; and engage communities, non-governmental organizations, and governmental agencies.

During its first two decades, BES developed in several ways. First, it advanced beyond the shortcomings of the Chicago School (Grove et al. 2015b). Prior to BES and CAP, the most widely recognized school of urban ecology had been the Chicago School from the 1920s. For BES and CAP, the Chicago School was an encumbrance because individuals new to urban ecology often assumed that the Chicago School's vintage research agenda and approaches typified the urban ecology agenda of BES and CAP. The Chicago School, developed in the early twentieth century by sociologist Robert Park and colleagues at the University of Chicago, had adopted a spatial model of concentric zones of human communities and interactions among communities. The Chicago School hypothesized that cities went through a deterministic life-cycle driven by immigration, invasion, and replacement of human communities in a rigid succession, drawing upon the ecologist Frederic Clements's theory of plant succession to a stable "climax" community (Light 2009; Cadenasso and Pickett 2013). This hypothesis combined idealized spatial form, pre-determined succession, and an equilibrium end point. Second, in contrast to the romanticized integration and order of rural communities, cities were conceived of negatively as dominated by processes of social dysfunction, vice, and disorder. These theories of urban sociological dynamics and plant community dynamics had been replaced by the time the two urban LTERs began. However, urban ecology was still sometimes tarred with that obsolete brush (e.g., Gottdiener and Hutchinson 2000).

BES sought to distinguish itself from the Chicago School. First, BES built its conceptual apparatus and research program upon contemporary ecological, social, and economic theory (Grove et al. 2015b; Pickett et al. 2020). The use of ecological succession ideas that had been challenged as early as 1916 were poor materials from which to construct a viable urban ecological theory for the twenty-first century (Cadenasso and Pickett 2013). Further, a deterministic and negative view of cities in contrast to rural life, both socially *and* environmentally, was a prejudicial approach for an urban ecology research program.

BES and CAP expanded the intellectual foundations of urban ecology. Since the 1970s, urban ecology deployed three increasingly inclusive paradigms (Cadenasso and Pickett 2013; Childers et al. 2014). One is the ecology *in* the city. The second is the ecology *of* the city, which originated with BES and CAP (Pickett et al. 1997, Grimm et al. 2000). The third is the ecology *for* the city, which emerged with the growing development, geographic spread, and co-production of practical applications from urban ecological science throughout the world (Box 10.1; Childers et al. 2015; Pickett et al. 2016; Zhou et al. 2017).

The contrast between an ecology *in* and the ecology *of* cities was introduced in an editorial in 1997 in the first issue of the journal *Urban Ecosystems* (Pickett et al. 1997). That short piece expanded urban ecological science from its familiar biotic foundations and suggested how the two urban LTER programs could add to ecological theories and approaches by examining cities as spatially complex and extensive social-ecological systems. In part, this was necessary to overcome many American ecologists' skepticism about the value of research in urban areas. Shortly thereafter, the integrative contrast between the two paradigms was explored in greater depth in an article in *BioScience,* "Integrated approaches to long-term studies of urban ecological systems" (Grimm et al. 2000).

The "ecology *in* cities" approach had been led by biologically oriented ecologists who applied their rural or wildland toolkits to cities, suburbs, and towns to study habitats or ecosystem types that were familiar to them but were embedded in an urban or urbanizing matrix. This approach was pioneered in Europe and Asia (Numata 1977; Goode 1989; Wang and Lu 1994; Sukopp 2008; McDonnell 2011; Lachmund 2013; Wu 2014; Kowarik 2020). The habitat types chosen for ecological study in cities were often structural analogs to those outside of cities, including forested areas, parks, cemeteries, meadows, vacant lots, wastelands, streams, and wetlands. In its pure form, the ecology *in* the city paradigm used a binary representation of spatial heterogeneity in cities: built versus non-built. Although some researchers parsed the built environment to varying degrees (Fischer et al. 2013; Lehmann et al. 2014), this idealization helps to clarify the different approaches that an ecology *in* and an ecology *of* cities took to spatial heterogeneity.

The idealized theory of spatial heterogeneity for the ecology *in* cities paradigm can be called "patch/matrix" theory. The patch/matrix theory follows such biological precedents as island biogeography, where the two components in a binary landscape are hospitable non-built areas, comprising patches and corridors, embedded in a hostile or uninhabitable built matrix (Cadenasso et al. 2013). The patch/matrix approach in urban ecology had constrained integration with the social sciences and the development of social-ecological approaches. The patch/matrix framework also followed the "human impact" model (Fig. 10.1), so that the demographic and social drivers affecting urban ecological analogs were assumed to emerge from the built environments and the surrounding human populations. Descriptive variables in patch/matrix context under the ecology *in* cities paradigm included impervious cover, roads, buildings, human population density, or economic activity, and were usually taken as external to the patches of interest.

The ecology *of* the city focused on entire urban areas, not just the rural or wild analogs used by the ecology *in* the city with its binary, patch/matrix conception of social ecological systems and spatial heterogeneity. This altered focus changed the perspectives of ecological research in two ways. First, the spatial conception of heterogeneity shifted from a patch/matrix view to a mosaic view: All patch types in an urban area were within the scope of ecology of the city. Furthermore, ecological research in the city was required to go beyond the pure analogs of natural areas outside the city.

Second, all elements of heterogeneity were included and potentially consequential to social-ecological relationships. In other words, an ecology *of* cities theorized urban systems as spatially complex and inclusive mosaics of interacting patch types. Under this theory, all measurable features of patches could be ecologically important. Patch features consist of the types (i.e., content), numbers, frequencies, boundaries, changes in, and interactions among patches. These features could all drive system structure and function. Further, the patches themselves could be hybrids, that is, they could include both biotic and human-derived elements (Cadenasso et al. 2007). With this new view, interactions between social and biogeophysical structures and processes were pervasive, intertwined, reciprocal, multi-scale, and dynamic (McPhearson et al. 2016; Rademacher et al. 2019). These additions to the ecology *in* the city approach greatly expanded the potential interactions with a variety of social sciences, economics, and urban designers. Explorations with engineers, complexity scientists, and science and technology studies have further broadened the scope of urban ecology (McPhearson et al. 2016).

Study of urban tree canopy is an example of how research changed between the ecology *in* and the ecology *of* the city approaches. In the patch/matrix view of streamside vegetation in cities, ecologists expected these habitats to help reduce nitrate pollution in streams. However, BES discovered that such riparian forest did not perform that function (Groffman et al. 2003). Given the desire of city managers to satisfy the demands to reduce nitrate pollution, collaboration between BES researchers, city agencies, and local non-profits shifted focus to the nitrate dynamics of a larger range of patch types in the city mosaic. Structure and function of all patch types in city watersheds were examined regardless of how "green" they were. Ecology *in* the city would have maintained focus on large, conspicuous green patches, and missed the opportunity to help mitigate nitrate pollution throughout the city. The broader approach is an example of ecology *of* the city.

The shift from a matrix to mosaic view of urban ecological systems required more robust theory and methods from both ecological and social sciences. Landscape

Box 10.1: Watershed Studies Have Been a Core Area of Focus for BES. Differences Between an Ecology *in, of, and for* Cities Can Be Illustrated with Examples of Watershed Questions

Ecology *in* Cities: How do forested areas affect fluxes of water and nutrients in urban watersheds?

Ecology *of* Cities: How do forested areas compare to the other types of land uses—residential, commercial, and industrial—in the urban mosaic in terms of fluxes of water and nutrients in urban watersheds; and how does land management, spatial arrangement, and spatial connectivity of these land uses affect fluxes of water and nutrients? How do long-term legacies of residential, racial segregation and environmental inequities affect fluxes of water and nutrients in urban watersheds?

Ecology *for* Cities: How can urban stream restoration projects be designed and implemented to minimize flooding, improve nutrient processing, and enhance aesthetics while minimizing the loss of trees and introduction of invasive species? At the same time, are stream restoration projects more effective for minimizing fluxes of water and nutrients than increasing tree canopy cover on upstream areas of the urban watershed mosaic? How do watershed revitalization projects affect property values, risk of gentrification, and residents' sense of place?

ecology, the study of spatial pattern and its functional implications, became critical for understanding ecosystem fluxes, population dynamics, community interactions, and biodiversity within patches and across the urban mosaic. From a social science perspective, the shift from an ecology *in* to an ecology *of* cities required greater social science sophistication than simply including the contextual parameters such as human population density or built infrastructure around an urban forest. Humans and their institutions were considered essential parts of the urban ecological system, not simply external and allegedly negative influences. Further, the role of humans at multiple scales of social organization—from individuals, households, and neighborhoods to complex and persistent agencies—was linked to the biophysical scales of urban systems. Thus, an ecology *of* cities opened the way toward understanding the dynamic feedbacks among biophysical and human components of the system; their spatial and temporal contexts; and their effects on ecosystem inputs and outputs at various social scales (Grove et al. 2015b).

The third paradigm that emerged from the two urban LTERs was an ecology *for* the city. Ecology *for* the city is *transdisciplinary* in that it involves urban scientists from many research disciplines, along with local communities, associations, agencies, designers, and decision-makers in the planning, design, and co-production of research and societal solutions (Childers et al. 2015). With an ecology *for* cities, urban ecology moved from "knowledge to action," meaning that research was conducted in cooperation and often led by the problems identified by communities or agencies. Ecology for the city resides in a realm of environmental stewardship, where the scientific knowledge about an urban area as a social-ecological system is integrated with decision-making processes of all sorts (Chapin et al. 2011). Examples of areas in ecology *for* cities include environmental restoration, social revitalization, economic vitality, public health, and environmental justice.

An ecology *for* cities often engages with societally wicked problems. Wicked problems have several characteristics. They occur in situations that are composed of complex systems of interacting and inter-dependent parts (Liu et al. 2007). Actors often hold diverse and conflicting values and perspectives. Solutions are often uncertain and sub-optimal (Simon and Schiemer 2015; Duckett et al. 2016). Urban stream restoration projects are an example of a wicked problem. Actors from government agencies, environmental groups, communities, and businesses may disagree, leading to conflicts over preferred solutions when large trees and forested areas must be removed to re-engineer and restore the stream channel and riparian areas. Both the value and scientific validity of the stream restoration may be called into question. Others may question whether such expenditures are even a priority given other obvious social, economic, or environmental concerns.

Wicked problems can also have different temporal dimensions. Wicked problems can be immediate, short term, or long term (Collins et al. 2011). A stream restoration is an example of a short-term event. Longer term wicked problems include the needs to mitigate or adapt to changes in climate, urbanization and reconfiguration of urban regions, and disruptions of food, energy, and water systems. Wicked problems are also evident in real-time responses to emergencies (Machlis and McNutt 2010) such as catastrophic storm events, epidemics, heatwaves, droughts, fires, disease outbreaks, and toxic spills. BES has contributed to solutions of wicked problems, for example, in streamside vegetation function (Groffman et al. 2003), neighborhood restoration (Hager et al. 2013), increase of urban tree canopy (Locke et al. 2013), environmental justice relative to polluting industry (Boone 2009), and downsides of different kinds of stormwater management interventions (Irwin et al. 2017).

BES and CAP represented a major initiative to build a new urban ecology based upon contemporary ecological and social sciences. During this time, urban ecology grew from a marginal interest to a widely pursued, theoretically motivated social-ecological field. The shifts in paradigms, led by BES and CAP (e.g., Cadenasso and Pickett 2013, Childers et al. 2015, Zhou et al. 2017) have proven to be a useful organizing lens for the continued growth and maturation of urban ecological science and have become familiar enough to be used in urban ecology textbooks as a framing device (e.g., Gaston 2010, Adler and Tanner 2013, Douglas and James 2015).

10.4 Three Strategies for Advancing Transdisciplinary Socio-Ecological Trajectory

Over its first 20 years, BES developed several strategies to pursue an ecology *of* and *for* cities, and to support interdisciplinary and transdisciplinary collaborations over the long term. These strategies consist of 1) boundary objects, 2) teams of teams, and 3) management of boundaries and boundary organizations. We present these three strategies in turn.

10.4.1 Boundary Objects as Conceptual Tools and Processes

We first introduce a rigorous conception and typology of boundary objects (Star 1988; Star and Griesemer 1989). Susan Leigh Star (1954–2010) was a sociologist and historian of science whose ethnographic studies included information infrastructures and how diverse groups found ways to collaborate in scientific projects. Star's initial conception of boundary objects was intended to elucidate the "nature of cooperative work in the absence of consensus" (Star 2010, p. 604). In particular, Star and Griesemer (1989) observed that scientific problems and their solutions can often appear to be ill-structured, inconsistent, ambiguous, illogical, and complex. These characteristics are akin to those of wicked problems.

At the same time, science often requires cooperation among actors to create common understandings, to ensure reliability across scientific domains, and to gather information that retains its integrity across time, space, and context. These requirements represent a fundamental conflict between reconciling divergent viewpoints and the desire to produce generalizable findings. Exacerbating this dilemma, many researchers see cooperation as a form of compromise and a loss of scientific identity. Thus, a key question in science, particularly when addressing ill-structured and complex problems, is how to manage diversity and cooperation among actors.

Star observed that many models of scientific cooperation assume that consensus must occur before cooperation can begin. Her own ethnographic research on contemporary science projects and historical analyses indicated that this was not true; instead, she observed that teams developed strategies for cooperation without first requiring consensus (Star 2010). The sequence of conceptual frameworks that BES has used are boundary objects, as we will detail later. Star proposed the idea of boundary objects based upon the cooperation she observed within teams. Star wrote that boundary objects "sat in the middle between different groups" (Star 2010, p. 608).

In this usage, objects are flexible and shared intellectual or material structures that were manifest by how they functioned to allow scientists or groups to work cooperatively. Thus, the construction of boundary objects was a "community phenomenon, requiring at least two sets of actors with different viewpoints" (Star 1989, p. 52). That is, boundary objects do not exist prior to or independent from cross-disciplinary interactions. For example, one of the initial boundary objects of BES was the human ecosystem framework, but the version used to help establish BES differed from the source framework (Machlis et al. 1997) in that it reflected additional "hooks" for biological and physical scientists.

Star argued that developing boundary objects is an essential process in achieving cooperation and managing diversity across intersecting social worlds to address ill-structured or complex problems. Boundary objects are useful in several ways. They "allow scientists to cooperate and work [collectively] 1) without having good models of each other's work; 2) while employing different units of analysis, methods of aggregating data, and different abstractions of data; and 3) having different goals, time horizons, and audiences to satisfy (Cash et al. 2003)." Boundary objects may be abstract or concrete. An important test of boundary objects is their ability to encompass, change, and adapt to multiple points of view while increasing communication across viewpoints (Star and Griesemer 1989). This is essential for different actors to conceive of and negotiate transdisciplinary problems and to conceptualize how they fit in and identify appropriate roles for their participation (Cash et al. 2003; Barry et al. 2008).

Star and Griesemer (1989) proposed four types of boundary objects (Fig. 10.4): 1) repositories, 2) ideal types, 3) coincident boundaries, and 4) standardized forms and practices. If Star had studied BES, we believe she would have found that her conditions for the utility of these boundary objects matched the situation that the URGE and URI teams faced as they developed the first, interdisciplinary BES proposal. We were tackling an ill-structured and complex problem: the urban ecological dynamics of the Baltimore metropolitan region. We had diverse disciplines and goals "without good models of each other's work" from ecosystem, community, and landscape ecology, hydrology, biogeochemistry, sociology, geography, and economics. We needed to find ways to combine data and work on different units of analysis. And we had different audiences to satisfy, from academic institutions to government agencies. Over time, the boundary objects BES has employed have evolved (Pickett et al. 2020). Further, some of these newer boundary objects had been identified through LTER social science or interdisciplinary workshops and meetings mentioned earlier. Importantly, we have found that these four types of boundary objects can be combined into an interacting system.

Repositories are ordered "piles" of things that are indexed in a standardized fashion. For instance, Star and Griesemer (1989) examined the University of California Berkeley's Museum of Vertebrate Zoology as an example of a repository with its

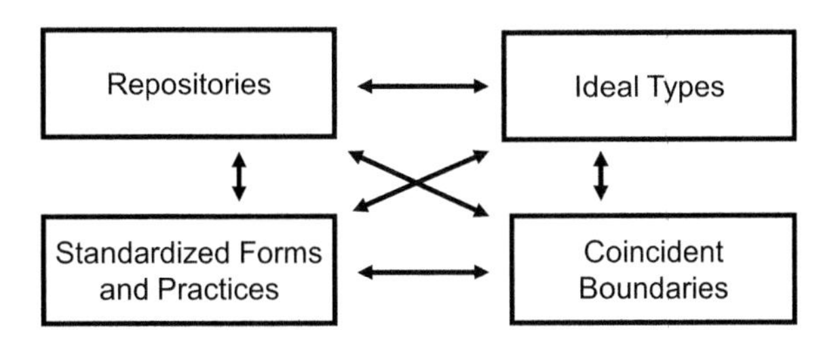

Fig. 10.4 Star and Griesemer's (1989) typology of boundary objects. Our experience indicates the value of each type of boundary object as well as the value of using them together as a system

collection of specimens and associated standardized documentation. Repositories have sub-components and individuals can use and re-use these sub-components for their own purposes.

To meet LTER Network requirements, BES developed an information management system—*a repository*—of diverse materials that included field samples, interviews, and remote sensing data. These data were analyzed and documented with metadata. The BES information management system spans a broad range of disciplines; academic, governmental, non-profit, and community sectors; and scales of time, space, and social-ecological nested hierarchies. The BES information system is designed and managed to be user-oriented and accessible over the long term by multiple users for multiple purposes.

Ideal types include diagrams and other forms of symbolic abstraction. They are not intended to describe precisely the details of a place or a thing. They are abstractions from the relevant domains and may be fairly vague. An ideal type "serves as a means of communicating and cooperating symbolically—a 'good enough' road map for all parties" (Star and Griesemer 1989, p. 410). In its initial proposal (1997), BES used two types of conceptual diagrams as ideal types. The first conceptual framework (Fig. 10.5), proposed the integration of physical, ecological, and social components by pursuing an ecology of cities approach using patch structures, functions, dynamics, causes of change, and education activities. This diagram provided an overarching view of how BES proposed to implement an urban patch dynamics approach that would enable cooperation and collaboration among biophysical and social scientists and educators.

In its 2016 incarnation, BES evolved the original framework to account for exogenous drivers and explicitly to address the dynamic feedbacks among different urban ecosystem structures, including built patches, and functions over the long

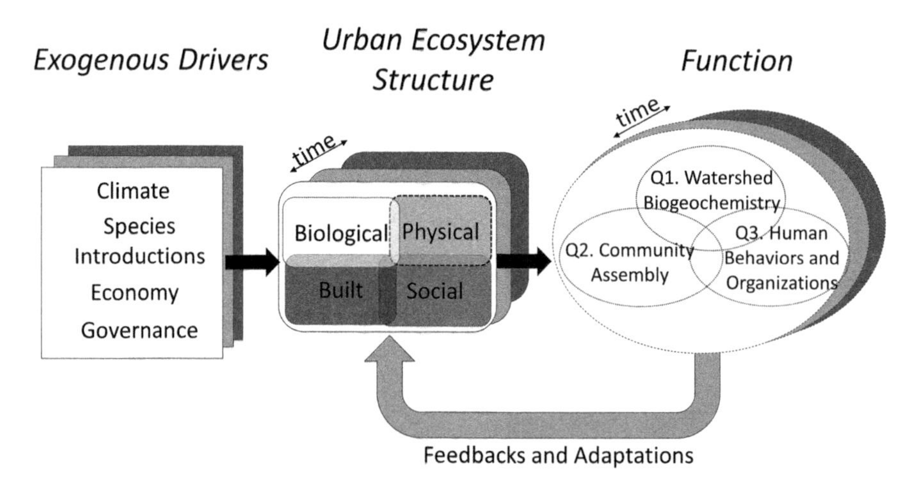

Fig. 10.5 Conceptual framework for BES LTER showing how exogenous drivers play out on complex urban ecosystem structure to produce functional responses in three focal areas; with internal feedbacks

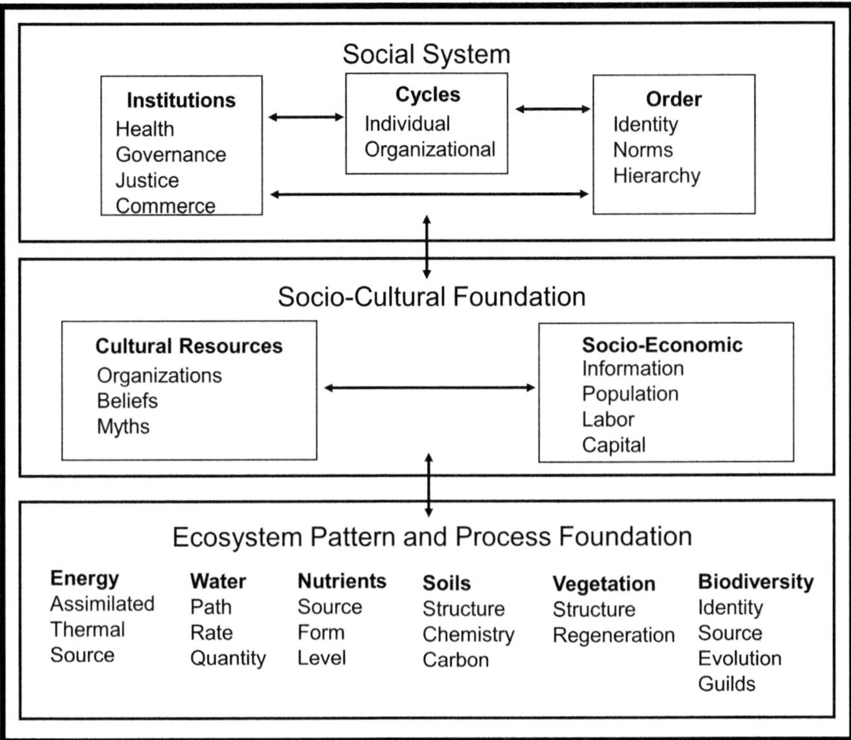

Fig. 10.6 The Human Ecosystem Framework as it appeared in the original BES proposal. (Adapted from Machlis et al. 1997)

term (Fig. 10.5). In both ideal types, the conceptual frameworks remain abstract and inclusive to facilitate collaborative work in diverse disciplines.

BES's urban patch dynamic ideal type is complemented by a human ecosystem framework (Fig. 10.6), which also appeared in the first proposal. This framework expanded upon the items for a patch dynamics approach (Fig. 10.2) and indicated an integrative approach for research teams and potential participants. Further, it provided guidance for the data to be included in the BES information management system (repository).

Coincident boundaries are often relatively familiar locales that have the same boundaries but whose contents will appear different to different teams. These objects may be multi-scaled, nested places or different types, such as the Baltimore region, municipalities, and neighborhoods, or climatic zones, geologic formations, or watersheds. An example of how various people may see the same locale differently is how those charged with regional stormwater management may see the city differently from those interested in air quality, or in social justice. The original BES

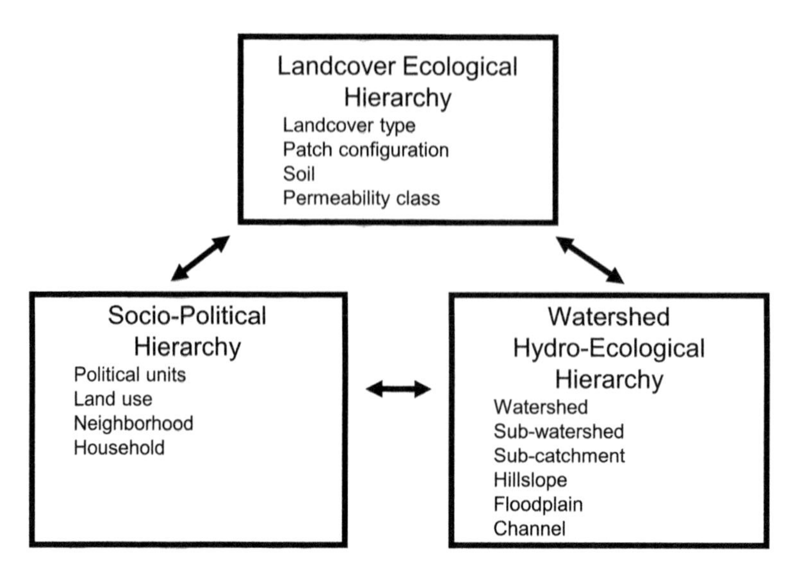

Fig. 10.7 An interdisciplinary, nested set of interacting place-based hierarchies

proposal identified a nested set of interacting place-based hierarchies that included watersheds, socio-political units, and land cover (Fig. 10.7).

Standardized forms and practices are methods of common communication across dispersed work groups over space and time to facilitate cooperation and collaboration. BES developed such standardized forms and practices in order to facilitate cooperation and collaboration. For instance, BES developed a common practice of interdisciplinary teams going to a place to engage in "a walkabout," during which they discussed – and often argued about – what might be the present, past, and future physical, ecological, social, demographic, economic, and political drivers and responses for the set of households along the street, the neighborhood, the neighborhood in the watershed, the watershed in the City and County, and the City and County in relation to the State and the Chesapeake Bay.

Furthermore, participants in a walkabout would discuss plans and organize data collection teams so that physical, ecological, and social data were co-located and synchronized. BES created strategies to link and integrate diverse data types such as field data, remotely-sensed data, and administrative data with "data hooks" that included latitude/longitude, address, time, and scale (Grove et al. 2015b). For example, a common practice for our household social survey has been to always document the address and latitude/longitude of respondents so that the survey results can be linked to and enhanced with other social, economic, physical, and ecological data.

BES combined the practices for collecting and integrating diverse data with different approaches or ways of knowing (Fig. 10.8). The ways of knowing can be diagrammed as a BES "Data Temple," which shows how some of the different kinds of boundary objects can be combined in a project strategy. The base is the information management system; the pillars are forms and practices for how data are

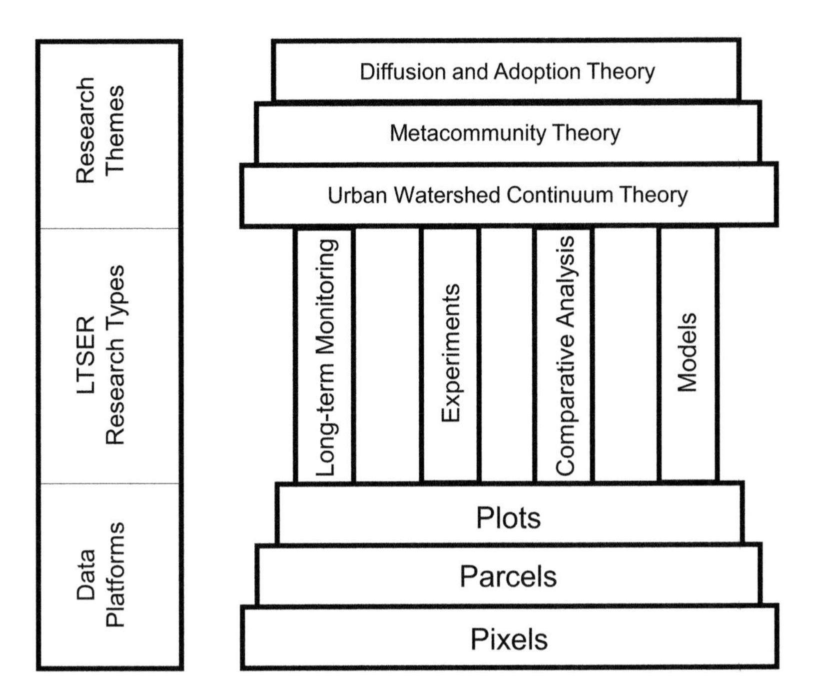

Fig. 10.8 The BES "Data Temple" integrates the data platform, research types, and research themes. It represents an evolution of how we approach the *standardized forms and practices* for collaboration among BES scientists and teams

integrated and then applied through long-term monitoring, experiments, comparative analyses, and modeling to support the roof: the specific research themes shown in Fig. 10.5. The temple is an example of an ideal type, and is an important shared view of the structure of BES.

Star did not suggest that her four types of boundary objects be used interactively as a system. BES has found that the integration of boundary objects was crucial for advancing its social-ecological work. However, the development and use of these boundary objects takes time and learning. Over time, BES teams learned to use these boundary objects simultaneously and interactively. During a walkabout, individuals in the group contribute their disciplinary perspectives on what is happening in that place (*coincident boundary).* At the same time, they are framing their discussion in terms of their conceptual diagram (*ideal type*), the data they need to answer the question (*repository*), and the different ways to collect and integrate their data and ways to answer the question (*standardized forms and practices*). Conversely, when the group is sitting in a room in front of a white board working on their conceptual diagram (*ideal type*), they are also thinking about how it would be operationalized in the context of one neighborhood versus another, or this watershed versus that watershed (*coincident boundaries*) and the data that are needed (*repository*). The disciplined use of boundary objects, as an interactive system, creates a structure for learning and collaboration among researchers from disparate disciplines.

10.4.2 A Team of Teams as a Scientific Social Practice

A second strategy for collaboration is how teams are organized to solve a problem. When BES began, several obstacles to team science were immediately clear. Biophysical scientists had a history of doing team science, but those teams were limited to working with other, familiar types of biophysical scientists. Biophysical scientists were apparently uncomfortable working with social scientists. The social scientists had neither a tradition of working in teams in general nor with biophysical scientists. Further, we needed to determine how to organize as teams in ways that matched the complexity of an urban system (e.g., Cadenasso et al. 2006).

We rely on General Stanley A. McChrystal's conception of a "Team of Teams" (McChrystal et al. 2015) to explore the social practices that BES eventually developed. General McChrystal developed his ideas around a "team of teams" as the Director of the Joint Staff of the Joint Special Operations Command of the U.S. military's fight against Al-Qaeda from 2003–2008 in Iraq. During this time, McChrystal and his staff recognized that the U.S. military was organized hierarchically, which was appropriate to solve complicated problems. Complicated problems may have many parts, but they have stable structures, and the interactions among the parts are deterministic and predictable (Allen et al. 2018). Reductionist approaches and siloed organizations can be highly effective at solving complicated problems.

However, Al-Qaeda was a networked and decentralized, complex adversary, and the task of defeating Al-Qaida presented a complex, wicked problem. Such problems have many interdependent parts and the interactions are unpredictable. Networked systems often need to be designed to solve complex problems (Fig. 10.9). Just as Star noted the value of boundary objects to build cooperation to tackle ill-structured, ambiguous, and complex problems, McChrystal and his staff realized that they needed to develop the necessary social organization and teamwork for a complex rather than a complicated problem.

There are several critical features for teams, or a team of teams, to solve complex problems. Such features include particular organizational structures, cultures, and interactions among teams. Organizationally, there is a need to shift to using small teams and from an emphasis on efficiency to an emphasis on adaptability. Culturally, what makes small teams adaptable are trust, common purpose, shared awareness, and the empowerment of individuals to act.

These adaptive features are critical because they invigorate teams with an ability to solve problems that could never be foreseen by a single leader. Ideas and innovations often emerge through the bottom-up result of interactions, rather than from top-down directions. Trust is crucial within a team and among teams. Strong lateral ties are essential for developing trust and the construction of shared awareness.

Insights from Philosophy of Science to Facilitate a Team of Teams BES has developed many of the attributes for a team of teams approach. In BES, there have been philosophical, cultural, and functional aspects at the overall project level that have facilitated a team of teams approach. (We use the term culture to signify both

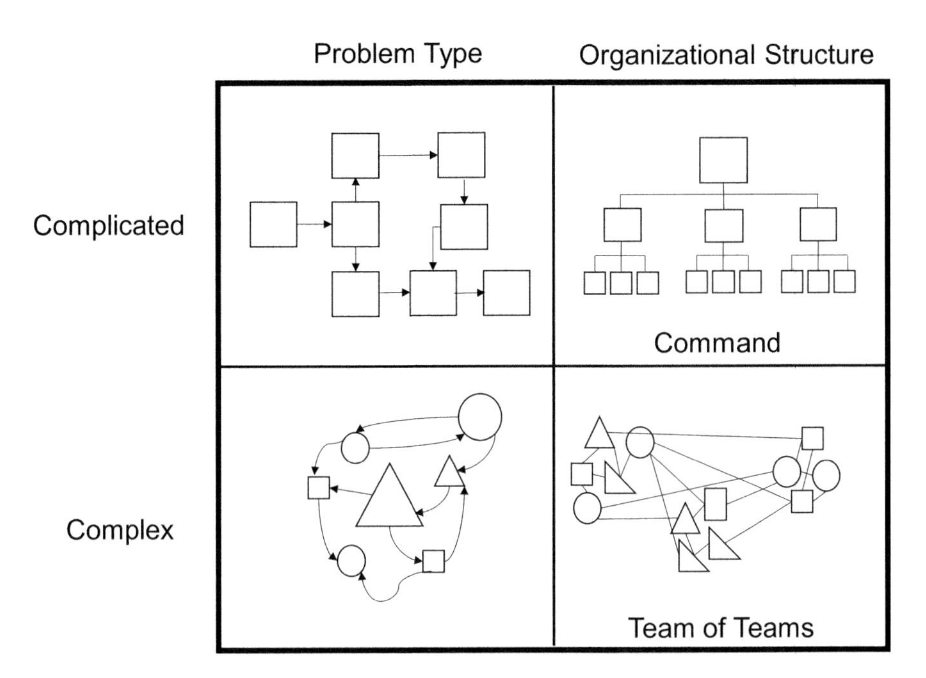

Fig. 10.9 The type of problem to be solved has important implications for the organizational structure adopted to solve the problem. This is true for both scientific and societal problems. Complicated problems can be broken down into its subcomponents without regard for interactions among subcomponents. In contrast, solutions to complex problems have to account for interactions among subcomponents. (Inspired by McChrystal et al. 2015)

specialized training and "the totality of socially transmitted behavior patterns, arts, beliefs, institutions, and all other products of human work and thought" (Pickett 1999).)

First, several of the founders of BES articulated a philosophy of exploration across disciplinary boundaries (Pickett et al. 1999). This culture emphasized the creative and open-minded impulse in contrast to the more conservative critical motivation that is so often emphasized in disciplinary science. Furthermore, the project identified "habits of mind" focused on synthesis, on overcoming narrow, defensive disciplinary differentiation, and on avoiding strictly reductionist stances to hypothesis generation and testing. In particular, the integrative habits of mind include 1) long-term commitment to integration, 2) use of analogy and radical juxtaposition of seemingly disparate ideas; 3) vanquishing the "eureka myth" of instantaneous discovery; and 4) improved use of diverse types of human knowledge and perspectives in the research community (Pickett 1999). These practices are not intended to be efficient, but rather, they must be adaptable.

This final attribute of adaptability has been particularly evident in BES. Talent and skills were sought broadly, and participants from many institutions were involved from the beginning. The institutional and disciplinary diversity have increased over time. For example,[5] BES social science expertise initially resided at Yale and the USDA Forest Service, and grew to include historical geography and social justice experts from Ohio University. Hydrology expertise came first from the University of Maryland, Baltimore County, University of North Carolina, and USGS, but has grown to include researchers from Penn State and University of Florida. Stream studies initially focused on water flow, and pollution by nitrate and phosphorus, but has now grown to include focus on ecological effects of contamination by pharmaceuticals and personal care products. Such flexibility existed because participation in the project was not limited to particular departments or institutions. The project became recognized as a "big tent," embracing members from many states in the U.S. and indeed several foreign countries. Budgetary constraints meant that no contributors could be fully supported by project funds. Consequently, a certain level of commitment has been intellectual rather than financial.

In BES, we found that attention to team building and nurturing was fundamental. In particular, tools such as standards that provide the "rules of the game," and philosophy have been important (Pickett 1999; Pickett et al. 1999; Holzer et al. 2018). Opportunities for individuals and teams to have social interactions and to develop group identity, trust, and bonding were key. Because new participants entered, and continue to enter, with different experiences and expectations, formal and informal coaching and support for interdisciplinary, team, and applied science are vital functions. Finally, participation involved discussion of incentives and rewards (Pickett et al. 2019).

Getting to know colleagues from other disciplines and acquiring an awareness of the whole can take significant time and effort. Cooperation and collaboration can create overlaps and redundancy. However, these inefficiencies are often what instills teams with high levels of adaptability and efficacy. Implementation of a team of teams approach requires constant maintenance. It involves transparent information sharing and decision making. Solving complicated problems requires teams that resemble machines; solving complex problems involves teams that are more like evolving organisms.

10.4.3 Management of Boundaries and Boundary Organizations

The use of boundary objects and the development and maintenance of teams of teams do not happen spontaneously. Both Cash et al. (2003) and McChrystal et al. (2015) emphasize the importance of organizations that act as intermediaries among teams, disciplines, and sectors. Called "boundary organizations," they may consist

[5] Not all institutions or disciplines that have been or are part of BES are listed in these examples. See https://baltimoreecosystemstudy.org/people/ for more information.

of diverse organizational arrangements and procedures to manage functions of communication, translation, and mediation at the boundaries among disciplines.

BES has adopted strategies to support a culture in which all participants were aware of the whole project and are encouraged to participate as members of the team of teams. BES holds quarterly project meetings that include all disciplines as well as different types of participants, including researchers, practitioners, and students. Its annual, two-day meeting is organized so that presentations are inter-mixed by topic. This means that there is no grouping of all the hydrologists, then the wildlife ecologists, then the economists, and then the geographers. This mixing tends to engage meeting participants and expose them to a diversity of concepts, methods, data, and findings from the diversity of BES disciplines.

While there is an emphasis on sharing and interaction at the project level as a whole, BES has teams organized around its three project areas: 1) Watershed biogeochemistry; 2) Ecological communities and sentinel systems; and 3) Human environmental perceptions, behaviors, and organizations. Each team always includes at least one member from each of the other teams. Thus, similar to McChrystal et al.'s (2015) approach, every member of BES does not need to know everyone else on the project, but everyone on one team knows at least one person on the other teams. Finally, our BES walkabouts build social affinity and trust.

10.5 Shaping New Trajectories: Changing Conceptions of Cities and Roles for Urban Ecology

BES enters its third decade facing several challenges. BES started with external support as a long-term study from the National Science Foundation's LTER Program. It is now intellectually ready to shift more fully to a science platform that is part of the Baltimore region and a continuing actor in the region's future (Grove and Pickett 2019). In doing so, BES must explore how to transition from an NSF LTER project to a long-term, transdisciplinary platform for Baltimore. At the same time, BES needs to address the changing conception of social-ecological science and the changing nature of cities themselves.

10.5.1 Changing Conceptions of Cities

A classic 1997 article by ecologists Peter Vitousek, Harold Mooney, Jane Lubchenco and Jerry Melillo, "Human domination of Earth's ecosystems," brought to prominence the need to understand the role of humans in local, regional, and global ecosystems (Vitousek et al. 1997). The paper provided scientific justification for addressing the fact that the world's population was becoming predominantly urban.

With ecological attention focused on cities, there have been several progressive phases of how cities have been conceived.

The first phase tended to see cities as complicated systems, composed of many predictably interconnected parts, and to focus on their negative impacts on the environment. A second phase consolidated early in the 2000s (Batty 2009), recognizing that cities are actually complex systems, with many interacting and interdependent parts that include positive and negative feedbacks, thresholds, multi-scalar dynamics, and path dependencies or historical legacies (cf. Allen et al. 2018). Often, this complex urban system was assumed to exist in a static environment. The third phase recognizes that not only are cities complex, but they exist in environments that are themselves spatially and temporally complex. For example, while cities were typically designed for relatively stationary climates, technologies, and economies, today they need to be understood in the context of rapid global and regional change and uncertain futures.

These changing conceptions of cities create scientific and practical challenges. Science is progressing from a focus on human impacts, through coupled "natural-human" systems, to the simultaneity of co-produced[6] social-ecological systems (Rademacher et al. 2019). Co-production, in this sense, implies that urban ecology needs to address urban systems that are simultaneously "a biophysical entity, a territory, a commodity, a habitat for nonhuman species, a resource for productive activities, [a place of consumption], and a buffer for absorbing pollutants. These systems are allocated, regulated, and administrated by various laws, norms, and rules and a source of meaning and sense of place, a landscape component, and symbolically loaded" (Meyfroidt et al. 2018, p. 63). There is also the simultaneity of time. To quote Faulkner (1951), "The past is never dead. It's not even past." One might also observe that in urban systems the future is never realized, but it is always present.

This progression in how we conceive of cities and their environments, beginning from a complicated system in a stable context, to a complex system in a changing world, will require both enhanced and new scientific and professional capacities. There is a need for diverse scientific perspectives on how an urban system works, the nature of its multi-stranded past, and the shape of its likely futures. There may be incomplete data, uncertain knowledge, or varying levels of confidence in the data and knowledge. Analytical models may not exist or may be insufficient to deal with the complexity of urban ecological problems. Finally, there may not be resources for monitoring and evaluating the long-term impacts of societal interventions. It is critical for cities to have these scientific capacities to address these challenges, and so it is vitally important for professionals to have practical and scientific training in interdisciplinary and synthetic solutions.

---

[6]Previously in this chapter, co-produced has been used to refer to transdisciplinary production of actionable scientific knowledge jointly by researchers and other stakeholders. Here, we use the same term for a distinct idea that urban systems are themselves co-produced by the interaction of biophysical and social processes.

10.5.2 A Shift from Transdisciplinary Projects to Platforms

To meet the challenges of complex urban problems, enhanced and new scientific and professional capacities will require changes in how transdisciplinary science is organized and supported (Gordon et al. 2019). We propose that an adaptation is to expand from transdisciplinary science projects to platforms (Grove and Pickett 2019). BES has been tending toward this kind of expansion for some time. However, its scope has been somewhat constrained by the need to satisfy a particular funding body. There is the possibility for continued growth as a platform and to contribute to local needs. Individual, short-term projects in research, innovation, monitoring, and assessment can certainly contribute to the sustainability and resilience of cities. But the full utility of such individual contributions depends on developing integrated knowledge systems that can persist and evolve as problems, societal needs, and scientific research and analytical technologies also evolve. A long-term social-ecological science platform for Baltimore can be such an adaptive, integrated knowledge system.

The contrast between short-term projects and long-term platforms falls along five dimensions: boundary objects; capacities; types and number of problems; collaboration; and time (Table 10.1). This list may not be exhaustive and the

Table 10.1 The difference between Projects and Platforms can be conceived of along five dimensions: boundary objects, capacity, problems, collaboration, and time. Projects can exist on their own or interact with other projects and build upon existing platforms. As societies face long-term, complex, and uncertain futures, social-ecological platforms may become increasingly important to address wicked problems

Dimensions	Projects	Platforms
Boundary objects	One-time use Short-term use and availability	Adaptation and reuse for multiple and unexpected applications over time Maintenance over the long term
Capacity	Teams developed for pre-identified problem. Team limited to length of 3–5 year funding	Teams adaptable as: New expertise needed, or New generations emerge Teams persist with diverse sources of long-term support
Problems	One problem Phenomenon over months or years Long-term historic change	Many problems "Real-time" responses to crises and emergencies over days Long-term future change Problems not yet identified or defined
Collaboration	Team is pre-identified and limited. Hierarchical structure.	Teams adapt as problems change or new problems emerge. Transboundary organizations and management. Network structure
Time	Closed span (often 3–5 years)	Open span (often decades)

differences among the dimension are in degree and not absolute. Projects and platforms will both develop and use boundary objects and employ teams, or a team of teams. However, the maintenance, reuse, and adaptation of boundary objects for multiple and unexpected applications over time is an important feature of long-term platforms.

There are differences in the types of phenomena most appropriate for projects compared to platforms. Projects may tend to focus on events occurring over months and years. Platforms, in contrast, are particularly aimed at extensive, pervasive, and subtle changes occurring over several decades. The standing capacity and competencies of long-term platforms will enable them to be persistent, available, nimble, and more effective for providing real-time, immediate responses to unfolding emergencies. Long-term platforms can also be a "place" for the discovery, exploration, and prototyping of science and applications that are not pre-identified and may be initially ill-defined. This allows long-term platforms to adapt and evolve as a whole in contrast to engaging in a series of new projects, each of which must reinvent the cultures and mechanisms of transdisciplinarity. Further, the long-term nature of platforms also allows new actors to join over time. Thus, platforms need to be able to acculturate, assimilate, and adapt to new participants, and changes in capacity, leadership, and relevant decision makers.

It is important to recognize that transitioning to platforms as a more effective knowledge system for transdisciplinary urban ecology will take time, continuing effort, and intentionality. Strategies to promote such systems require a sufficiently long-term perspective that accounts for the generally slow impact of ideas on practice, the need to learn from field experience, and the time scales involved in enhancing human and institutional capital necessary for doing all these things. Cash and colleagues observe that a decade or more may be the minimal period for transdisciplinary projects to be planned, implemented, and evaluated (Cash et al. 2003). This is relevant to individual studies and to transdisciplinary team projects. Indeed, such projects might span entire professional careers. In the case of Baltimore, the investments over the past 20 years provide a significant foundation for BES to transition to a transdisciplinary science platform.

10.5.3 From Baltimore Ecosystem Study to Baltimore Ecosystem Alliance

To expand to a long-term, transdisciplinary science platform in Baltimore, we anticipate several changes to the current structure of BES. The support and maintenance of boundary objects and boundary management are essential to the transdisciplinary capacity of the platform. To ensure the capacity for such activities, we anticipate that we will need to build an alliance from multiple sectors and multiple disciplines. A diverse alliance is valuable for engaging and solving complex problems because such problems often involve multiple boundary organizations to facilitate discussion among parties with multiple interests regarding differences in perspective, methodology, preferences, values, and desired outcomes (Cash et al. 2003). In particular, a transdisciplinary platform can play an important role in the science-to-action process by continually engaging stakeholders in all phases of the work: defining the problem, specifying the data and knowledge needs, constructing alternative solutions, and negotiating policy, planning, or management decisions. Such collaboration creates a process more likely to produce salient scientific information for decision making. It can also increase credibility by bringing multiple types of expertise to the table and enhance legitimacy by providing transparency to multiple stakeholders throughout the information production process (Cash et al. 2003).

An alliance anticipates an expansive view of transdisciplinary science and for co-producing data and knowledge. Cities, in the broadest sense, are substantially new things in the twenty-first century (McHale et al. 2015), requiring a science that can adapt to the changing reality of the urban world. This reality is complex, in both the figurative and the technical senses. It would seem ideally suited to and require both the integration of numerous ways of generating knowledge beyond a hypothetico-deductive conception of science (Meyfroidt et al. 2018; Pickett et al. 2019; Kamarainen and Grotzer 2019). It also entails a transition from coupled understandings of social and ecological systems to the simultaneity of co-produced social-ecological systems (Rademacher et al. 2019).

Many changes may be anticipated in a platform built to address complex or wicked problems in the long term. Several are implied by the contents of Table 10.1. For the sake of brevity, we present two changes to exemplify the nature of this new approach.

The Alliance will need to consider whether the existing boundary organizations are sufficient to support an increasingly diverse and multi-sector community. Additional organizations may need to be added. Further, existing approaches to support a team of teams approach such as annual, quarterly, and ad hoc meetings may be insufficient or ineffective. Supplemental or alternative approaches may need to be developed.

The Alliance will need to embrace its role and the use of its platform for training in place-based STEM science for diverse populations and transdisciplinary science and practice. This will require working with local programs from elementary through graduate school and developing common educational and training resources for Alliance members to use. It is important to stress here that this would be about more than producing more scientists. It would be about bolstering current and future generations of professionals in government, private-sector, non-governmental organizations, and communities. An important feature of the Alliance world be the opportunity to enhance students' education and training with experiences from local, relevant, real-life situations.

Finally, the evaluation of success would shift from an external set of NSF LTER program officers and panelists to local Baltimore Ecosystem Alliance members and their communities. An important goal would be to develop both an "authentic" urban ecology science on topics of interest at local and regional levels while continuing its place as a national and international leader in urban ecology. Further, the "effectiveness" of such a transdisciplinary alliance would be gauged not only in terms of solutions and outcomes, but in terms of who participated, how inclusively issues were defined and framed, and on the diversity of options considered.

10.6 Conclusion: A Way Forward

Over its first twenty years, BES, along with CAP-LTER, played a significant role in scientists' and practitioners' experience with transdisciplinary science for urban ecological dynamics (Börner et al. 2010; Zscheischler and Rogga 2015; Grove et al. 2016; Pearce et al. 2018). BES developed expertise and provided training in boundary objects, teams of teams, boundary organizations and boundary management. It demonstrated that transdisciplinary science could both advance scientific knowledge and be useful to practitioners. While short-term, urban transdisciplinary projects have generated valuable experience and training programs for scientists and students[7] BES is one of the only social-ecological science programs that has developed the experiences, expertise, and multi-generational training that comes with passing the 10-year milestone suggested by Cash et al. (2003) to characterize a platform.

During this time, NSF made numerous calls for interdisciplinary, transdisciplinary, and convergence science (e.g., Ramaswami et al. 2018). Ironically, during the first 20 years of BES and CAP, program elements associated with advancing convergence science were the same elements that created unresolved tensions and struggles with the NSF LTER program. The NSF LTER program failed to adopt the recommended core areas from LTER social scientists and to fully fund long-term social science data and social-ecological science. It did not support either of the social-ecological frameworks (Redman et al. 2004; Collins et al. 2007) that emerged from the LTER community. It insisted on reductive theory building and experimental approaches and dismissed the multiple pathways to generate theory and understanding of complex, social-ecological systems (Pickett et al. 2019; Kamarainen and Grotzer 2019).

Looking forward, it is clear that the United States and the world will continue to urbanize and that urban areas will be home to a growing proportion of the world's population. Many of these urban areas are most vulnerable to complex,

---

[7] For example: *Social-Environmental Immersion Program for Post-Graduate Fellows*, https://www.sesync.org/for-you/educator/programs/immersion; National Science Foundation, 2018, *Dear Colleague letter: Growing convergence research*, https://www.nsf.gov/pubs/2018/nsf18058/nsf18058.jsp; Belmont Forum, *Transdisciplinary science funding opportunities*, http://www.belmontforum.org/opportunities/. Accessed 4 January 2019.

social-ecological problems such as climate change, economic re-structuring; re-distributions of population, and technological changes in energy, transportation, and buildings. Many of these problems will be associated with conflict and few will have optimal solutions. Given these ongoing changes, Risser and Lubchenco's first decadal analysis of the LTER network was prescient. It articulated the value of studying human-dominated ecosystems, including urban systems, with long-term social-ecological approaches. This value was addressed with the urban LTER request for proposals in 1997. In the same report, Risser et al. (1993) also suggested that there might be other sources of support for LTER projects. Although they considered only other federal agencies for potential support, perhaps it is time to consider multi-sector support and enable broader scale investments and benefits. In this case, there may be the need for new alliances, models, and approaches not based upon NSF and the LTER program. This would require conceptualizations, strategies, and investments for a new transdisciplinary urban ecological science.

The bold history of transdisciplinary science in BES over the past 20 years should be complemented by an equally bold future for its next 20 years. Core institutional partners that have shaped and promoted BES are also committed to this unique project and its contributions to the science required for the rapidly changing global urban reality. Transitioning the Baltimore Ecosystem Study as an alliance, the Baltimore Ecosystem Alliance, may be the next step. And we also need to update the form of our transboundary objects and transboundary organizations to better reflect the need for transdisciplinary, urban ecology platforms (Sternlieb et al. 2013).

Over the past twenty years, BES, CAP, and the growing transdisciplinary urban ecology community have achieved intellectual, methodological, and practical success. What remains to be seen is how well this community can develop in the future to understand and support shrinking cities; expanding urban megaregions; cities not yet born; cities of consumption; cities of service; sustainable cities; resilient cities; and many more. The dimensions of urban change in the Baltimore region echo those occurring in many other towns and cities worldwide that share some of the attributes that now exist or are set to appear in Baltimore. These attributes, changes, and novelties are the raw material for producing the sustainable, resilient approach to the urbanizing, globalizing world of the twenty-first century. Nothing could be a more compelling justification for the continued investigation of humankind's most intimate and dynamic habitat, already so well understood in Baltimore.

Acknowledgments This material is based upon work supported by the National Science Foundation under grants # DEB-1637661, DBI-1052875, DBI-1639145, the USDA Forest Service, and the University of Maryland. Any opinions, findings, and conclusions or recommendations expressed in this material are those of the author(s) and do not necessarily reflect the views of the National Science Foundation, the USDA Forest Service, or the University of Maryland. We thank Miriam Avins for her thoughtful review and suggestions. We thank Scott Collins and remember the late Henry Gholz for visionary leadership and support for social-ecological research in the LTER network. None of this would have been possible without them.

References

Adler, Frederick R., and Colby J. Tanner. 2013. *Urban ecosystems: Ecological principles for the built environment*. Cambridge: Cambridge University Press.
Allen, T.F.H., P. Austin, M. Giampietro, Z. Kovacic, E. Ramly, and J. Tainter. 2018. Mapping degrees of complexity, complicatedness, and emergent complexity. *Ecological Complexity* 35: 39–44.
Barry, A., G. Born, and G. Weszkalnys. 2008. Logics of interdisciplinarity. *Economy and Society* 37 (1): 20–49.
Batty, M. 2009. Cities as complex systems: Scaling, interaction, networks, dynamics and urban morphologies. In *Encyclopedia of complexity and systems science*, ed. R.A. Meyers, 1041–1071. New York: Springer.
Boone, C.G., M.L. Cadenasso, J.M. Grove, K. Schwarz, and G.L. Buckley. 2009. Landscape, vegetation characteristics, and group identity in an urban and suburban watershed: Why the 60s matter. *Urban Ecosystems* 13: 255–271. https://doi.org/10.1007/s11252-009-0118-7.
Börner, Katy, N. Contractor, H.J. Falk-Krzesinski, S.M. Fiore, K.L. Hall, J. Keyton, B. Spring, D. Stokols, W. Trochim, and B. Uzzi. 2010. A multi-level systems perspective for the science of team science. *Science Translational Medicine* 2 (49): 49cm24. https://doi.org/10.1126/scitranslmed.3001399.
Cadenasso, M.L., and S.T.A. Pickett. 2013. Three tides: The development and state of the art of urban ecological science. In *Resilience in ecology and urban design: Linking theory and practice for sustainable cities*, ed. S.T.A. Pickett, M.L. Cadenasso, and B. McGrath, 29–46. New York: Springer.
Cadenasso, M.L., S.T.A. Pickett, and J.M. Grove. 2006. Dimensions of ecosystem complexity: Heterogeneity, connectivity, and history. *Ecological Complexity* 3: 1–12. https://doi.org/10.1016/j.ecocom.2005.07.002.
Cadenasso, M.L., S.T.A. Pickett, and K. Schwarz. 2007. Spatial heterogeneity in urban ecosystems: Reconceptualizing land cover and a framework for classification. *Frontiers in Ecology and the Environment* 5: 80–88.
Cadenasso, M.L., S.T.A. Pickett, B. McGrath, and V. Marshall. 2013. Ecological heterogeneity in urban ecosystems: Reconceptualized land cover models as a bridge to urban design. In *Resilience in ecology and urban design: Linking theory and practice for sustainable cities*, ed. S.T.A. Pickett, M.L. Cadenasso, and B. McGrath, 107–129. New York: Springer.
Cash, D.W., W.C. Clark, F. Alcock, N.M. Dickson, N. Eckley, D.H. Guston, J. Jäger, and R.B. Mitchell. 2003. Knowledge systems for sustainable development. *Proceedings of the National Academy of Sciences* 100 (14): 8086–8091.
Chapin, F.S., III, M.E. Power, S.T.A. Pickett, A. Freitag, J.A. Reynolds, R.B. Jackson, D.M. Lodge, et al. 2011. Earth stewardship: Science for action to sustain the human-Earth system. *Ecosphere* 2: 89. https://doi.org/10.1891/ESA11-00166.1.
Childers, Daniel L., S.T.A. Pickett, J. Morgan Grove, Laura Ogden, and Alison Whitmer. 2014. Advancing urban sustainability theory and action: Challenges and opportunities. *Landscape and Urban Planning* 125: 320–328. https://doi.org/10.1016/j.landurbplan.2014.01.022.
Childers, D.L., M.L. Cadenasso, J.M. Grove, V. Marshall, B. McGrath, and S.T.A. Pickett. 2015. An ecology for cities: A transformational nexus of design and ecology to advance climate change resilience and urban sustainability. *Sustainability* 7 (4): 3774–3791.
Collins, J.P., A. Kinzig, N.B. Grimm, W.F. Fagan, D. Hope, J.G. Wu, and E.T. Borer. 2000. A new urban ecology. *American Scientist* 88: 416–425.
Collins, S., S.M. Swinton, C.W. Anderson, B.J. Benson, J. Brunt, T.L. Gragson, N. Grimm, J.M. Grove, D. Henshaw, G. Kofinas, J.J. Magnuson, W. McDowell, J. Melack, J.C. Moore, L. Ogden, L. Porter, J. Reichman, G.P. Robertson, M.D. Smith, J. Vande Castle, and A.C. Whitmer. 2007. *Integrated Science for Society and the Environment: A strategic research initiative*. Albuquerque, New Mexico: LTER Network Office.

Collins, S.L., S.R. Carpenter, S.M. Swinton, D.E. Orenstein, D.L. Childers, T.L. Gragson, N.B. Grimm, J.M. Grove, S.L. Harlan, J.P. Kaye, A.K. Knapp, G.P. Kofinas, J.J. Magnuson, W.H. McDowell, J.M. Melack, L.A. Ogden, G.P. Robertson, M.D. Smith, and A.C. Whitmer. 2011. An integrated conceptual framework for social-ecological research. *Frontiers in Ecology and the Environment* 9 (6): 351–357.

Collins, S.L. 2019. Foreword. In *Science for the sustainable city: Empirical insights from the Baltimore School of Urban Ecology*, ed. S.T.A. Pickett, J.M. Grove, E.G. Irwin, E.J. Rosi, and C.M. Swan, xi–xiii. New Haven: Yale University Press.

Cressey, D. 2015. Ecologists embrace their urban side. *Nature* 524: 399–400.

Douglas, Ian, and Philip James. 2015. *Urban ecology: An introduction*. New York: Routledge.

Duckett, D., D. Feliciano, J. Martin-Ortega, and J. Munoz-Rojas. 2016. Tackling wicked environmental problems: The discourse and its influence on praxis in Scotland. *Landscape and Urban Planning* 154: 44–56.

Faulkner, W. 1951. *Requiem for a nun*. New York: Random House.

Fischer, Leonie K., Moritz von der Lippe, and Ingo Kowarik. 2013. Urban land use types contribute to grassland conservation: The example of Berlin. *Urban Forestry & Urban Greening* 12: 263–272. https://doi.org/10.1016/j.ufug.2013.03.009.

Gaston, Kevin, et al. 2010. *Urban ecology*. New York: Cambridge University Press.

Goode, D.A. 1989. Urban nature conservation in Britain. *Journal of Applied Ecology* 26: 859–873. https://doi.org/10.2307/2403697.

Gordon, I.J., K. Bawa, G. Bammer, C. Boone, J. Dunne, D. Hart, J. Hellmann, A. Miller, M. New, J. Ometto, S. Pickett, G. Wendorf, A. Agrawal, P. Bertsch, C.D. Campbell, P. Dodd, A. Janetos, H. Mallee, and K. Taylor. 2019. Forging future organizational leaders for sustainability science. *Nature Sustainability* 2: 647–649.

Gottdiener, Mark, and Ray Hutchison. 2000. *The new urban sociology*. 2nd ed. New York: McGraw Hill.

Gragson, T.L., and J.M. Grove. 2006. Social science in the context of the long term ecological research program. *Society and Natural Resources* 19 (2): 93–100.

Grimm, N.B., J.M. Grove, S.T.A. Pickett, and C.L. Redman. 2000. Integrated approaches to long-term studies of urban ecological systems. *Bioscience* 50 (7): 571–584.

Groffman, Peter M., Daniel J. Bain, Lawrence E. Band, Kenneth T. Belt, Grace S. Brush, J. Morgan Grove, Richard V. Pouyat, Ian C. Yesilonis, and Wayne C. Zipperer. 2003. Down by the riverside: Urban riparian ecology. *Frontiers in Ecology and the Environment* 1: 315–321.

Grove, J.M., and J. Carrera. 2019. Lessons learned from the origin and design of long-term socio-ecological research. In *Science for the sustainable city: Empirical insights from the Baltimore school of urban ecology*, ed. S.T.A. Pickett, J.M. Grove, E.G. Irwin, E.J. Rosi, and C.M. Swan, 289–306. New Haven: Yale University Press.

Grove, J.M., and S.T.A. Pickett. 2019. From transdisciplinary projects to platforms: Expanding capacity and impact of land systems knowledge and decision making. *Current Opinion in Environmental Sustainability* 38: 7–13. https://doi.org/10.1016/j.cosust.2019.04.001.

Grove, J.M., R. Chowdhury, and D. Childers. 2015a. Co-design, co-production, and dissemination of social-ecological knowledge to promote sustainability and resilience: Urban experiences from the U.S. long term ecological research (LTER) network. *Global Land Project News* 11 (April): 6–11.

Grove, J.M., M.L. Cadenasso, S.T. Pickett, G.E. Machlis, and W.R. Burch, eds. 2015b. *The Baltimore school of urban ecology: Space, scale, and time for the study of cities*. New Haven: Yale University Press.

Grove, J.M., D.L. Childers, M. Galvin, S. Hines, T. Muñoz-Erickson, and E.S. Svendsen. 2016. Linking science and decision making to promote an ecology for the city: Practices and opportunities. *Ecosystem Health and Sustainability* 2 (9): e01239.

Hager, G.W., K.T. Belt, W. Stack, K. Burgess, J.M. Grove, B. Caplan, M. Hardcastle, D. Shelley, S.T.A. Pickett, and P.M. Groffman. 2013. Socioecological revitalization of an urban watershed. *Frontiers in Ecology and the Environment* 11: 28–36.

Holzer, J.M., M.C. Adamescu, R. Díaz-Delgado, J. Dick, J.M. Grove, R. Rozzi, and D.E. Orenstein. 2018. Negotiating local vs. global needs in the International Long Term Ecological Research (ILTER) Network's socio-ecological research agenda. *Environmental Research Letters* 13 (10): 105003.

Irwin, Nicholas B., H. Allen Klaiber, and Elena G. Irwin. 2017. Do stormwater basins generate co-benefits? Evidence from Baltimore County, Maryland. *Ecological Economics* 141: 202–212. https://doi.org/10.1016/j.ecolecon.2017.05.030.

Kamarainen, A.M., and T.A. Grotzer. 2019. Constructing causal understanding in complex systems: Epistemic strategies used by ecosystem scientists. *Bioscience* 69: 533–543.

Kingsland, S.E. 2005. *The evolution of American ecology, 1890–2000*. Baltimore: Johns Hopkins University Press.

———. 2019. Urban ecological science in America: The long march to cross-disciplinary research. In *Science for the sustainable city: Empirical insights from the Baltimore school of urban ecology*, ed. S.T.A. Pickett, M.L. Cadenasso, J.M. Grove, E.G. Irwin, E.J. Rosi, and C.M. Swan, 24–44. New Haven: Yale University Press.

Kowarik, Ingo. 2020. Herbert Sukopp – an inspiring pioneer in the field of urban ecology. *Urban Ecosystems* 23 (3): 445–455.

Lachmund, Jens. 2013. *Greening Berlin*. Cambridge MA: MIT Press.

Lehmann, Iris, Juliane Mathey, Stefanie Roessler, Anne Braeuer, and Valeri Goldberg. 2014. Urban vegetation structure types as a methodological approach for identifying ecosystem services - Application to the analysis of micro-climatic effects. *Ecological Indicators* 42: 58–72. https://doi.org/10.1016/j.ecolind.2014.02.036.

Light, J.S. 2009. *The nature of cities: Ecological visions and the American urban professions, 1920–1960*. Baltimore: Johns Hopkins University Press.

Likens, G.E. 1989. *Long-term studies in ecology: Approaches and alternatives*. New York: Springer.

Liu, J., T. Dietz, S.R. Carpenter, M. Alberti, C. Folke, E. Moran, and E. Ostrom. 2007. Complexity of coupled human and natural systems. *Science* 317 (5844): 1513–1516.

Locke, Dexter H., J. Morgan Grove, Michael Galvin, Jarlath P.M. O'Neil-Dunne, and Charles Murphy. 2013. Applications of urban tree canopy assessment and prioritization tools: Supporting collaborative decision making to achieve urban sustainability goals. *Cities and the Environment* 6. http://digitalcommons.lmu.edu/cgi/viewcontent.cgi?article=1132.

Machlis, G.E., and M.K. McNutt. 2010. Scenario-building for the deepwater horizon oil spill. *Science* 329 (5995): 1018–1019.

Machlis, G.E., J.E. Force, and W.R. Burch. 1997. The human ecosystem 1. The human ecosystem as an organizing concept in ecosystem management. *Society & Natural Resources* 10: 347–367. https://doi.org/10.1080/08941929709381034.

McChrystal, G.S., T. Collins, D. Silverman, and C. Fussell. 2015. *Team of teams: New rules of engagement for a complex world*. London: Portfolio.

McDonnell, M.J. 2011. The history of urban ecology: An ecologist's perspective. In *Urban ecology: Patterns, processes, and applications*, ed. J. Niemela, 5–13. New York: Oxford University Press.

McDonnell, M.J., and A.K. Hahs. 2008. The use of gradient analysis studies in advancing our understanding of the ecology of urbanizing landscapes: Current status and future directions. *Landscape Ecology* 23: 1143–1155.

McDonnell, M.J., and S.T.A. Pickett, eds. 1993. *Humans as components of ecosystems*. New York: Springer.

McHale, M.R., S.T.A. Pickett, O. Barbosa, D.N. Bunn, M.L. Cadenasso, D.L. Childers, M. Gartin, G.R. Hess, D.M. Iwaniec, T. McPhearson, M.N. Peterson, A.K. Poole, L. Rivers, S.T. Shutters, and W. Zhou. 2015. The new global urban realm: Complex, connected, diffuse, and diverse social-ecological systems. *Sustainability* 7: 5211–5240.

McPhearson, Timon, Steward T.A. Pickett, Nancy B. Grimm, Jari Niemelä, Marina Alberti, Thomas Elmqvist, Christiane Weber, Dagmar Haase, Jürgen Breuste, and Salman Qureshi.

2016. Advancing urban ecology toward a science of cities. *BioScience* 66: 198–212. https://doi.org/10.1093/biosci/biw002.

Meyfroidt, P., R. Roy Chowdhury, A. de Bremond, E.C. Ellis, K.-H. Erb, T. Filatova, R.D. Garrett, J.M. Grove, A. Heinimann, T. Kuemmerle, C.A. Kull, E.F. Lambin, Y. Landon, Y. le Polain de Waroux, P. Messerli, D. Müller, J.Ø. Nielsen, G.D. Peterson, V. Rodriguez García, M. Schlüter, B.L. Turner, and P.H. Verburg. 2018. Middle-range theories of land system change. *Global Environmental Change* 53: 52–67.

National Research Council. 1999. *Human dimensions of global environmental change: Research pathways for the next decade*. Washington, DC: National Academies Press.

———. 2000. *Grand challenges in environmental sciences*. Washington, DC: National Academies Press.

Numata, M., et al. 1977. *Tokyo project: Interdisciplinary studies of urban ecosystems in the metropolis of Tokyo*. Chiba: Chiba University.

Pearce, B., C. Adler, L. Senn, P. Kruthi, M. Stauffacher, and C. Pohl. 2018. Making the link between transdisciplinary learning and research. In *Transdisciplinary theory, practice and education: The art of collaborative research and collective learning*, ed. D. Fam, L. Neuhauser, and P. Gibbs, 167–183. Cham: Springer.

Pickett, S.T.A. 1999. The culture of synthesis: Habits of mind in novel ecological integration. *Oikos* 87: 479–487.

Pickett, Steward T. A., W. R. Burch Jr., Shawn E. Dalton, and Timothy W. Foresman. 1997. Integrated urban ecosystem research. *Urban Ecosystems* 1: 183-184. https://doi.org/10.1023/A:1018579628818.

Pickett, S.T.A., W.R. Burch, and J.M. Grove. 1999. Interdisciplinary research: Maintaining the constructive impulse in a culture of criticism. *Ecosystems* 2: 302–307.

Pickett, S.T.A., J. Kolasa, and C.G. Jones. 2007. *Ecological understanding: The nature of theory and the theory of nature*. 2nd ed. San Diego: Academic Press.

Pickett, S.T.A., M.L. Cadenasso, D.L. Childers, M.J. McDonnell, and W. Zhou. 2016. Evolution and future of urban ecological science: Ecology in, of, and for the city. *Ecosystem Health and Sustainability* 2 (7): e01229.

Pickett, S.T.A., J.M. Grove, E.G. Irwin, E.J. Rosi, and C.M. Swan, eds. 2019. *Science for the sustainable city: Empirical insights from the Baltimore school of urban ecology*. New Haven: Yale University Press.

Pickett, S.T.A., J.M. Grove, S.L. LaDeau, and E.J. Rosi. 2020. Urban ecology as an integrative science and practice. In *Urban ecology-its nature and challenges*, ed. Pedro Barbosa. Wallingford: CABI Publishing.

Rademacher, A., M.L. Cadenasso, and S.T.A. Pickett. 2019. From feedbacks to coproduction: Toward an integrated conceptual framework for urban ecosystems. *Urban Ecosystem*. https://doi.org/10.1007/s11252-018-0751-0.

Ramaswami, A., L. Bettencourt, A. Clarens, S. Das, G. Fitzgerald, E. Irwin, D. Pataki, S. Pincetl, K. Seto, and P. Waddell. 2018. *Sustainable urban systems: Articulating a long-term convergence research agenda*. Washington, DC: National Science Foundation. https://www.nsf.gov/ere/ereweb/ac-ere/sustainable-urban-systems.pdf. Accessed 18 Mar 2020.

Redman, C.L., J.M. Grove, and L.H. Kuby. 2004. Integrating social science into the long-term ecological research (LTER) network: Social dimensions of ecological change and ecological dimensions of social change. *Ecosystems* 7 (2): 161–171.

Risser, Paul G., Jane Lubchenco, Norman L. Christensen, Philip L. Johnson, Peter J. Dillon, Pamela Matson, Luis Diego Gomez, Nancy A. Moran, Daniel J. Jacob, Thomas Rosswall, and Michael Wright. 1993. Ten-year review of the National Science Foundation's Long-Term Ecological Research program. https://lternet.edu/wp-content/uploads/2010/12/ten-year-review-of-LTER.pdf. Accessed 15 June 2020.

Simberloff, D. 1980. Succession of paradigms in ecology - essentialism to materialism and probabilism. *Synthese* 43: 3–39.

Simon, D., and F. Schiemer. 2015. Crossing boundaries: Complex systems, transdisciplinarity and applied impact agendas. *Current Opinion in Environmental Sustainability* 12: 6–11.
Southgate, E.W.B. 2019. *People and the land through time: Linking ecology and history*. 2nd ed. New Haven: Yale University Press.
Spirn, A.W. 1984. *The granite garden: Urban nature and human design*. New York: Basic Books.
Star, S.L. 1988. The structure of ill-structured solutions: Boundary objects and heterogeneous distributed problem solving. In *Distributed artificial intelligence*, ed. M. Huhns, vol. 2, 37–54. San Francisco: Morgan Kaufmann.
Star, S.L. 1989. *Regions of the mind: Brain research and the quest for scientific certainty*. Stanford: Stanford University Press.
———. 2010. This is not a boundary object: Reflections on the origin of a concept. *Science, Technology and Human Values* 35 (5): 601–617.
Star, S.L., and J.R. Griesemer. 1989. Institutional ecology, translations and boundary objects: Amateurs and professionals in Berkeley's Museum of Vertebrate Zoology, 1907–39. *Social Studies of Science* 19 (3): 387–420.
Sternlieb, F., R.P. Bixler, and H.H. Huber-Stearns. 2013. A question of fit: Reflections on boundaries, organizations and social–ecological systems. *Journal of Environmental Management* 130: 117–125.
Sukopp, H. 2008. On the early history of urban ecology in Europe. In *Urban ecology: An international perspective on the interaction between humans and nature*, ed. E. John Marzluff, W. Shulenberger, M. Endlicher, G. Alberti, C. ZumBrunne Bradley, and U. Simon, 79–97. New York: Springer.
Vajjhala, S., A. Krupnick, E. McCormick, M. Grove, P. McDowell, C. Redman, L. Shabman, and M. Small. 2007. Rising to the challenge: Integrating social science into NSF environmental observatories. *Resources for the Future Report*. https://media.rff.org/documents/NSFFinalReport.pdf. Accessed 7 May 2020.
Vitousek, P.M., H.A. Mooney, J. Lubchenco, and J.M. Melillo. 1997. Human domination of earth's ecosystems. *Science* 277: 494–499.
Wang, Rusong, and Lu. YongLong. 1994. *Urban ecological development: Research and application*. Beijing: China Environmental Science Press.
Wu, Jianguo, Wei-Ning Xiang, and Jingzhu Zhao. 2014. Urban ecology in China: Historical developments and future directions. *Landscape and Urban Planning* 125: 222–233. https://doi.org/10.1016/j.landurbplan.2014.02.010.
Zhou, W., S.T.A. Pickett, and M.L. Cadenasso. 2017. Shifting concepts of urban spatial heterogeneity and their implications for sustainability. *Landscape Ecology* 32: 15–30.
Zimmerman, J.K., F.N. Scatena, L.C. Schneider, T. Gragson, C. Boone, and J.M. Grove. 2009. Challenges for the implementation of the decadal plan for long-term ecological research: Land and water use change. Available at LTER Network Documents Archive, 2009 Working Groups: http://lternet.edu/wp-content/uploads/2011/12/Workshop Report Zimmerman_Land and Water Use Change_ implementation of decadel plan (ISSE).pdf.
Zscheischler, J., and S. Rogga. 2015. Transdisciplinarity in land use science – a review of concepts, empirical findings and current practices. *Futures* 65: 28–44. https://doi.org/10.1016/j.futures.2014.11.005.

Chapter 11
Integration of the Arts and Humanities with Environmental Science in the LTER Network

Mary Beth Leigh, Michael Paul Nelson, Lissy Goralnik, and Frederick J. Swanson

Abstract A broad spectrum of arts and humanities activities has emerged organically within the Long Term Ecological Research (LTER) program including disciplines such as philosophy and ethics, creative writing, and the visual, multimedia, musical, and performing arts. The majority of LTER sites now hosts activities that integrate the environmental sciences with the arts and humanities (eSAH). These programs serve important functions central to the LTER mission, including, but not limited to, public engagement, outreach, and education. Some LTER eSAH programs additionally consider these activities as steps toward the aspirational goal of helping society address grand social-ecological challenges of the twenty-first century, challenges that science alone cannot overcome. The arts and humanities can offer critical dimensions to this mission and to outreach, education, and general edification, such as awakening and engaging ethics, values, empathy, and wonder in individuals and societies. In this chapter, we reflect upon eSAH efforts across the LTER network, including their history, value to LTER's mission, challenges, and aspirations and share case studies of eSAH activities from several LTER programs, including their objectives, organizational models, audiences, and outcomes.

M. B. Leigh (✉)
Institute of Arctic Biology, University of Alaska Fairbanks, Fairbanks, AK, USA
e-mail: mbleigh@alaska.edu

M. P. Nelson
Department of Forest Ecosystems and Society, Oregon State University, Corvallis, OR, USA
e-mail: mpnelson@oregonstate.edu

L. Goralnik
Department of Community Sustainability, Michigan State University, East Lansing, MI, USA
e-mail: goralnik@msu.edu

F. J. Swanson
Pacific Northwest Research Station, US Forest Service, Corvallis, OR, USA
e-mail: fred.swanson@oregonstate.edu

R. B. Waide, S. E. Kingsland (eds.), *The Challenges of Long Term Ecological Research: A Historical Analysis*, Archimedes 59,
https://doi.org/10.1007/978-3-030-66933-1_11

Keywords LTER program · Long-term ecological research · Science-arts collaborations · Science-humanities collaborations · Environmental ethics · Ecological outreach · Ecological education

11.1 Introduction

Arts and humanities activities are components of the majority of Long-Term Ecological Research (LTER) programs. These activities cut across all the relevant areas of the humanities and creative arts, and include consideration of philosophical and ethical themes, creative writing, visual arts and film-making, multimedia exhibits, and musical and performing arts. They have arisen organically from within LTER programs in the service of a variety of missions, including public engagement, outreach, and education. These arts and humanities efforts, together with the environmental science activities within LTER programs, complement each other as methods for understanding the world. Like the ecosystems in which they evolved and the scientific emphases of each site's community of researchers, the LTER network's multiple arts and humanities programs vary in terms of the disciplines involved, the organizational models employed, the goals of their integrative efforts, and the audiences engaged. While the majority of programs focus on integrating the environmental sciences, arts, and humanities (eSAH) for the purposes of engagement, outreach, and/or education, some programs now share a common aspirational goal of pursuing interdisciplinary integration for the benefit of society, whether for enrichment, understanding, community building, and/or developing problem-solving capacity (Goralnik et al. 2017).

In this chapter, we reflect upon eSAH efforts across the LTER network, including their history, value to LTER's mission, challenges, and aspirations. To provide context for this work and highlight some of the contributions, we also share case studies of eSAH activities from several LTER programs, including their objectives, organizational models, and audiences. These programs have become interwoven with LTER's mission and goals, especially as related to outreach, education, and public understanding of environmental problems and ecological science. We argue that this growing network of eSAH programs represents important linkages between science and society that contribute to the aspirational goal of addressing major social-ecological challenges.

11.2 LTER eSAH Activities: Origins

Although the arts and humanities have since emerged within LTER, they were not specifically mentioned in the National Science Foundation's (NSF) establishment of the LTER network in 1980 (Swanson 2015). But, following the establishment of LTER, NSF Program Officer James T. Callahan (1984) emphasized the importance of long-term research not only for providing continuity in research, but for using this research to understand and address anthropogenic disturbance, thereby situating problem-solving and outreach as central to LTER objectives. Callahan credits the creation of the National Environmental Policy Act of 1969 (NEPA), as well as other federal agency initiatives, for laying the foundation for LTER's mission, by mandating that "it is the continuing policy of the Federal Government...to use all practicable means and measures...to create and maintain conditions under which man and nature can exist in productive harmony."

Since prescriptive conservation is inherently a fusion of understanding how the world works (science) and why the world is valuable (humanities/ethics), LTER's articulated origin as one focused on environmental problem solving opens the door to eSAH efforts. An additional factor contributing to the presence of eSAH activities within LTER is the fact that such activities were already underway within research sites that existed prior to the establishment of LTER (in 1980) that were later incorporated into the LTER (Swanson 2015). Starting in the 1950s, NSF sponsored an Antarctic artist and writer residency program that continues to be active today, including in association with McMurdo Dry Valleys LTER. In other cases, arts and humanities activities were initiated by the sites themselves. At Harvard Forest (HFR) in the 1930s, large and detailed dioramas were created to depict landscape change in Central New England from 1700 to 1930. These dioramas still serve as a central feature of Harvard Forest's visitor center, where the arts continue to help communicate ecosystem science to the public. Harvard Forest also hosts a long-running artist-in-residence program that continues to the present day.

The interest in eSAH in the LTER network is situated in a much larger national and international trend of interactions between the environmental sciences, arts, and humanities. In the twentieth–twenty-first centuries, environmental science, arts, and humanities interactions and collaborations have emerged in academic, nonprofit, and federal agency contexts (for current examples, see: artists-in-labs[1]; Brown University[2]; Cape Farewell[3]; Climarte[4]; IHOPE[5]; SciArt Center[6]; SymbioticA[7]; The Institute for Figuring[8]; Dixon et al. 2011b; Ingram 2011; Jacobson et al. 2007; Muchnic 2013). Other formalized networks of ecologically oriented research sites (e.g. biological field stations and marine laboratories) have hosted arts and humanities activities, while US National Park properties have also established formalized artist residency programs focusing on eSAH engagement with place. Collectively, the presence of these many programs has contributed to the groundswell of eSAH

[1] https://www.transartists.org/air/artists_in_labs_projects.7686.html

[2] https://arts.brown.edu/theme/arts-environment-2017-2020

[3] https://capefarewell.com/

[4] https://climarte.org

[5] http://ihopenet.org/

[6] https://www.sciartinitiative.org/

[7] http://www.symbiotica.uwa.edu.au/

[8] http://www.theiff.org/

activities of which the LTER eSAH activities are a part. While the applications and objectives differ across programs, scholars argue that bridging disciplines in this way facilitates emotional engagement or care for the natural world in ways that can impact appreciation, respect, and inspiration (Demaray 2014; Dixon et al. 2011a, b; Houtman 2012; Jacobson et al. 2007; Kimmerer 2016; Muchnic 2013; Patterson 2015; Root-Bernstein 2003; Swanson et al. 2008).

11.3 LTER eSAH Activities: Growth

Since the establishment of the LTER network, eSAH activities have emerged and gained traction widely across the network. Our 2013 survey of LTER Principal Investigators (PIs) found that 21 out of the total of 24 LTER sites hosted some kind of arts and/or humanities activities at that time (Goralnik et al. 2015). This network-wide activity reflects a wider trend nationally and internationally in engaging arts and humanities alongside scientific endeavors.

Modern scientific enterprises are focused narrowly on their respective disciplinary methods and practices. Engaging with other disciplines and integrating scientific work with other disciplines, especially disciplines different from the sciences, such as the arts and humanities, both benefit from and require active cultivation. A common feature across many programs is the presence of one or more multidisciplinary LTER researcher(s) already working in the sites, who might identify as or already collaborate with musicians, artists, writers, or dancers. These contributors find value in the work beyond their disciplines and have a passion for advancing eSAH work (Swanson 2015). Often these individuals create and build eSAH programs, serving as bridges or translators of this work for those working in disparate disciplines, building multidisciplinary communities and engaging broader audiences in the local/regional area. Successful programs are often characterized by alliances and partnerships with other institutions with strong existing memberships and outreach capabilities, such as museums, journals, and arts associations, which may enhance sustainability and extend networks.

Connections among active LTER-based eSAH programs have served to facilitate the emergence and growth of new programs across the network. In some cases, LTER-based eSAH programs were developed under the mentorship of more established LTER eSAH programs. For example, H.J. Andrews Experimental Forest LTER (AND) created a writer's residency in 2004 through collaboration with the Spring Creek Project for Ideas, Nature, and the Written Word at Oregon State University. As one example of growth facilitation, eSAH leaders from the existing writer's program at AND LTER visited Bonanza Creek LTER (BNZ) to co-facilitate that site's first field excursion for artists and writers in 2007, which then grew into BNZ's integrative *In a Time of Change* program.

To introduce a wider public to LTER eSAH activities, as well as provide opportunities for LTER programs to learn from each other, representatives from active programs have hosted workshops and panels at LTER All Scientists Meetings since 2006, and at several Ecological Society of America (ESA) and the American

Geophysical Union (AGU) annual meetings. LTER eSAH organizers also presented exhibits of cross-network eSAH work in the National Science Foundation's headquarters in Arlington, Virginia, in 2012 and 2013. The 2012 exhibit *Long Term Ecological Reflections: Bridging Science, the Arts, and Society* was also displayed at the ESA Meeting in Portland, Oregon, and at Oregon State University (in 2012). Additionally, organizers regularly share digital exhibits at the LTER All Scientists Meetings prior to plenary lectures and/or in poster halls.

Many eSAH activities associated with LTER sites are captured under the umbrella title *Ecological Reflections*, a heading that also includes non-LTER-based programs (e.g., biological field stations) with related eSAH structures and goals, and provides an additional resource for more broadly exploring the diversity of activities that have emerged. The Ecological Reflections website (www.ecological-reflections.com) communicates LTER eSAH work with a wider public and facilitates community-building within active programs, including beyond-LTER sites that share similar interests. The website includes information about eSAH programs and goals, as well as links to websites of participating projects.

Finally, LTER eSAH organizers have advanced scholarship on eSAH more broadly. In 2015, LTER-affiliated eSAH organizers convened an NSF-funded workshop in collaboration with organizers within the Organization of Biological Field Stations (OBFS) and the National Association of Marine Laboratories (FSMLs), Sagehen Creek Research Station, and the Nevada Museum of Art in Reno, Nevada. The 2.5-day workshop hosted 25 invited participants, including eSAH organizers, artists, scientists, humanists, and funding agency program officers, for a series of presentations, discussions, and a field trip to Sagehen Creek Research Station. The meeting was documented with a blog post (http://artsciconverge.blogspot.com/2015/07/artsciconverge-nsf-workshop-in-reno-nv.html). The workshop led to the development of the ArtSciConverge network, which describes the collaboration of LTER Ecological Reflections, OBFS, and other programs that wish to promote eSAH activities and scholarship.[9]

11.4 Value of eSAH to LTER Programs, Ecological Science, and Society

11.4.1 Perceived Value of eSAH in the LTER Network: Empathy and Observation

In summer 2013, we surveyed LTER Principal Investigators (PIs) to explore the nature and extent of arts and humanities engagement that their sites had hosted, and to understand the perceived benefits and challenges of this work (Goralnik et al. 2015). Of the 21 sites that had hosted some kind of eSAH inquiry (out of the 24 total

[9] Some of these activities can be viewed through the associated social media hashtag, #ArtSciConverge

LTER sites at that time), 19 agreed or strongly agreed that this inquiry was both relevant for and important to LTER sites. Respondents were offered 13 potential values[10] that arts and humanities might contribute to LTER sites, derived from the literature (see Demaray 2014; Dieleman 2008; Dixon et al. 2011b; Houtman 2012; Jacobson et al. 2007; Kimmerer 2016) and interpreted through the lens of general academic goals (e.g., plays a role on grants). Five values consistently ranked the highest: (1) *Fosters outreach and public engagement*, (2) *Is good in and of itself*, (3) *Inspires creative thinking*, (4) *Provides opportunities for education*, and (5) *Broadens our understanding of the natural world.*

While several of these outcomes are instrumental to the LTER mission and goals, (e.g., outreach and education), others hint at an emergent, less tangible contribution of this work, including its capacity to inspire creative thinking and broaden our understanding of the natural world, as well as the intrinsic quality of being good in and of itself. These suggested values allow for a less scripted, more open role for creative inquiry in the LTER Network, whereby arts and humanities might contribute something new and worthwhile that has its own shape and purpose, in addition to or rather than just supporting the science in a service capacity. This perspective is perhaps akin to science driven by discovery and wonder, whereby we create space for unexpected outcomes rather than aim for particular objectives. That the survey participants, most of whom were scientists and LTER PIs, were open to values like this was somewhat surprising and encouraging, given the trend within art-science interactions to employ the arts merely as a means to scientific ends (Jeffreys 2018; Stevens and O'Connor 2017), e.g., as an instrument for the purposes of illustration, outreach, or education.

Respondents were also provided with a list of 12 statements[11] drawn from explicit mission statements and goals that were stated on the LTER website, and from implicit objectives of the LTER network drawn from commitments to long-term, place-based inquiry. Participants were asked to rank the relative contribution of arts and humanities inquiry to these objectives of the LTER Network. In line with the earlier question about perceived value, respondents associated arts and humanities inquiry most closely with Network goals related to (1) *Outreach* and (2) *Communication*. But what was surprising was the relatively high ranking of (3) *Relationship Building – To develop empathetic relationships with the natural world and stimulate inspiration, awe, and wonder*, and (4) *Human Dimensions – To understand human drivers on natural systems, investigate the impacts of ecosystems on humans, and explore human perceptions of and attitudes about the natural world.*

---

[10] (1) Markets the science, (2) Stimulates collaboration, (3) Develops observational skills, (4) Contributes to environmental problem-solving, (5) Fosters outreach, (6) Provides opportunities for education, (7) Plays a role on grants, (8) Is good in and of itself, (9) Enables interdisciplinary scholarship, (10) Broadens our understanding of the natural world, (11) Stimulates empathy, (12) Enhances the science, (13) Inspires creative thinking.

[11] (1) Understanding, (2) Synthesis, (3) Information, (4) Legacies, (5) Education, (6) Outreach, (7) Conservation, (8) Communication, (9) Environmental Impact, (10) Relationship building, (11) Long-term Ecological Research, (12) Human dimensions.

Eleven sites (of 24), nearly half, ranked these responses >80%, above the fifth-ranking contribution, *Education* (Goralnik et al. 2017).

In 2015, we followed up the original survey with phone interviews with 14 LTER PIs and two outreach and education coordinators (see Goralnik et al. 2017). These interviews aimed to investigate the suggestion that LTER PIs recognized non-utilitarian value in the contributions of arts and humanities inquiry across the network or at their own sites, specifically related to empathy. Empathetic relationships with the natural world imply responsibility, a kind of reciprocity embedded in deep respect; they also suggest imagination, wonder, and awe, shown in the literature to provide entrance to empathy (Lorkowski and Kreinovich 2015; Piff et al. 2015), which are certainly qualities that can be associated with ecological research, but are not often the kinds of outcomes ecologists promote about their research. The interviews probed how PIs thought (a) empathy was relevant for Network activities, and (b) how arts and humanities might contribute this kind of value.

Results showed that awe and wonder are powerful drivers of many of the scientists' own careers. One explained:

> I think that a lot of environmental scientists get into what they do because of a sense of wonder, and we, as educators, can do well to think more deeply about that when offering [educational] opportunities ... because we tend to forget that the root of what we are doing is deeply rooted [in a] sense of curiosity and appreciation for the natural world that I know for myself was really the reason why I chose to do what I do.

Another summed up what he understands is a central motivation for many scientists:

> [P]eople ... want to understand how this unbelievably terrific system works, and what it does and what can cause it to unravel and what are the responses to disturbance and resiliency .... I think that's very important. To me it is fundamentally an emotional thing.

Several interviewees were uncomfortable with the language of empathy related to their own work, primarily because they worried about the appearance of advocacy interfering with the objectivity of their work. But all participants told stories about the beautiful, curious, amazing things they studied. These stories all imply a kind of awe, wonder, or empathy for the natural world that guides and nurtures their research.

Nearly all respondents also explained that inspiration, awe, and empathy are outcomes their LTER sites were either already accomplishing or should be doing. In some cases, respondents went as far as to say that these outcomes are a hidden LTER objective, a duty to the public, and an important part of education. Most agreed that arts and humanities can facilitate inspiration, awe, and wonder, especially by sharing LTER science with people beyond the site. One PI explained that "I see the value of the humanities [as] inspiring people that maybe don't have that natural curiosity or natural wonder [...who might be] more likely to get inspired by a beautiful sculpture or painting than by a Ph.D. scientist espousing the virtue of ecological theory." Another explained: "[Y]ou can throw a bunch of data at people all you want, but they're not going to change their minds just because they have the information, right? They need to be touched in some way.... Sometimes art does that."

There was also a sense in the interviews that something more impactful, or perhaps less outward facing, than outreach was happening when scientists engage with artists and humanists in LTER sites. One PI explained:

> With scientists, we're looking at the nuts and bolts of what's going on in the environment. Most scientists still conceive of the environment as a machine, and it's hard to talk about the environment as a holistic entity, which is the kind of thing you gain empathy with. That's, I suppose, a role for the arts and humanities, to educate scientists about the environment.

This comment is striking because often in these kinds of eSAH collaborations we find artists benefitting from the interaction with the scientists, and then using their medium to share science stories with other audiences. There are fewer examples of the ways the arts and humanities are nurturing the science or scientists directly. Providing new avenues of observation and attentiveness is one way arts and humanities interactions can benefit scientists and their practice. Another PI echoes this suggestion:

> One of the things I see in linking arts and humanities is it ... enhances and sharpens the power of observation of scientists, for example, that it gets us out of our box and our computers filled with data and it really gets us out there and appreciate the beauty of nature and the perception of the natural surroundings.

These PIs echo the wider conversation about arts, humanities, and science collaborations related to emotion and observational capacity (see Burns 2015; Goralnik and Nelson 2015; Goralnik and Nelson 2017; Mann 2017; Reilly et al. 2005). Beyond being a direct benefit to scientific practice and public engagement, there is also the chance that when we see with new eyes, we potentially acquire new problem-solving capabilities, as well.

11.4.2 Impacts of LTER eSAH Work in the Public Sphere

Data from participant surveys collected at the BNZ LTER site's *In a Time of Change* eSAH gallery exhibits demonstrate that eSAH activities can achieve the kinds of public engagement outcomes PIs identified as worthwhile contributions of eSAH activities. In August 2013, we surveyed audiences at *In a Time of Change: Trophic Cascades*, an eSAH exhibit documenting the yearlong collaboration of artists, storytellers, and writers with scientists in Denali National Park and BNZ LTER in Fairbanks, Alaska. Opening night attendance at this show was 280 visitors, and total attendance during the 3-week run was 1820. We collected 94 surveys, most on opening night. In addition to questions about the exhibit's impact on participant knowledge and attitudes related to predators and ecosystem health, participant motivations, and overall participant reactions to the show, we asked two questions about the role of art in building awareness about environmental issues.

Of the 91 people who responded to the survey, more than 97% agreed or strongly agreed that art can be an effective way to understand ecosystems and the role of humans in the natural world. They also agreed that art can be an effective mechanism for building public awareness and understanding of important issues. While this does not yet describe the problem-solving capability of these kinds of collaborations, the audience's reactions to this exhibit reflect the potential of this work to engage a broad public in social-ecological issues. Additionally, open-ended responses about the most thought-provoking elements of the show reflect PI interview responses about empathy and awe, as well as the artist capacity to capture a holistic perspective that can be inspiring or eye opening for viewers.

One participant explained that the show "helped me...empathize with the hunted," while another shared "the artist perspective brings a wholeness to the world." A third explained that she, "stood in awe, surrounded by the power of words and images to evoke deep feelings of amazement at the precarious balance of the natural world." Participants regularly invoked words like awe, amazement, unexpected, and fascinated. All of these qualities open doors to empathetic awareness.

We also collected audience surveys at a second *In a Time of Change* eSAH multimedia exhibit in 2017 in Fairbanks, Alaska, "Microbial Worlds." There were over 900 visitors at this opening, a notably well-attended Fairbanks art event, and we received 109 survey responses. Again, we asked about knowledge, attitudes, and motivations; we were also curious about the impact of these eSAH collaborations on problem-solving capacity. In an open-ended question, analyzed with an emergent coding protocol (Vaismoradi et al. 2013), we asked participants: *What is the role of arts (if any) in capturing your attention about important issues?* Respondents identified arts and humanities as an effective tool for capturing attention about issues because of its immediate impact, ability to make the familiar unfamiliar, emotional resonance, and approachability. They described these impacts as an opportunity to invite broader audiences to the table, bring people together, educate, and create awareness. One participant shared, "[eSAH collaborations let] the idea behind the art sink in more because you feel emotion when viewing art, and thus you feel emotion toward the issue being addressed."

We also asked participants: *What role (if any) do arts-humanities-science collaborations play in environmental problem-solving?* Respondents explained that eSAH collaborations demand engagement, exercise mental muscles, and provoke innovation. These collaborations can impact creativity and connection, inspire wonder, make issues personally felt, broaden the audience for the issues, and change viewers' perceptions. "They bring environmental problems into the home to make an abstraction out of a concrete problem and make a concrete object out of an abstract concept," shared one participant.

In general, these collaborative exhibits demonstrate that eSAH work gets people in the room to engage with LTER science and environmental issues. There are some self-identified learning gains, and a good deal of emotional engagement with the pieces and the ideas. There is wonderful conversation and relationship building, the kinds of social interactions that are necessary to engage in broad scale problem solving. And there is potential for awareness, attitude shifts, and even empathy. More research is necessary to assess these outcomes directly.

11.5 Challenges and Future Directions

While arts and humanities in the LTER network is largely perceived to be a value-adding contribution to site activities, supporting eSAH programs and collaborations is not without challenge. The 2013 survey asked participants to rank 11 provided potential challenges[12] of hosting arts and humanities inquiry in LTER sites. Participants consistently ranked three challenges at the top: (1) funding, (2) time or available labor, and (3) available expertise. These were anticipated responses, because LTER sites are grant-funded and most research, regardless of discipline, requires similar resources, including funding, labor, and time. What was encouraging about these responses is that they are, in many ways, surmountable hurdles, not specific to arts and humanities inquiry, therefore requiring skills most researchers already possess. Grants exist, experts are available, schedules and appointments can be restructured.

These kinds of challenges, though, also open the door to potentially deeper challenges. For example, the transformative capacity of these collaborations can potentially be hindered by forcing work into certain boxes in response to financial pressures or grant structures (Bieler 2014), or by requiring deep interdisciplinary work to function within traditional disciplinary research outcomes and institutional expectations (Holm et al. 2013). Prescribing the shape, scale, and outcomes of these collaborations with overly structured objectives likely limits their potential. The greater goal of these kinds of deep interdisciplinary collaborations, scholars contend, is to transcend horizontal knowledge sharing between the disciplines to allow for something new and emergent – something potentially transformational – to arise (Brown et al. 2017; Gabrys and Yusoff 2012; Jones et al. 2010).

11.5.1 Potential Value of eSAH: Social-Ecological Problem Solving

We suggest that many environmental challenges are the product of a relationship between humans and the natural world characterized by declines not so much in knowledge about nature, but rather, empathy with, humility in the face of, and sense of wonder about our world. We propose that an emergent property can result from

[12] (1) Funding, (2) Time, available labor, (3) Available expertise, (4) Clear vision or goals, (5) Lack of alignment with research, (6) Scheduling, (7) Limited relationships, (8) Limited space, (9) Challenging to find collaborators, (10) Little on-site interest, (11) Not in LTER purview.

the intersection of the environmental sciences, arts, and humanities: a holistic understanding of nature and environmental change that incorporates scientific knowledge as well as values, emotions, and aesthetics. It is through such shared understanding that societies can begin to address environmental challenges that lie ahead. The current paradigm for addressing grand environmental challenges recognizes the need to bridge the biological, physical, social, and political sciences to tackle these complex problems (Berkes et al. 2002; Ostrom 2009), but commonly overlooks the role of the arts (writ large to include the visual and performing arts, and other media) and humanities (writ large to include creative writing, philosophy, ethics, etc.). The arts and humanities, however, offer critical dimensions such as ethics, values, empathy, and wonder.

If we desire fully formed positions on how we ought to respond to grand social-ecological challenges, it is only from the combination of facts (science) and values (arts and humanities) that society can reach conclusions about what should be done to address those challenges. There is a growing recognition that a new paradigm is needed; one in which authentic relationships among the environmental sciences, arts, and humanities are revitalized to bring together the combined expertise and ways of knowing of these diverse disciplines to connect humans more deeply with their environment (Brown et al. 2017). This reintegration has the potential to generate a more unified approach to solving the ecological and social crises of the twenty-first century (Snow 1959; Frodeman 2010; Nisbet et al. 2010). However, methods for achieving interdisciplinarity across such large epistemological divides have yet to be well developed (Wilson 1999). Given the existence of a well-established scientific community, and a burgeoning community of artists and humanities scholars in their midst, the LTER program is poised to lead this critical eSAH effort.

Swanson et al. (Chap. 8, this volume) note that, although arts and humanities programs were not embedded in the programs with the intent of affecting policy outcomes, public perception may have been shaped in part by eSAH work in creative writing and the visual arts. Since early in the twentieth century, for example, Harvard Forest has offered public expression of change and legacies of land use in the New England landscape through dioramas and text. This appreciation of landscape forms a basis for public support of the Wildlands and Woodlands regional forest conservation strategy. The acid rain/air pollution issue deeply informed by work at Hubbard Brook came to the public eye in part through photographic work in dead and dying forests in New England. The awe-inspiring beauty and complexity of old-growth forests in the Pacific Northwest captured the public's imagination during the ca. 1990s "forest war" over the future of federal forestry through photography and compelling writings, some of which were set in the Andrews Experimental Forest where the science basis for understanding that ecosystem was based. Going forward, it will be interesting to see how arts and humanities embedded in LTER programs, coupled with the science, will be posed to contribute to society's ability to cope with looming environmental challenges.

11.6 Highlights of LTER Site-Based Programs and Projects

H.J. Andrews Experimental Forest and LTER The Andrews Forest LTER is located in Oregon and involves both Oregon State University and the U. S. Forest Service. Its arts and humanities program commenced in 2002 through collaboration with the privately-endowed Spring Creek Project for Ideas, Nature, and the Written Word, and with support from the U. S. Forest Service and private donors. Echoing the LTER science model, the Long-Term Ecological Reflections program brings creative writers, visual artists, composers, and humanities scholars into the forest with the objectives of conducting place-based arts/humanities inquiry and outreach to the public. The residents are encouraged to engage with the place, reflect on ecological and human change over a 200-year period of time, and archive and share their works through *The Forest Log* on the Spring Creek website.[13] Over the course of 100 residencies, works have appeared in *Orion* and *The Atlantic*, as a chapter in a finalist for the 2019 Pulitzer nonfiction book prize, in an image-essay in the online journal *terrain.org*, and through art exhibits and performances in eight Oregon cities, among other outlets.

A compilation of works by 35 writers and scientists from the Andrews Forest Reflections program is represented in *Forest Under Story* (Brodie et al. 2016). Charles Goodrich introduces the book in his essay "Entries into the Forest":

> Trying to comprehend what these forests *mean* can be bewildering. We may intuit and celebrate the wholeness of the forest, but we know it in pieces and threads, by its species and cycles, its products and processes. We come to know the forest via the paths laid down in stories, stories told in anecdotes, photographs, essays, and poems, or in hypotheses, data, and graphs. All these stories are entries into the forest, paths that others have made and which we may follow, perhaps to discover new insights and entice others to enter too.

The essays and poems that follow, including Robin Wall Kimmerer's "Interview with a Watershed" (written in 2004, see excerpt below), explore mysteries and surprises in the forest, our kinship with other beings, the soul of the science enterprise. Writers give voice to feelings scientists do not articulate – empathy, hope, love. In a "gratitude duet" the poet thanks the scientist for persistent, deep inquiry into how one particular facet of the world works; the scientist thanks the poet for approaching the world with such an open mind. This resonance is just beginning to unfold.

Excerpt from "Interview with a Watershed" by Robin Wall Kimmerer, in *Forest Under Story* (Brodie et al. 2016):

> It's a hopeful thing when scientists look to the land for knowledge, when they try to translate into mathematics the stories that water can tell. But it is not only science that we need if we are to understand. Lewis Thomas identified a fourth and highest form of language. That language is poetry. The data may change our minds, but we need poetry to change our hearts.

[13] https://liberalarts.oregonstate.edu/centers-and-initiatives/spring-creek-project/programs-and-residencies/long-term-ecological-reflections/forest-log

> Rich though they are, conversation, mathematics, and poetry are but human languages. And I think there is another language, the forgotten language of the land. Its alphabet is the elements themselves, carbon, hydrogen, oxygen, nitrogen. The words of this language are living beings, and its syntax is connection. There is a flow of information, a network of relationship conveyed in the rising sap of cedars, in tree roots grafted to fungi, and fungi to orchids, orchids to bees, bees to bats, bats to owls, owls to bones, and bones to the soil of cedars. This is the language we have yet to learn and the stories we must hear, stories that are simultaneously material and spiritual. The archive of this language, the sacred text, is the land itself. In the woods, there is a constant stream of data, lessons on how we might live, stories of reciprocity, stories of connection. Species far older than our own show us daily how to live. We need to listen to the land, not merely for data, but for wisdom.

Bonanza Creek Experimental Forest and LTER Bonanza Creek Experimental Forest and LTER involves collaboration between the University of Alaska and the U. S. Forest Service, and is located in interior Alaska. Established in 2007, the *In a Time of Change* program provides opportunities for competitively selected artists to engage with scientists on programmatic themes (e.g., climate change, wildfire, predator control, microorganisms), with each one-to-two-year project culminating in the production of original art exhibits, performances, and/or literary readings for the public. Through a series of group field trips, lectures, and hands-on activities, the participating artists and scientists exchange knowledge and perspectives and build cross-disciplinary relationships. Analyses of audience survey data revealed that exhibit attendance led to increased knowledge about the ecosystem, changed attitudes, and the cultivation of empathy for the natural world that can enable pro-environmental behavior (Goralnik et al. 2017). Participating artists and writers have credited their participation in *In a Time of Change* programs with long-term impacts on their work and with the formation of lasting collaborative relationships with scientists and other artists.

In the collaborative arts-humanities-science exhibit *Microbial Worlds* (itoc.alaska.edu), fourteen artists and writers magnified the physical beauty of microbes and illuminated the many roles they play in human and environmental health. The artistic media represented includes collage (Fig. 11.1), painting, sculpture, tile, printmaking, textile art, artist books, writing, and multimedia works. The exhibit was the culmination of over 1 year of monthly interactions between artists and a total of over 30 scientists through lectures, hands-on lab and field activities, as well as independent research and one-on-one collaborations. Since premiering in February, 2017, in Fairbanks, Alaska, the exhibit has actively toured within and beyond the state of Alaska.

Harvard Forest LTER Hosted by Harvard University and located in Petersham, Massachusetts, the Harvard Forest has a deep history of incorporating environmental literature, history, photography, and fine art with scientific research to characterize past landscapes, depict future scenarios, communicate complex scientific concepts, and educate a range of audiences. The detailed, 1930s-era dioramas of the Fisher Museum transcend typical boundaries among art, history, and science and effectively convey the conceptual foundation for much of the Forest's work. They

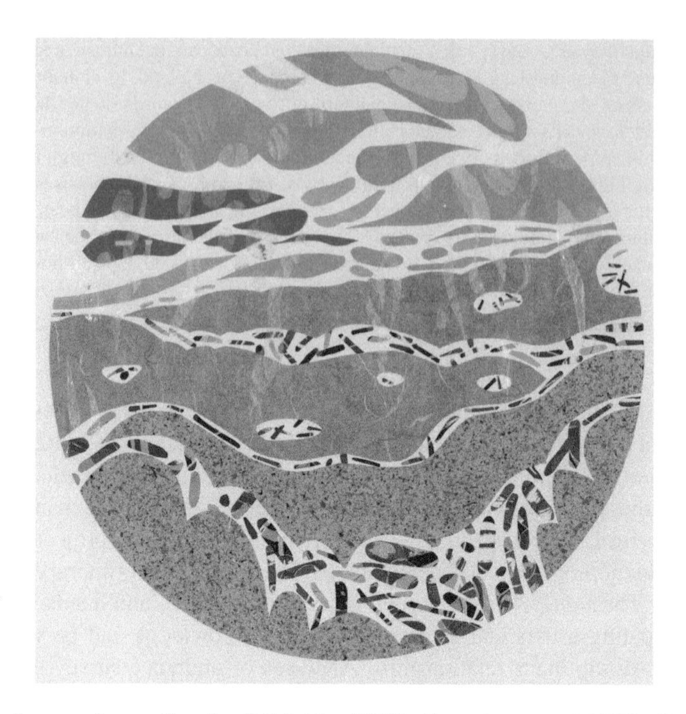

Fig. 11.1 *Impermafrost,* collage by Gail Priday (2017). Created as part of BNZ's "*In a Time of Change: Microbial Worlds*" project. Additional information and works by the fourteen participating artists can be viewed at itoc.alaska.edu. (Photograph by Todd Paris)

are visited by thousands of students, scientists, decisionmakers, land managers, and members of the public each year – and seamlessly provide rich footing in landscape-based inquiry for scientists and non-scientists alike. In the modern era, the Forest community incorporates artists, writers, and humanities scholars in three main ways: (1) interdisciplinary workshops and exhibits for students and the public, which provide creative entry points into complex global change topics; (2) the highly competitive Bullard Fellowship Program, endowed to support advanced work by any practitioner (including artists, writers, and historians) who shows promise to make an important contribution to forest-related subjects; and (3) an emerging effort to archive and analyze site-based artistic products as data.

Hemlock Hospice (Fig. 11.2) was a 2017–2018 art-science collaboration between Bullard Fellow David Buckley Borden and Harvard Forest Senior Ecologist Aaron Ellison. An 18-sculpture public art series embedded for a year into one of the Forest's main research tracts, it focused viewers on the death of eastern hemlock, a foundation tree in eastern forests now slowly vanishing from North America as it is weakened and killed by an invasive insect. In creating the sculptures (made primarily from discarded material from past ecological experiments) and in collaborating with filmmakers on a companion documentary, Borden and Ellison sought to encourage empathetic conversations among all caregivers for our forests—ecologists and artists, foresters and journalists, students, naturalists and citizens—while

Fig. 11.2 *Exchange Tree*, one of eighteen Hemlock Hospice field installations at Harvard Forest that highlights the death and disappearance of eastern hemlock trees from North America due to invasive insects. 8×10×12.5 feet. Wood and acrylic paint. 2017. Collaborators: David Buckley Borden, Aaron Ellison, Salvador Jiménez-Flores, and Saluo Rivero. Additional works and information at https://harvardforest.fas.harvard.edu/hemlock-hospice. (Courtesy of David Buckley Borden)

fostering social cohesion around ecological issues. The exhibit was featured in the *Boston Globe*, *Living on Earth*, *Orion*, and *SciArt Magazine*; in addition to leading guided tours of the installation, the creators have given dozens of presentations at universities, galleries, and conferences. Visitors' written responses to the sculptures in the field provided long-term data now being analyzed.

Konza Prairie Biological Station Konza Prairie Biological Station, part of Kansas State University located at Manhattan, Kansas, is one of the six initial LTER sites, and includes both intact and restored tallgrass prairie. This ecosystem type has high conservation value since most tallgrass prairie was converted to production agriculture and existing landscapes are threatened by global change, woody encroachment and fragmentation. Arts and humanities activities benefit from and contribute to the research station's mission of long-term ecological research, education, and conservation. Primary among these activities are the creation of original creative work, especially painting and drawing, photography, and literature. Both university-based and independent artists explore the prairie's complexity, natural history, climatic vulnerability, and other topics. Some from outside the region come for short-term artist residencies; others conduct longer-term projects. In addition, artists and writers offer public outreach through presentations of their work, including art exhibitions and readings and lectures by resident authors and writers.

Fig. 11.3 Kansas State University Associate Professor of Art Erin Wiersma develops drawings for "Konza Drawings" by dragging paper through recently burned prairie at Konza Prairie LTER. (Photograph by David Mayes)

Illustrative of the long-term art activities at the Biological Station is "Konza Drawings," a series of land drawings using burned plant matter during the seasonal prescribed-burn treatments of the prairie's watersheds (Fig. 11.3). In conjunction with the experimental and conservational burn regime, Erin Wiersma, Associate Professor of Art at Kansas State University enters the watersheds and works with the charred vegetative matter by pulling, rubbing, dragging, and lifting paper through the recently-burned prairie. The images she creates record the landscape patterns resulting from the varying burn treatments, fire intensities, and vegetative composition from each locale. Her intention is to "bridge art and conservation and create a greater public awareness of the land," she says. Her work has been featured in a lecture at the Mariana Kistler Beach Museum of Art on the Kansas State University campus, the Robischon Gallery in Denver, Colorado, and through public programming. Upcoming exhibitions include Galerie Fenna Wehlau in Munich, Germany, and Mid-America Art Alliance in Kansas City, Missouri.

Virginia Coast Reserve LTER The *Art & Ecology* program at Virginia Coast Reserve LTER, which involves the University of Virginia and studies Virginia's coastal ecosystems, provides professional development for K-12 art and science teachers through a weekend immersion in the coastal systems of Eastern Virginia. Since its founding in 2012, the program has used the shared dependence on observation to interweave art techniques with ecology lessons. Observational Drawing is taught in the spring; Plein Air Painting is taught in the fall. Contributed artists' statements reveal a heightened sense of natural dynamics and environmental fragility gained from "participating" in an ecosystem during the artistic endeavor to capture its image. Virginia Coastal Reserve artists and scientists have also begun closing their eyes to focus on the emerging theme of listening experiences and sounds as data. Ecoacoustic listening and reflective activities have captivated researchers with deep experience in the systems of the coastal reserve. Similar listening experiences are being envisioned as a new approach to outreach on changing coastline issues. Recent collaborations between music and environmental science graduate students contributed to studies of oyster reef productivity and resulted in musical compositions using reef sounds.

Donna Dixon Aliff, a 2012 *Art & Ecology* program participant and a teacher at York County Middle school, shared her experience: "Artists and scientists both harbor a passion for what they see and record as changes take place in their world". Elementary school teacher and 2014 participant Laura McGowan echoes her, "The world needs scientists to find out how things work, but it also needs artists to interpret the workings, and the beauty, of our world". Thanks to the ongoing *Art & Ecology* program, comparing paintings (e.g., Fig. 11.4) from the same marsh site over nearly a decade is beginning to reveal signs of marsh migration and forest die off with sea level rise and salt-water intrusion. Together, artists and scientists are capturing the process.

11.7 Conclusions

The two-pronged mission of LTER is infused with aspirational language focused on conserving and protecting the natural world (https://lternet.edu/vision-mission/):

- LTER envisions a society in which exemplary science contributes to the advancement of the health, productivity, and welfare of the global environment that, in turn, advances the health, prosperity, welfare, and security of our nation
- Thus, LTER's mission is to provide the scientific community, policy makers, and society with the knowledge and predictive understanding necessary to conserve, protect, and manage the nation's ecosystems, their biodiversity, and the services they provide.

The environmental challenges we currently face, however, are not just problems of science. That is, they are not just problems concerning the empirical facts (biophysical or social) of the working world now and into the future. Our challenges are as much the result of certain philosophical and ethical assumptions about what humans are, what the world is, and what an appropriate relationship between the two looks like. And they are just as much problems of the imagination as well. We seem tragically unable to imagine our way out of our current challenges. It seems easier, in fact, to imagine a complete cultural apocalypse than it does to imagine a world without fossil fuels. The arts and humanities prompt us not only to understand the origin of our grand environmental challenges, but they also serve as the wellspring of our imagining a future that fulfills the mission of LTER.

Fig. 11.4 Plein Air oil painting of a seaside marsh by Donna Dixon Aliff of York County Middle School. This work was created in 2012, the first year of *Art & Ecology*. It is part of a collection that has been displayed at local, regional, and national venues, including UVA's Science Library and the Eastern Shore's Barrier Islands Center cultural museum

Acknowledgements We would like to thank Clarisse Hart (HFR), Cora Johnston (VCR), and Elizabeth Dodd (KNZ) for their assistance in preparing the highlights from their programs and our editors Sharon Kingsland and Robert Waide for valuable feedback throughout the writing process. The Andrews Forest Reflections program gratefully acknowledges support of the Shotpouch Foundation, Pacific Northwest Research Station of the USDA Forest Service, and NSF LTER (grant 14-40409). The *In a Time of Change* program (BNZ) thanks NSF LTER (grant DEB-1636476), the USDA Forest Service, Pacific Northwest Research Station (RJVA-PNW-01-JV-11261952-231), and NSF DEB-1257424 for support of its activities.

References

Berkes, F., J. Colding, and C. Folke. 2002. *Navigating social-ecological systems: Building resilience for complexity and change*. Cambridge: Cambridge University Press. https://doi.org/DOI. https://doi.org/10.1017/CBO9780511541957.

Bieler, A. 2014. *Exhibiting climate change: An examination of the thresholds of arts-science collaborations in the context of learning for a sustainable future*. Ph.D. dissertation, York University, Toronto.

Brodie, N., C. Goodrich, and F.J. Swanson. 2016. *Forest under story: Creative inquiry in an old-growth forest*. Seattle: University of Washington Press.

Brown, K., N. Eernstman, A.R. Huke, and N. Reding. 2017. The drama of resilience: Learning, doing, and sharing for sustainability. *Ecology and Society* 22 (2). https://doi.org/10.5751/ES-09145-220208.

Burns, H.L. 2015. Transformative sustainability pedagogy: Learning from ecological systems and indigenous wisdom. *Journal of Transformative Education* 13 (3): 259–276.

Callahan, J.T. 1984. Long-term ecological research. *Bioscience* 34 (6): 363–367.

Demaray, E. 2014. Work samples from the field of art and science collaboration. *Journal of Environmental Studies and Sciences* 4: 183–185.

Dieleman, H. 2008. Sustainability, art and reflexivity: Why artists and designers may become key change agents in sustainability. In *Sustainability: A new frontier for the arts and cultures*, ed. S. Kagan and V. Kirchberg, 108–146. Frankfurt, Germany: Verlag fur Akademische Schriften.

Dixon, D., H. Hawkins, and M. Ingram. 2011a. Blurring the boundaries. *Nature* 472: 417.

Dixon, D., E. Straughan, and H. Hawkins. 2011b. When artists enter the laboratory. *Science* 331: 860.

Frodeman, R., J.T. Klein, and C. Mitcham. 2010. *The Oxford handbook of Interdisciplinarity*. 1st ed. Oxford, UK: Oxford University Press.

Gabrys, J., and K. Yusoff. 2012. Arts, sciences and climate change: Practices and politics at the threshold. *Science as Culture* 21 (1): 1–24. https://doi.org/10.1080/09505431.2010.550139.

Goralnik, L., and M.P. Nelson. 2015. Empathy and agency in the Isle Royale field philosophy experience. *Journal of Sustainability Education* 10. http://www.jsedimensions.org/wordpress/content/empathy-and-agency-in-the-isle-royale-field-philosophy-experience_2015_12/. Accessed 20 June 2020.

Goralnik, L. and M.P. Nelson. 2017. Field philosophy: Environmental learning and moral development in Isle Royale National Park. *Environmental Education Research* 23 (5): 687–707. https://doi.org/10.1080/13504622.2015.1074661.

Goralnik, L., M.P. Nelson, H. Gosnell, and L. Ryan. 2015. Arts and humanities efforts in the US LTER Network: Understanding perceived values and challenges. In *Earth stewardship: Linking ecology and ethics in theory and practice*, ed. R. Rozzi, F.S. Chapin, J.B. Callicott, S.T.A. Pickett, M.E. Power, J.J. Armesto, and R.H. May Jr., 249–268. Berlin: Springer.

Goralnik, L., M.P. Nelson, H. Gosnell, and M.B. Leigh. 2017. Arts and humanities inquiry in the long-term ecological research network: Empathy, relationships, and interdisciplinary collaborations. *Journal of Environmental Studies and Sciences* 7 (2): 361–373.

Holm, P., M.E. Goodsite, S. Cloetingh, M. Agnoletti, B. Moldan, D.J. Lang, R. Leemans, J.O. Moeller, M.P. Buendía, W. Pohl, R.W. Scholz, A. Sors, B. Vanheusden, K. Yusoff, and R. Zondervan. 2013. Collaboration between the natural, social and human sciences in global change research. *Environmental Science and Policy* 28: 25–35.

Houtman, N. 2012. Forms from the sea. *Terra* 8:20–25. http://terra.oregonstate.edu/2012/10/forms-from-the-sea/. Accessed 13 Jan 2020.

Ingram, M. 2011. Eliciting a response through art. *Nature Climate Change* 1: 133–134.

Jacobson, S.K., M.D. Mcduff, and M.C. Monroe. 2007. Promoting conservation through the arts: Outreach for hearts and minds. *Conservation Biology* 21 (1): 7–10.

Jeffreys, T. 2018. Experiments in the field: Why are artists and scientists collaborating? *Frieze*: https://frieze.com/article/experiments-field-why-are-artists-andscientists-collaborating. Accessed February 21, 2018.

Jones, P., D. Selby, and S. Sterling, eds. 2010. *Sustainability education: Perspectives and practice across higher education*. London: Earthscan.

Kimmerer, R.W. 2016. Interview with a watershed. In *Forest under story: Creative inquiry in an old growth forest*, ed. N. Brodie, C. Goodrich, and F.J. Swanson, 41–49. Seattle: Washington University Press.

Lorkowski, J., and V. Kreinovich. 2015. *Why awe makes people more generous: Utility theory can explain recent experiments*. University of Texas El Paso Departmental Technical Reports. Paper 927. http://digitalcommons.utep.edu/cs_techrep/927. Accessed 13 Jan 2020.

Mann, S. 2017. *Focusing on arts, humanities to develop well-rounded physicians*. Association of American Medical Colleges (AAMC), 15 (Aug 15). https://www.aamc.org/news-insights/focusing-arts-humanities-develop-well-rounded-physicians. Accessed 12 Jan 2020.

Muchnic, S. 2013. Under the microscope. *ARTnews* 112 (3): 70–75.
Nisbet, M.C., M.A. Hixon, K.D. Moore, and M.P. Nelson. 2010. Four cultures: New synergies for engaging society on climate change. *Frontiers in Ecology and the Environment* 8 (6): 329–331.
Ostrom, E. 2009. A general framework for analyzing sustainability of social-ecological systems. *Science* 325 (5939): 419–422. https://doi.org/10.1126/science.1172133.
Patterson, B. 2015. *Al Gore inspires 'CO2,' an opera; A physicist lectures on climate change with a string quartet*. http://www.eenews.net/stories/1060019537. Accessed 13 Jan 2020.
Piff, P.K., M. Feinberg, P. Dietze, D.M. Stancato, and D. Keltner. 2015. Awe, the small self, and prosocial behavior. *Journal of Personality and Social Psychology* 108 (6): 883–899.
Reilly, J.M., J. Ring, and L. Duke. 2005. Visual thinking strategies: A new role for art in medical education. *Family Medicine* 37 (4): 250–252.
Root-Bernstein, R.S. 2003. Sensual chemistry: Aesthetics as a motivation for research. *HYLE* 9: 33–50.
Snow, C.P. 1959. *The two cultures and the scientific revolution*. Cambridge: Cambridge University Press.
Stevens, C. and G. O'Connor. 2017. When artists get involved in research, science benefits. The Conversation (Aug. 16, 2017). theconversation.com/when-artistsget-involved-in-research-science-benefits-82147. Accessed February 21, 2018.
Swanson, F.J. 2015. Confluence of arts, humanities, and science at sites of long-term ecological inquiry. *Ecosphere* 6 (8): 1–23.
Swanson, F.J., C. Goodrich, and K.D. Moore. 2008. Bridging boundaries: Scientists, creative writers, and the long view of the forest. *Frontiers in Ecology and the Environment* 6: 499–504. https://doi.org/10.1890/070076.
Vaismoradi, M., H. Turunen, and T. Bondas. 2013. Content analysis and thematic analysis: Implications for conducting a qualitative descriptive study. *Nursing and Health Sciences* 15 (3): 398–405.
Wilson, E.O. 1999. Consilience: The Unity of knowledge. In *New York*. New York: Vintage Books.

Part IV
The Importance of Community in the Evolution of a Research Network

Chapter 12
History of Comparative Research and Synthesis in the LTER Network

John J. Magnuson and Robert B. Waide

Abstract We present the historical development of cross-site/inter-site comparisons and synthesis by the Long Term Ecological Research (LTER) Network from 1980–2015. Comparative ecosystem research began to flourish after John Brooks of the National Science Foundation (NSF) challenged the LTER community in 1986 to function as a collaborating network of sites rather than as a collection of independent sites. We traced the history over five periods defined by key events and changes that occurred. These periods were: First Steps in Founding the LTER Network from 1980–1986; Responding to Brooks' Challenge from 1987–1995; Dealing with Institutional Instability from 1996–2000; Efforts to Plan for the Future from 2001–2010; and Implementing Parts of the Plan from 2011–2015. Early examples of cross-site research in the 1990s and 2000s were described. Comparisons of similar types of ecosystems that were characterized by similar properties such as grasslands with grassland or lakes with lakes were more common and proved easier to accomplish than did comparisons of different types of ecosystems such as lakes with deserts or forests with coastal waters. Analyses of the network's publication history of cross-site research and synthesis pointed out that LTER sites required 15 years to work as a network but that cross-site research and synthesis increased strongly over the next 20 years. Finally, we posit that the LTER Network and NSF played an important role in helping move ecosystem sciences toward a new level of comparative analyses and syntheses across time and space.

Keywords LTER program · Long-term ecological research · LTER network · Ecological networks · Comparative ecology · Cross-site experiments · Ecological synthesis · Networking

J. J. Magnuson (✉)
Center for Limnology, University of Wisconsin-Madison, Madison, WI, USA
e-mail: john.magnuson@wisc.edu

R. B. Waide
Department of Biology, University of New Mexico, Albuquerque, NM, USA
e-mail: rbwaide@unm.edu

R. B. Waide, S. E. Kingsland (eds.), *The Challenges of Long Term Ecological Research: A Historical Analysis*, Archimedes 59,
https://doi.org/10.1007/978-3-030-66933-1_12

12.1 Introduction

The initial 1979 announcement from the National Science Foundation (NSF) for the Long Term Ecological Research (LTER) program had stated goals to (1) "initiate the collection of comparative data at a network of sites representative of major biotic regions of North America," (2) "evaluate the scientific, technical, and managerial problems associated with such long-term comparative research," and 3) "involve groups of investigators working at representative sites located over the continent or within geographic regions." Investigators were asked to focus on a series of core research topics, coordinate their studies across sites, utilize documented and comparable methods, and commit to continuation of work for the required time.[1]

Despite the emphasis on comparative research, the concept of competing to create a network was alien to the culture at NSF and led to significant controversy.[2] Eloise ("Betsy") Clark, who was Assistant Director for Biological, Behavioral, and Social Sciences, was skeptical of the value of long-term ecological research but approved the initial Request for Proposals with the stipulation that the sites would be selected individually and not as part of a network.[3] However, once the program was created, NSF staff in Biotic Systems and Resources quickly realized the importance, not only of comparative measurements, but also of a functioning LTER Network to facilitate cross-site research. In doing so, NSF anticipated the evolution of the larger environmental community toward "big biology".[4]

Our chapter reviews initiatives, challenges, and research by the evolving network from the first call for proposals by NSF in 1979 through 2015, some 36 years later. We consider the development of the LTER Network, the comparative, cross-site, research accomplished, and the dynamic interactions between NSF and the Network. We detail several early cross-site research projects that began in the late 1980s, developed in the 1990s, and continued to publish through the 2000s. Documentation for the nature and growth of cross-site research during the second 20 years of the LTER program are largely from Johnson et al. (2010) and Huang et al. (2020). We trace this history through five periods that we defined by key events or changes that occurred during the creation and evolution of the Network and its interactions with NSF. A final section assesses the success achieved in doing comparative ecosystem research. The periods are:

- 1980–1986: First steps in founding the LTER Network
- 1987–1995: Responding to John Brooks' challenge
- 1996–2000: Dealing with institutional instability

[1] 1979 "A New Emphasis in Long-term Research" announcement from Long-term Research, Division of Environmental Biology, National Science Foundation.

[2] Jerry Franklin and Frank Harris, personal communication.

[3] Jerry Franklin, personal communication.

[4] Frank Harris, personal communication.

- 2001–2010: Efforts to plan for the future
- 2011–2015: Implementing parts of the plan

At the onset we note that over the years of changing priorities and levels of support from NSF, cross-site research developed and expanded, as judged by published papers and interaction between sites. We posit that pressures to act as a network had a greater influence on LTER activities than any specific set of scientific questions such as, for examples, climate change or biodiversity. Much of the scientific planning in LTER over the years has searched for unique comparative analyses that could be explored and addressed by a network of diverse sites representing a broad range of aquatic, marine, and terrestrial biomes.

12.2 1980–1986: First Steps in Founding the LTER Network

In October 1980, soon after the first group of six Long-Term Ecological Research sites were funded, the NSF invited principal investigators to Washington D.C. for a first network meeting (Magnuson et al. 2006b). Two people came from each site.[5] Most of the participants had never met or worked together before. They were from different communities of scientists who identified themselves with forests or prairies or deserts or lakes or rivers or estuaries. At the meeting they were reminded that they were a network of ecological research sites. John Brooks, Tom Callahan, and Frank Harris represented NSF at the meeting and announced that substantial effort would be devoted to comparative ecosystem analysis and cross-site synthesis, including comparability of methods, information management, and cross-site experimentation.[6]

Participants began to organize and get acquainted (Fig. 12.1). They chose G. Richard (Dick) Marzolf (Konza Prairie) as chair and Richard H. Waring (Andrews Forest) as vice chair (Magnuson et al. 2006b). They shared the networking tools of the day (phone numbers and postal addresses) and scheduled the first Coordinating Committee meeting at Konza Prairie for December, 1980. It was not clear, at least to John Magnuson, who represented the North Temperate Lakes LTER in Wisconsin, that the sites really knew what being part of a network meant or what might be expected from that aspect of their new grants. The North Temperate Lakes' initial proposal did not develop the idea of comparative, cross-site research but did mention in a short paragraph that they intended to develop closely correlated "data sets

[5]Arthur McKee and Richard Waring (H. J. Andrews), Dac Crossley and Wayne Swank (Coweeta), Pat Webber and Nel Caine (Niwot Ridge), John Vernberg and Dennis Allen (North Inlet Estuary), John J. Magnuson and Carl Bowser (North Temperate Lakes), and John Zimmerman and Richard Marzolf (Konza Prairie). Source: D Marzolf plenary talk, "Konza prairie and the origin of the LTER network," 2006 LTER All Scientists Meeting.

[6]G. R Marzolf, plenary talk, "Konza prairie and the origin of the LTER network," 2006 LTER All Scientists Meeting.

Fig. 12.1 LTER researchers getting to know each other over dinner after meeting with NSF staff at their first meeting in 1980. Full faces from left to right: Nelson Caine (Niwot Ridge), Carl Bowser (North Temperate Lakes), Pat Webber (Niwot Ridge), Dennis M. Allen (North Inlet Estuary), G. Richard Marzolf (Konza Prairie), and John J. Magnuson (North Temperate lakes). (Photo by R. Marzolf; from Magnuson et al. 2006b, on p. 304)

with a wider array of lake ecosystems in other geochemical and climatic regions."[7] NSF required the lake site to write an addendum to their proposal, but again the addendum request did not mention cross-site research nor did the addendum address it. We were asked to include modest travel funds so that we could attend Coordinating Committee meetings. The lake site and, Magnuson presumed, other sites were focused on the major responsibility of getting a Long-Term Ecological Research site up and running.

At LTER Coordinating Committee meetings in the early 1980s, we were reminded about comparative cross-site research, but complained that we did not have sufficient resources. Each site attended Coordinating Committee meetings held annually that rotated from site to site. Carl Bowser and Magnuson usually attended for the North Temperate Lakes site, at least initially. We used the meetings, not only to participate in network business and discussion, but also to learn more about our colleagues and research at their sites. For example, at an early Coordinating Committee meeting at Niwot Ridge, we (Bowser and Magnuson) extended our stay for an additional night and day to camp out on the ridge. The overnight was valuable and even inspirational for the two of us as we experienced the site, the city lights, and the reflections at dawn of the many reservoirs on the distant flatlands below. We began to appreciate our colleagues, their sites, and the research they did.

As required by NSF, Dick Marzolf as well as other network staff routinely compiled an annual description of the sites with a map on the cover of their locations. In the 1982 report Marzolf, in his role as chair of the Steering Committee (later the Coordinating Committee), explicitly pointed out the values of establishing a

[7] North Temperate Lakes LTER proposal "1981 Comparative Studies of a Suite of Lakes in Wisconsin", p. 15, in archives of the LTER at the Center for Limnology, Steenbock Library, Records and Archives, University of Wisconsin, Madison.

collection of LTER sites (Halfpenny et al. 1982). These important bullets about the need for an established set of long-term ecological research sites were fully recognized and greatly appreciated:

1. Investigation of ecological phenomena occurring at time scales of decades and centuries were not normally supported by NSF funding in ecology.
2. Ecological experiments were conducted with little recognition of the high interannual variability in the studied systems.
3. Long-term trends were not being systematically monitored in ecological systems with the consequence that unidirectional changes could not be distinguished from more cyclic variation.
4. The absence of a coordinated network of ecological research sites inhibited comparative research and its benefits.
5. Natural ecosystems where research was being conducted were being lost to other uses.
6. Ecological research was often done on only a selected component of the system, and multilevel, integrated data were not available at intensive research sites.

Only bullet 4 explicitly referred to the importance of being a network. The other bullets were just as relevant to long term research at each of the sites as they were to a functioning network of sites.

In 1981 and 1983 NSF provided additional support for planning cross-site research to the Coordinating Committee. The work proposed in 1981 focused on identifying a set of principles common to multiple ecosystems that would lead to hypotheses that could be the basis of further proposals. The 1983 award provided funds for workshops to investigate the general principles that had been identified, and expanded the network to include six new sites selected in 1982.

Even so, the cross-site research products expected from the LTER Network did not suddenly materialize in the early 1980s. NSF decided to help change that reality. In January 1985, as part of the guidance for the renewals of the first cohort of sites, Tom Callahan directed each site to spend at least 15% of their budgets on cross-site experiments and synthesis.[8] This was not popular across the sites.

Expectations of NSF for comparative, cross-site research became very clear when John Brooks (Fig. 12.2), Director of NSF's Division of Environmental Biology, attended the November 1986 LTER Network Coordinating Committee meeting in Denver. He pointed out that the LTER program had begun five years earlier with the implicit goal to "develop integrated comparative study of ecosystems, going one step beyond what was established during the International Biological Program (IBP)."[9] Brooks indicated that by 1990 the LTER Network needed to

[8] See Dec 19, 1984 Guidelines for Renewal Proposals revised, JJ Magnuson LTER 1, Box 20E9B, in LTER Network, Limnology archives, Steenbock Library, Records and Archives, University of Wisconsin, Madison.

[9] Minutes, LTER Coordinating Committee Meeting, Nov 8–9, 1986, Denver Colorado, Box 20E8C, in LTER Network, Limnology Archives, Steenbock Library, Records and Archives, University of Wisconsin, Madison.

Fig. 12.2 John Brooks, NSF Director of the Division of Environmental Biology in winter 1976 at NSF in Washington D.C. (Photo by J. Magnuson; from Magnuson et al. 2006a, on p. vi)

demonstrate what NSF got by spending $15 million on 11 sites. John Brooks' high hopes were to move ecosystem science to a new level by developing approaches to comparative ecosystem analysis. LTER needed to incorporate greater time spans and to demonstrate comparative capabilities, including working at larger spatial scales as well as at the landscape level. The importance of climate and climate change was mentioned specifically. The hope was that the LTER sites could be Biospheric Observatories on the North American continent. LTER was expected to make a presentation to NSF in 1987 and then prepare major publications by the ten-year review.

Magnuson's personal notes from the above meeting confirmed what was in the minutes: that the intention was basically to achieve a quantum jump in the science to create a new comparative ecosystem science.[10] Magnuson took John Brooks' presentation as a pep talk with a "carrot and stick" approach. The carrots were to reach high for a new kind of ecosystem science and to provide $60,000 to the LTER Coordinating Committee to catalyze the effort. The stick (which Magnuson remembered, although it was not in the minutes or his personal notes) was a warning to the effect that if they did not conduct comparative research, the LTER sites would no longer be a network of ecosystem sites, but would return to being individual ecosystem sites as far as NSF was concerned. Without producing unique results based on comparative research, the LTER program would be in jeopardy at the ten-year

[10] Magnuson personal notes, LTER Coordinating Committee Meeting, Nov 8–9, 1986, Denver Colorado, Box 20E8C, in LTER Network, Limnology archives, Steenbock Library, Records and Archives, University of Wisconsin, Madison.

review. Moreover, Brooks informed the group that the National Science Board, which established the policies for NSF, felt that NSF did not have enough control over LTER projects and their coordination. Another statement that Magnuson recalls from about the same time was that Brooks believed that the long-term data sets on a suite of ecosystems would be more valuable to science than the papers we might produce in the short term.

12.3 1987–1995: Responding to Brooks' Challenge

The newly emerging network of LTER sites had been convinced; we prepared during the next 12 months for the 1987 presentation to NSF.[11] The 12 presentations were: Jerry Franklin (the importance of long-term studies in ecology); Walter Whitford (site and research overview); John Magnuson (time lags and temporal variation); Frederick Swanson (geomorphic perspectives on landform, soil, and vegetation relationships); William Parton (regional soil and organic matter dynamics); Larry Ragsdale (biotic regulation of material flows); Mark Harmon (coarse woody debris in ecosystems); Timothy Seastedt (the First International Satellite Land Surface Climatology Project Field Experiment program on Konza Prairie); Thomas Frost (interactions between LTER and the Little Rock Lake experimental acidification project at North Temperate Lakes); David Tilman (biological diversity and bioengineered organisms); Robert Woodmansee (responses to global change); and Jerry Franklin (cross-site comparisons).

Jerry F. Franklin, Chair of the LTER Coordinating Committee, ended our presentations with the statement: "Ecological science has suffered severely from investigator myopia. Comparative research is essential to understanding how ecological processes vary across major spatial and temporal gradients." The LTER scientists were committed and largely on the same page as NSF. So, what did we do, and how well did we do it?

NSF did its part to encourage comparative research, creating the position of LTER Research Coordinator in the Division of Environmental Biology in 1988 (Coleman 2010). One of the principal responsibilities of this position was to facilitate cross-site research. Caroline Bledsoe was recruited to the position in 1988 and worked with the LTER Coordinating Committee to organize semi-annual meetings to foster cross-site synthesis. She helped establish the informal "Gang of X" at NSF that included John Brooks, Jerry Melillo, Tom Callahan, Bob Robbins and others (X meant that the number in the gang fluctuated).[12] This group met regularly to discuss the direction of the LTER program and worked to develop support for LTER both within NSF and the ecological community. Bledsoe played an important role in the

---

[11] LTER Coordinating Committee Meeting Nov 1987 and Washington briefing session on Long-Term Ecological Research Program, Box 20E8C, in LTER Network, Limnology archives, Steenbock Library, Records and Archives, University of Wisconsin, Madison.

[12] LTER Network News 1997, Issue 20.

development of the first LTER directory and all-site bibliography (Chinn and Bledsoe 1997). She worked with other Federal agencies on expanding the LTER concept into a "network of networks" (Bledsoe 1993) and helped build relationships with the Environmental Protection Agency, Department of Defense, and NASA. From an LTER Network perspective, the Coordinating Committee did allocate the $60,000 to five-to-six cross-site projects and did group exercises to prioritize what topics should be considered.[13]

Cross-site research did emerge from the LTER Network in response to the Brooks challenge; studies began to appear in the late 1980s and early 1990s. Typically, these efforts continued to produce comparative cross-site publications into the late 1990s and 2000s, well past the ten-year and twenty-year reviews. Using both LTER sites and non LTER sites, researchers did comparative analyses among ecosystems of the same type and less commonly across different ecosystem types. Not surprisingly, researchers found it easier to compare within a single ecosystem type than between dissimilar ecosystems; likewise, comparisons of the same measure were more straightforward than comparisons among a diversity of measures. However, explorations in comparative study among dissimilar ecosystem types and measures did occur. These issues were noted in the Ten-Year Review, which was conducted by a committee chaired by Paul Risser and Jane Lubchenco and was submitted to NSF in 1993. The executive summary of the report noted that "Some intersite comparisons have been conducted, but the power of the network of coordinated research sites has not yet been fully realized." The report also noted that the criteria used to select sites and participate in the program "have not uniformly included specific expectations for developing a comprehensive, representative network of sites" (Risser et al. 1993, p. 2).

At a 1989 Coordinating Committee meeting at Harvard Forest, the Network undertook the first of several forward-looking planning efforts aimed in part at enhancing cross-site research and synthesis (Table 12.1). The 1989 plan focused largely on steps needed to establish a functioning research network including guidelines for cross-site experimental science.[14] These guidelines addressed how to identify tractable scientific issues and how the network could address such issues. Establishing guidelines was important because sites were still trying to define the appropriate allocation of effort between site and cross-site research. John Brooks, who retired from NSF June 1989, revealed his continuing interest in the LTER by attending some of the sessions. Participants mostly expressed forward looking and collaborative views about the LTER Network; several expressed negative views about the formality of the planning process among the sites.

---

[13] Coordinating Committee at Andrews LTER site in April 1988. Rankings summarized in June 13, 1988 memo from Caroline Bledsoe to the LTER Coordinating Committee, Box 20E8B, in LTER Network, Limnology archives, Steenbock Library, Records and Archives, University of Wisconsin, Madison.

[14] A long-range strategic plan for the Long-Term Ecological Research Network. https://lternet.edu/wp-content/uploads/2014/11/Strategic%20Plan.pdf. Accessed 31 May 2020.

Table 12.1 Five principal planning efforts that shaped cross-site and comparative research in the LTER Network

Name	Year initiated & completed	Leadership	Short rerm result	Long term result
A long-range strategic plan for the long-term ecological research network	1989	LTER coordinating committee	First formal description of synthesis at the network level	Defined the LTER Network as "a collaborative effort of LTER sites to facilitate and extend the capabilities of the individual sites and to promote synthesis and comparative research"
LTER 2000: Creating a Global Environmental Research Network	1992	LTER Executive committee	Informed the Ten-Year Review of the LTER Network	Expanded the vision for comparative research and synthesis to a global network of sites
LTER 2000–2010: A decade of synthesis	2000/2001	LTER Executive committee	Set network priorities in anticipation of the 20 Year review	Established six interrelated goals of the LTER network: Understanding, synthesis, information, legacies, education, and outreach
The decadal plan for LTER Integrative science for society and the environment: A plan for research, education, and cyberinfrastructure in the U.S. long term ecological research network	2003/2007	Science task Force (five co-investigators of the planning award from NSF)	Defined a role for the LTER Network in a research environment that integrates biophysical and social sciences	Created a vision for long-term socioecological research that is synthetic across multiple sites and described a research framework to achieve that vision
Strategic and implementation plan	2010/2011	LTER Executive board	Achieved a common vision for future development of the LTER network	Provided detailed guidance for LTER committees and the network office to enhance network research, education, communication, information management, and outreach

In 1992, Dr. Mary Clutter, Assistant Director of NSF's Biological Sciences Division, invited the LTER Network to brainstorm the kinds of new science that could be accomplished by LTER if site budgets were increased to $2.5–3.0 million. This invitation resulted in an expansive new vision for LTER (Table 12.1) that included the enhancement of network-level science, creation of a global network of research sites, development of centers for research and synthesis, partnerships with research centers funded by NSF and other agencies, and development of new technologies to advance the field of environmental science. These ideas that resulted from Clutter's request were incorporated into the LTER 2000 (Table 12.1) document that provided input into the Ten-Year Review.[15]

The Ten-Year Review of the LTER made the expectation of cross-site research strongly and clearly. However, this expectation was clothed in an expanded vision for LTER that included doubling the number of sites, expanding the geographic extent of research, incorporating physical and social sciences, and increasing funding by an order of magnitude. Significantly, the Ten-Year Review noted that "Reviewers of the LTER program…. have not always placed consistent emphasis or value on cross-site comparisons", resulting in mixed messages to LTER researchers (Risser et al. 1993). Not surprisingly, site renewal proposals varied in their emphasis on cross-site research. The LTER Coordinating Committee, in a letter to Mary Clutter responding to the recommendations of the Ten-Year Review, embraced the challenge of increased cross-site and network research but cautioned that success would depend on strong support from NSF.[16]

NSF did demonstrate increasing support for the LTER Network during the period from the ten-year review to 1995. NSF sponsored open competitions for cross-site research in 1994 and 1995. In these competitions, twenty-two awards to LTER and non-LTER sites ranged from $109,000–200,000 and provided the means to conduct new comparative research.

NSF also announced a special competition to augment two existing LTER sites to expand regionally, create comprehensive site histories, and increase disciplinary breadth. Proposals from North Temperate Lakes and Coweeta sites were selected in 1994. By creating winners and losers, these competitions went counter to the spirit of cooperation that LTER had fostered, but the Network hoped that additional competitions would restore parity across the eighteen LTER sites.

The 1993 All Scientists Meeting (ASM) provided important opportunities for the LTER Network to expand comparative research with partners outside of the Network. An invited workshop at the ASM led to the formation of the International LTER Network, expanding opportunities for cooperation across national boundaries (see Waide and Vanderbilt, Chap. 16 in this volume). NSF signed memoranda of

[15] LTER 2000: Creating a Global Environmental Research Network, 1992. https://lternet.edu/wp-content/uploads/2010/12/LTER2000.pdf. Accessed 31 May 2020.

[16] Letter from Jerry Franklin to Mary Clutter based on Coordinating Committee discussions on July 30, 1993. https://lternet.edu/wp-content/uploads/2010/12/10year-review-response.pdf. Accessed 31 May 2020.

understanding with the U.S. Department of Agriculture (Forest Service) and the National Biological Service to encourage collaboration with the LTER Network.

The degree of support from NSF during this period encouraged the LTER Network to plan for a significant expansion of comparative research and synthesis. By 1994, the Network was deeply engaged in discussions about the kinds and extent of comparative ecological research that could be conducted using LTER data. In 1994, the LTER Network created a Synthesis Committee to help guide these discussions.

12.4 1996–2000: Dealing with Institutional Instability

Beginning in 1995 NSF began to anticipate shortfalls in the LTER budget resulting from across the board funding reductions. In 1996, site budgets were reduced by $20,000 and NSF predicted a flat budget scenario for LTER through 2000, with the alternative being a reduction in the number of sites. No further augmentation competitions took place and competitions for cross-site research were not held until 2000.[17] No All Scientists Meeting was supported in 1996 or subsequent years until 2000. Ironically, at the same time, NSF issued a special competition for two new urban LTER sites.

The LTER Network had entered a period where resources for cross-site comparisons were no longer readily accessible through the LTER Program at NSF. Sites were encouraged to seek funding for cross-site work through other programs at NSF or other agencies and to pay particular attention to new initiatives for which we might be well positioned to compete.

This guidance coincided with several important changes in the management of LTER. Scott Collins replaced Tom Callahan as LTER program officer in 1995. Jerry Franklin retired as Coordinating Committee Chair, and Jim Gosz replaced him. NSF announced an open competition for the LTER Network Office (LNO) and declined a renewal proposal from the University of Washington. The LTER Network encouraged a joint proposal for the Network Office between the University of New Mexico and the University of Washington, but NSF again declined to fund the Washington request. In the move from Seattle to Albuquerque in 1997, the position of Executive Director of the LNO was created and subsequently filled by Robert Waide.

During this period of significant administrative transition for the LTER program, the Network of sites needed to identify new approaches to achieving comparative research and synthesis. These approaches built on strategic goals for comparative research that had been advanced for nearly a decade by the LTER Network.

These new approaches began with a re-examination of LTER research priorities. The US LTER National Advisory Board (see Zimmerman and Groffman, Chap. 15 in this volume) identified the need for priority-setting in their 1998 review of

[17] Scott Collins, personal communication.

LTER. Similar recommendations arose from NSF's review of the LNO in 2000. A third stimulus for priority setting was the need to prepare for the 20-year review of the LTER Program by NSF in 2001. To address these recommendations, the LTER Executive Committee (Zimmerman and Groffman, Chap. 15, this volume) drafted a white paper that described the priorities of the LTER Network for the first decade of the new millennium. The resulting "LTER 2000-2010: A Decade of Synthesis" (Table 12.1), describes how the LTER Network would refresh and refocus its core aims and mission "to place its diverse current activities in a clear and consistent context and to develop clear priorities for the future."[18]

LTER 2000–2010 describes the activities needed to attain two goals relating to cross-site research: (1) increase the pace of synthesis and (2) increase experimental and comparative cross-site research. The document establishes specific targets to achieve the two goals. Synthesis entails the integration of existing information across disciplines and projects to create new understanding. Synthesis at individual sites would be accomplished, in part, through a series of comprehensive research volumes published by Oxford University Press. The LTER Network set a goal for a complete 24 volume set by 2010; however, by 2020 only 11 had been completed. Synthesis at the network level would be accomplished, in part, through All Scientists Meetings every three years, annual exploration of topics selected by the Coordinating Committee, writing books and journal special features addressing comparative research or methods, and forming multisite working groups. In addition, individual investigators, students, and post-docs were to become more engaged in synthesis of LTER data.

LTER 2000–2010 proposed creation of a standing Committee on Scientific Initiatives charged with identifying and helping to develop new opportunities for LTER synthesis. At the same time, logistical support for synthesis would be provided through database development and informatics techniques optimized for ecological research.

Support to increase the pace of synthesis came from a variety of sources. The LTER Network Office provided support for Coordinating Committee meetings and six to eight synthesis working groups per year. Books, special features, and individual synthesis projects were accomplished through regular research activities of the sites, sabbaticals from academic institutions, and support from the National Center for Ecological Analysis and Synthesis (NCEAS) or other sources. The Network proposed a new program to fund student and post-doctoral synthesis projects. Big ticket items like the All Scientists Meeting and database development would require investments from NSF and perhaps other agencies. Most of these synthesis activities were carried out through coordinated efforts among sites, the Network Office, and NSF. Unfortunately, sabbatical, student, and post-doctoral projects did not receive support by NSF despite repeated joint efforts by NCEAS and LNO.

---

[18] LTER 2000–2010: A Decade of Synthesis. https://lternet.edu/wp-content/uploads/2010/12/lter_2010.pdf. Accessed 31 May 2020.

The goal of increasing experimental and comparative cross-site research presented a different set of design and funding challenges. New cross-site research in an existing network of sites required joint decisions about units of measurement, definitions of variables, and experimental treatments. Such decisions inevitably led to modifications in existing research approaches and additional investments. The more sites involved in the cross-site research, the larger the overall investment. For a network of 24 long-term research sites operating on a fixed budget, new collaborative research was not affordable. Additional sources of funding were critical to stimulate increased levels of cross-site research. Unfortunately, NSF eliminated cross-site research competitions after 2000 and thus removed an important source of funding for cross-site projects.

Despite the instability caused by changes in LTER and NSF, the Network continued to promote a broad range of cross-site research activities. Waide made the All Scientists Meeting (ASM) a high priority and was able to obtain support for a meeting in 2000 in association with the Ecological Society of America. The LNO received follow up proposals for 39 working groups from the ASM and was able to fund 24 of these, many of which led to successful synthesis products. Research topics covered the range of LTER core areas as well as information management, education, and social science. Funds to continue funding new comparative cross-site research, however, remained scarce; few future projects could be implemented.

12.5 2001–2010: Efforts to Plan for the Future

Stimulated by recommendations of the Twenty-Year Review, NSF once again began to look for ways to provide more funding for the LTER program (see Collins, Chap. 14 in this volume). Joanne Roskowski, Deputy Assistant Director for Biological Sciences, signaled the possibility of increased funding through LTER Program Director Scott Collins. The possibility of additional funds stimulated the LTER Executive Committee to consider the development of a research plan that would provide a vehicle for new funding. With support from a planning grant from NSF, the LTER Network engaged in an intensive research planning effort that took place from 2004–2007 (Table 12.1), (Collins et al. 2011) followed by the development of a broad Strategic and Implementation Plan in 2010 that described activities necessary to enhance research, education, communication, information management, and coordination with other networks.[19] Specific research objectives included:

1. To increase site-based capacities for cross-site research;
2. To conduct synthetic and cross-site research that builds upon existing long-term, site-based data, experiments, and models across the Network;

[19] Strategic and Implementation Plan: Long Term Ecological Research Network. https://lternet.edu/wp-content/uploads/2010/12/LTER_SIP_Dec_05_2010.pdf. Accessed 2 June 2020.

3. To perform transformative research at regional to continental scales that expands upon existing LTER infrastructure and human and intellectual capital and capitalizes on emerging observatory networks and technologies.

Strategies proposed to achieve these goals focused on using existing resources and expanding funding "by means of conventional, existing, and new funding models at NSF and elsewhere."

NSF's response to the Strategic and Implementation Plan was lukewarm (Collins, Chap. 14, this volume), leaving few options to fund research involving most or all of the 26 LTER sites. The discrepancy between resources needed to conduct network research and those available through NSF is nowhere clearer than in the recommendations of the Thirty-Year Review[20] and NSF's response to those recommendations. The report from the Thirty-Year Review repeated concerns raised in the Ten- and Twenty-Year Reviews:

> Resources are a key limiting factor for the future of the network. The network should: 1) make realistic prioritizations within the existing resource base to create more science per dollar, and 2) engage with NSF and others to pro-actively develop new resources.

NSF's response to this recommendation focused on doing more science at sites for the same money and downplayed any possibility of additional support to achieve the cross-site research goals of the Strategic and Implementation Plan:

> NSF considers it essential that LTER sites prioritize their research, data management, and education and outreach efforts. These priorities should highlight sites' scientific strengths and strengthen sites' abilities to address compelling research questions. NSF acknowledges that sites cannot excel in all of the areas identified in the Thirty-Year Review and emphasizes the importance of developing priorities that advance long-term research. Priorities must be based on resources currently available. As the LTER community moves forward with clear and wise allocation of existing resources, NSF will consider seeking additional sources when they are available. At the same time, individual sites and the Network Office should examine other funding initiatives within NSF as opportunities to enhance the LTER research portfolio.[21]

12.6 2011–2015: Implementing Parts of the Plan

The LTER Network entered its fourth decade with a strong new conceptual framework (Collins et al. 2011), a strategic plan with prioritized goals for research, education, cyberinfrastructure, and management, a detailed plan for implementation of the steps needed to achieve strategic goals (Strategic and Implementation Plan 2011, Table 12.1), and an effective governance structure (Zimmerman and Groffman, Chap. 15, this volume) designed to meet our strategic goals. However, the budgets

[20] Long-Term Ecological Research Program: A Report of the 30 Year Review Committee. https://lternet.edu/wp-content/uploads/2011/12/bio12001.pdf. Accessed 31 May 2020.

[21] https://lternet.edu/wp-content/uploads/2012/05/Recommendations_30_year_review.pdf. Accessed 31 May 2020.

at LTER sites and the LNO were limited and could not accommodate the new costs inherent in our planning.

With the advice and consent of the LTER Network's Executive Board (formed as part of the new governance structure), the LNO attempted to address this need in its renewal proposal for the 2009–2015 time period by requesting an increase from 9 to 15 million dollars. Additional funds were directed at carrying out LTER strategic goals for comparative research and synthesis, cyberinfrastructure, and outreach (including education). Our request for additional funding was not approved, and we were directed to submit a budget with only a cost of living increase.

However, the economic crisis of 2008 led to the American Recovery and Reinvestment Act of 2009; NSF submitted our requested increase to this new funding source. The American Recovery and Reinvestment Act provided an additional $6.7 million beginning in 2010, and the LTER Network suddenly had the resources to implement several of our strategic goals. The bulk of these funds went to the development of the Network Information System ($2.2 million) and support for cross-site research ($1.2 million).

To increase synthesis and comparative research, the LTER Network adopted both broad- and small-scale approaches. For broad-scale projects involving many sites engaging in new studies or experiments, LTER invested in developing ideas that addressed goals of existing programs at NSF or other agencies. For example, the LTER Science Council identified four research themes that together could involve a majority of LTER sites and that had potential to attract additional funding: inland climate change, the disappearing cryosphere, future scenarios of landscape change, and vulnerability to climate change in coastal systems. Planning for these research themes took place at workshops at the 2009 All Scientists Meeting and subsequent meetings funded by the LNO. However, the research projects that emerged from these planning efforts were not well matched with new or existing programs at NSF, few of which offered awards sufficiently large to support multi-site projects. Thus, none of these efforts were able to move forward at the scope that was originally planned.

The Science Council also invested in smaller scale synthesis studies that could be carried out with existing resources. Most of these projects involved existing data and did not require additional field studies. This kind of synthesis requires one or more dedicated leaders (researchers and information managers) willing to volunteer their time. Other costs are minimal, usually limited to a few meetings of participating sites. Resources that were brought to bear on comparative studies of this type included LTER ASMs, working groups sponsored by the LNO, Science Council research themes, LTER annual mini-symposia at NSF, ad hoc working groups at professional meetings, and working groups sponsored by NCEAS, the National Socio-Environmental Synthesis Center, or other organizations. Online meetings were also used successfully to carry out comparative studies.

A prerequisite for all LTER comparative research was the development of standardized, comparable databases. Although seemingly simple in concept, the process of developing algorithms to homogenize diverse data, creating cyberinfrastructure to make those data accessible, and analyzing the data was costly both in dollars and human effort (Stafford, Chap. 13 in this volume). A vision for an LTER Network Information System (NIS) had existed since the first decade of LTER but developing operational plans to achieve that vision was extremely difficult. Two interacting efforts began to take shape in the mid-2000s. The Ecotrends Project was an effort to provide access to comparable ecological data in a way that would promote synthesis of long-term data (e.g., Carpenter et al. 2009; Peters 2010; Peters et al. 2011).[22] Arising from a conversation between two LTER scientists (Debra Peters and Ariel Lugo) at a Science Council meeting, the project aimed at simplifying and homogenizing diverse data sets to facilitate comparison across a range of LTER, experimental forest, and rangeland research sites (Peters et al. 2013). The Ecotrends Project represented an important commitment on the part of LTER scientists to develop comparable data. Integrating Ecotrends with the developing Network Information System[23] did not occur owing to lack of funds, and as a result data in Ecotrends have remained static. However, development of the Ecotrends data portal by LNO software developers and LTER information managers did provide a base on which to construct the current Network Information System.

The Decadal Plan for LTER (Table 12.1; Collins, Chap. 14, this volume) included an LTER Cyberinfrastructure Strategic Plan that formalized the vision for the Network Information System. While the steps to achieve the Network Information System were well understood, support for developing the necessary cyberinfrastructure was not available. In was not until the LNO was able to obtain funding for the NIS as part of the NSF American Recovery and Reinvestment Act award that work was able to proceed. The NIS was made operational in 2013 and allowed access to data from all LTER sites through a single data portal. Since then, the NIS has expanded its scope to include non-LTER data as the Environmental Data Initiative, which took over the management of the NIS in 2016.[24] At present, this program allows access to 43,868 unique data packages including LTER and Ecotrends data. The availability of data through the Ecological Data Initiative facilitated comparative research among LTER sites and the incorporation of LTER data in broader syntheses (Dornelas et al. 2014; Xu 2015; Blowes et al. 2019).

12.7 LTER Cross-Site Research

As we have discussed, planning for network cross-site research began in the first decade of the LTER in the 1980s (Table 12.1). NSF support for such activities reached a peak in the early 1990's following Brooks' Challenge and Mary Clutter's support for LTER, but then waned from 1996–2000 during a period of stagnant

[22] Ecological data on demand. At: https://ecotrends.info/. Accessed 2 June 2020.

[23] Envisioned as early as 2007 in The Decadal Plan for LTER: Integrative Science for Society and the Environment: https://lternet.edu/wp-content/uploads/2010/12/TheDecadalPlanReformattedForBook_with_citation.pdf. Accessed 19 June 2020.

[24] EDI data portal: https://portal.edirepository.org/nis/home.jsp

budgets. Beginning with the 2000 All Scientists Meeting, NSF did support small-scale cross-site projects through the LNO, direct supplements to sites, and a final cross-site research competition in 2000. However, by 2014 NSF made cross-site research optional in its guidance to sites and eliminated all reference to cross-site research in 2016 (Jones and Nelson, Chap. 3 in this volume).

Despite shifting attitudes toward cross-site research at NSF, participation by LTER scientists in cross-site studies increased steadily over four decades (Johnson et al. 2010; Huang et al. 2020). Increasing cross-site research required commitment, persistence and creativity on the part of the LTER Network and individual groups of researchers. Individuals, often not recognized, were instrumental both for the ups and the downs in cross-site research. Yet, enthusiasm for a functioning network for comparative ecological research among a diversity of ecosystems at longer and broader scales survived and flourished.

Below we highlight a few among the early projects catalyzed by the Brooks challenge. Then we review metanalyses of the network's cross-site research that reveals the character of the growing stream of cross-site publications over LTER's first 40 years of this grand experiment.

Climate Variability and Ecosystem Response Comparing climate variability and ecosystem response provided examples of comparing the same common measures across the entire LTER Network. The LTER Network's Climate Committee successfully functioned as an early network-wide research group. Each site had a member or two on the Climate Committee. The initial goals were to (a) prepare a standard for meteorological measurements, (b) publish a monograph (Greenland 1987) describing the climates of each LTER site, and (c) stimulate bioclimatology studies in the LTER program. Network cross-site research on climate was not specified, per se, but Greenland did initiate several cross-site syntheses (Greenland and Swift 1990; Greenland 1994; Greenland et al. 2003a, b). These reports and papers were creative examples of network scientists responding to Brooks' 1986 challenge from NSF and extended for 16 years from 1987 to 2003.

Their first product (Greenland 1987) provided robust, climate data summaries for each of the 11 sites, with comprehensive tables comparing the climates of the sites, and graphs of the temperature and precipitation clusters and the moisture relations. An interesting observation for the Konza Prairie was that the climate classification systems suggested that it could have supported a forest ecosystem if fire were absent.

By the ten-year review, Greenland and Swift (1990) had published a technical report titled "Climate Variability and Ecosystem Response." Eleven of the 15 sites participated in the workshop and 10 presented papers, all of which were based on their individual site data. Topics ranged from short- to long-term changes, climate trends and variability, spatial variability in microclimates, weather data quality, hydrology, stream flow, and lake ice phenology. The best of these were edited and illustrated in Greenland et al. (2003a).

Discussions of the climate committee were summarized by Greenland and Swift (1991) and Greenland et al. (2003b); they were insightful and productive. Major discussion related to the definition of terms like climate variability, events, and episodes; the importance of scale in both time and space, indices for cross-site comparisons, and the value of considering similarities and dissimilarities. The paper was full of great advice; discussions of scales in time and space and similarity and dissimilarity were useful in cross-site research done at North Temperate Lakes and other sites.

The framework (Greenland and Swift 1991; Greenland et al. 2003b) for climate variability and ecosystem response first pointed out the need to identify the type of climate variability apparent in the climate time series. Variability based on Karl (1985) ranged from a long-term trend, to a step or abrupt change, to regular oscillations. Once the kinds of variability were identified, Greenland et al. (2003b) recommended a set of useful questions to evaluate the data. This synthesis was a major cross-site contribution of the LTER Network that worked. The analyses and ideas presented have been useful to LTER sites and should be in future decades.

Lake Ice Seasonality: Comparing Similar Ecosystems Measuring lake ice seasonality in time and space used common measures (ice cover seasonality) to make comparisons among the same ecosystem type. Reaching out and including many non-LTER sites and locations were essential because few lakes were found at other LTER sites and long-term data on ice seasonality were absent. This project began in the earliest years of the LTER with doctoral research by Dale M. Robertson[25] (Magnuson 1990), was significantly enhanced when the North Temperate Lakes site was expanded regionally, and continued to evolve through 2020. Supplements by NSF over the years initially funded the research. (See Magnuson, Chap. 9 in this volume, for details of this study).

Comparing Primary Production Measurements This was an early unsuccessful attempt to develop a common measure to compare across the entire network. The successes within the climate committee encouraged similar approaches in other core areas of LTER research. Sometimes we stumbled as we attempted to develop common measures across the sites. An ad hoc committee set up by the Coordinating Committee had the goal of determining a common measure for cross-site research on primary production.

Interest in comparing primary production by plants was high among LTER sites; after all it was important enough to be one of the five core areas for each LTER site to measure. As with the climate committee a first need was presumed to be to standardize measures of primary production so that comparative analyses could be conducted. Yet, what measure would work across the range of primary producers of forest, prairie, desert, river, and lake ecosystems? Leaf area index or litter fall might work for a forest, clipping above ground plant biomass for prairie and tundra, and C_{14} primary production and chlorophyll concentration might for algae in lakes. The

[25] D. M. Robertson, 'The Use of Lake Water Temperature and Ice Cover as Climatic Indicators." Ph.D. diss., 1989, University of Wisconsin, Madison.

committee met and concluded that there was no measure of primary production that could be used at all sites.

Clearly, another approach than standardizing a common measurement would be required to compare dissimilar ecosystems. This would be the case, not only for primary production, but for many ecosystem measures. This effort challenged us to identify common measures that could be used and to explore approaches to comparisons even when common measures, per se, did not exist.

Comparing Variability Among North American Ecosystems (VARNAE) Comparisons of variability analyzed dissimilar LTER sites that had dissimilar physical, chemical, and biological measures. VARNAE synthesized or generalized to the character of all ecosystems such as lakes, deserts, forests, or any other ecosystem type. Partial support came from the $60,000 provided by NSF to the Coordinating Committee in 1986; results were published from 1991 to 2006. The challenge was that few common measures existed across the diversity of ecosystem types. For example (as discussed above), we had been unable to find a common measure for primary production across the LTER sites. This would have been true for many other site-specific measures. We had to find common measures to compare between and among ecosystems.

The project was breaking new ground from the beginning. No site was to be excluded owing to its particular features and measures. Timothy K. Kratz and John Magnuson chose to ask questions about ecosystem variability across the network of LTER sites. This was a pragmatic decision. We thought that the coefficient of variation[26] could provide a useful comparative metric. What we requested from each LTER site for each variable was measurements in a grid of years and locations within each LTER site. From these data we could compute the relative variance (coefficient of variation) among years and among locations at all LTER sites.

Magnuson visited most of the sites and explained the study and what we needed from them. We had no network policy about data sharing at that time; some were wary about sharing. Sites were more willing to share when they realized we were going to publish analyses from derived data rather than the original measurements. The network had no centralized cross-site database; data arrived in the mail on floppy discs, or in a briefcase when Magnuson visited a site; some arrived by email. Literature on ecosystem variability was not abundant and the discussion sections of our first papers were brief.

Importantly, we had access to scientists who were thinking and writing about complexity and scale. Both Tim Kratz and Barbara J. Benson had recently received their Ph.D. degrees with Timothy F. H. Allen at Wisconsin who had published *Hierarchy: Perspectives for Ecological Complexity* (Allen and Starr 1982). At that time Tim Kratz served as the LTER field-site scientist at the Trout Lake Station and Barbara Benson as the LTER data manager in Madison. Their backgrounds and

[26] The coefficient of variation for a particular measure such as "leaf area index" or "fish abundance" is relative to the mean of that same measure at that LTER site. It is a relative measure of variation not dependent on the units of the measure and is expressed as a percent or proportion.

creativity significantly shaped the questions and the approach we took with VARNAE. Thomas M. Frost also led, especially on matters of scales of taxonomic aggregation (Frost et al. 1992).

We ran a network-wide workshop in 1988 at the Trout Lake Station in the Northern Highlands Lake District with 11 LTER sites and 1 non-LTER site; 18 researchers participated (Fig. 12.3). Tim Kratz did all the calculations of the 448 data matrices received in Excel 1.5 on an early Macintosh computer. The derived data were all available at the workshop and we self-selected questions on which to work and the subgroups analyzed the data on the spot. Magnuson found it to be an

Fig. 12.3 Photo of VARNAE workshop participants at Trout Lake Station in April 1988. 1st row left to right: Randy Dahlgren (Hubbard Brook), Thomas M. Frost (North Temperate Lakes), Dennis Heisey (North Temperate Lakes); 2nd row: Caroline S. Bledsoe (National Science Foundation), Stephen R. Carpenter (North Temperate Lakes), John J. Magnuson (North Temperate Lakes), John Yarie (Bonanza Creek); 3rd row: Barbara J. Benson (North Temperate Lakes), James C. Halfpenny (Niwot Ridge), Elisabeth R. Blood (North Inlet Estuary), Arthur McKee (Andrews Forest); 4th row: Donald W. Kaufman (Konza Prairie), Peter B. Bayley (Illinois River), Cory W. Berish (Coweeta), Gary L. Cunningham (Jornada); 5th row: John Anderson (Short Grass Steppe), Timothy K. Kratz (North Temperate Lakes), Richard C. Inouye (Cedar Creek). (Photo by C. Bowser; from Riera et al. 2006, on p. 111)

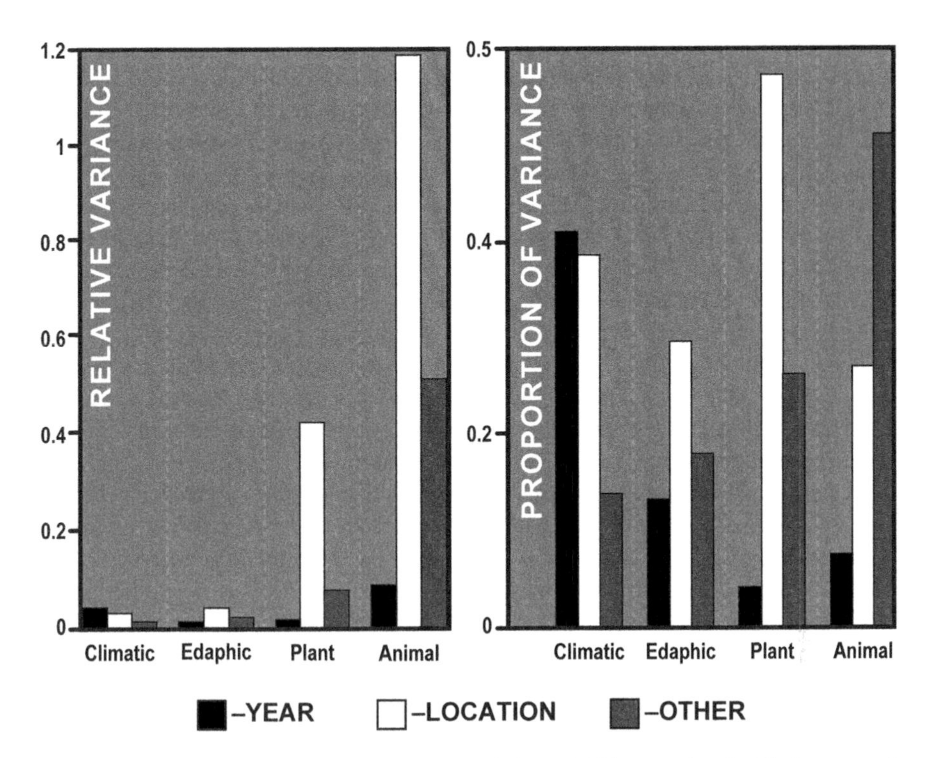

Fig. 12.4 Relative variance and the proportion of the variability associated with the years of record for that site, the locations within an LTER Site, and other sources (namely the statistical interaction between locations and years and measurement error). Values are compared for climatic (physical) variables, edaphic (chemical) variables, plant variables, and animal variables. Interaction measured the fact that the influence of location was not consistent among years. (Modified from Magnuson et al. 1991, on p. 57, and Riera et al. 2006, on p. 115)

exciting activity. The resulting papers included Kratz et al. (1991), Magnuson et al. (1991), Kratz et al. (1995), Magnuson and Kratz (2000), and Riera et al. (2006).

Measures of relative variance for animals was greatest, followed by plants; the variability was much less for edaphic and climatic measures (Fig. 12.4). Perhaps the most surprising result, considering the emphasis of LTER research on the years and long term, was that the variation among years was much less than variation among locations within an LTER for all three types of variability except for climatic variables (Fig. 12.4). Time clearly was an important component for the LTER research, but so were differences in location within LTER sites. Animals had the greatest variability resulting from the interaction between year and location. This suggested that animals could react more rapidly by moving in response to changing conditions and also were more difficult to measure precisely. Most specific conclusions made common sense as well. It made sense that lakes had a greater variability among locations than did deserts because lakes are island-like water bodies embedded in the landscape and would appear to differ more than locations down a continuous hillside in the desert.

Response of Forests to Hurricanes Comparing the response of forests to hurricanes was one attempt to understand the response of ecosystems to disturbance. LTER sites play an important function in understanding the role of disturbance because ongoing measurements exist before, during, and after a disturbance, rather than just after a major disturbance. Moreover, the long-term nature of LTER research allows comparison of multiple disturbance events and analysis of their interactions.

Hurricane Hugo struck Puerto Rico in 1989, one year after establishing the Luquillo LTER site. The same storm caused severe damage at the North Inlet LTER site in South Carolina and lesser damage at the Harvard Forest LTER site in Massachusetts. David Foster, principal investigator at Harvard Forest, was also a member of the research team at Luquillo, which facilitated comparative studies of hurricane damage in temperate and tropical forests (Boose et al. 1994; Boose 2003).

The goal was to understand the long-term ecological role of hurricanes by considering three related questions. How does wind damage: 1) vary in space and time, 2) change with disturbance intensity, and 3) change with repeated disturbance events? Boose (2003) analyzed historical hurricanes and reported field studies of ecosystem response in recent years. Interestingly, the most informative analyses were from the historical sets of hurricanes over the last 500 years – 52 hurricanes for New England and 143 for Puerto Rico. Overall impacts were greater for Luquillo Experimental Forest than Harvard Forest. The forests of both sites were remarkably resilient to wind damage, but visible effects persisted longer at Harvard Forest than at Luquillo.

Hurricanes were more frequent at Luquillo than Harvard Forest; strong storms occurred in 1997 (Georges) and 2017 (Irma and Maria). As a result, Luquillo has engaged in comparative studies of hurricanes with other LTER sites (North Inlet and Florida Coastal Everglades) as well as sites around the globe. An informal research working group produced special features in *Biotropica* (1991, 1996) and *Ecosphere* (2018). These special features highlighted research from the U.S. Virgin Islands, Martinique, Guadeloupe, Dominican Republic, Jamaica, Mexico, Nicaragua, Taiwan, and Australia as well as multiple sites in Puerto Rico. Collaboration with researchers working in ecosystems affected by hurricanes continues.

Comparing the Signal and Response to El Niño-Southern Oscillation (ENSO) This study analyzed whether the El Niño climatic driver influenced the air temperature and precipitation for each of the LTER sites. Climatologist Bruce Hayden from the Virginia Coast Reserve LTER organized a question-based workshop at the 1993 All Scientists Meeting in Estes Park, Colorado. This Climate Committee effort constituted an early consideration of El Niño's influences outside of the area in the Pacific Ocean where it is measured as the Southern Oscillation Index (SOI) (Greenland 1994, 1999). Hayden provided a primer on El Niño, with time series data on the Southern Oscillation Index. Six LTER sites presented papers, and Greenland provided a general discussion.

Later Greenland (1999) with climate data from 17 LTER sites analyzed correlations (1957 to 1990) between both monthly mean temperature and precipitation at the sites and El Niño and La Niña events indicated by the SOI; correlation analyses were lagged by 0–11 years. Major results were:

- Correlations were higher when a site was responding to both El Niño and La Niña events.
- Only 3 sites had strong correlations (r > 0.2) with SOI: Andrews in Oregon, Luquillo in Puerto Rico, and Palmer Station in Antarctica; 15 had moderate correlations; and 10 even had weak or no response. The correlations, while small, often were significant statistically owing to the large number of data points (years).
- Highest correlations occurred at lags greater than 1 month and up to a year.
- Significant correlations were more common for temperature than for precipitation.

Litter Decomposition Experiment in Terrestrial Ecosystems This experiment compared rates of decomposition across a wide range of temperature and precipitation. The Long-term Inter-site Decomposition Experiment Team (LIDET) examined decomposition and nitrogen accumulation in above- and below-ground litter (leaves and fine roots) in terrestrial ecosystems as controlled by substrate quality (pine vs hardwoods) and climate (broadly distributed sites) (Long-term Inter-site Decomposition Experiment Team 1995). One goal of the study was to improve the accuracy of estimates of global carbon dynamics in terrestrial ecosystems. All sites could participate provided that there were terrestrial soils on the site. Decomposition of leaves, needles, and roots was planned to continue from 1990 to 1999 at 27 sites of which 17 were LTER sites (Fig. 12.5). A few new LTER sites were added and some data were collected as late as 2007.[27]

An important early achievement was that the 27 sites rather rapidly settled on the questions to be addressed and were comfortable with a common set of measurements of decomposition. This differed starkly from the LTER Network's first attempts to standardize measures of primary production. The reason, of course, was that the sites compared, even though very different from each other, all had areas of terrestrial soils and the same measures could be made at all of them for this study.

The cross-site project was initiated by Mark Harmon at Andrews Forest with grants from the National Science Foundation. The project's first workshop took place at Woods Hole in May 1989. The 35 individuals divided into three working groups of field collaborators, laboratory analysts, and modelers. After each year of the experiment, each site was to make its data available to the overall project. They had a year to write site specific papers and after that the data was available for use by the entire LIDET group. A second workshop was held at Sevilleta Field Station in March 1996, to pull the analyses and papers together.

---

[27] Long-term Cross Site Decomposition Experiment, 1990–2002: http://andlter.forestry.oregonstate.edu/data/abstract.aspx?dbcode=TD023

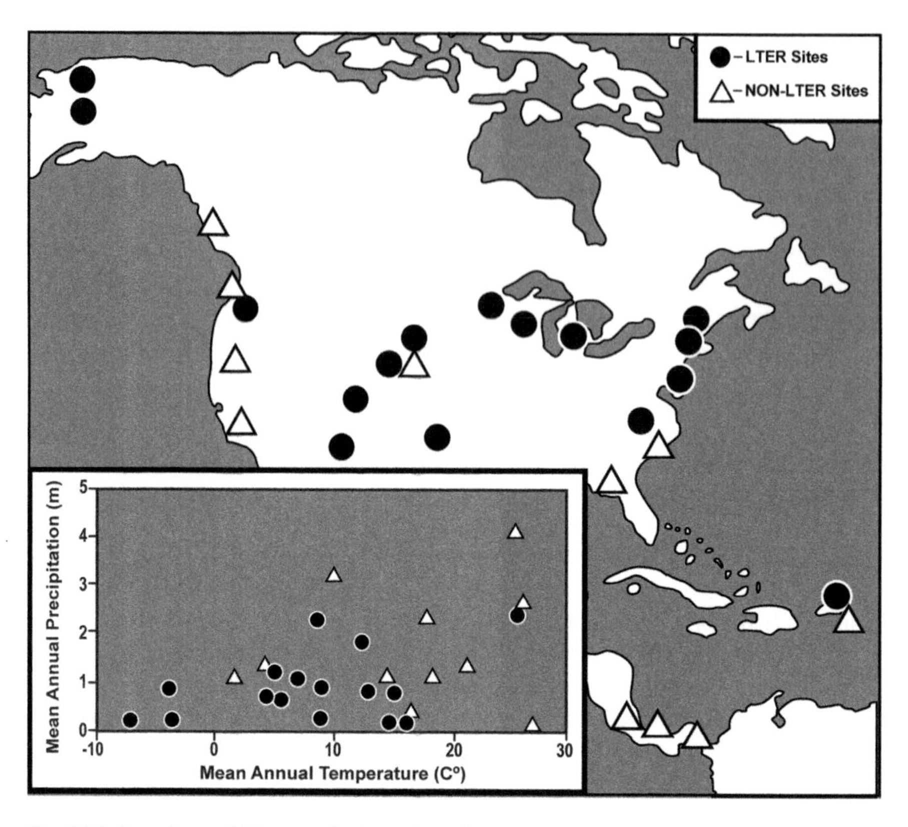

Fig. 12.5 Locations of 27 research sites where the same measures were made to do the Long-Term Inter-site Decomposition Experiment (LIDET) that tested the importance of climate on litter decomposition in terrestrial (usually forested) ecosystems. Analyses included 16 LTER sites and 11 non-LTER sites. Insert: the mean annual precipitation (Y) plotted against the mean annual temperature at each site (X). (Modified from Long-term Intersite Decomposition Experiment Team (LIDET) on pp. 7 and 9)

We make no attempt to review the total number of papers related to LIDET.[28] The publication by Gholz et al. (2000) summarizing the first 5 years provides an early look at the results. Two surprising conclusions were: 1. neither annual air temperature nor annual precipitation alone explained the global pattern of decomposition; derived variables that combined both temperature and moisture were better predictors of decomposition and 2. broadleaf litter decomposed much more rapidly than pine litter at sites with broadleaf trees compared to pine forest sites; authors called this a "home-field advantage."

Many of the later papers delved into global patterns of decomposition, nitrogen release, modeling global patterns of decomposition and nitrogen release, and the

[28] LIDET related publications, as of March 30, 2011: https://andrewsforest.oregonstate.edu/sites/default/files/lter/pubs/webdocs/reports/lidet/pub_list.htm

climate-litter quality paradigm of decomposition. Ten years of data from seven biomes found predictable global-scale patterns in net N release during decomposition driven by initial tissue N concentration and mass remaining rather than climate, edaphic conditions, or biota (Parton et al. 2007). In grasslands, temperature did not explain variation in root and leaf decomposition, but precipitation partially explained variation in root decomposition (Bontti et al. 2009). Models of decomposition and nitrogen loss varied in the accuracy of predictions for different biomes, suggesting that models must incorporate greater resolution of climate and litter quality controls as well as finer resolution of the relationship between carbon and nitrogen dynamics during decomposition (Moorhead et al. 1999).

The project was a direct response to John Brooks' 1986 challenge to the LTER Network and investigators. The experiment was designed to provide important scientific information, as well as, demonstrate the usefulness of long-term, broad-scale research. The use of a common measure in a cross-site experiment worked across the Network and beyond. It provided a grand example of comparative ecological science by the LTER Network.

Comparing Greenness Heterogeneity Using Satellite Images This study of greenness heterogeneity in LTER landscapes used satellite images to compare dissimilar ecosystems with a common measure. Sites had gained some experience with remote sensing at a 2-week, network training workshop on "Remote Sensing for Landscape Ecology" held in January 1987 in Albuquerque, New Mexico.[29] Later Magnuson used an NSF mid-career fellowship in 1992 from the Division of Environmental Biology to analyze the images with John Vande Castle at the LNO in Seattle.

The question was not "that now we have a new tool and how should we use it." Rather, the motivation was to be able to observe a diverse set of LTER sites with the same measurement tool. Thirteen Landsat images in which LTER sites were imbedded were observed with the same sensor, Landsat, and with a common parameter (greenness). Greenness was calculated for each 30-by-30 m pixel and the variation in greenness among pixels was calculated for the entire image (Vande Castle et al. 1995; Riera et al. 1998; Riera et al. 2006).

The heterogeneity (variation) in greenness within the 13 LTER Landsat images was compared with surface water pixels included or not included in the calculations. With Landsat, water has greenness value near zero. Thus, when water pixels were included, two of the sites, North Temperate Lakes and the Seattle area (site of the LNO at that time) became the most heterogeneous owing to Wisconsin's many lakes and Seattle's Puget Sound. This pointed out that looking at landscapes with both land and water in the analyses revealed dramatic information on the role of water areas contributing to the heterogeneity of a landscape.

---

[29] LTER January 1987 Remote Sensing Workshop. Limnology archives, Box 20E8B and 20E9C, in LTER Network, Limnology archives, Steenbock Library, Records and Archives, University of Wisconsin, Madison.

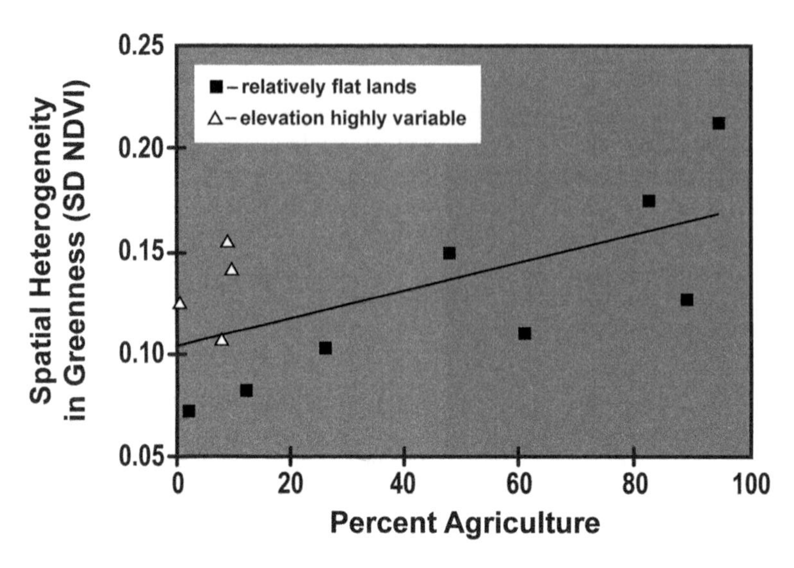

Fig. 12.6 Spatial heterogeneity in greenness is related to topography (variability in elevation) and landuse (% agriculture) measured from full Landsat satellite scenes containing an LTER site. (Water pixels had been filtered out before this analysis.) Spatial heterogeneity in greenness was measured by the standard deviation (SD) among pixels of an index of greenness (NDVI, Normalized Difference Vegetation Index). (Modified from Riera et al. 1998, on p. 276)

Finer patterns among the land pixels became more evident when the waters were removed from the analysis. For example, spatial heterogeneity increased with the percent of agriculture lands (Riera et al. 1998, 2006) (Fig. 12.6). For lands with little agriculture, greenness heterogeneity was greater for Landsat images that were less flat. Heterogeneity also differed with the spatial scale of that greenness; for agricultural lands the heterogeneity was at a fine spatial scale among fields with different crops, whereas for the mountainous areas it was greatest at the larger spatial scales of topography.

We thought that the variability we observed would have implications for the processes occurring on the LTER sites and landscapes. If so, these linkages have not yet been made. One thing Magnuson learned from this satellite-based study was that the technology was so complex that it required remote sensing professionals and ecological scientists, one to deal with the complexity of the tool and the other to ask and answer questions about ecological systems.

Species Diversity and Productivity The relationship between species diversity and productivity was compared among a diverse set of ecosystem types. In their 1995 meeting at Virginia Coast Reserve, the LTER Coordinating Committee (CC) discussed three topics that could be addressed at all sites: controls on biodiversity, trophic interactions, and food webs. A subsequent workshop on biodiversity at Cedar Creek LTER chose the relationship between diversity and productivity as a common project. With funds from the newly formed National Center for Ecological Analysis and Synthesis in Santa Barbara, California, Michael Willig and Bob Waide organized three meetings of a working group to examine theoretical predictions about the relationship between productivity and diversity in both producers and consumers using data from 18 LTER sites.

Products included 10 publications, two of which (Waide et al. 1999 and Mittelbach et al. 2001) have been cited more than 1000 times, and 12 presentations. The idea that there is a relationship between species richness and productivity (defined as the rate of conversion of resources to biomass per unit area (or volume) per unit time) goes back at least to the 1960s. One commonly held view was that the relationship of biodiversity and productivity was unimodal or hump-shaped across scales. That is, increases in productivity were associated with increases in species diversity only up to a certain point, and further increases in productivity were associated with decreased diversity. Unimodal patterns were by the 1990s understood to be among the few valid generalizations in ecology, and were considered to be textbook examples of productivity-diversity relationships. Studies by the LTER working group examined, and in some cases cast doubt, on those claims of universality.

A thorough survey of the literature found that the relationship between biodiversity and productivity in a broad range of ecosystems was highly variable and scale-dependent with respect to geographic extent, ecological extent, taxonomic hierarchy, and energetic basis of productivity (Waide et al. 1999). A meta-analysis showed that a positive monotonic relationship was pervasive in studies that comprise data from two or more biomes, whereas the relationship was most commonly unimodal (humped) when data derived from two or more communities within a biome (Mittelbach et al. 2001).

The relationship between species diversity and primary productivity at six grassland LTER sites was also scale dependent (Gross et al. 2000). Gough et al. (2000) found that nutrient enrichment in herbaceous communities decreased species density by a constant amount regardless of initial levels of productivity, suggesting that the unimodal pattern is unlikely to exist at these scales, in part because of the absence of the opportunity for an evolutionary response by the biota. Scheiner et al. (2000) posited the *pattern accumulation hypothesis* in which the between-community pattern of the diversity-productivity relationship is a simple accumulation of local effects and patterns (Fig. 12.7). If true, this hypothesis indicates that there is a general functional relationship between diversity and productivity that applies at both local and regional scales. Alternatively, if the relationship changes across scales, different mechanistic explanations of the pattern must operate at different scales.

Dodson et al. (2000) synthesized information on the interrelationships between species richness, area, and productivity for 33 well-studied lakes around the world, including eight from the North Temperate Lake LTER. Analyses of phytoplankton, rotifers, cladocerans, copepods, macrophytes, and fish uniformly revealed a unimodal response of productivity to richness when controlling for lake area. Mittelbach et al. (2001) suggested that the prevalence of unimodal responses in lakes resulted from the fact that lake studies tend to measure the species richness of an entire trophic level (e.g., primary producers, grazers, carnivores) whereas terrestrial studies generally measure species richness of taxa. The species richness of an entire trophic

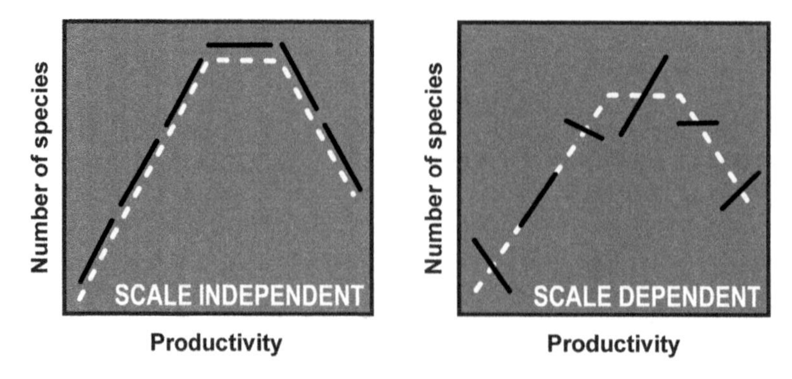

Fig. 12.7 Scale dependence in the relationship between species richness and productivity for a region. Solid line segments represent the relationships within a landscape whereas the dotted line shows the regional pattern. The pattern accumulation hypothesis posits that the regional pattern is simply the sum of the local patterns (scale independent). The regional-scale pattern differs from the local-scale patterns (e.g., Simpson's Paradox), suggesting that patterns at different scales arise from different processes (scale dependent). (From Scheiner et al. 2000)

level should relate more closely to changes in productivity than should a taxonomic group (e.g., birds) that spans multiple trophic levels.

In addition to improved understanding of diversity-productivity relationships, this coordinated effort led to insights about the challenges of comparative research. While sites with short stature vegetation (e.g., grasslands, deserts, and savannas) used similar techniques to measure productivity and diversity, forested sites employed a range of different methods, making comparison difficult if not impossible. Comparisons of productivity between terrestrial and aquatic sites was also problematic, and the existence of only a single lake site in the network required Dodson et al. (2000) to find comparative data from non LTER sites.

Financial and other support from NCEAS was a key factor in the successful completion of the biodiversity-productivity project. Moreover, LTER and NCEAS continued to pursue the diversity-productivity theme in two other joint projects. The Knowledge Network for Biocomplexity (KNB) and the Science Environment for Ecological Knowledge (SEEK) both used the diversity-productivity framework as a case study for developing information management techniques. As a result, many more investigators and graduate students were able to participate in working groups investigating the relationship between diversity and productivity with access to a much more extensive data base. This work has generated many additional publications and is an example of the importance of generous, long-term financial support in stimulating comparative ecological research.

Physical, Chemical, and Biological Features of Lakes The physical, chemical, and biological features were compared for lakes from high to low in the landscape down a hydrologic flow path. Timothy K. Kratz and Magnuson had noted that the LTER sites intuitively captured the spatial variability within their sites by sampling positions from high to low in the landscape. In some cases, the differences were from alpine to lowland forests, in others they were from locations down a catena in the desert, or from lakes down a groundwater flow path.

For the North Temperate Lakes site, the decision to sample lakes in respect to the groundwater flow paths from lakes high in the flow path to those low in the flow path was explicit (Magnuson et al. 2006b). Carl Bowser and Magnuson in 1980 had initially chosen lakes that differed greatly in color, depth, and area in the Northern Highlands Lake District of northern Wisconsin, but they were not in the same groundwater flow path. In June 1981, the First Advisory Committee for our site made a key recommendation on lake selection. In particular, Thomas C. Winter, a groundwater geologist, helpfully recommended that we alter our lake choice somewhat from a heterogeneous set of lake ecosystems scattered across the lake district to a heterogeneous set within the same, well-defined groundwater flow path. We did just that in choosing our primary LTER lakes. It was a good decision.

Measurements and analyses of our primary lakes soon characterized major differences sequentially down the hydrological flow path; lake area, ion content, and fish species richness all increased from high in the flow path (landscape) to low in the flow path (landscape) (Kratz et al. 1997). We did not view position in the landscape as a cause, but as a context for the processes that differed in effect down the flow path. This finding ended up being the case not only for the groundwater system we chose for our LTER primary sampling locations, but more generally for lakes across the entire Northern Highland Lake District (Fig. 12.8).

Landscape position became an important organizing context that we used as we evaluated differences in characteristics, coherence (synchrony), and variability at our LTER site. The same context persisted for lakes and reservoirs more broadly than for our LTER site (Soranno et al. 1999).

12.8 Meta-analyses of Cross-Site Research

The evolution of LTER research towards a new way of doing ecological science became apparent even during the first 20 years at the North Temperate Lakes site (Magnuson et al. 2006a, b). The number of authors per publication increased as did the number not in Wisconsin or North America. The changes were well underway during the second decade. Perhaps more significantly, the time and space scales increased markedly in extent as judged by number of years and the number of lakes analyzed per publication. Again, these changes occurred by the last half of the second decade. The collaboration with sites outside of Wisconsin and North America was the result of reaching out to lake researchers around the Northern Hemisphere.

This evolution in the way the LTER program at NSF changed ecological research was apparent across the entirety of LTER sites and persisted through the fourth decade of its existence (Johnson et al. 2010; Huang et al. 2020). The most commonly used metric to assess success of network-based science has been collaboration, measured by the number of joint publications and proposals among researchers affiliated with different sites (Johnson et al. 2010). The expectation is that

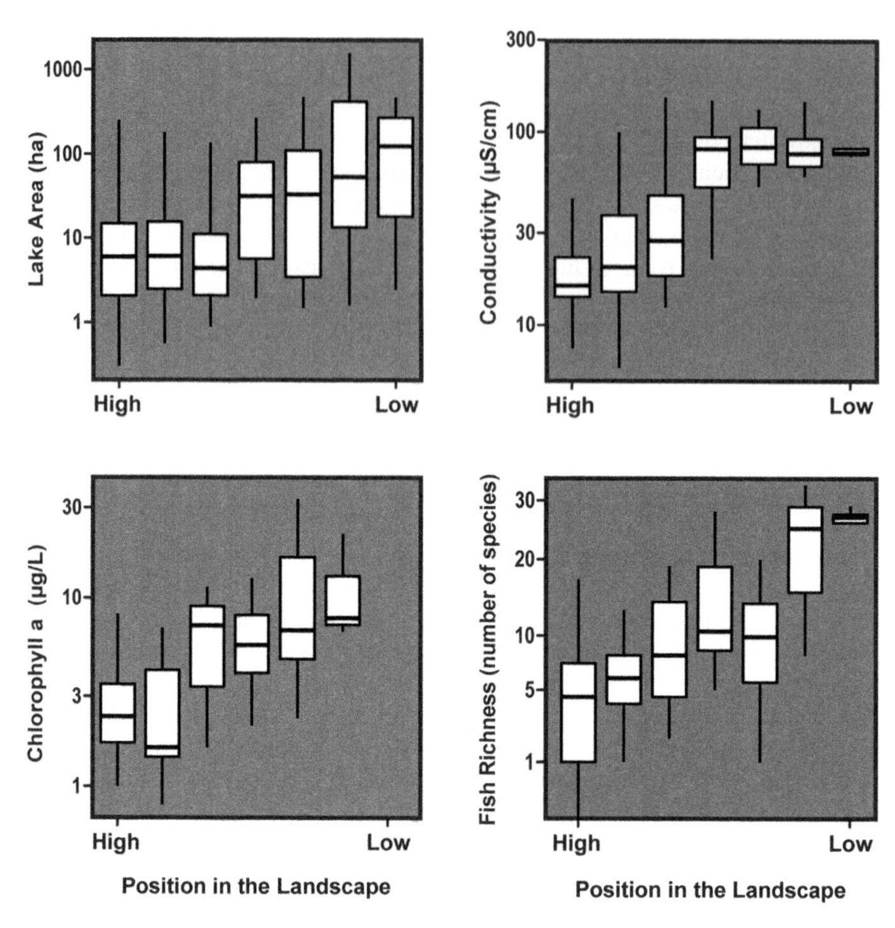

Fig. 12.8 Lake properties change sequentially from high in the landscape or groundwater flow system to low in the landscape or groundwater flow system. Lakes high in the landscape are smaller and have lower conductivity (dissolved solids), less chlorophyll (phytoplankton), and fewer fish species compared to those low in the landscape. Data on 556 lakes were from four catchments in Wisconsin's Northern Highlands Lake District. Individual boxes give the median and the 25th and 75 percentiles; the vertical lines for each box extend to include about 90% of the observations. The overall difference in values in each panel are statistically significant between high and low in the landscape, but not between adjacent pairs. (Modified from Kratz et al. 2006, on p. 59, and from Riera et al. 2000, on pp. 309, 310, and 313)

participation in a network will result in greater standardization of approaches, sharing of ideas, and collaborative research, and thus more joint publications.

Social network analysis showed that the LTER Network evolved from a loose collection of individual sites to a densely connected network (Johnson et al. 2010). Key measures in this analysis included the number of network components (where a completely connected network has only a single component), the number of collaborations with other sites (degree centrality), and the tendency of sites to associate with sites in similar ecosystems. At the beginning of the LTER program, the number of components was equal to the number of sites, but evidence of collaboration was seen beginning in 1985. The number of components declined over time, reaching complete connectivity for the first time in 2000. Thus, it took the LTER Network two decades to form a fully collaborating network, in part because new sites were being added to the network. On average the sites had 6 joint publications annually with other sites by 2000, compared with zero joint publications 20 years earlier. Surprisingly, the similarity of ecosystem was not a driving factor in the development of collaboration at the network level.

Collaboration (as measured by co-authorship) is influenced by many factors, the most important of which is interest in common scientific problems. Common scientific interest can be fostered by social, financial, or familial relationships, but in the LTER Network an important factor is membership in the clans defined by the scientific communities at each site. At new LTER sites, these communities undergo a period of adjustment during which interdependence develops and intra-site collaboration among individuals grows. Once sites are established (and generally after they have undergone a successful renewal), cross-site collaborations are more likely to develop. Cross-site collaborations are promoted by the LTER core areas, which define a different set of clans based on common interests and membership in professional societies. Cross-site collaboration is facilitated by organizational links (e.g., among Forest Service scientists at LTER sites), geographic proximity of sites, association of individuals with multiple sites, shared institutional affiliations, and collegial relationships developed as students or as supervisors of students. Another important factor is the dispersal of students from one LTER site to another after graduation. Some sites attempt to build cross-site relationships from the beginning of their programs. For example, the Luquillo site included scientists from the Andrews, Harvard Forest, and Coweeta sites in its proposed research team.

Recently, Huang et al. (2020) conducted an analysis of social networking in LTER using papers from 158 ecological journals published between 1980–2018. They focused on whether participation in LTER increased collaboration among research institutions rather than among LTER sites. Despite this difference in focus, results were similar to Johnson et al. (2010). LTER-related publications involved more collaborators and more institutions than other publications in the field of ecology. LTER collaborations also persisted for longer and covered greater distances. Network connectedness was initially high once the initial cohort of sites formed collaborations but declined as new sites joined the network. In 2005, after the number of sites had stabilized, nearly all sites were connected in a network.

Huang et al. (2020) found that collaboration among institutions did not increase until the second decade of LTER, like the pattern found by Johnson et al. (2010) for 25 sites, and by Magnuson et al. (2006a, b) for the lake site. A close examination (Fig. 12.9) confirms the suggestion made above that a period of intra-site collaboration precedes the development of cross-site collaboration. Between 1980 and 1995, the average number of collaborating authors on LTER papers averaged between two and three and was consistently greater than the mean number of authors on non-LTER papers. However, the average number of institutions is slightly above 1 for the same period, suggesting the dominance of intra-site collaboration among

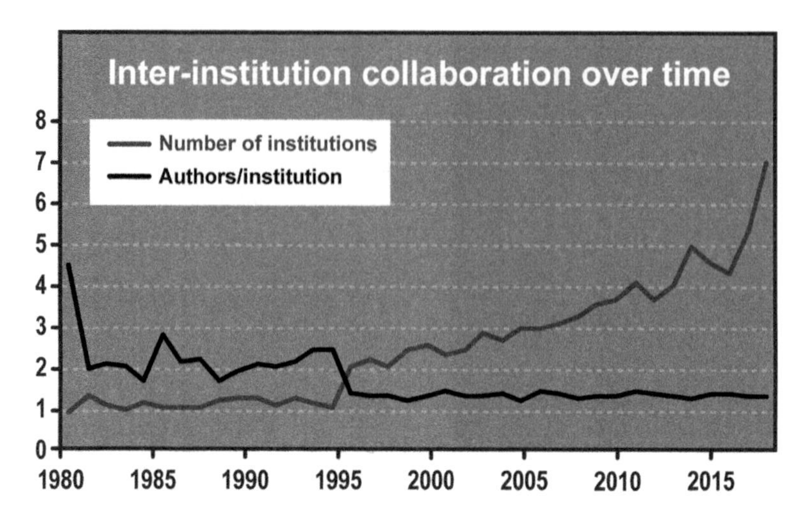

Fig. 12.9 Change in collaboration among LTER institutions from 1981 to 2018. Black = number of authors/institutions per article and Grey = number or institutions per article. The inflection point between 1995 and 1996 indicates a shift from intra-institution collaboration to inter-institution collaboration. This shift may be related to cross-site research competitions held by NSF in 1994 and 1995

colleagues at the same institution. Beginning in 1996, the average number of authors and institutions both increase more strongly for LTER papers than for non-LTER papers, indicating an increase in the proportion of cross-site collaborations.

Complex, interacting factors have enabled the evolution of the LTER Network from an aggregation of loosely connected sites to a highly integrated, collaborative network (Johnson et al. 2010; Huang et al. 2020). The long-term nature of the LTER program allowed the initiation and maturation of collaborative relationships between individuals and sites, but external pressure to form a network provided a consistent focus for LTER development. However, the formation of cooperative relationships between individuals has always been the intellectual force underlying the development of the network. Without the strange attraction that brings individuals from different backgrounds together to address problems of common interest, no amount of pressure from above could have created LTER as it exists today. Other intangibles stimulate collaboration, but their relative importance is hard to assess.

The emergence of grand challenges like climate change and loss of biodiversity has created a trend towards long-term, large-scale, multidisciplinary research. The challenge of forging a successful LTER research proposal established enduring partnerships aimed at achieving the goals of decades-long studies and experiments. These partnerships have survived even under the trend toward mobility between institutions. The longevity of LTER sites allows individuals to maintain research relationships over the course of their entire academic careers. Finally, the existence of a close-knit community of LTER scientists leads to what Huang et al. (2020) call the "friendship factor", which establishes the trust that allows scientists to share ideas, data, and credit.

12.9 Discussion

Since 1980 when the LTER originated, grouping of scientists into formal and informal networks has become more common, and for good reason. Collaboration in networks provides insights and generality not visible from site-based research. In addition, new technologies have provided improved mechanisms to make distance collaboration more efficient. Short courses are now offered over the web on how to form and work in a network setting. The comparative approach works and allows the creative use of minds, data, and tools that together create new information and knowledge and solve problems.

The impetus for comparative analysis and synthesis is emerging from the entire ecological community, but the LTER Network has played a key role in facilitating comparative ecosystem science. More frequently, data generated by LTER researchers are used in cross-site studies conducted by other communities of researchers. In such studies the data received from LTER sites are over longer time series on a greater diversity of measures than are available for most non-LTER sites, and they are backed up with group of committed LTER researchers. Clearly, the LTER Network has advanced comparative studies through its commitment to producing, documenting, and sharing long-term data.

Similarly, LTER researchers wishing to do a cross-site study on any particular ecosystem type typically necessitates networking with non-LTER sites. This is the case because the LTER network of 26 active sites in 2020 has only a few from any one ecosystem type. LTER sites include forests, grasslands, deserts, agricultural lands, urban areas, lakes, wetlands, coral reefs, and coastal and open water marine areas that are scattered across polar, temperate, boreal, and tropical regions.

Comparative analyses and syntheses are now common in the ecosystem sciences, especially of the same measures in the same ecosystem type. Typically, it has been easier for us to visualize the utility of comparing similar ecosystem types to broaden the temporal or spatial scales of analyses through networking among sites or data sources. However, comparisons or syntheses across a diversity of ecosystem types characterized by a diversity of measures remain rare and intellectually challenging both in the LTER Network and in ecological sciences more generally. Interestingly, the ecosystem types in the LTER Network do not always differ from each other in any absolute sense. For example, are forests a single ecosystem type or is a tropical forest a different type than a temperate forest? Or are lakes a single ecosystem type or is a brownish bog lake a different ecosystem type than a clearwater lake? On the other hand, a forest is clearly a different ecosystem type than a lake. This suggests that comparing differences and similarities among ecosystem types could provide an interesting and perhaps fruitful activity.

As we have learned through 40 years of LTER research, the relative importance of different ecosystem processes may shift over long temporal (White et al. 2006; Smith et al. 2009) and broad spatial scales (Scheiner et al. 2000). Moreover, processes operating at different scales may interact, further complicating cross-site comparisons. To better understand processes that vary across both sites and scales, researchers must broaden both the temporal and spatial focus of observations and analyses. This expanded focus is particularly challenging for a network dedicated to understanding the long-term dynamics of different ecosystems and requires a different set of comparative approaches. The search to find common measures across a diversity of ecosystem types has been a persistent challenge for LTER researchers.

The network of LTER sites and NSF deserve credit for helping move ecosystem sciences towards a new level of comparative analyses and syntheses across time and space. Organizations like the LTER Network, the NCEAS, the Global Lake Observatory Network, the International LTER Network, and other networks have provided data and approaches that facilitate cross-site analyses and syntheses. We believe that the values of long-term ecological research sites listed in 1982 by Marzolf and the urgent need for the LTER emphasized by Franklin, as discussed earlier in this chapter, have been borne out over the first 40 years of the LTER and remain true. Even given the uneven history of support for cross-site research described in this chapter, the LTER has steadily advanced comparisons and syntheses across broader space and longer time dimensions (Johnson et al. 2010; Huang et al. 2020). Through it all and because of it all, the LTER has become a key leader in advancing comparative long-term ecological research and synthesis through a network of research sites.

12.10 Epilogue: A Few Points to Keep in Mind

Comparisons and syntheses should be more than a shelf of books on dissimilar ecosystems. The same point can be made for the chapters in a book. We should connect and meaningfully compare and analyze the set of books or, similarly, interleaf the pages and the analyses in a synthesis book. We need to think of broader and perhaps less specific questions than those which we have been comfortable with the system we know best.

LTER researchers have advanced comparative ecosystem science over the last 40 years. Our success has been clear in comparisons of similar ecosystem types characterized by similar measures; these have been and will continue to be important. Our success in comparisons among different ecosystem types that are characterized by different features and measures has been less clear. Some approaches that we have used were presented above and others exist in the literature. Early in the LTER venture, several LTER scientists contributed to an analysis of comparative ecosystem methods (Cole et al. 1991), including a careful and creative analysis of comparative studies of "apples and oranges" ecosystems by Downing (1991).

We believe that intellectual and practical opportunities to explore a larger variety of approaches across dissimilar ecosystems exist, especially if we focus on major issues of the day such as climate change, species diversity and invasions, water shortages and contamination, sustainability and resilience, to name but a few. We should challenge ourselves to identify and ask the same questions across the diversity of ecosystem types from urban and agricultural systems to natural areas and wilderness or from desert to ocean.

Finally, we note that the LTER enterprise is intrinsically altruistic. By embarking on long term studies that will continue for decades, ecologists know that many questions they are asking may not be answered within their lifetimes and some important questions may not even be identified yet. Likewise, investigators whose contribution to a publication is buried among several or tens or even hundreds of other authors (e.g., Dornelas et al. 2018) may not receive the recognition that reflects their individual effort in making that contribution. Long-term comparative research and synthesis requires good leaders and dedicated participants who live by the idiom that you can accomplish almost anything if you do not need the credit.[30] Clearly, participants in the goals of the Long Term Ecological Research Network share this credo.

Acknowledgements We thank the National Science Foundation for supporting the LTER sites and the LTER Network beginning in the early 1980s. We recognize investments of time, talent, and data from the many colleagues and students across the LTER Network. NTL LTER program supported final figure preparation. We thank Jerry Franklin, Frank Harris, and Scott Collins for their insight on development of the LTER program, and Gina Rumore for sharing notes and interviews from her research. We also thank Sharon Kingsland and Emily Stanley for their many suggestions.

References

Allen, T.F.H., and T. Starr. 1982. *Hierarchy: Perspectives for ecological complexity*. Chicago: University of Chicago Press.

Bledsoe, C. 1993. *Ecological network of networks: Creating a network to study ecological effects of global climate change*. Washington, DC: U.S. National Committee on Man and the Biosphere.

Blowes, S.A., and 22 others. 2019. The geography of biodiversity change in marine and terrestrial assemblages. *Science* 366: 33945.

Bontti, E.E., J.P. Decant, S.M. Munson, M.A. Gathany, A. Przeslowska, M.L. Haddix, S. Owens, I.C. Burke, W.J. Parton, and M.E. Harmon. 2009. Litter decomposition in grasslands of Central North America (US Great Plains). *Global Change Biology* 15: 1356–1363.

Boose, E.R. 2003. Hurricane impacts in New England and Puerto Rico. In *Climate variability and ecosystem response at long-term ecological research sites*, ed. D. Greenland, D. Goodin, and R.C. Smith, 25–42. Oxford: Oxford University Press.

Boose, E.R., D.R. Foster, and M. Fluet. 1994. Hurricane impacts to tropical and temperate forest landscapes. *Ecological Monographs* 64: 369–400.

Carpenter, S.R., E.V. Armbrust, P.W. Arzberger, F.S. Chapin III, J.J. Elser, E.J. Hackett, A.R. Ives, P.M. Kareiva, M.A. Leibold, P. Lundberg, M. Mangel, N. Merchant, W.W. Murdoch, M.A. Palmer, D.P.C. Peters, S.T.A. Pickett, K.K. Smith, D.H. Wall, and A.S. Zimmerman. 2009. Accelerate synthesis in ecology and environmental sciences. *BioScience* 59: 699–701.

Chinn, H., and C. Bledsoe. 1997. Internet access to ecological information: The US LTER All-Site Bibliography Project. *BioScience* 47: 50–57.

[30]An alternate version of the original from an 1863 diary by the Jesuit Priest, Father Strictland, "A man may do an immense deal of good, if he does not need the credit.

Cole, J., G. Lovett, and S. Findlay, eds. 1991. *Comparative analyses of ecosystems: Patterns, mechanisms, and theories*. New York: Springer.

Coleman, D.C. 2010. *Big ecology: The emergence of ecosystem science*. Berkeley: University of California Press.

Collins, S.L., S.R. Carpenter, S.M. Swinton, D.E. Orenstein, D.L. Childers, T.L. Gragson, N.B. Grimm, J.M. Grove, S.L. Harlan, J.P. Kaye, A.K. Knapp, G.P. Kofinas, J.J. Magnuson, W.H. McDowell, J.M. Melack, L.A. Ogden, G.R. Robertson, M.D. Smith, and A.C. Whitmer. 2011. An integrated conceptual framework for long-term social-ecological research. *Frontiers in Ecology and the Environment* 9: 351–357.

Dodson, S.I., S.E. Arnott, and K.L. Cottingham. 2000. The relationship in lake communities between primary productivity and species richness. *Ecology* 81: 2662–2679.

Dornelas, M., N.J. Gotelli, B. McGill, H. Shimadzu, F. Moyes, C. Sievers, and A.E. Magurran. 2014. Assemblage time series reveal biodiversity change but not systematic loss. *Science* 344: 296–299.

Dornelas, M., L.H. Antao, F. Moyes, A.E. Bates, A.E. Magurran, and BioTIME consortium (200+ authors). 2018. BioTIME: A database of biodiversity time-series for the Anthropocene. *Global Ecology and Biogeography* 27: 760–786.

Downing, J.A. 1991. Comparing apples with oranges: Methods of interecosystem comparison. In *Comparative analyses of ecosystems*, ed. J. Cole, G. Lovett, and S. Findlay, 24–45. New York: Springer.

Frost, T.M., S.R. Carpenter, and T.K. Kratz. 1992. Choosing ecological indicators: Effects of taxonomic aggregation on sensitivity to stress and natural variability. In *Ecological indicators*, ed. D.H. McKenzie, D.E. Hyatt, and V.J. McDonald, vol. 1, 215–227. New York: Elsevier Science Publishers.

Gholz, H.L., D.A. Wedin, S.M. Smitherman, M.E. Harmon, and W.J. Parton. 2000. Long-term dynamics of pine and hardwood litter in contrasting environments: Toward a global model of decomposition. *Global Change Biology* 6: 751–765.

Gough, L., C.W. Osenberg, K.L. Gross, and S.L. Collins. 2000. Fertilization effects on species density and primary productivity in herbaceous plant communities. *Oikos* 89: 428–439.

Greenland, D. ed. 1987. *The climates of the Long-Term Ecological Research Sites*. Occasional paper No. 44. Boulder, Colorado: Institute of Arctic and Alpine Research, University of Colorado.

———. ed. 1994. *El Niño and Long-Term Ecological Research sites*. Publication No. 18, LTER Network Office. Seattle: US LTER Network Office, College of Forest Resources, University of Washington.

———. 1999. ENSO- related phenomena at Long-Term Ecological Research sites. *Physical Geography* 20: 491–507.

Greenland, D., and L.W. Swift. 1991. Climate variability and ecosystem response: Opportunities for the LTER network. *Bulletin of the Ecological Society of America* 72: 118–126.

Greenland, D., and L.W. Swift Jr. eds. 1990. *Climate variability and ecosystem response. Proceedings of a Long-Term Ecological Research workshop, Niwot Ridge/Green Lakes Valley LTER Site, Mountain Research Station, University of Colorado, Colorado, August 21–23. 1988.* General Technical Report SE-65. Asheville, NC: USDA Forest Service., Southeastern Forest Experiment Station.

Greenland, D., D.G. Goodin, and R.C. Smith, eds. 2003a. *Climate variability and ecosystem response at long-term ecological research sites*. New York: Oxford University Press.

Greenland, D., B.P. Hayden, J.J. Magnuson, S.V. Ollinger, R.A. Pielke, and R.C. Smith. 2003b. Long-term research on biosphere-atmosphere interactions. *Bioscience* 53: 33–45.

Gross, K.L., M.R. Willig, L. Gough, R. Inouye, and S.B. Cox. 2000. Patterns of species density and productivity at different spatial scales in herbaceous plant communities. *Oikos* 89: 417–427.

Halfpenny, J.C., K. Ingraham, and J. Hardesty, eds. 1982. *Long-term ecological research in the United States: A network of research sites*. Albuquerque: Long-Term Ecological Research (LTER) Network.

Huang, T., M.R. Downs, J. Ma, and B. Shao. 2020. Collaboration across time and space in the LTER network. *Bioscience* 70 (4): 353–364.

Johnson, J.C., R.R. Christian, J.W. Brunt, C.R. Hickman, and R.B. Waide. 2010. Evolution of collaboration within the US Long Term Ecological Research Network. *Bioscience* 60: 931–940.

Karl, T.R. 1985. Perspective on climate change in North America during the twentieth century. *Physical Geography* 6: 207–229.

Kratz, T.K., B.J. Benson, E. Blood, G.L. Cunningham, and R.A. Dalgren. 1991. The influence of landscape position on temporal variability in four north American ecosystems. *American Naturalist* 138: 355–378.

Kratz, T.K., J.J. Magnuson, P. Bayley, B.J. Benson, C.W. Berish, C.S. Bledsoe, E.R. Blood, C.J. Bowser, S.R. Carpenter, G.L. Cunningham, R.A. Dahlgren, T.M. Frost, J.C. Halfpenny, J.D. Hansen, D. Heisey, R.S. Inouye, D.W. Kaufman, A. McKee, and J. Yarie. 1995. Temporal and spatial variability as neglected ecosystem properties: Lessons learned from 12 North American ecosystems. In *Evaluating and monitoring the health of large-scale ecosystems*, ed. D.J. Rapport, C.L. Gaudet, and P. Calow, 359–383. New York: Springer.

Kratz, T.K., K.E. Webster, J.L. Riera, D.B. Lewis, and A.I. Pollard. 2006. Making sense of the landscape: Geomorphic legacies and the landscape position of lakes. In *Long-term dynamics of lakes in the landscape: Long-term ecological research on North Temperate Lakes*, ed. J.J. Magnuson, T.K. Kratz, and B.J. Benson, 49–166. New York: Oxford University Press.

Kratz, T.K., K.E. Webster, C.J. Bowser, J.J. Magnuson, and B.J. Benson. 1997. The influence of landscape position on lakes in Northern Wisconsin. *Freshwater Biology* 37: 209–217.

Long-term Inter-site Decomposition Experiment Team (LIDET). 1995. *Meeting the challenge of long-term, broad-scale ecological experiments.* Publication No. 19, LTER Network Office. Seattle: US LTER Network Office, College of Forest Resources, University of Washington.

Magnuson, J.J. 1990. Long-term ecological research and the Invisible Present: Uncovering the processes hidden because they occur slowly or because effects lag years behind causes. *BioScience* 40 (7): 495–501.

Magnuson, J.J., and T.K. Kratz. 2000. Lakes in the landscape: Approaches to regional limnology. *Internationale Vereinigung für Theoretische und Angewandte Limnologie* 27: 74–87.

Magnuson, J.J., T.K. Kratz, T.M. Frost, B.J. Benson, R. Nero, and C.J. Bowser. 1991. Expanding the temporal and spatial scales of ecological research and comparison of divergent ecosystems: Roles for the LTER in the United States. In *Long term ecological research: An international perspective*, ed. P.G. Risser, 45–70. Chichester: Wiley.

Magnuson, J.J., T.K. Kratz, and B.J. Benson, eds. 2006a. *Long-term dynamics of lakes in the landscape: Long-term ecological research on North Temperate lakes.* New York: Oxford University Press.

Magnuson, J.J., B.J. Benson, T.K. Kratz, D.E. Armstrong, C.J. Bowser, A.C.C. Colby, T.W. Meinke, P.K. Montz, and K.E. Webster. 2006b. Origin, operation, evolution, and challenges. In *Long-term dynamics of lakes in the landscape: Long-term ecological research on North Temperate lakes*, ed. J.J. Magnuson, T.K. Kratz, and B.J. Benson, 280–322. New York: Oxford University Press.

Mittelbach, G.G., C.F. Steiner, S.M. Scheiner, K.L. Gross, H.L. Reynolds, R.B. Waide, M.R. Willig, S.I. Dodson, and L. Gough. 2001. What is the observed relationship between species richness and productivity? *Ecology* 82: 2381–2396.

Moorhead, D.L., W.S. Currie, E.B. Rastetter, W.J. Parton, and M.E. Harmon. 1999. Climate and litter quality controls on decomposition: An analysis of modeling approaches. *Global Biogeochemical Cycles* 13: 575–589.

Parton, W., W.L. Silver, I.C. Burke, L. Grassens, M.E. Harmon, W.S. Currie, J.Y. King, E.C. Adair, L.A. Brandt, S.C. Hart, and B. Fasth. 2007. Global-scale similarities in nitrogen release patterns during long-term decomposition. *Science* 315: 361–364.

Peters, D.P.C. 2010. Accessible ecology: Synthesis of the long, deep, and broad. *Trends in Ecology and Evolution* 25: 592–601.

Peters, D.P.C., A.E. Lugo, F.S. Chapin III, S.T.A. Pickett, M. Duniway, A.V. Rocha, F.J. Swanson, C. Laney, and J. Jones. 2011. Cross-system comparisons elucidate disturbance complexities and generalities. *Ecosphere* 2 (7): art81. https://doi.org/10.1890/ES11-00115.1.

Peters, Debra P.C., Christine M. Laney, Ariel E. Lugo, Scott L. Collins, Charles T. Driscoll, Peter M. Groffman, J. Morgan Grove, Alan K. Knapp, Timothy K. Kratz Mark D. Ohman, Robert B. Waide, and Jin Yao. 2013. *Long-term trends in ecological systems: A basis for understanding responses to global change*. Technical Bulletin No. 1931. N.P.: USDA Agricultural Research Service.

Riera, J.L., J.J. Magnuson, J.R. Vande Castle, and M.D. MacKenzie. 1998. Analysis of large scale spatial heterogeneity in vegetation indices among North American landscapes. *Ecosystems* 1: 268–282.

Riera, J.L., J.J. Magnuson, T.K. Kratz, and K.E. Webster. 2000. A geomorphic template for the analysis of lake districts applied to the Northern Highland Lake District, Wisconsin, USA. *Freshwater Biology* 43: 301–318.

Riera, J.L., T.K. Kratz, and J.J. Magnuson. 2006. Generalization from cross-site research. In *Long-term dynamics of lakes in the landscape: Long-term ecological research on North Temperate lakes*, ed. J.J. Magnuson, T.K. Kratz, and B.J. Benson, 107–120. New York: Oxford University Press.

Risser, Paul G., Jane Lubchenco, Norman L. Christensen, Philip L. Johnson, Peter J. Dillon, Pamela Matson, Luis Diego Gomez, Nancy A. Moran, Daniel J. Jacob, Thomas Rosswall, and Michael Wright. 1993. Ten-year review of the National Science Foundation's Long-Term Ecological Research program. https://lternet.edu/wp-content/uploads/2010/12/ten-year-review-of-LTER.pdf. Accessed 31 May 2020.

Scheiner, S.M., S.B. Cox, M. Willig, G.G. Mittelbach, C. Osenberg, and M. Kaspari. 2000. Species richness, species-area curves and Simpson's paradox. *Evolutionary Ecology Research* 2: 791–802.

Smith, M.D., A.K. Knapp, and S.L. Collins. 2009. A framework for assessing ecosystem dynamics in response to chronic resource alterations induced by global change. *Ecology* 90: 3279–3289.

Soranno, P.A., K.E. Webster, J.L. Riera, T.K. Kratz, J.S. Baron, P.A. Bukaveckas, W. Kling, D.S. White, N. Cane, R.C. Lathrop, and P.R. Leavitt. 1999. Spatial variation along lakes within landscapes: Ecological organization along lake chains. *Ecosystems* 2: 395–410.

Vande Castle, J.V., J.J. Magnuson, and J Riera. 1995. Regional Ecosystem Comparison using a standardized NDVI Approach. In *The next step: Ninth annual symposium on Geographic Information Systems in natural resources management, Vancouver, British Columbia, Canada. Symposium proceedings,* vol. 2. Fort Collins: GIS World, Inc.

Waide, R.B., M.R. Willig, C.F. Steiner, G. Mittelbach, L. Gough, S.I. Dodson, G.P. Juday, and R. Parmenter. 1999. The relationship between productivity and species richness. *Annual Review of Ecology and Systematics* 30: 257–300.

White, E.P., P.B. Adler, W.K. Lauenroth, R.A. Gill, D. Greenberg, D.M. Kaufman, A. Rassweiler, J.A. Rusak, M.D. Smith, J.R. Steinbeck, R.B. Waide, and J. Yao. 2006. A comparison of the species-time relationship across ecosystems and taxonomic groups. *Oikos* 112: 185–195.

Xu, M. 2015. Ecological scaling laws link individual body size variation to population abundance fluctuation. *Oikos* 125: 288–299.

Chapter 13
A Retrospective of Information Management in the Long Term Ecological Research Program

Susan G. Stafford

Abstract This chapter describes the evolution of information management protocols for Long Term Ecological Research, starting with work conducted at the Andrews Forest Long Term Ecological Research site in Oregon in the 1980s. This early work involved the design, testing, and implementation of a data and information management system that helped establish standards and protocols across the Long Term Ecological Research Network. Following this initial work, a growth period ensued. The chapter discusses the creation of Ecological Metadata Language, explores the impact of the internet on ecological data management and shows how other countries adopted the same model for their long-term research networks. The Network Information System that was developed had broad applications to other projects, such as EcoTrends, the Environmental Data Initiative, and DataONE. Innovations in information management represent major and far-reaching accomplishments of the Long Term Ecological Research Program, and have influenced the entire field of interdisciplinary ecological research. They have helped to change the culture of scientific collaboration by making it both feasible and fair for scientists to share their data, and have in general promoted greater data literacy within the long-term ecological network.

Keywords LTER program · Long-term ecological research · Ecological networks · Data management · Data literacy · Data sharing · Information management · Cyberinfrastructure · Ecological Metadata Language

---

S. G. Stafford (✉)
Department of Forest Resources, University of Minnesota, St Paul, MN, USA
e-mail: stafford@umn.edu

R. B. Waide, S. E. Kingsland (eds.), *The Challenges of Long Term Ecological Research: A Historical Analysis*, Archimedes 59,
https://doi.org/10.1007/978-3-030-66933-1_13

13.1 Introduction

The establishment of the first Long-Term Ecological Research (LTER) site in 1980 coincided with the beginning of my professional career as a tenure-track, Assistant Professor in the Department of Forest Science, College of Forestry, at Oregon State University (OSU), 40 years ago. At that time, investments of time and resources in data and information management were almost non-existent. As the first Information Manager for the Andrews Experimental Forest (AND) LTER, this fortuitous timing provided a blank canvas on which to design, test and implement a data and information management system that helped establish standards and protocols across the LTER Network. My job was to help faculty, researchers and graduate students design statistically sound experiments, to manage their experimental data, to assist with analyses, and to ensure that statistically sound results were reported. Today, data management is taken very seriously and plays a very significant role within the LTER Network and in all proposals submitted to the National Science Foundation (NSF).

This chapter is organized into three parts. The first describes what I refer to as the evolutionary period in the development of data management protocols. The second describes the growth period during which our initial work in this field took hold and grew across the entire field of interdisciplinary ecological research. In the third section I recount my personal journey over the past four decades and conclude with some thoughts on what the future may bring.

13.2 The Evolutionary Period – How We Established Data Management Protocols

13.2.1 History of the Andrews LTER

The Andrews Experimental Forest had been in the data generation business for a long time before the Andrews LTER (AND LTER) site was first established. The AND LTER benefitted from the strength of the relationship between the OSU College of Forestry and the US Forest Service. The Andrews Forest was a recognized Man and the Biosphere (MAB) site and a key player (along with some of the other early LTER sites) in the International Biological Program (IBP).

NSF, the LTER network, and the scientific community learned many valuable lessons from the IBP about the importance and value of collecting, documenting, and managing data from long-term studies (Aronova et al. 2010). For example, I remember walking past a vault of IBP data (punched cards in those days) that had not been properly documented and archived. Consequently, that data had limited value to others unfamiliar with the original projects. These types of shortcomings seriously reduced the benefit derived from the initial investment – both financially and scientifically. From the beginning of the LTER program in 1980, there was a new policy that 15% of a site's budget should be dedicated to data and information management (Porter and Callahan 1994).

Fallout from IBP created high expectations for LTER data availability and documentation (later to be called "metadata") completeness on the part of both NSF and the scientific community. Although individual Principal Investigators complied with the new 15% budgetary requirement, each site's data and information management protocols were not standardized and coordinated to facilitate sharing of the data among all the sites.

For the same reasons, early progress towards standardization of methods among the first cohort of sites was stymied when the second cohort of sites was chosen in 1982. The rapid growth of the LTER program from 0 to 17 sites in 8 years complicated efforts to create a networked approach to information management.

The early LTER Network had a serious game of catch-up managing an enormous amount of legacy data at each site. Such data had been acquired and archived using a wide range of approaches. Sites collected data differently depending on the kinds of ecosystems being studied, and they used different terms to describe these data. Even within a site, the same process might be labelled differently by investigators studying streams versus forests. Much of this data was collected before "metadata" was even a word, yet the inclusion of these legacy datasets was invaluable and irreplaceable for both current and future generations of students and researchers. Historical practices and legacy data had to be integrated into a system consistent with the new information management protocols (Karasti et al. 2010).

13.2.2 Forest Science Data Bank and the Quantitative Sciences Group

To manage more effectively new and existing LTER data at the AND LTER, we took several innovative steps. First, we created the Forest Science Data Bank (FSDB) (Stafford 1998; Stafford et al. 1984, 1988) and established the Quantitative Sciences Group (QSG). In addition, we developed protocols (Stafford et al. 1986) for managing and archiving data that soon became prototypes for other LTER sites.

My goal was to fully integrate sound data management practices into the research process. To do so, we needed to include protocols from the beginning of a research study in a proactive manner, rather than retroactively (Stafford et al. 1986, 1994). The guiding principle of the FSDB was that documentation about the data was just as important as the data itself. Working alongside the researchers, at the beginning of a study, allowed for the co-design of data management solutions in parallel with the scientific process. This practice helped create an early trust in data stewardship and reinforced the importance of sound data management practices from the onset of a project. My experience in developing an information management system for the AND LTER site was repeated by Information Managers (IMs) at each of the sites.

The members of the QSG were nearly equally split between OSU and U.S. Forest Service (USFS) employees. This closely mirrored the productive relationship between the OSU College of Forestry and the USFS at the AND LTER. The first LTER Network Office was housed in the OSU Department of Forest Science. Jerry Franklin, a USFS scientist, held a courtesy faculty appointment in our Department and chaired the first Network Office. The integration of OSU and USFS personnel gave me, as QSG director, greater leverage in maximizing the benefit from our pooled resources. This was not necessarily the case at other LTER sites.

In 1994, Mary Clutter, Assistant Director of NSF's Biological Sciences Directorate (BIO), invited me to serve as visiting Division Director of Biological Instrumentation and Resources (Stafford 1996). During that time, I invited John Porter, Information Manager from the Virginia Coast Reserve (VCR) LTER site, to serve as a rotator Program Officer for Database Activities, a role for which he was very well suited and in which he was highly effective.

In my opinion, this Division was the best kept secret in the BIO Directorate, housing programs that related to data and research infrastructure for programs in the other Divisions within BIO, and was eventually renamed the Division of Biological Research Infrastructure. This interdisciplinary leadership opportunity was invaluable to me in many ways (Stafford 2016), bringing more attention to LTER IM as well as providing a springboard for me to serve on various Advisory Committees going forward. These included chairing the BIO Advisory Committee (BIOAC) as well as the Advisory Committee for Environmental Research and Education (ACERE). The ACERE served all of NSF and reported directly to the Director of NSF (Arden Bement at the time.)

13.2.3 Forging an Identity for Information Managers

One of the early challenges among the various LTER sites was building a community and a culture of collaboration. Bill Michener, Data Manager of the former North Inlet (NIN) LTER, and I wrote the first proposal to secure funds for an annual meeting of LTER Data Managers. These meetings were highly successful in bringing cohesion and a common vision to the initially small and disparate group of Data Managers (who would later be known as Information Managers). Michener and I emerged as the *de facto* leaders of the Information Management Committee (IMC). With NSF support, we institutionalized the annual Information Managers meeting. These meetings have persisted to this day (LTER Network News, http://news.lterent.edu) and are used as a model to create cohesion and singleness of purpose within a diverse cadre of researchers who value collegiality and professionalism (Stafford 2016).

These meetings became community forums for discussion of issues that were not being broadly addressed elsewhere. DataBits, first a printed then an electronic newsletter (https://lternet.edu/?taxonomy=document-types&term=databits),

chronicles these discussions. By the end of 2017, there had been approximately 35 annual meetings of the IMC (Henshaw 2018).

Because LTER IMs have come from various backgrounds and have varying responsibilities, there is no standard description of the position. Individual sites have managed IM positions in many different ways; as faculty (as in my case), as research associates, as technical staff, as computer scientists, or as other positions. Some IMs have advanced degrees, including PhD's and doctorates that make it easier for IMs to have faculty positions. Predictable backgrounds include ecology, statistics, or computer science, but IMs have also been drawn from the field of archaeology (Peter McCartney, former IM at the Central Arizona-Phoenix LTER and now Program Officer at NSF) and civil engineering (IM Tim Whiteaker at the Beaufort Lagoon Ecosystem LTER) to name only a few. The IMs are an eclectic bunch!

Moreover, because the technical expertise required of IMs often dictates high salaries, it is difficult to fit IM positions into the standard academic human resources model. I recall a situation at the AND, when I was able for the *first* time to secure NSF funding for a person (a UNIX System administrator) rather than software or hardware. An interesting call from OSU Human Resources ensued because I had posted a salary on the position description far in excess of what an incoming Assistant Professor would make. I explained that I had to pay this person that salary because that was what she could easily make on the open market. Our success set a new precedent for hiring positions and helped pave the way for other sites to write similar grants with similar success.

A perennial challenge for data managers has been their drive to get things done in the short-term rather than considering their own career trajectory. Within Universities, the role of data management typically was considered ancillary to existing job categories rather than as a wave of the future. The IM liaison roles that coordinated science, data and technology were difficult to convey in a market of specialists (Baker and Millerand 2007). Data managers have few colleagues (other than themselves) to teach them how to be project managers within existing power structures. They are, after all, in charge of data production, an endeavor distinct from but complementary to site-based knowledge production (Baker and Millerand 2010). It has been challenging for the LTER IM community to become more outward facing where their work would become more visible outside the Network.

13.2.4 Data Managers vs. Information Managers

In the early days, we were known as Data Managers. Over time, our title evolved to Information Managers. The term "information manager" first shows up in a 1992 report (https://lternet.edu/wp-content/uploads/2010/12/im_1992_report.pdf). This was a gradual transition and occurred for several reasons. First, the term "data manager" had menial connotations for many researchers who thought that managing data was simply a routine task, rather than the complex mix of tasks that we knew to be the case. Using the term "information manager" let us better define ourselves to the PIs rather than using an older, misleading label.

Second, this was near the advent of the use of Internet Information Servers (e.g. Gopher and shortly thereafter the World Wide Web). This meant that we were not dealing with data in the traditional sense of merely columns of numbers but rather text, images, bibliographies, and personnel databases, to mention only a few! Consequently, "information" was a far better fit than "data" for describing what we were dealing with. And lastly, the term "information manager" was showing up in other organizations.[1]

13.2.5 Building Collaboration

A collaborative approach emerged at the LTER IMC annual meetings that created a place of inclusiveness. Exposure to such diversity cultivated an understanding of and a sensitivity to a broad array of issues – issues that were not necessarily being addressed elsewhere. The genial attitudes of participants also fostered agreement on general standards and practices that were not as evident in cross-site scientific efforts. A sharing mentality developed for applications and data that eventually changed a very proprietary view of principal investigators' data to a more open and sharing perspective. This has now led to most of the LTER core data sets from all sites being available in federated systems such as DataONE (see below). It was through work within the LTER Information Management Committee that data managers began to learn that what might initially be perceived as an "individual trouble" may well be a "community issue" (Millerand et al. 2013).

It was always my intent to piggy-back the Annual IM meetings on larger scientific meetings (e.g., Ecological Society of America) to not only minimize cost, but equally important, to increase the visibility of IM efforts. Holding our meetings where large groups of ecologists were already assembling allowed us to open our discussions to a larger swath of the community. Guests were always invited and welcomed. The relevance of these meetings was evident in the number of requests for outside participation from the early 1990s into the 2000s (Henshaw 2018).

Co-locating these meetings in larger venues provided more opportunities for LTER IMs to present papers at larger conferences. To me, this was a way to provide a platform to encourage IMs to produce publications – the coin of the realm in academia – thus increasing their stock in the eyes of their PIs at their home sites. In the first 25 years, LTER IMs organized several successful symposia, resulting in several books and publications. Since the initial cohort of IMs were predominantly ecologists who also enjoyed working with data and computers, it made sense going to ESA and ESA-like meetings. Over time, as the IMs became more tech-savvy, conferences that were oriented towards computer science began to be more appro-

[1] J. Porter, personal communication, 22 September 2018.

priate. Regardless of the venue, the annual IM meetings created a level playing field among the great diversity of IMs. On more than one occasion, IMs have said that the single most important meeting they attended every year was the annual IM meeting.

The annual meetings helped the IM community to bridge the challenging divide between responsibility to the site versus responsibility to the Network. The generosity of spirit and openness that became a hallmark of the annual IM meetings facilitated efforts to work together toward common goals and solutions. More than any other group of LTER scientists, the IMs have consistently found ways to work together across site boundaries to forge viable partnerships and develop durable solutions to common network-wide challenges. The LTER IMs know how to play well together.

Technical advances in the LTER Network provide an example of how infrastructure for local site-based research can be configured as a distributed network in contrast to a centralized venue remote from where field data were generated (Karasti and Baker 2008). Parallel with advances in infrastructure, LTER forged a new kind of identity for scientists responsible for data stewardship. With an embedded data manager at each site, LTER grew its own workforce, one that took what was initially perceived by others as mundane work and unpacked it into a multi-faceted new kind of position shaped by an understanding of how to support both hypothesis–driven scientific inquiry and long-term data stewardship. In recognizing data's importance to science in the digital age, LTER data managers devoted their time, energy, and innovative thinking to data care (Baker and Karasti 2018).

The early data managers were faced with analyzing everyday data practices and carrying out the work of collective data management before the concepts of data repositories, data curation, and open data became part of the digital data scene (Karasti et al. 2006; Baker and Bowker 2007; Baker and Chandler 2008). The role of data management was emergent at a time when technologists and computer scientists thought in terms of standard technical solutions rather than designing processes adapted to the science and the times. For example, an LTER IM describing the data system at the Sevilleta LTER site complained:

> This solution has been called uninspiring, yet the fact remains it is a functional system that recognizes the way scientists work; it does not try to control the way they work. What scientists need from software and database engineers is fewer 'omnipotent' database packages and more tools to integrate existing software. (Brunt 1994)

I have described the great enthusiasm that IMs have for their responsibilities, but to be fair, that enthusiasm has not always been shared by all PIs. Some view the requirement for information management as a tax on their research dollars, and others consider the details of IM boring. At Coordinating Committee meetings, some PIs would roll their eyes and joke about going into a "data coma"[2] as the agenda turned to more technical topics. Part of the problem in this case was

[2] P. Groffman, personal communication, 14 June 2018.

communication, as IMs are accustomed to talking about their work in technical terms, and often have had a hard time translating their ideas into plain language.

Sometimes I questioned whether the true wealth of experience and technical knowledge represented by the LTER IM community was recognized and valued by the PIs and other LTER researchers. Today the LTER Network is at a point where the expectation for archiving data has become routine. However, individual IMs have had to devise their own approaches and ad hoc solutions for many sites because the industry and eco-informatics community investment in ready-to-use cyberinfrastructure for front-line environmental data management (sensor data management, Quality Assurance/Quality Control, metadata creation and management, etc.) has been sparse. IMs have created an on-line document library, all with the same general purpose of improving access to site information and facilitating easy navigation to information and data at the site, at other sites and across the LTER Network. Collectively, these resources provide a rich set of tutorial materials for incoming IMs to LTER as well as other information management professionals from other networks. In many ways, the IMs have served, and continue to serve, as a vanguard for new ideas and developments in an evolving technology.

13.2.6 The Challenge of Rapidly Changing Technology

It is easy to forget just how unsophisticated technology was in 1980, the year the LTER program was first funded. We didn't have many of the capabilities we take today for granted. It was "before WiFi, before the internet, before generic email, even before the first IBM PC. When LTER started, GenBank did not exist. When Amazon sold its first book and when Microsoft first shipped an operating system with built-in support for networking, LTER was already 15 years old. It was 20 years old when the DOT-COM bubble popped." (Robbins 2011).

Before LTER sites could function as a network though, they needed the technological capabilities to *be* a network. The establishment of a fully functioning network required equivalent infrastructure for communication, internet connectivity, and web access. Yet the technological capabilities across the Network were uneven, and NSF realized this. The North Inlet site and the Virginia Coast Reserve site were perhaps the most technologically advanced,[3] yet the overall strength of the Network was only as strong as its weakest link.

In 1988, NSF and the LTER Network defined the level of technology that needed to be available at all sites to allow robust interactions. This Minimum Standard Installation (MSI) included compatible Geographic Information Systems (GIS), local and wide area networks, and high capacity data storage systems. NSF awarded supplemental grants to sites to achieve the MSI across the Network. Moreover, NSF provided support to establish high throughput internet connections at field sites,

[3] R. Robbins, personal communication, 16 June 2018.

where connectivity was poor (Brunt et al. 1990). For the first time, NSF decided to provide collective support to sites for equivalent hardware and software platforms so that there would be comparable capability across the Network.

To achieve the MSI, NSF made available a pot of "new money." As a result, sites saw the mandate for data management as a *source* rather than as a drain of resources. This was a good example of using social engineering to accomplish things. By calling it "new" money and making it available as supplements to existing LTER sites, it transformed the role of data managers within the sites from cash sinks into a possible cash source. This required direct leadership from both within the Biological, Behavioral, and Social Sciences Directorate (the precursor to the Biological Sciences Directorate) from key individuals like David Kingsbury (Assistant Director), John Brooks (Division Director) and Tom Callahan (Program Officer) in concert with strong advocacy from LTER PIs, specifically Jerry Franklin and John Magnuson. Kingsbury was instrumental in creating the data-management supplement awards within LTER and those awards were crucial in shaping the improvement of LTER data management from 1987 onwards.[4]

This coordinated approach from both the NSF and key PIs helped guide the Network during its earliest years. NSF appeared to understand that they needed to exercise patience as LTER sites learned how to function successfully as a Network rather than as a collection of independent, strong-minded PIs. "Patience" was defined as allowing more time for the demonstration of results. As the nascent Network developed, interest was building in developing capability for obtaining spatially explicit data in the form of GIS. Early meetings between NSF and IMs determined that Arc Info would be the best software platform. A pattern was developing where NSF, in concert with IMs and domain scientists, collaborated to build the capabilities across the network in terms of computing, data storage, and analysis.

13.2.7 The Internet Impact

The development of the internet provided the opportunity for rapid communication via e-mail. In 1988, it became possible – with only a few keystrokes – to communicate with anyone and everyone within the LTER network if you knew their first initial and last name. Today the creation of an email alias system seems rather trivial, but the power of "sstafford@lternet.edu" or "im@lternet.edu" revolutionized how the LTER network functioned and helped facilitate a feeling of connectedness that heretofore had only existed for a few domain-specific groups.

The internet also provided the means for individuals and sites to share data, and the LTER Network embraced this opportunity by adopting a network-wide policy of making data accessible. This forward-looking approach anticipated the current requirement for data sharing by funding agencies and publishers and demonstrated

[4] R. Robbins, personal communication, 16 June 2018.

the leadership role that LTER assumed in providing open access to all publicly-funded data. It's been said "data sharing is not a natural state" (Robbins 2011), yet I submit that the LTER Network and early data sharing guidelines and policy have set the stage for the rest of biology. Guidelines for NSF's programs in BIO now require that there be a data management plan.

The LTER has been a leader in devising both technologies and policies to drive environmental data sharing (Porter 2010). The LTER Network supported long-term interdisciplinary projects and data sharing long before it became mandatory to do so, and LTER efforts pre-dated the data sharing mandate in the United States by 30 years.[5] This puts the remarkable vision of the LTER Information Management enterprise in greater perspective.

13.3 The Growth Years – Watching Our Work Take Hold

13.3.1 Challenges in Creating an Information System for Ecological Data

To assess the conceptual framework underlying long-term research at each site and to direct future research efforts, the LTER Network was charged with collecting, managing, and making accessible long-term ecological data collected over many sites using many different collecting techniques. These data needed to be described in detail, archived in perpetuity, and made discoverable by a broad scientific community. When the first LTER sites were selected, neither the approach nor the technology to achieve such goals existed. To address this challenge, the LTER IM community focused first on data collected at individual sites that needed to be shared and synthesized among investigators only at their individual sites. In many cases, site information systems were created from scratch, and as a result, a variety of information management solutions arose among the sites; see Brunt (1994) for an example. In the early stages of development of information systems, the IMs viewed this diversity as a strength because it allowed them to test and choose among different technical approaches to addressing common goals.

From 1990 to 2000 the emphasis began to shift from an IM strategy focusing on individual sites to a strategy that encompassed the entire LTER Network (Brunt 1999). Homogenizing technology, defining IM standards, and creating shared databases such as the Core Dataset Catalog (Michener et al. 1990) characterized IM efforts beginning in 1990. The development of the World Wide Web in 1991 and the first graphical web browser in 1993 (Porter and Brunt 2001) provided the tools to

[5] See memo of John P. Holdren, Director, Office of Science and Technology Policy, Executive Office of the President, to Heads of Executive Departments and Agencies, Feb. 22, 2013, on "Increasing Access to the Results of Federally Funded Scientific Research": https://www2.icsu-wds.org/files/ostp-public-access-memo-2013.pdf. Accessed 21 July 2020.

share data online. In 1994, the LTER Coordinating Committee agreed that all sites should post data sets online. The existence of online data facilitated comparison and synthesis across LTER sites and provided important input to the IMs. These collaborations encouraged the development of a more ambitious approach to building an LTER Network Information System (NIS) beginning in 1996.

13.3.2 The Decade of Synthesis (2000–2010)

The LTER Network announced a plan for a Decade of Synthesis from 2000–2010. This plan formalized the goal of sharing data and metadata broadly to facilitate "regional, national and global syntheses, thus providing a resource for the broader scientific community" (http://lternet.edu/wp-content/uploads/2010/12/lter_2010.pdf). This bold step envisioned collaboration and data sharing at an expanded scale, and this vision set the parameters for the developing NIS. Expanding the scope and goals of the NIS, coupled with rapid changes in technology and cyberinfrastructure capabilities, required close cooperation among IMs, domain scientists, and data engineers both at the LTER Network Office (LNO) and at collaborating institutions.

13.3.3 Data and Metadata Standards and the Long-Term Preservation of Data

Integration of data from different sources is one of the most common and frustrating challenges in ecology, and much thought and effort have gone into addressing this challenge. The decision to share data implied that scientists must also provide descriptions of their data and the methods used to collect them. Descriptions of data, or metadata, were more useful if they conformed to recognized standards, but at that time such metadata standards did not exist for ecological data. The next challenge for LTER IMs, therefore, was to develop standard approaches to formatting and describing data so that new users could interpret those data.

Issues with metadata content and data structure inconsistency and comparability potentially inhibit data discovery and re-usability. From the earliest days of LTER, the IMs have emphasized "standards" rather than "standardization." Some ask why we need such high standards for metadata. The answer is simple. If one knows how data were collected, one can later replicate a study and produce data to be compared to the earlier data. Strong metadata is a value-added component to the data. High standards for metadata help insure that data will be able to be used in perpetuity by future researchers totally unfamiliar with the original data collection effort and not associated with the original research project that generated the data in the first place. Reproducibility is particularly important for the LTER program because long-term studies are conducted by successive generations of scientists.

Metadata is one of those elements that requires the researcher and IM to "pay it forward", i.e. the more time and effort that is invested up front, the greater the value and payoff in the future. This asymmetry of initial cost vs. future benefit, however, can be problematic. Michener et al. (1997) have noted "although increasing metadata structure reduces the burden on data re-users, it significantly increases the burden on the data originator." The maximum benefits accrue to the future user. In this context, one could ask the question: "What does the present owe the future?" In my opinion, the failure to make this initial investment answers the question incorrectly.

13.3.4 Homogeneous Data

One simple approach is to standardize data formats and units to simplify comparison. Thus, instead of constantly transforming measurements from English to metric units, a better solution is to establish a standard so that all data are provided in the same units.

LTER IMs used this approach to create common databases of meteorological (named CLIMDB) and hydrological (named HYDRODB) data in which each site provided data and metadata in standard formats. A similar approach was effective in creating databases on site personnel (PERSDB) and site characteristics (SITEDB) as well as a basic catalog of data collected at each site, Data Table of Contents (DTOC) (Brunt 1999). IMs worked with domain scientists to develop these early databases, which used different technological approaches for comparison. These databases served as prototypes for the development of the LTER Network Information System (NIS) (Brunt 1999).

13.3.5 Heterogeneous Data

Other kinds of ecological data are more heterogeneous and less amenable to standardization. For example, measurement of primary productivity uses completely different methods in forests, grasslands, and lakes and the data are expressed in different formats. Because of the great diversity of formats, the focus therefore was on developing metadata standards that would facilitate the development of software tools to produce data in required formats (Servilla et al. 2016) to foster the exchange of data between sites.

The development of metadata exchange standards was a challenge that reached beyond the boundaries of the LTER Network. The Ecological Society of America convened a committee on the Future of Long-term Ecological Data (FLED) (Gross and Pake 1995) that addressed similar issues.

At the 1992 IM meeting, discussions of data and metadata standards resulted in articulating a vision for using machine-readable metadata to facilitate data manipulation and sharing. If data are described with detailed metadata that can be interpreted by computer programs, then software can be written that can analyze data stored in any format. To achieve this vision, metadata that were mostly written in plain text would need to be converted to something that a computer could interpret. This challenge required the development of a descriptive language that made sense to both humans and computers.

13.3.6 Ecological Metadata Language and Partnership for Biodiversity

The development of such a language was undertaken by the Partnership for Biodiversity Informatics (PBI; http://pbi.ecoinformatics.org/), a consortium of five institutions including the National Center for Ecological Analysis and Synthesis (NCEAS), the LTER Network Office, the San Diego Supercomputer Center, the Natural History Museum and Biodiversity Research Center at the University of Kansas, and the California Institute for Telecommunications and Information Technology. The goal of PBI was to enable scientists and other users to deploy vast amounts of ecological, biodiversity and environmental information in research, education and public service in order to help society achieve the means to safeguard our future and a sustainable planet. Ecological Metadata Language (EML), based on work initiated by FLED and described in Michener et al. (1997), was formalized through the Knowledge Network for Biocomplexity (KNB), a PBI project funded by NSF. LTER IMs and domain scientists played an important role in the development of version 1.0 of EML (Fegraus et al. 2005) by helping to define data needs of LTER research projects associated with KNB.

The PBI partners initiated two other projects aimed at improving the software and infrastructure available to support ecological and biodiversity science. SEEK (Science Environment for Ecological Knowledge) was a comprehensive knowledge management project for ecological and biodiversity science. RDIFS (Resource Discovery Initiative for Field Stations) initiated collaboration on software infrastructure for field stations and informatics training for field station personnel. LTER IMs and domain scientists continued to play a leading role in these projects by providing training, setting priorities, and testing developing cyberinfrastructure.

EML thus addressed a critical challenge in fulfilling the vision put forth by LTER IMs in 1992. EML was designed specifically for use with ecological data through a collaboration among ecologists, IMs, and software engineers. EML provides a common structure to describe ecological data so that subsequent generations of scientists can accurately interpret the data. Because EML is a machine-readable language, it allows researchers to develop software applications to search for, acquire, manipulate, integrate, and analyze data distributed through data archives connected via the internet.

Some researchers believe EML should establish a base-level of metadata standards while others advocate the benefits of having a higher level that captures the semantic elements to fully automate integration and analysis. A new version of EML may address some of these issues – but this would make EML metadata even more time consuming to prepare. Not surprisingly, LTER IMs tend to be more supportive of the idea that we need more metadata, not less, contending that high-quality, archival data will be used by researchers in the future.

EML has become the standard for documenting and describing in detail ecological data and their characteristics not only for the LTER Network, but also for many other national and international programs. Researchers are now able to find and use LTER data for their research using the Network Information System and accompanying EML metadata. By having more ready-access to data archives connected via the internet, researchers world-wide can now work together more effectively (see Wolkovich et al. 2012, Dornelas et al. 2014, Zhang et al. 2017, Collins et al. 2018, Hautier et al. 2018, Rodriguez et al. 2018, Sillett et al. 2018, and Song et al. 2018 for examples). In addition, Servilla et al. (2016) lists over 50 articles which cite LTER data as a basis for their research.

At the AND LTER site alone, an educated guess is that 20% of publications from that site come from work that they didn't participate in but which were made possible by having the data available on-line.[6] As ways are implemented to reward data creators and incentivize data publication (Kratz and Strasser 2015), the re-use of data should only increase. While some may contend that researchers find data from publications, not data banks, regardless of the route to the data, data can now be found with requisite metadata and can be used by researchers not associated with the original research.

13.3.7 Drupal Ecological Information Management System

A good example of the collaborative spirit of LTER was the development of DEIMS (Drupal Ecological Information Management System). DEIMS is a grass-roots, Drupal-based system that provides a unified framework for ecological information management for LTER sites, Biological Field Stations and other similar groups. Drupal is a free software package that allows an individual or community of users to easily publish, manage and organize a wide variety of content on the web. One needn't be a database expert to use DEIMS.

Marshall White at LNO recognized that the Drupal Content Management System represented an opportunity to build a form-based EML editor and thus to simplify the conversion of text metadata to EML and the ability to integrate content of different types of data (e.g., people and publications). In the words of one LTER IM, Kristin Vanderbilt, "It was slick." The success of the original DEIMS system (DEIMS 1) interested other IMs, and a group of IMs decided to pool their IM supplement grants to hire a team of Drupal developers to transfer DEIMS into Drupal

[6] J. Jones, personal communication, 7 October 2018.

7. DEIMS 2 is now being used at eight LTER sites. It has been adopted at a few field stations, as well, and has also been used by national networks in other countries. The aim was to develop an information management system that could be used to manage all the disparate kinds of information that a site generates.

Besides the technical aspects of this accomplishment, the behind-the-scenes collaboration that occurred cannot be overlooked or understated. Given that the aim of DEIMS was to develop an information management system that could be used to manage all the disparate kinds of information that a site generates, three sites and the LNO pooled their resources for a project that benefitted many others, representing an effort to harmonize data management across several sites. Although it is still something that needs work, it would not have been possible without the generosity of spirit and collaboration between IMs and the LTER Network Office.[7]

13.3.8 Managing Disparity Between Expectations and Resources: Development of the Network Information System (NIS)

Any LTER PI will tell you that the cost of maintaining a long-term ecological program in the face of increasing expectations is underestimated. The perennial tension between resources and expectations is particularly keen in the area of information management. Infrastructure is a funny thing – when it works, nobody notices and when it doesn't, everyone knows. The Report of the Twenty-Year Review of LTER (Harris and Krishtalka 2002) was quite emphatic about this tension when it stated:

> Increased NSF investment in the LTER informatics infrastructure is a particularly critical need. In addition to fostering synthesis science, *it will help reverse the perception that informatics is an "add-on" rather than a fundamental component of ecological research.* According to statistics from NSF's Division of Biological Infrastructure, research projects in biology allocate an average of 5% of resources to informatics when the actual need is between 35–40% of total project costs. LTER science, which is data intensive, exceeds this average, allocating approximately 10%–20% of total project costs to informatics depending on the site. Still, this short-changes informatics, which has diminishing returns in the long run, resulting in information that is less capable of integration, analysis, synthesis and prediction by LTER and other scientists. (Emphasis added)

> An appropriate level of investment in an informatics infrastructure for the entire LTER community will be cost-effective and achieve economies of scale for NSF and the LTER program. Part of the increased investment in informatics should target the management and maintenance of LTER data, an irreplaceable asset for current and future research and applications. LTER data are no different in this respect from federal census data, remote sensing and genomic data, taxonomic and culture collections and other national archives.

The report of the Twenty-Year Review highlighted the most significant problem standing in the way of creating a centralized LTER data system. The development

[7] Kristin Vanderbilt, personal communication, 28 March 2018.

of such a system would require significant inputs of time from an already overtaxed community of IMs as well as the participation of software engineers and programmers that did not exist in the LTER Network at that time. The task of defining the scope and capabilities of the Network Information System (NIS) had already been initiated by the IMC in collaboration with domain scientists, but progress was slow because of the complexity of the project and the absence of dedicated personnel. The philosophy behind the early development of the NIS was to use requirements for research synthesis to define NIS capabilities (Brunt 1999). This philosophy required IMs to participate in LTER cross-site and synthesis working groups and to gather requirements for the new NIS from participants. Then IMs needed to propose means of addressing these requirements and develop and test prototype solutions. Because of the technical complexity of these tasks, disproportionate responsibility fell on the most technically-capable IMs.

Two fundamental principles guided early development of the NIS prior to the Twenty-Year Review (Brunt 1999). The LTER NIS would focus on integrating site information systems, not replacing them. The NIS would be a dynamic system, evolving to incorporate technological advances and improvements in understanding of how scientists most efficiently use data. With these principles in mind, the IM community defined three goals for work on the NIS at the turn of the millennium: (1) the LNO would work to improve the utility of the existing components of the NIS (e.g., the data catalog); (2) an NIS working group composed of IMs and staff from the LNO would improve access and query capabilities for inter-site data and (3) diverse solutions to problems in design and implementation would be encouraged to take advantage of the variety of approaches developed at sites.

In preparation for the Twenty-Year Review, the LTER Network underwent an intensive self-analysis "to refresh and to update the overall aims and mission of the LTER Network". This analysis led to a document titled "LTER 2000–2010: A Decade of Synthesis" that laid out six goals for LTER:

- Understanding: Gaining ecological understanding of a diverse array of ecosystems at multiple spatial and temporal scales
- Synthesis: Using the network of sites to create general ecological knowledge through the synthesis of information gained from long-term research and development of theory
- Information: Creating well-designed, documented data bases that are accessible to the broader scientific community
- Legacies: Leaving a legacy of well-designed and documented long-term observations, experiments, and archives of samples and specimens
- Education: Using the uniqueness of the LTER programs and network to promote training, teaching, and learning about long-term ecological research and the earth's ecosystems
- Outreach: Providing knowledge to the broader ecological community, general public, resource managers, and policy makers to address complex environmental challenges.

These broad goals provided guidance for the further development of the NIS, but more specific strategic direction was needed. In 2002, the Executive Committee, responding to a recommendation from the IMC, created a NIS Advisory Group (later Committee; known as the Network Information System Advisory Committee (NISAC)) to draft a long-term strategic plan for the development of the NIS. Membership in this group was drawn from the Coordinating Committee, the IMs, and the LNO. Planning for the NIS was deliberate, with emphasis on prototyping and evaluation of different technologies. This strategy arose both from the recognition that it would produce a stronger NIS in the long-term, but also from the reality that IMs had only limited amounts of time given their other responsibilities. The planning effort was inclusive and long-term, and encouraged joint planning by information managers, domain scientists and LNO staff. All groups engaged in iterative cycles of software development and assessment across the network with teams of domain scientists, IMs, and graduate students. The process emphasized information exchange among groups to inform the next steps of development, thus building a "community of practice" (Karasti and Baker 2004).

The LNO provided reinforcements to the effort in 2003 by hiring a software engineer and a programmer, but the amount of work to be accomplished was still daunting. Researchers, IMs, and LNO staff formed a series of working groups focused on designing, funding, and implementing key components of the NIS. These working groups addressed issues of data standardization that arose from synthesis efforts, methods of data integration, interoperability with other national standards, guidelines for data and metadata longevity, and other topics. As issues were resolved, the LNO and IMs integrated solutions into the developing NIS infrastructure. In 2005, the LTER Coordinating Committee approved the LTER Network Information System Strategic Plan produced by NISAC (http://lternet.edu/wp-content/uploads/2010/12/ApprovedNISStrategicPlanVersion2.9.pdf). The 2007 LTER Network Cyberinfrastructure Strategic Plan projected the infrastructure needed to achieve the goals of the NIS Strategic Plan (http://lternet.edu/wp-content/uploads/2010/12/LTER_CI_Strategic_Plan_4.2.pdf). Finally, the 2011 LTER Strategic and Implementation Plan formalized the steps necessary to implement plans for the NIS (http://lternet.edu/wp-content/uploads/2010/12/LTER_SIP_Dec_05_2010.pdf).

By 2008, planning for the NIS was largely completed, but progress on implementation of the plans was still slow. Bob Waide, Executive Director of the LNO, made the decision to request additional resources to complete work on the NIS as part of the LNO's renewal proposal in 2009. NSF was amenable to the request, but eventually declined the proposal for additional funds because of budgetary constraints. The economic downturn and the subsequent stimulus package, however, provided another opportunity for funding. Ultimately, the LNO received an additional $2 million over 4 years to advance the NIS from funds made available under the American Recovery and Reinvestment Act of 2009 (ARRA).

This surge in funding allowed the LNO to recruit additional software engineers, to accelerate the work of information managers responsible for key components of the NIS, to simplify the process of encoding metadata in EML, to acquire key cyberinfrastructure, and to support working groups looking for solutions to outstanding technical problems. In January 2013, a functional version of the NIS that incorporated the major capabilities required by LTER scientists went live (Servilla et al. 2016).

Subsequently, a team of NIS developers including IMs and staff from the LNO continued to improve the software package underlying the NIS, dubbed the Provenance Aware Synthesis Tracking Architecture (PASTA). These improvements added requested features to the software, some of which were only identified once the NIS was in use. The basic functions of the NIS are to harvest data and metadata from each LTER site on a regular basis and to archive these data in a repository at the University of New Mexico. Data users can then browse or search for data from all LTER sites through a single data portal. Each package of data and associated metadata are assigned a unique Digital Object Identifier (DOI) so that they can be distinguished from every other data package in existence. Updates to a data/metadata package are assigned a new DOI, so different versions of the same data stream are identifiable. The PASTA software records the relationship between original and updated data packages (Provenance Aware). New data packages derived from the integration of one or more data packages during analysis are given a separate DOI, and the software keeps track of the data packages used in analysis (Synthesis Tracking). Thus, the metadata for each synthetic data package has all the information needed to replicate the analysis.

LTER developed a controlled vocabulary (http://vocab.lternet.edu/vocab/vocab/index.php) but the lack of uniformity of keywords across LTER sites made the ability to conduct more sophisticated keyword analyses problematic. In 2016, data were described by over 6000 unique keywords. As with all keywords, the hope is that a small number can be used to describe and organize a collection of items, in this case datasets. Unfortunately, this has not been the case. Of the roughly 6000 keywords, 2498 have been used *to describe only one data set* (Servilla et al. 2016). In hindsight, limiting the number of keywords to be used would have improved the usefulness of using a controlled vocabulary.

Other features of the software facilitate data management or use (Servilla et al. 2016). For example, each data package submitted is subject to a series of checks that determine whether the metadata accurately describes the structure of the data. Any mismatch is reported to the data provider. An open programming interface allows data users to write software to download data in desired formats.

Some LTER scientists were concerned that they would not get credit for re-use of their data, and this concern prevented some individuals from contributing data to the NIS. This concern was addressed by using DOIs to identify all data in the NIS, which provided a way of tracking data through citations in new publications. The widespread and continued contribution of data into the NIS from all active LTER sites has demonstrated that attribution concerns have largely abated within the LTER community (Servilla et al. 2016).

Data in the NIS is described by EML which makes the data amenable to manipulation and analysis by LTER IMs and other researchers who want to use the data. EML allows users to successfully combine disparate data sets and where appropriate, create integrated datasets for larger spatial and/or temporal scales. None of this progress would have been possible without the leadership from a very engaged, proactive and productive IM and LNO LTER community, their partners and domain scientists themselves.

The NIS addresses three of the primary goals of the LTER Network. It makes well-designed and documented databases accessible to scientists and the public. The NIS promotes synthesis by facilitating the process of data discovery across all sites and making data easier to reuse. Finally, the NIS provides a legacy of LTER research by describing the observations and experiments that form the core of the LTER Network. As envisioned at the beginning of the LTER program, the NIS integrates research across the Network and provides a resource that is greater than the sum of the individual site contributions. The significance of having a NIS is that now the community has a centralized LTER data system. The development of the NIS came at great expense, both in time and money, and represents another example of how the LTER Network has honored its debt to the future.

13.3.9 Information Management in the International LTER Network (ILTER)

Over 40 other countries have initiated long-term research networks based on the model of the U.S. LTER program. One of the strongest areas of collaboration among these international networks has been in information management. The ILTER grew from a satellite of the U.S. LTER to a self-sustaining and vibrant entity in its own right. The IM component of the ILTER was part of what helped cement it together. While the scientists in the ILTER were still figuring out how to collaborate at an international scale, the information managers were already doing so. For example, John Porter, IM VCR, introduced EML to information managers at the Taiwanese LTER Network (TERN) and the Taiwanese took the EML model all over Asia – China, Malaysia, Thailand, Vietnam, Mongolia. It is widely used in Asian LTER sites today.

The U.S. LTER contributed significantly to the development of the ILTER information management systems, including the dissemination of the DEIMS information management framework mentioned above. David Blankman, formerly of the LNO, is now an Israeli citizen and heavily involved with LTER Europe. David knew about DEIMS from attending U.S. LTER All Scientists Meetings, and he convinced LTER Europe to use it to develop a website where the whole ILTER can enter site information and describe datasets. This was a major undertaking for ILTER, and it was built on the shoulders of the U.S. LTER collaboration to create DEIMS. Although the U.S. is no longer the IM leader in the ILTER, the LTER IM efforts and success enabled the evolution of the ILTER information management strategy.

In 2006 the ILTER adopted a new governance model that called for standing committees, and the ILTER IM Committee was formed. As a committee activity, IM experts from around the ILTER were brought together in China in 2008 to consider all the IM systems within the ILTER and to choose a path forward for the entire ILTER. EML was chosen because of its maturity and because there were tools to support it. To their collective credit and foresight, the LTER IMs also recognized that there were challenges in making data discoverable in a multilingual network, and co-authored a paper on this topic (Vanderbilt et al. 2015). This was followed by a second workshop (Vanderbilt et al. 2017) during which they considered how to make the ILTER information management system accept input languages other than English and return relevant data. Participants from around the ILTER attended this workshop and co-authored a paper (Vanderbilt and Gaiser 2017). These were among the earliest collaborations that were branded as ILTER.[8] This collaboration is another profound example of how the U.S. LTER IM community embraced the larger international IM challenges. Working collaboratively with their international colleagues, they facilitated and enabled durable IM solutions for future generations.

13.3.10 Broader Applications of the NIS Concept

The intent was always to design the NIS in such a way that it could continue to evolve in tandem with the growth and evolution of the LTER program. One of the successes of the NIS has been the way that it has led to new applications of the concepts that define it.

EcoTrends EcoTrends is an excellent example of a productive collaboration among the PIs, researchers, IMs, and the LNO with other ecoinformatics partners that helped inform and guide a significant landmark accomplishment of the LTER IM enterprise, namely the NIS. The EcoTrends project started in 2004 as a simple idea between two scientists (Debra Peters and Ariel Lugo) asking the question: how do we make long-term data easily accessible to a large group of people who may not be familiar with the raw data? A committee of scientists and technical experts from several agencies and sites was formed to ensure that different kinds of data (e.g., population, community, ecosystems) from a variety of ecosystems (e.g., lakes, grasslands, marine, polar, alpine) would be well-represented, documented, and made accessible through a common web page. This committee provided the guidance and determination to pull together over 1200 datasets from 50 sites into the EcoTrends project (www.ecotrends.info).

Although the approach to data management was much different in EcoTrends than it is in NIS, the experience provided invaluable insights that guided the development of the NIS. The EcoTrends project manipulated LTER data sets *by hand* to standardize their formats for comparison. Because this work was labor intensive, fewer data sets could be presented than in the NIS, and data were never updated

[8] K. Vanderbilt, personal communication, 28 March 2018.

after the initial analysis. The challenges faced by the EcoTrends project in sorting though and cleaning up a multitude of datasets, however, contributed significant insight into the issues involved in standardizing data that informed and guided the development of the NIS. In addition, software developers in the LNO helped design and develop the cyberinfrastructure for the EcoTrends Project. Many of the approaches they used in this work were incorporated into the NIS.

The Environmental Data Initiative The Environmental Data Initiative (EDI) is derived from the LTER NIS, but EDI serves the broader ecological community as well as the LTER Network. EDI is funded by the National Science Foundation to provide support, training, and resources to help archive and publish high-quality environmental data and metadata, particularly data from projects funded by the NSF's Division of Environmental Biology (DEB). Programs served include, but are not limited to, Long Term Research in Environmental Biology (LTREB), Organizaton for Biological Field Stations (OBFS), Macrosystems Biology (MSB), and Long Term Ecological Research (LTER). The EDI data repository uses an improved version of the PASTA software to serve these communities (as discussed in Sect. 13.3.8).

To fully understand the emergence of the EDI, it is helpful to relate a bit of history. In 2009, NSF made the decision to hold an open competition for the LNO in 2015. The call for this competition split the communications and data management components of the LNO into two entities to be funded separately. Concurrent with the LNO competition, a few key IMs had been invited by NSF to create a plan for a more comprehensive data center that would include funds for collaborative software development, cross-site technology transfer and a mechanism for cross-site work by informatics specialists (i.e., centers of expertise within the network). That work on a Network Information Management Office (NIMO) formed the centerpiece of a proposal headed by University of Wisconsin (UW), specifically the North Temperate Lakes (NTL) LTER site. Scientists and data engineers from the LNO submitted a parallel proposal to maintain the operations of the NIS and to continue development of the PASTA software framework (dubbed PASTA+). The LNO proposal, submitted from the University of New Mexico (UNM), aimed at serving the whole ecological community, including LTER. Both the NIMO and PASTA+ teams originated from the LTER NIS community and had some overlap of personnel. NSF decided to link the two proposals into a single entity, the EDI. In the process, some goals were re-assigned between University of Washington and University of New Mexico, and support for LTER cross-site collaboration work on shared IM solutions was eliminated.

Thus, EDI is a re-branding of the LTER NIS aimed at a broader audience. The EDI includes the full archive of LTER data packages and uses PASTA+. This new configuration resulted in pulling the EDI away from direct LTER influence – the EDI would now serve *all* environmental biology with LTER as only one of many client projects. The PASTA+ portion of the EDI was awarded to University of New Mexico (the old LTER Network Office) and the NIMO portion went to North Temperate Lakes at University of Wisconsin. In some ways this made sense, as there is nothing about PASTA+ functions or good IM practice that is exclusive to LTER, and there are great needs for improved IM in Environmental Biology.[9] NSF saw value in the separation of EDI and LTER to get the value of EML, PASTA+, and what LTER had learned to the wider community.

The creation of the EDI as an entity separate from LTER resulted in several potential issues. Support for cross-site, collaborative work on shared IM solutions for LTER, originally part of the NIMO proposal, was not funded in either EDI award (NTL or LNO). Thus, there is no support for information technology or IM development across sites in the LTER Network other than what can be squeezed and justified from site budgets. Moreover, neither the EDI nor the new National Communications Office (NCO) at the University of California-Santa Barbara has allocated funds to support meetings of the IMC. These two issues threaten to affect the strong collaborative spirit that has developed in the LTER Network that led to the development of the NIS and ultimately the EDI. In addition, support for *scientific* synthesis in LTER comes from the NCO while support for *data* synthesis resides with EDI. LTER IMs have always bridged the gap between information management and synthesis, but with their diminished role, this function seems less likely to be successful. Despite its strong, competent leadership and a highly capable but small team, given its budget, it will be a tall order for the EDI to meet the full range of NSF expectations and researcher needs.

DataONE Data Observation Network for Earth (DataONE) (Michener et al. 2012) is a digital metadata catalog that provides links to data with shared characteristics across the whole ecological community (Waide et al. 2017). DataONE provides access to data from multiple member repositories, including the EDI. DataONE serves as a high-level aggregator of metadata, allowing domain and data scientists to discover and access ecological data from over 40 repositories (Waide et al. 2017). EDI is one of the three principal coordinating nodes of DataONE, and as such provides additional search and display tools beyond those available at EDI to cooperating scientists.

The LTER Network played an important role in the development of DataONE along with our PBI partners. Bill Michener, principal investigator of DataONE, was a member of the LNO when he developed the initial proposal to create DataONE. The ideas behind DataONE are drawn in part from the KNB and SEEK projects, both of which had strong participation from LTER IMs and the LNO. Thus, the existence of DataONE as the most important source of ecological data in the country is a direct offshoot of work from the original PBI collaboration.

[9] J. Porter, personal communication, 29 March 2018.

13.4 A Retrospective and Where We Go From Here

Information management in the LTER Network evolved from: (a) an initial approach in which data were archived and accessed directly from each site, to (b) a more centralized approach in which data contributed by all sites could also be accessed through a single LTER portal, to (c) the present configuration, where data from the LTER Network are also discoverable, along with data from 39 other entities, through DataONE (Michener et al. 2012).

NSF Program Officer, Tom Callahan noted in 1984: "Neither NSF nor the LTER investigators intend to make LTER data the exclusive province of scientists associated with the LTER projects. In fact, the intent is exactly the opposite, and it is hoped that the scientific community at large will come to regard the data sets as valuable resources" (Callahan 1984). After nearly 40 years, it's fair to say that this box has been checked: mission accomplished!

13.4.1 Accomplishments

The early emphasis NSF placed on sound IM practices resonated across the Network from the very beginning. Ecologists gained a new appreciation for long-term data and in effect improved their data literacy. Willig and Walker (2016) chronicled the careers of over three dozen LTER-affiliated individuals. In their individual essays, most mention an aspect of IM as being a valuable component of their skill set and an important take-home lesson from their LTER experience.

The accomplishments of the IM Enterprise embody the success of the LTER Network. Our significant achievements, of which Tom Callahan would be proud, include the following:

- Created a functioning computer network (hardware and software) between and among all LTER sites
- Forged an identity for and an awareness of Information Management as an integral part of scientific research
- Promoted "data literacy" within the LTER network
- Established a collaborative and inclusive working environment among LTER IMs for other domain scientists to emulate
- Developed EML to describe datasets so that they can be personnel-independent and used into perpetuity by researchers unfamiliar with the original research
- Created the NIS as a robust community-driven information system
- Changed the culture of scientific collaboration by developing PASTA which addressed the attribution issue of researchers not wanting to share their data for fear that they wouldn't get credit when others re-used their data
- Incorporated the use of DOIs in PASTA, to provide a way of tracking data through citation in new publications that didn't infringe upon individual accomplishments

- Created the platform upon which the EDI and DataONE emerged thus expanding the principles of information management across disciplines and fields of science
- Facilitated the establishment and growth of the ILTER IM program thus expanding the LTER geographically.

The adoption and implementation of EML by the entire LTER and ILTER community deserves special mention. The IM community worked collectively to make that happen, and for IMs from the U.S. LTER, that represents a remarkable ability to work together. Most of the impetus for this project came from the IMs themselves.[10] The collective effort involved in structuring legacy text metadata to meet the technological demands of this complex metadata standard was an enormous accomplishment.[11] The LTER IM enterprise facilitated research far and beyond the LTER Network. Researchers and students are now able to find and use LTER data for their research that simply wouldn't have been possible without the NIS and the accompanying EML metadata.

13.4.2 Issues Moving Forward

We have shown how LTER IM is replete with examples of how sound information management practices, while taking time and expense early, accrue benefit for future generations of researchers and ecologists. Yet, despite these significant accomplishments issues persist. Cheruvelil and Soranno (2018) describe the future of science as being more data intensive and characterized by more open and more team-based approaches.

To support synthesis science today and into the future, researchers are going to need more high-level integrated datasets. Because support for synthesis and data management were decoupled in the creation of the EDI and the NCO, however, a greater coordinated effort is now required to capture these new data sets. Historically, LTER IMs have always bridged the gap between information management and synthesis but since the support for cross-site, collaborative work on shared IM solutions was dropped in the descoping of the EDI, IMs have fewer opportunities to contribute to this process. As a result, the LTER IM enterprise is at a tipping point. This issue needs to be addressed and will require NSF support for LTER scientists to continue to produce these value-added data sets as part of the synthesis working groups sponsored by the NCO.

If the strengths of the past accomplishments are an indication of the promise of the future, I'm very optimistic that a strategy can be found. Modelling a behavior of collaboration, IMs working productively with data scientists, domain scientists, cyberinfrastructure specialists, and data engineers from the LNO to solve problems

[10] K. Vanderbilt, personal communication, 11 April 2018.

[11] W. Sheldon, personal communication, 28 March 2018.

and work toward common goals has served ecology and ecologists well and will continue to do so going forward.

13.5 Conclusion

I have been associated with the LTER Information Managers for nearly four decades. I stayed actively involved even when I left Oregon State University to become Department Head of Forest Sciences at Colorado State University (CSU) in 1998. At that time, CSU was the home of the former Short Grass Steppe (SGS) LTER site. The SGS was administratively housed in my Department. It was only after I became Dean of the College of Natural Resources at the University of Minnesota in 2002 that I stepped away from my role of chairing the LTER Information Management Committee and IM Executive Committee (IMEXEC).

I was very fortunate to begin my association with the LTER at the AND. This experience taught me many invaluable skills and helped me develop a managerial style that served us well within the IM community. The culture at the AND site tended to attract other like-minded researchers with an *a priori* tendency to trust others already in the group (Grier 2007). Over time, I came to realize and value the true collaborative culture at the AND site.

As was the case at several LTER sites, long-term research at the AND was conceptually positioned at the interface of basic and applied science. Communicating research results across a wide spectrum of stakeholders – from state, federal, international and non-governmental organizations – was commonplace. Watching the AND leadership cultivate, nurture and build enduring partnerships provided invaluable lessons for effective communication with diverse and numerous constituencies. I learned first-hand how to work towards successful conflict resolution by tackling issues directly to find common ground, an approach we honed within our Annual IM meetings. Direct participation in a collaborative and openly inclusive atmosphere has served as an excellent working model for the LTER IMC, domain scientists, data engineers and cyber specialists working together successfully and productively for the last four decades. The rapid growth and maturation of the infrastructure available to manage ecological data engendered a parallel evolution in the culture of data sharing and collaborative science. The LTER Network has been at the forefront of this cultural shift, and the LTER IM community led the way. I am forever grateful to have had the opportunity to be a part of this organization and to have worked with so many of my esteemed colleagues.

Acknowledgements I want to thank Bob Waide, John Porter, Wade Sheldon, Kristen Vanderbilt, Corinna Gries, Karen Baker, Don Henshaw, Bob Robbins, Dave Schimel, and Bill Michener for providing their respective insights and candid thoughts on the successes and challenges facing the LTER IM enterprise.

References

Aronova, E., K.S. Baker, and N. Oreskes. 2010. Big science and big data in biology: From the international geophysical year through the international biological program to the Long Term Ecological Research (LTER) network, 1957–present. *Historical Studies in the Natural Sciences* 40 (2): 183–224.

Baker, K.S., and G.C. Bowker. 2007. Information ecology: Open system environment for data, memories, and knowing. *Journal of Intelligent Information Systems* 29 (1): 127–144.

Baker, K.S., and C. Chandler. 2008. Enabling long-term oceanographic research: Changing data practices, information management strategies and informatics. *Deep Sea Research II* 55: 2132–2142.

Baker, K.S., and H. Karasti. 2018. Data care and its politics: Designing for local collective data management as a neglected thing. In *PDC '18: Proceedings of the 15th participatory design conference, August 20–24, Hasselt and Genk, Belgium*, vol. 1. https://doi.org/10.1145/3210586.3210587.

Baker, K.S., and F. Millerand. 2007. Articulation work supporting information infrastructure design: Coordination, categorization, and assessment in practice. In *Proceedings of the 40th Hawaii international conference on Systems Sciences*.

———. 2010. Infrastructuring ecology: Challenges in achieving data sharing. In *Collaboration in the new life sciences*, ed. J.N. Parker, N. Vermeulen, and B. Penders, 111–138. Farnham: Ashgate.

Brunt, J.W. 1994. Research data management in ecology: A practical approach for long-term projects. In *Seventh international working conference on Scientific and Statistical Database Management*, 272–275. Charlottesville: IEEE. https://doi.org/10.1109/SSDM.1994.336941.

———. 1999. The LTER network information system: A framework for ecological information management. In *North American science symposium: Toward a unified framework for inventorying and monitoring forest ecosystem resources, 2–6 Nov 1998, Guadalajara, Mexico*, ed. C. Aguirre-Bravo and C.R. Franco, 435–440. Fort Collins: US Department of Agriculture, Forest Service, Rocky Mountain Research Station.

Brunt, J.W., J. Porter, and R. Nottrott. 1990. *Internet connectivity in LTER: Assessment and recommendations*. Seattle: LTER Network Office, University of Washington, College of Forest Resources.

Callahan, J. 1984. Long-term ecological research. *Bioscience* 34: 363–367.

Cheruvelil, K.S., and P.A. Soranno. 2018. Data-intensive ecological research is catalyzed by open science and team science. *Bioscience* 68 (10): 813–822.

Collins, S.L., M.L. Avolio, C. Gries, L. Hallett, S.E. Koerner, K.J. LaPierre, A.L. Rypel, E.R. Sokol, S.B. Fey, D.F.B. Flynn, S.K. Jones, L.L. Ladwig, J. Ripplinger, and M.B. Jones. 2018. Temporal heterogeneity increases with spatial heterogeneity in ecological communities. *Ecology* 99 (4): 858–865.

Dornelas, M., N.J. Gotelli, B. McGill, H. Shimadzu, F. Moyes, C. Sievers, and A.E. Magurran. 2014. Assemblage time series reveal biodiversity change but not systematic loss. *Science* 344: 296–299.

Fegraus, E., S. Andelman, M.B. Jones, and M.P. Schildauer. 2005. Maximizing the value of ecological data with structured metadata: An introduction to the Ecological Metadata Language (EML) and principles for metadata creation. *Bulletin of the Ecological Society of America* 86: 158–168.

Grier, M.G. 2007. *Necessary work: Discovering old forests, new outlooks, and community on the H.J. Andrews Experimental Forest, 1948–2000*, General Technical Report PNW-GTR-687. Portland: U.S. Department of Agriculture, Forest Service, Pacific Northwest Research Station.

Gross, K.L., and C.E. Pake. 1995. *Final report of the Ecological Society of America Committee on the Future of Long Term Ecological Data (FLED). Volume II: Directories to sources of long-term ecological data*. Washington, DC: Ecological Society of America.

Harris, F., and L. Krishtalka. 2002. *Long Term Ecological Research Program twenty-year review: A report to the National Science Foundation*. http://lternet.edu/wp-content/uploads/2014/01/20_yr_review.pdf. Accessed 22 May 2020.

Hautier, Y., F. Isbell, E.T. Borer, E.W. Seabloom, W.S. Harpole, E.M. Lind, A.S. MacDougall, C.J. Stevens, P.B. Adler, J. Alberti, J.D. Bakker, L.A. Brudvig, Y.M. Buckley, M. Cadotte, M.C. Caldeira, E.J. Chaneton, C. Chu, P. Daleo, C.R. Dickman, J.M. Dwyer, A. Eskelinen, P.A. Fay, J. Firn, N. Hagenah, H. Hillebrand, O. Iribarne, K.P. Kirkman, J.M.H. Knops, K. la Pierre, R.L. McCulley, J.W. Morgan, M. Pärtel, J. Pascual, J.N. Price, S.M. Prober, A.C. Risch, M. Sankaran, M. Schuetz, R.J. Standish, R. Virtanen, G.M. Wardle, L. Yahdjian, and A. Hector. 2018. Local loss and spatial homogenization of plant diversity reduce ecosystem multifunctionality. *Nature Ecology & Evolution* 2 (1): 50–56.

Henshaw, D. 2018. *A history of LTER Information Management Committee meetings: Venues and participation. LTER Databits: Information Management Newsletter of the Long Term Ecological Research Network*. Spring 2018 Issue. https://lternet.edu/wp-content/uploads/2018/03/2018DatabitsSpringIssue-web.pdf. Accessed 20 May 2020.

Karasti, H., and K.S. Baker. 2004. Infrastructuring for the long-term: Ecological information management. In *Proceedings of the 37th Hawaii international conference on system sciences*. Washington, DC: IEEE Computer Society. https://doi.org/10.1109/HICSS.2004.10014.

———. 2008. Digital data practices and the long-term ecological research program growing global. *International Journal of Digital Curation* 3 (2): 42–58.

Karasti, H., K.S. Baker, and E. Halkola. 2006. Enriching the notion of data curation in e-science: Data managing and information infrastructuring in the Long Term Ecological Research (LTER) Network. *Computer Supported Cooperative Work* 15 (4): 321–358.

Karasti, H., K.S. Baker, and F. Millerand. 2010. Infrastructure time: Long term matters in collaborative development. *Computer Supported Cooperative Work* 19: 377–415.

Kratz, J., and C. Strasser. 2015. Making data count. *Scientific Data* 2: 150039. https://doi.org/10.1038/sdata.2015.39.

Michener, W.K., A.B. Miller, and R. Nottrott, eds. 1990. *Long-Term Ecological Research Network core data set catalog*. Columbia: Belle W. Baruch Institute for Marine Biology and Coastal Research, University of South Carolina.

Michener, W.K., J.W. Brunt, J.J. Helly, R.B. Kirchner, and S.G. Stafford. 1997. Nongeospatial metadata for the ecological sciences. *Ecological Applications* 7: 330–342.

Michener, W.K., S. Allard, A.E. Budden, R. Cook, K. Douglass, and M. Frame. 2012. Participatory design of DataONE—Enabling cyberinfrastructure for the biological and environmental sciences. *Ecological Informatics* 11. https://doi.org/10.1016/j.ecoinf.2011.08.007.

Millerand, F., D. Ribes, K. Baker, and G.C. Bowker. 2013. Making an issue out of a standard: Storytelling practices in a scientific community. *Science, Technology &Human Values* 38 (1): 7–43.

Porter, J.H. 2010. A brief history of data sharing in the U.S. Long Term Ecological Research Network. *Bulletin of the Ecological Society of America* 91: 14–20.

Porter, J.H., and J.W. Brunt. 2001. Making data useful: The history and future of LTER metadata. *LTER Network News* 14 (2): 12–13. http://lternet.edu/wp-content/uploads/2010/12/fall01.pdf. Accessed 22 May 2020.

Porter, J.H., and J.T. Callahan. 1994. Circumventing a dilemma: Historical approaches to data sharing in ecological research. In *Ecological information management*, ed. W.K. Michener, S. Stafford, and J.W. Brunt, 193–203. Bristol: Taylor and Francis.

Robbins, R.J. 2011. *Data management for LTER: 1980–2010*. http://lternet.edu/wp-content/uploads/2012/01/report-RJR-fin-revised.pdf. Accessed 22 May 2020.

Rodriguez, N.B., K.J. McGuire, and J. Klaus. 2018. Time-varying storage-water age relationships in a catchment with a Mediterranean climate. *Water Resources Research* 54 (6): 3988–4008.

Servilla, M., J. Brunt, D. Costa, J. McGann, and R. Waide. 2016. The contribution and reuse of LTER data in the Provenance Aware Synthesis Tracking Architecture (PASTA) data repository. *Ecological Informatics* 36: 247–258.

Sillett, S.C., R. Van Pelt, J.A. Freund, J. Campbell-Spickler, A.L. Carroll, and R.D. Kramer. 2018. Development and dominance of Douglas-fir in North American rainforests. *Forest Ecology and Management* 429: 93–114.

Song, C., W.K. Dodds, J. Ruegg, A. Argerich, C.L. Baker, W.B. Bowden, M.M. Douglas, K.J. Farrell, M.B. Flinn, E.A. Garcia, A.M. Helton, T.K. Harms, S. Jia, J.B. Jones, L.E. Koenig, J.S. Kominoski, W.H. McDowell, D. McMaster, S.P. Parker, A.D. Rosemond, C.M. Ruffing, K.R. Sheehan, M.T. Trentman, M.R. Whiles, W.M. Wollheim, and F. Ballantyne. 2018. Continental-scale decrease in net primary productivity in streams due to climate warming. *Nature Geoscience* 11 (6): 415–420.

Stafford, S.G. 1996. Finding leadership opportunities in an era of dual-career families. *Bioscience* 46: 52–54.

———. 1998. Issues and concepts of data management: The H.J. Andrews Forest Sciences Data Bank as a case study. In *Data and information management in the ecological sciences: A resource guide*, ed. W.K. Michener, J.H. Porter, and S.G. Stafford, 1–5. Albuquerque: Long-Term Ecological Research Network Office.

———. 2016. Data, data everywhere. In *Long-term ecological research: Changing the nature of scientists*, ed. M.R. Willig and L.R. Walker, 63–71. New York: Oxford University Press.

Stafford, S.G., P.B. Alaback, G.J. Koerper, and M.W. Klopsch. 1984. Creation of a forest science data bank. *Journal of Forestry* 82: 432–433.

Stafford, S.G., P.B. Alaback, K.L. Waddell, and R.L. Slagle. 1986. Data management procedures in ecological research. In *Research data management in the ecological sciences*, ed. W.K. Michener, 93–113. Columbia: University of South Carolina Press.

Stafford, S.G., G. Spycher, and M.W. Klopsch. 1988. The evolution of the Forest Science Data Bank. *Journal of Forestry* 86: 50–51.

Stafford, S.G., J.W. Brunt, and W.K. Michener. 1994. Integration of scientific information management and environmental research. In *Environmental information management and analysis: Ecosystem to the global scales*, ed. W.K. Michener, J.W. Brunt, and S.G. Stafford, 3–19. London: Taylor and Francis.

Vanderbilt, K., and E. Gaiser. 2017. The International Long Term Research Network: A platform for collaboration. *Ecosphere* 8 (2): e01697. https://doi.org/10.1002/ecs2.1697.

Vanderbilt, K.L., C.-C. Lin, S.-S. Lu, A.R. Kassim, H. He, X. Guo, I. San Gil, D. Blankman, and J.H. Porter. 2015. Fostering ecological data sharing: Collaborations in the International Long Term Ecological Research Network. *Ecosphere* 6 (10): 1–18.

Vanderbilt, K., J.H. Porter, S.-S. Lu, N. Bertrand, D. Blankman, X. Guo, H. He, D. Henshaw, K. Jeong, E.-S. Kim, C.-C. Lin, M. O'Brien, T. Osawa, E.O. Tuama, W. Su, and H. Yang. 2017. A prototype system for multilingual data discovery of International Long-Term Ecological Research (ILTER) Network data. *Ecological Informatics* 40: 93–101.

Waide, R.B., J.W. Brunt, and M.S. Servilla. 2017. Demystifying the landscape of ecological data repositories in the United States. *Bioscience* 67: 1044–1051.

Willig, M.R., and L.R. Walker, eds. 2016. *Long-term ecological research: Changing the nature of scientists*. New York: Oxford University Press.

Wolkovich, E.M., B.J. Cook, J.M. Allen, T.M. Crimmins, J.L. Betancourt, S.E. Travers, S. Pau, J. Regetz, T.J. Davies, N.J.B. Kraft, T.R. Ault, K. Bolmgren, S.J. Mazer, G.J. McCabe, B.J. McGill, C. Parmesan, N. Salamin, M.D. Schwartz, and E.E. Cleland. 2012. Warming experiments underpredict plant phenological responses to climate change. *Nature* 485: 494–497.

Zhang, M., N. Liu, R. Harper, Q. Li, K. Liu, X. Wei, D. Ning, Y. Hou, and S. Liu. 2017. A global review on hydrological responses to forest change across multiple spatial scales: Importance of scale, climate, forest type and hydrological regime. *Journal of Hydrology* 546: 44–59.

Chapter 14
Network Level Science, Social-Ecological Research and the LTER Planning Process

Scott L. Collins

Abstract This chapter provides a personal perspective and history of the LTER Planning Process that took place from 2004 through 2007 with support from the National Science Foundation (NSF). Decadal reviews of LTER in 1990 and 2000 commissioned by NSF emphasized the need for interdisciplinary science, greater cross-site synthesis and the desire for a network-level research agenda. The purpose of the Planning Process was to develop the scientific basis and conceptual framework for network-level science that would facilitate synthesis and integration from the start. Many researchers from the biophysical and social sciences were involved in the process, which resulted in a conceptual framework for integrated, long-term, social-ecological research that has been widely embraced globally. Although the LTER Network did not get to implement its Network-level science initiative, the process demonstrated that LTER scientists could work together across sites to develop a research agenda essential for understanding how global environmental change will affect the dynamics of social-ecological systems during the Anthropocene.

Keywords LTER program · Long-term ecological research · Ecological strategic planning · Network science · National Science Foundation · Integrative science · Conceptual framework · Socio-ecological systems

---

S. L. Collins (✉)
Department of Biology, University of New Mexico, Albuquerque, NM, USA
e-mail: scollins@unm.edu

R. B. Waide, S. E. Kingsland (eds.), *The Challenges of Long Term Ecological Research: A Historical Analysis*, Archimedes 59,
https://doi.org/10.1007/978-3-030-66933-1_14

14.1 Introduction

The goal of this chapter is to provide some background and context for the Long-term Ecological Research (LTER) Planning Process, an NSF-funded activity that took place from 2004 through 2007. I want to emphasize that this chapter will contain a highly personal perspective on what happened and why. I will provide some background relevant to the start of the process, the overarching objectives of the process, a summary of the process itself, the ultimate outcomes and products, and their reception by NSF management and the broader scientific community. A lot of people put a lot of time and energy into this process over a 3 year time span, with some important, concrete outcomes for the LTER Network and beyond. Like the International Biological Program (IBP) that is widely but wrongly criticized for not achieving its primary goal of modeling net primary production globally (Golley 1993), we did not accomplish our overly ambitious goal of establishing a new, long-term, cross-site, fully integrated, social-ecological research program. But also like the IBP, several long-lasting positive outcomes emerged from the planning process, including a new conceptual framework for social-ecological research, a new governance structure for the LTER Network, and an enhanced web portal to manage and deliver LTER data. Despite these important success stories, one might ask whether or not these outcomes were worth the time and money invested in the Planning Process? I will return to this nagging question under lessons learned at the end of the chapter.

14.2 Background and History

From September 1992 to February 2003 I was a Program Director in the Division of Environmental Biology (DEB) at the National Science Foundation (NSF). From 1995 to 2000 I served as the Program Director (PD) for LTER, before being reassigned to be the first PD for the National Ecological Observatory Network (NEON). Despite that administrative move, I remained actively engaged in LTER management, as well as a long-time researcher at the Konza Prairie LTER site in northeastern Kansas. In February 2003, I left NSF to take a faculty position at the University of New Mexico and to become the lead Principal Investigator on the Sevilleta LTER program. On one of my last days as a Program Director at NSF, I was having a meeting with Dr. Joann Roskoski who was the Deputy Assistant Director for Biological Sciences (BIO) at the time. During that meeting, much to my great surprise and pleasure, she said, "we need to find a way to get more money to LTER." LTER was on her mind because the program had just been through the 20-year review (Krishtalka 2002) commissioned by the BIO Directorate. That review called for more resources, as well as a greater emphasis on synthesis research and cross-site coordination.

In the early years of LTER, staff at NSF realized that LTER would not be successful or justifiable from an agency standpoint without the program acting more like a network. As was typical of NSF through much of the evolution of the LTER Program, managers at NSF would identify what they considered to be an important direction (e.g., act like a network, or develop a data management system) for LTER to move and then ask the scientists to figure out how to make it happen. These mandates often, but not always, came with extra resources. Indeed, the LTER Network Office was conceived early on as a facility to support and encourage network-level activities, such as annual meetings among the site PIs to promote collaboration. The first "LTER All Scientists Meeting" occurred in 1985 hosted by the University of Minnesota and since then these meetings have been held approximately every 3 years. Thus, network integration and coordination were goals from the start of the LTER Network (LTER Network Office 1989), but it was not always clear how to achieve these goals given how the Network was established through multiple competitions for site-based science.

The need for cross-site and synthesis activities was further reinforced by the LTER Ten Year Review (Risser 1993) which concluded that although, "...intersite comparisons have been conducted...the power of the network of coordinated research sites has not yet been fully realized." The LTER Twenty Year Review continued that theme (Krishtalka 2002) noting that, "...missing is a clear exposition of what synthesis science LTER should accomplish - what should the scientific focus, niche and priorities of the LTER program be for the next decade? Despite...accomplishments, some of the critical recommendations of the Ten-Year Review for LTER science have yet to be fully realized. The transition from individual site-based research and science projects to a broader, more integrative research platform has not been sufficient to address large-scale, interdisciplinary environmental issues."

Synthesis can be achieved in two ways. The first is to integrate across disciplines within a site. Long-term, integrated, site-based research was and still is the essential ingredient in LTER science. Indeed, most sites have a long history of blending biophysical perspectives from the start, and the addition of urban sites provided yet another level of integration that included the social, behavioral and economic sciences. Cross-site synthesis, on the other hand, was slower to materialize within LTER, increasing gradually as the LTER Network matured (Johnson et al. 2010). In some cases, multi-site research projects were generated externally, with funding provided by various programs in the Division of Environmental Biology (DEB) (e.g., LIDET, Gholz et al. 2000; Parton et al. 2007), and others were established through two NSF-sponsored LTER cross-site competitions open to researchers within and outside the LTER Network. Both competitions included funds contributed by other programs in DEB and Biological Oceanography in the Directorate for Geosciences. Cross-site competitions were designed to generate multi-site LTER research, as well as to attract non-LTER scientists to work at LTER sites and to facilitate research between LTER and non-LTER sites. Although these competitions were popular, because of constant budget constraints, no permanent internal funds were earmarked to keep them going.

In the mid-1990s, while James Gosz was DEB Division Director, NSF received an unexpected budget windfall from which DEB held a competition within the LTER Network to expand site based research regionally and to increase disciplinary breadth. The North Temperate Lakes (NTL) and Coweeta (CWT) LTER sites were selected following peer review to receive budget increases from ~$560,000 per year (the Network standard at the time) to $1,000,000 per year. The ultimate plan was to repeat this competition periodically so that more sites could expand their research programs. In truth, the budget windfall was intended for other federal agencies (NASA, USDA, Department of Energy), not NSF. This funding bonanza occurred because NSF had room in its budget request for additional global change research funds through its annual request to Congress. These funds were directed to NSF by the Office of Management and Budget with the intention of NSF participating in a cross-site competition for global change research. As a consequence, rather than continuing to expand LTER site science, most of these funds were used for NSF's contribution to the Terrestrial Ecology and Global Change interagency competition, known as TECO. That effectively ended the plan to use these funds to expand the scale and scope of sites in the LTER Network.

As the LTER Network grew, there was a clear need for a governance structure to promote cross-site interactions. The Coordinating Committee (CC) meeting initially served in that capacity. Starting in the mid-1990s research symposia at the CC meetings were used to explore interconnections among LTER sites. For example, one highly successful CC workshop hosted by Dave Tilman (Cedar Creek LTER) resulted in an LTER working group led by Bob Waide and Mike Willig and supported by the National Center for Ecological Analysis and Synthesis (NCEAS). That working group resulted in several impactful cross-site publications (e.g., Waide et al. 1999; Dodson et al. 2000; Gough et al. 2000; Gross et al. 2000; Mittelbach et al. 2001). This was one of the first of numerous cross-site efforts, many of which were funded by resources provided through NCEAS and, more often, the LTER Network Office.

In fact, the LTER strategic planning at the time of the Twenty-Year Review identified the third decade of LTER science as one of cross-site research and synthesis that would lead to a better understanding of complex environmental problems and result in knowledge that serves science and society. Despite the increase in synthesis and cross-site research that had occurred by that time, most such activities were *ad hoc*, somewhat idiosyncratic, and relatively uncoordinated, thus preventing the LTER Network from achieving its full potential. This deficiency called for a coordinated, organized approach to Network-level science, collaboration and synthesis driven from the bottom-up by the LTER research community. Network level science to address Ecological Grand Challenges, a list of urgent research priorities identified by the National Research Council (National Research Council 2001), was incorporated into the LTER Network's vision, mission, and scientific priorities. In addition, Network-level science required improvements in governance and organizational structure, infrastructure needs, advanced informatics and integration with education and policy initiatives all built around a strong science-driven research agenda.

In addition to cross-site research, the LTER Network formed a partnership with Oxford University Press to publish site-based (e.g., Knapp et al. 1998; Bowman and Seastedt 2001; Magnuson et al. 2005; Havstadt et al. 2006; Chapin et al. 2000; Lauenroth and Burke 2008; Brokaw et al. 2012; Hobbie and Kling 2014; Swank and Webster 2014; Hamilton et al. 2015; Childers et al. 2019), methods-oriented (e.g., Robertson et al. 1999; Greenland et al. 2003; Fahey and Knapp 2007), and topical (e.g., Greenland et al. 2003; Shachak et al. 2014; Redman and Foster 2008; Willig and Walker 2016) synthesis volumes. The complete list of LTER books can be found at https://lternet.edu/books/. These syntheses provided a means to summarize years of site-based research, and they promoted standardized measurement and analysis protocols both across the Network and for ecological research in general. Finally, the triennial LTER All Scientists Meetings (ASM) increasingly acted as a catalyst for cross-site synthesis and coordination both nationally and internationally. Activities at the ASM led to proposals submitted to the LTER Network Office, and again several of these LNO funded meetings resulted in important publications (e.g., Redman et al. 2004; Suding et al. 2005; Houlahan et al. 2007; Peters et al. 2008; Fountain et al. 2012; Robertson et al. 2012; Bestelmeyer et al. 2012; Alber et al. 2013; Hallett et al. 2014; Kaushal et al. 2014; Smith et al. 2015). Thus, interest in synthesis was growing within and across the network. Like most syntheses, integration across LTER sites was often challenging because of variable time frames, and different methods and measurements across systems. What was needed was a framework for integrated LTER science that would enable synthesis from the start.

The organization of the LTER network certainly facilitated communication and interactions, but it was not well-suited to conduct and coordinate network-level science. For many years, the primary form of governance for the LTER Network included an Executive Committee (EC) and the Coordinating Committee (CC). As noted earlier, the CC was made up of all the lead PIs as well as individuals representing information managers and the graduate students, whereas the EC was an elected subset of CC members. Essentially, the EC was the "business" arm of the LTER Network, including interacting with NSF staffers from time to time. The role of the CC was not particularly clear because early on there were no LTER Network bylaws in place that specified its role in network governance, nor was there any explicit mechanism to promote cross-site research.

14.3 The Planning Process

At the time I moved to UNM in March 2003, the Chair of the EC/CC was James Gosz, former PI of the Sevilleta LTER and a Professor in the UNM Biology Department. The LTER Network Office with Bob Waide as Executive Director was also located at UNM. As Chair, Gosz was notified by Henry Gholz, LTER Program Director at NSF, to prepare a proposal that would lead to a forward looking research plan for Network-level science. This plan was to build off recommendations of the Twenty-Year Review. The science should be built around the Environmental Grand

Challenges recently defined by the National Research Council (2001), as well as the recommendations of the Ecological Society of America (ESA) Visions Committee (Palmer et al. 2004; 2005). The Visions Committee was established by ESA to update the highly successful Sustainable Biosphere Initiative (Lubchenco et al. 1991) that included a forward looking research and education agenda for ecology. In addition, the LTER Planning Process needed to walk a fine line between integrating with existing networks, including the development of the National Ecological Observatory Network (NEON) (National Research Council 2003), while also clearly differentiating LTER from NEON. To fulfill this agenda, I worked with Gosz and Waide to design a bottom-up planning process that would gather input from a wide-ranging group of scientists from both within and outside of the LTER network. Our goal was to generate a scientifically-based action plan for network-level, integrative, long-term, social-ecological research, to recruit more scientists to the LTER Network, and to justify increased funding that would be needed to implement this plan.

The Planning Process had three specific objectives. The first was to develop a plan for LTER network-level science, technology, and training by (1) developing a new initiative in long-term thematic, regional, and network-scale science; (2) increasing cyberinfrastructure and technical expertise at each site; (3) embedding graduate and undergraduate training into Network-level science and synthesis; and (4) integrating LTER and non-LTER sites and networks into a comprehensive international network of networks for ecological research. We also believed that the governance structure of the LTER Network needed to change to accommodate this new vision for LTER. Therefore, the second objective of the planning process was to explore alternative governance, planning and evaluation structures for managing LTER Network-level science. The new model required a governance structure to serve and support a more highly coordinated scientific network, one that included (1) a structure for network-wide science planning and evaluation, (2) a process for seamless integration of new sites and collaborative networks, and (3) an implementation plan to achieve these objectives.

The third objective for the planning process was to envision a much more ambitious plan for education, training, outreach, and knowledge exchange activities to link LTER science with application needs. Specifically, this objective included (1) establishing priority areas and key targets for education and outreach activities, (2) exploring mechanisms to facilitate collaborative science, (3) enhancing the participation of groups underrepresented in the discipline, and (4) developing skills and mechanisms for better exchange of knowledge among scientists, policymakers, and resource managers.

These were ambitious objectives that would require substantial increases in resources for the LTER Network. We did not want existing LTER research funds to be shifted to our new agenda. Instead, our goal was to build off the existing strengths of the LTER Network by enhancing research activities at each site through a new set of activities that would be layered on to existing research programs, but one that would be more fully integrated across sites from the start. Funds for the planning process came largely from the Directorates for Biological Sciences, Geosciences, and Social, Behavioral and Economic Sciences (SBE). SBE at the time was providing some of the funding for the urban LTER sites in Baltimore and Phoenix, with the hope that social sciences could be integrated into other LTER sites. Also, there was a growing movement in the research community globally for conceptual and empirical research on social-ecological systems (Haberl et al. 2006; Haberl et al. 2007). Thus, the planning process began.

The planning process was organized by a Science Task Force made up of the Planning Grant PIs – Jim Gosz (LTER Network Chair), Scott Collins (Sevilleta LTER), Dan Childers (Florida Coastal Everglades LTER), Barbara Benson (North Temperate Lakes LTER Information Manager representative), Alison Whitmer (Santa Barbara Channel LTER and Education and Outreach representative), along with Bob Waide (LTER Network Office) (Fig. 14.1). Input was also provided by the LTER National Advisory Board (NAB), an advisory committee specific to the

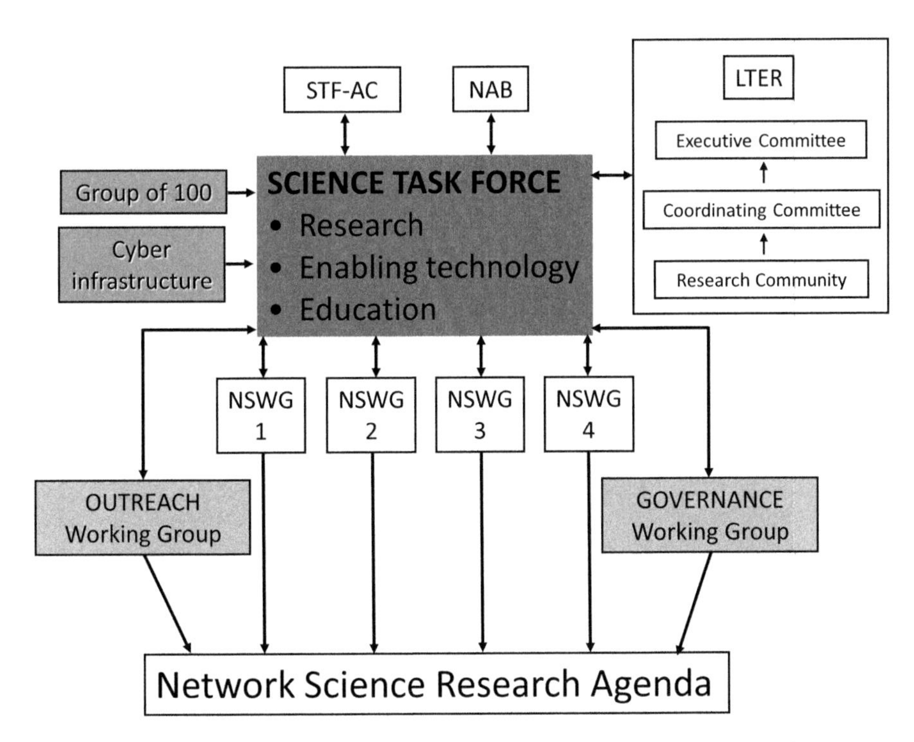

Fig. 14.1 A schematic overview of the LTER Planning Process that occurred from 2004–2007. The goal was to generate network-level science with input from as many participants and disciplines as possible. The Science Task Force was comprised of the Principal Investigators on the proposal to NSF that funded the process. The process started with a meeting of 100 participants from a wide range of disciplines to build a new research agenda based on the existing strengths of the LTER Network. Following the meeting of 100, four thematic working groups (NSWG 1-4) were formed to develop more focused activities. Researchers at All Scientists Meetings, the Coordinating Committee, and Advisory Committees (e.g., NAB – National Advisory Board; STF-AC – Scientific Task Force Advisory Committee) also provided input and guidance throughout the planning process

planning process (STF-AC) along with input from the broader LTER Network via the Executive Committee, Coordinating Committee and All Scientists Meetings. The goal was to start broad and then to narrow both the focus and the scientific team tasked with organizing the planning process. Shortly after the process got started, Jim Gosz retired from University of New Mexico, leaving me to take over as PI of the planning award.

The first step in the process began with the Meeting of 100, which was to be broadly inclusive, involving a number of social scientists (anthropologists, sociologists, economists, geographers) as well as biophysical scientists from within and outside the LTER Network. At one point during the initial Meeting of 100, I said to one of the resource economists at the workshop that we needed more sociologists at the next meeting, to which he replied, "oh, we don't need any more of those." I invited more sociologists anyway. The purpose of the Meeting of 100 was to focus the research themes, which ultimately resulted in four Network Science Working Groups (NSWGs). The themes for the four NSWGs were organized somewhat hierarchically (Fig. 14.2): at the broadest scale was climate variability and climate change. Embedded in that was coupled natural-human systems, which encompassed altered biogeochemical cycles and altered biotic structure. These themes were considered to represent the existing strengths of the LTER Network and provided a sound foundation for initiating network-level science. What followed was a series

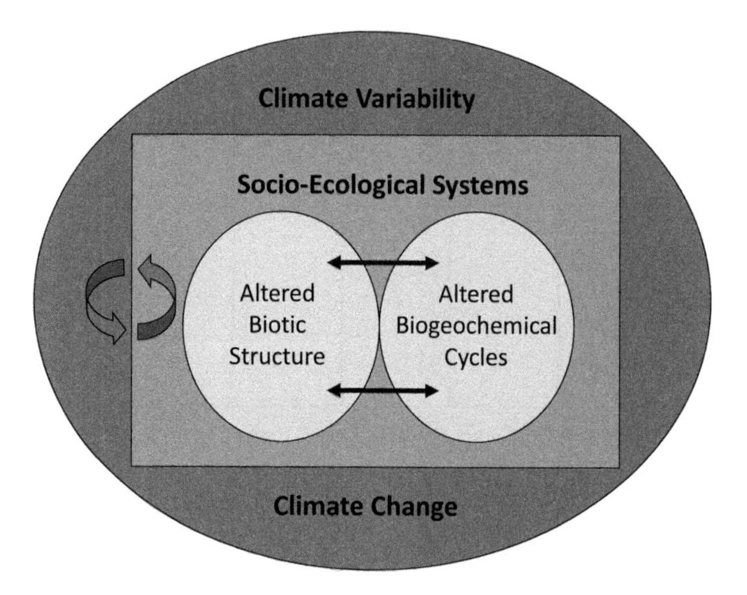

Fig. 14.2 A hierarchical schematic of the key strengths of the LTER Network research., which were the focus of four Network Science Working Groups. Altered biological structure and altered biogeochemical cycles were nested within social-ecological systems, all of which are affected by climate change. These research domains and their interactions are built around Environmental Grand Research Challenges (NRC 2001) and formed the basis of the expanded LTER Network research agenda

of meetings by Network Science Working Groups to fine tune their conceptual frameworks and research questions, and implementation plans. At the same time the Governance, Education and Outreach, and Cyberinfrastructure working groups also met to develop their ambitious plans for expanding the scale and scope of the LTER Network. Working Group meetings were often co-located to facilitate interaction and communication among all participants.

The input from Network Science Working Groups was then handed off to a Conference Committee, a smaller working group drawn from members of the NSWGs. It was the task of the Conference Committee to build the overarching scientific framework for network-level, integrated science based on the following premises. First, human activities are changing the abundance of key resources and other ecosystem drivers globally, such as elevated atmospheric CO_2, increased rates of nitrogen deposition, altered precipitation regimes and more extreme precipitation events, and sea level rise (Vitousek et al. 1997; Chapin et al. 2000). These changes can be classified as either *pulses* (e.g., discrete events, like wildfire) or *presses* (e.g., gradual increases in mean annual temperature) (e.g., Ives and Carpenter 2007; Smith et al. 2009). Many species traits (e.g., C_4 photosynthesis) result from evolutionary selection for scarce resources (e.g, atmospheric CO_2 concentrations, inorganic nitrogen) (Galloway et al. 2008; Edwards et al. 2010). Changes in resource availability or environmental drivers have significant consequences for species interactions, community structure and ecosystem functioning (Tilman et al. 2014; Komatsu et al. 2019; Clark et al. 2019). Moreover, human social systems are also spatially and temporally dynamic, and also respond to [and cause] pulse and press events (Grimm et al. 2017; Ripplinger et al. 2016). Social system drivers and dynamics (tax laws, regulations, preferences, behaviors) directly affect ecological processes (Millennium Ecosystem Assessment 2003; Carpenter et al. 2009; Larson et al. 2017), and changes in ecological processes have feedbacks that affect human social systems (Pace et al. 2015).

The conference committee determined that the overarching question for network-level science was, “How do changing climate, biogeochemical cycles, and biotic structure affect ecosystem services and dynamics with feedbacks to human behavior?” The infamous loop diagram (Fig. 14.3; Collins et al. 2011) was conceptualized to address this question, and to provide a common framework for site-based social-ecological research that could also facilitate cross-site integration. This loop diagram has four main components: biophysical systems and social systems that are linked explicitly via ecosystem services and press-pulse dynamics. Each of the major linkages is associated with a general question (see caption) that can be adapted for site based-applications.

The loop diagram had several important attributes for cross-site social-ecological research. This research agenda was designed to address *societally relevant* questions at regional and national scales. The process was *multivariate*. Cross-site research would expand beyond univariate-based understanding to study interactive effects of multiple stressors at multiple sites over long time frames and could identify commonalities in ecosystem and social system responses. The work was explicitly *interdisciplinary* and potentially *transdisciplinary*. Historically, people were

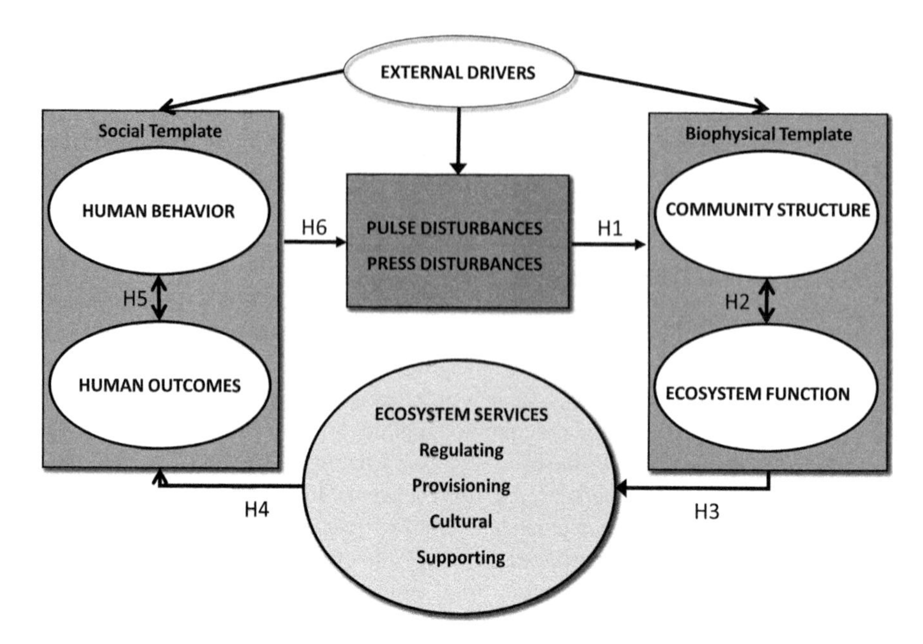

Fig. 14.3 The components of press-pulse dynamics that formed the basis for long-term, Network-level, social-ecological research. Each set of arrows in the diagram was associated with a generic hypothesis (H1-H6) that could be modified and applied to specific contexts. H1 – long-term press disturbances and short-term pulse disturbances interact to alter ecosystem structure and function; H2 – biotic structure is both a cause and a consequence of ecological fluxes of energy and matter; H3 – altered ecosystem dynamics negatively affect most ecosystem services; H4 – changes in vital ecosystem services alter human outcomes; H5 – changes in human outcomes, such as quality of life or perceptions, affect human behavior; H6 – predictable and unpredictable human behavioral responses influence the frequency, magnitude, or form of press and pulse disturbance regimes across ecosystems. (Modified from Collins et al. 2011)

typically viewed by ecologists as drivers of change, less frequently as response variables, but rarely as participatory actors as part of a research agenda, the goal of transdisciplinary science. The loop could be entered at any point, meaning projects could start with the social science drivers in some cases and the biophysical drivers in others. The conceptual framework facilitated research across sites and habitats. Multiple-site research would help to identify the most important underlying processes through a combination of observation, modeling and experimentation. The process would *integrate* education and outreach. Social-ecological research is participatory and thus requires full participation by citizens, educators, and policymakers.

Throughout the planning process we were well aware that the new and expanded research agenda for the LTER Network was not going to come cheap. At the same time, we hoped to expand this research agenda well beyond LTER. Quite simply the LTER Network was asking for a lot more money for the LTER Network, which seemed far too self-serving. Requesting large sums of new money just for this new agenda was unlikely to gain much support from NSF Program Directors or the

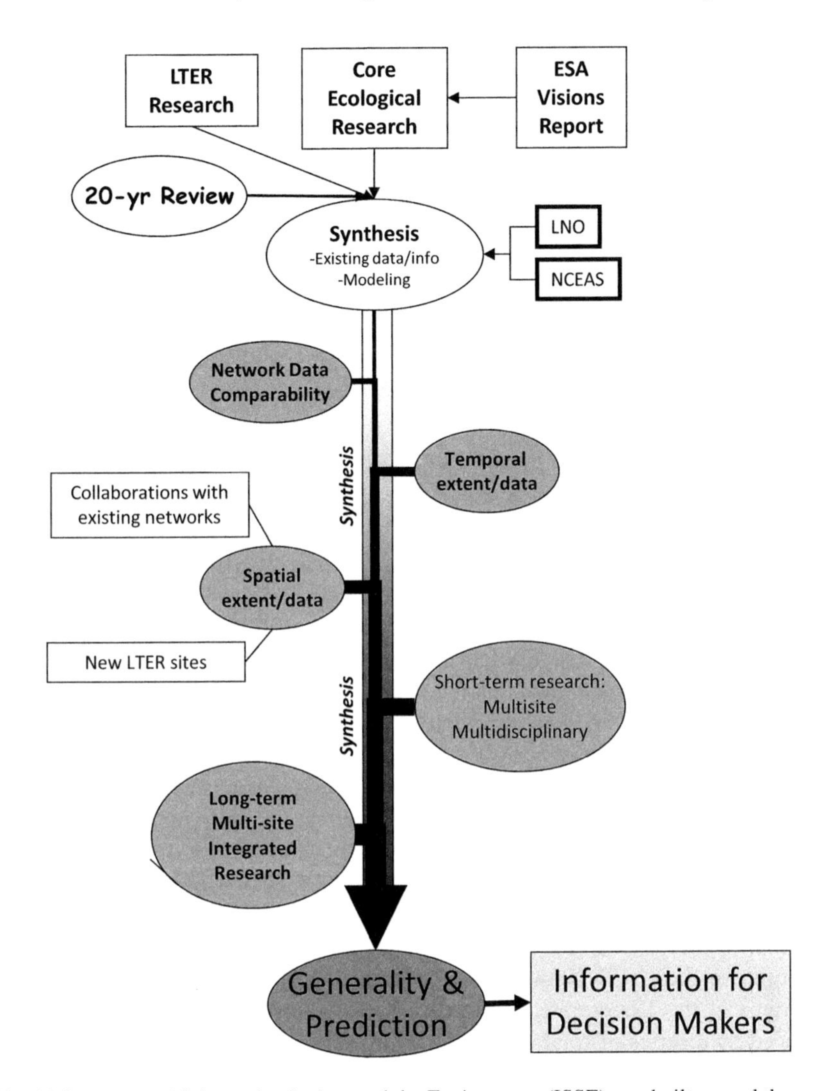

Fig. 14.4 Integrated Science for Society and the Environment (ISSE) was built around the premise that our ability to tackle challenging environmental problems and generate synthesis research over space, time, and disciplines is limited by impediments to data integration, the need for increased spatial coverage and additional long-term measurements, and coordinated, cross-disciplinary research which fully integrates social, geophysical, and ecological sciences. ISSE incorporated ideas from on-going LTER research programs, decadal reviews, and the Ecological Society of America's Visions Report (Palmer et al. 2004, 2005). Thus, ISSE recommended enhanced resources for existing as well as new funding opportunities for individual investigator and team-based long-term research, along with more resources for interdisciplinary research, more opportunities for synthesis of existing research, and a new network-scale, interdisciplinary, long-term research program for LTER. *LNO* LTER Network Office, *NCEAS* National Center for Ecological Analysis and Synthesis

broader scientific community. As a consequence, we put together a funding initiative directed at NSF, Integrated Science for Society and the Environment (ISSE; Collins et al. 2007), to justify a substantial increase in research funds that would be distributed across at least three research Directorates and multiple programs (Fig. 14.4). Therefore, when we approached NSF with our new plan for network-level science, we would also provide a scientifically based justification for a funding initiative that would broadly benefit and further integrate the biophysical and social sciences.

14.4 Outcomes of the Planning Process

It is safe to say that not all LTER scientists were enthusiastic about the goals of the planning process. The members of the Science Task Force did their best to communicate plans and progress to NSF and the LTER Network along the way. One All Scientists Meeting (2006) was dedicated to the planning process, many site scientists were involved in working groups throughout the process, and we regularly reported on progress at annual Science Council meetings and to the LTER Executive Board. Nevertheless, a few PIs felt that an unwanted research agenda was being forced on them. Others argued that human impacts were not that important at their sites, so they were concerned they would be punished for not being more engaged in social-ecological research. Still others just wanted more money for what they were already doing, which was simply not going to happen. And yet most sites and PIs fully embraced the planning process and the organizing framework, incorporating the loop diagram into their renewal proposals, with various degrees of success.

The planning process ran from 2004 to 2007. A lot can happen within a funding agency over a 3 year time span. In fact, during this period, Dr. Mary Clutter, Assistant Director (AD) for Biological Sciences, retired. Dr. Clutter was a strong supporter of the LTER Network and considered LTER to be one of the flagship programs in the Directorate. Dr. Clutter had been the AD since 1988. She was replaced by a series of rotators, all of whom had different interests and priorities. The BIO Directorate at NSF has a history of insularity from the research community. Although BIO occasionally reached out to the community (i.e., regarding the need for the national center to promote ecological synthesis), unlike other Directorates, BIO rarely sought advice about potential research-oriented funding initiatives from the community of active research scientists. But with new leadership, we hoped that the culture within BIO might have changed, and that the new management would be receptive to the social-ecological integration inherent in ISSE.

We were wrong. There was considerable skepticism expressed about ISSE and the plans for an expanded research agenda for the LTER Network. Although we regularly briefed NSF management on our progress and goals throughout the planning process, they were, in fact, completely unprepared for our initiative. Instead, Directorate-level management claimed that they were expecting a “strategic plan” for LTER, not a new research agenda. There is no mention of a strategic plan in the proposal that funded the planning process. At no time during the planning process or during our meetings with BIO Directorate management did the notion of a strategic plan come up. Instead of discussing the merits of our proposed research initiative, we were told to go back to the drawing board and develop a strategic implementation plan (SIP) for LTER. SIPs are the formal structure used by, for example, Science and Technology Centers funded by NSF. They include timelines, goals, how and when funds will be allocated. It is inappropriate to call for a SIP when no funds have been appropriated, because quite simply it is impossible to strategically implement funds you do not have.

Nevertheless, the LTER Network leadership developed an unfunded SIP as requested, which directly resulted in next to nothing. Essentially, the Directorate was not interested in our initiative nor did they have any intention of expanding and enhancing LTER science. The strategic plan and SIP felt very much like a make work program while BIO management pursued other priorities, especially NEON. Not surprisingly, we received more favorable receptions in other Directorates at NSF (GEO, SBE), which were more open to community input than BIO was. We were also invited to present ISSE to staff at USDA, NASA and on the Hill, and to other research networks (e.g., Consortium of Universities for the Advancement of Hydrologic Science; International LTER) where the initiative was well-received.

Despite our reception by the BIO Directorate staff, there were certainly some successes that emerged from the planning process. The Governance Working Group (GWG), led by Dr. Ann Zimmerman from the University of Michigan, and John Magnuson (North Temperate Lakes LTER) provided one of the most enduring outcomes of the Planning Process. They noted that the management structure and organization of LTER at the time was inadequate regardless of the plan to expand to network level science. Many lead PIs were avoiding the annual Coordinating Committee meetings because there was very little meaningful action and science at those meetings. The GWG proposed a new structure in which an Executive Board (EB) would conduct the day-to-day business of the Network. It would be made up of representatives from one third of the sites (hopefully the lead PI) and each representative would serve a 3 year term. That way, all sites would have representation on the EB every 9 years or so, and all sites would contribute to Network governance. The Chair of the EB would be elected and could serve at most two consecutive 3 year terms, assuring regular changes in Network-level leadership. The Coordinating Committee of lead PIs would then become the Science Council (SC) and the annual Science Council Meetings would focus on science and synthesis. These recommendations from the GWG, among others, were quickly adopted and implemented by the LTER Network, and they definitely led to re-engagement of PIs in network-level management through the EB, and participation in Science Council meetings, which now have an explicit science theme for synthesis each year.

The loop diagram was another success. The science behind it and the general framework, known as "press-pulse dynamics" or PPD was published in *Frontiers in Ecology and the Environment* (Collins et al. 2011). As of 1 April 2020, that paper has been cited 488 times according to Google Scholar. The framework has been widely referenced and incorporated into long-term social-ecological research programs, especially in Europe. I think that suggests that the intellectual contribution of the PPD was novel, important and useful. It would have provided a solid foundation for long-term, Network-level, integrated research. In that regard, I would also like to think that ISSE and the loop diagram provided some impetus to the US Forest Service in their efforts to establish Urban Long-term Research areas (ULTRAs), which required a strong integration of social and biophysical sciences. In addition to the *Frontiers* paper, we generated a second paper on the Hierarchical Response Framework (HRF), also built on press-pulse interactions that was published in *Ecology* (Smith et al. 2009) and as of 1 April 2020 has been cited 384 times. The HRF focuses on how global change presses, in particular, are driving long-term ecosystem dynamics, and how these presses can interact with pulse disturbances as potential drivers of state changes in ecological systems (e.g., Ratajczak et al. 2017).

An important obligation of long-term research, in general, and the LTER Network specifically, is a secure and perpetual data management system that facilitates data discovery, re-use and synthesis. For decades NSF pushed LTER to not only manage the data that were being collected, but to make those data and the metadata that describe the data freely available, discoverable and usable by anyone within or outside the Network, ideally through a single data portal. At the time most LTER data were accessed through websites hosted by individual LTER sites, which was highly inefficient for drawing together disparate datasets for cross-site synthesis. The Planning Process ended as the country was entering the 2008 financial crisis. To jumpstart the economy and preserve jobs, Congress passed the American Recovery and Reinvestment Act (ARRA), which allocated $787 billion for increased spending on education, health care, infrastructure and the energy sector. As part of ARRA, NSF received a one-time infusion of $1 billion to fund "shovel ready" research projects. Because NSF forced the LTER Network to develop a very detailed SIP, including plans for an advanced information management system to support synthesis, the LTER Network Office was poised to receive ARRA funding through NSF. The LNO then submitted a proposal for ARRA funding to support the development of PASTA (Provenance Aware Synthesis Tracking Architecture). Essentially, PASTA is a "one stop shop" for uploading, managing and discovering LTER data and metadata. ARRA funds were also used to complete the LTER Network Information System Data Portal, which provides public access to all open LTER data sets in PASTA. So, the benefits of the planning process allowed the LTER Network to achieve one of its long-standing goals, the development and implementation of an advanced information management system to facilitate data management, access and synthesis.

14.5 Lessons Learned

What are the lessons that were learned through the Planning Process and through our interactions with NSF? At the beginning of this chapter, I posed the question, "Despite some clear success stories, one might ask whether or not the benefits of the Planning Process were worth the costs in time and money?" Although we did not achieve our highly ambitious over-arching goal of establishing a long-term, multi-site, social-ecological research program within the LTER Network, solid research, management and infrastructure outcomes emerged from the planning process. In retrospect it seems as though staff at NSF had no intention of following through on our agenda and I remain deeply disappointed in how the BIO Directorate management dealt with our plan. Perhaps we were both naïve and too ambitious, and we certainly irritated BIO management by proposing a broadly based funding initiative let alone an expanded LTER research agenda.

These factors were further complicated by changes in NSF staff from the Assistant Director down to the LTER Program Director, individuals with dramatically different priorities than those who were in place when we started the Planning Process. Despite these roadblocks, we did everything we said we would in the funded planning proposal. Of significance, we clearly demonstrated that we could conceptualize and potentially carry out network-level, interdisciplinary science, which continues to be an aspirational goal for LTER in addition to maintaining and strengthening site-based, long-term research. Social-ecological research remains a solid core activity at a number of LTER sites. As we enter the Anthropocene, more and more interdisciplinary science will be needed to understand the dynamics of ecosystems increasingly influenced by human activities and decision making. I think the planning process has demonstrated that the LTER Network is ready, willing and able to lead such an important, vital and forward-looking research agenda.

References

Alber, M., D. Reed, and K. McGlathery. 2013. Coastal long term ecological research: Introduction to the special issue. *Oceanography* 26: 14–17.

Bestelmeyer, B.T., A.M. Ellison, W.R. Fraser, K.B. Gorman, S.J. Holbrook, C.M. Laney, M.D. Ohman, D.P.C. Peters, F.C. Pillsbury, A. Rassweiler, R.J. Schmitt, and S. Sharma. 2012. Analysis of abrupt transitions in ecological systems. *Ecosphere* 2: 129. https://doi.org/10.1890/ES11-00216.1.

Bowman, W.D., and T.R. Seastedt. 2001. *Structure and function of an alpine ecosystem*. Oxford: Oxford University Press.

Brokaw, N., T.R. Crowl, A.E. Lugo, W.H. McDowell, F.N. Scatena, R.B. Waide, and M.R. Willig. 2012. *A Caribbean forest tapestry: The multidimensional nature of disturbance and response*. Oxford: Oxford University Press.

Carpenter, S.R., H.A. Mooney, J. Agard, D. Capistrano, R.S. DeFries, S. Diaz, T. Dietz, A.K. Duraiappah, A. Oteng-Yeboah, H.M. Pereira, C. Perrings, W.V. Reid, J. Sarukhan, R.J. Sccholes, and A. Whyte. 2009. Science for managing ecosystem services: Beyond the

Millennium Ecosystem Assessment. *Proceedings of the National Academy of Sciences* 106: 1305–1312.

Chapin, F.S.I.I.I., E.S. Zavaleta, V.T. Eviner, R.L. Naylor, P.M. Vitousek, H.L. Reynolds, D.U. Hooper, S. Lavorel, O.E. Sala, S.E. Hobbie, M.C. Mack, and S. Diaz. 2000. Consequences of changing biodiversity. *Nature* 405: 234–242.

Childers, D.L., E. Gaiser, and L.A. Ogden. 2019. *The coastal Everglades*. Oxford: Oxford University Press.

Clark, C.M., S.M. Simkin, E.B. Allen, W.D. Bowman, J. Belnap, M.L. Brooks, S.L. Collins, L.H. Geiser, F.S. Gilliam, S.E. Jovan, L.H. Pardo, B.K. Schulz, C.J. Stevens, K.N. Suding, H.L. Throop, and D.M. Waller. 2019. Potential vulnerability of 348 herbaceous species to atmospheric deposition of nitrogen and sulfur in the U.S. *Nature Plants* 5: 697–705.

Collins, S.L., S.M. Swinton, C.W. Anderson, B.J. Benson, J. Brunt, T. Gragson, N.B. Grimm, M. Grove, D. Henshaw, A.K. Knapp, G. Kofinas, J.J. Magnuson, W. McDowell, J. Melack, J.C. Moore, L. Ogden, J.H. Porter, O.J. Reichman, G.P. Robertson, M.D. Smith, J. Vande Castle, and A.C Whitmer. 2007. Integrated science for society and the environment: A strategic research initiative. Publication #23 of the U.S. LTER Network. Albuquerque, New Mexico: LTER Network Office. http://www.lternet.edu/planning/. Accessed 20 Apr 2020.

Collins, S.L., S.R. Carpenter, S.M. Swinton, D.E. Orenstein, D.L. Childers, T.L. Gragson, N.B. Grimm, J.M. Grove, S.L. Harlan, J.P. Kaye, A.K. Knapp, G.P. Kofinas, J.J. Magnuson, W.H. McDowell, J.M. Melack, L.A. Ogden, G.R. Robertson, M.D. Smith, and A.C. Whitmer. 2011. An integrated conceptual framework for long-term social-ecological research. *Frontiers in Ecology and the Environment* 9: 351–357.

Dodson, S.I., S.E. Arnott, and K.L. Cottingham. 2000. The relationship in lake communities between primary production and species richness. *Ecology* 81: 2662–2679.

Edwards, E.J., C.P. Osborne, C.A.E. Strömberg, S.A. Smith, and C4 grasses consortium. 2010. The origins of C4 grasslands: Integrating evolutionary and ecosystem science. *Science* 328: 587–591.

Fahey, T.J., and A.K. Knapp. 2007. *Principles and standards for measuring primary production*. Oxford: Oxford University Press.

Fountain, A.G., J.L. Campbell, E.A.G. Schuur, S.E. Stammerjohn, M.W. Williams, and H.W. Ducklow. 2012. The disappearing cryosphere: Impacts and ecosystem responses to rapid cryosphere loss. *Bioscience* 62: 405–415.

Galloway, J.N., A.R. Townsend, J.W. Erisman, M. Bekunda, Z. Cai, J.R. Freney, L.A. Martinelli, S.P. Seitzinger, and M.A. Sutton. 2008. Transformation of the nitrogen cycle: Recent trends, questions, and potential solutions. *Science* 320: 889–892.

Gholz, H.L., D. Wedin, S. Smitherman, M.E. Harmon, and W.J. Parton. 2000. Long-term dynamics of pine and hardwood litter in contrasting environments: Toward a global model of decomposition. *Global Change Biology* 6: 751–765.

Golley, F.B. 1993. *A history of the ecosystem concept in ecology*. New Haven: Yale University Press.

Gough, L., C.W. Osenberg, K.L. Gross, and S.L. Collins. 2000. Fertilization effects on species density and primary productivity in herbaceous plant communities. *Oikos* 89: 428–439.

Greenland, D., D.G. Goodin, and R.C. Smith. 2003. *Climate variability and ecosystem response at long-term ecological research sites*. Oxford: Oxford University Press.

Grimm, N.B., S.T.A. Pickett, R.L. Hale, and M.L. Cadenasso. 2017. Does the ecological concept of disturbance have utility in urban social-ecological-technological systems? *Ecosystem Health and Sustainability* 3. https://doi.org/10.1002/ehs2.1255. Accessed 20 Apr 2020.

Gross, K.L., M.R. Willig, L. Gough, R. Inouye, and S.B. Cox. 2000. Patterns of species density and productivity at different spatial scales in herbaceous plant communities. *Oikos* 89: 417–427.

Haberl, H., V. Winiwarter, K. Andersson, R. Ayers, C. Boone, A. Castillo, G. Cunfer, M. Fischer-Kowalski, W.R. Freudenburg, E. Furman, R. Kaufmann, F. Krausmann, E. Langthaler, H. Lotze-Campen, M. Mirtl, C.L. Redman, A. Reenberg, A. Wardell, B. Warr, and H. Zechmeister. 2006. From LTER to LTSER: Conceptualizing the socio-economic dimension of long-term socio-ecological research. *Ecology and Society* 1: 13. http://www.ecologyandsociety.org/vol11/iss2/art13/. Accessed 20 Apr 2020.

Haberl, H., K.H. Erb, F. Krausmann, V. Gaube, A. Bondeau, C. Plutzar, S. Gingrich, W. Lucht, and M. Fischer-Kowalski. 2007. Quantifying and mapping the human appropriation of net primary production in earth's terrestrial ecosystems. *Proceedings of the National Academy of Sciences* 104: 12942–12947.

Hallett, L.M., J.S. Hsu, E.E. Cleland, S.L. Collins, T.L. Dickson, E.C. Farrer, L.A. Gherardi, K.L. Gross, R.J. Hobbs, L. Turnbull, and K.N. Suding. 2014. Biotic mechanisms of community stability shift along a precipitation gradient. *Ecology* 95: 1693–1700.

Hamilton, S.K., J.E. Doll, and G.P. Robertson. 2015. *The ecology of agricultural landscapes: Long-term research on the path to sustainability*. Oxford: Oxford University Press.

Havstadt, K.M., L.F. Huenneke, and W.H. Schlesinger. 2006. *Structure and function of a Chihuahuan Desert ecosystem*. Oxford: Oxford University Press.

Hobbie, J.E., and G.W. Kling. 2014. *Alaska's changing arctic*. Oxford: Oxford University Press.

Houlahan, J.E., D.J. Currie, K. Cottenie, G.S. Cumming, S.K.M. Ernest, C.S. Findlay, S.D. Fuhlendorf, U. Gredke, P. Legendre, J.J. Magnuson, B.H. McArdle, E.H. Muldavin, D. Noble, R. Russell, R.D. Stevens, T.J. Willis, I.P. Woiwod, and S.M. Wondzell. 2007. Compensatory dynamics are rare in natural ecological communities. *Proceedings of the National Academy of Sciences* 104: 3273–3277.

Ives, A.R., and S.R. Carpenter. 2007. Stability and diversity of ecosystems. *Science* 317: 58–62.

Johnson, J.C., R.R. Christian, J.W. Brunt, C.R. Hickman, and R.B. Waide. 2010. Evolution of collaboration within the US Long Term Ecological Research Network. *Bioscience* 60: 931–940.

Kaushal, S.S., W.H. McDowell, and W.M. Wolheim. 2014. Tracking evolution of urban biogeochemical cycles: Past, present, and future. *Biogeochemistry* 121: 1–21.

Knapp, A.K., J.M. Briggs, D.C. Hartnett, and S.L. Collins. 1998. *Grassland dynamics: Long-term ecological research in Tallgrass prairie*. Oxford: Oxford University Press.

Komatsu, K.J., M.L. Avolio, N.P. Lemoine, F. Isbell, E. Grman, G.R. Houseman, S.E. Koerner, D.S. Johnson, K.R. Wilcox, J.M. Alatalo, J.P. Anderson, R. Aerts, S.G. Baer, A.H. Baldwin, J. Bates, C. Beierkuhnlein, R.T. Belote, J.M. Blair, J.M.G. Bloor, P.J. Bohlen, E.W. Bork, E.H. Boughton, W.D. Bowman, A.J. Britton, J.F. Cahill Jr., E. Chaneton, N. Chiariello, J. Cheng, S.L. Collins, J.H.C. Cornelissen, G. Du, A. Eskelinen, J. Firn, B. Foster, L. Gough, K. Gross, L.M. Hallett, X. Han, H. Harmens, M.J. Hovenden, A. Jentsch, C. Kern, K. Klanderud, A.K. Knapp, J. Kreyling, W. Li, Y. Luo, R.L. McCulley, J.R. McLaren, J.P. Megonigal, J.W. Morgan, V. Onipchenko, S.C. Pennings, J.S. Prevéy, J. Price, P.B. Reich, C.H. Robinson, F.L. Russell, O.E. Sala, E.W. Seabloom, M.D. Smith, N.A. Soudzilovskaia, L. Souza, K.N. Suding, K.B. Suttle, T. Svejcar, D. Tilman, P. Tognetti, R. Turkington, Z. Xu, L. Yahdjian, Q. Yu, P. Zhang, and Y. Zhang. 2019. Global change effects on plant communities are magnified by time and the number of global change factors imposed. *Proceedings of the National Academy of Sciences* 116: 17867–17873.

Krishtalka, L. 2002. *Long-term ecological research program twenty-year review*. Report to the National Science Foundation. https://lternet.edu/wp-content/uploads/2014/01/20_yr_review.pdf. Accessed 20 Apr 2020.

Larson, K.L., J. Hoffmann, and J. Ripplinger. 2017. Legacy effects and landscape choices in a desert city. *Landscape and Urban Planning* 165: 22–29.

Lauenroth, W.K., and I.C. Burke. 2008. *Ecology of the shortgrass steppe*. Oxford: Oxford University Press.

LTER Network Office. 1989. *A long-range strategic plan for the long term ecological research network*. https://lternet.edu/wp-content/uploads/2013/04/1989StrategicPlan.pdf. Accessed 20 Apr 2020.

Lubchenco, J., A.M. Olsen, L.B. Brubaker, S.R. Carpenter, M.M. Holland, S.P. Hubbell, S.A. Levin, J.A. MacMahon, P.A. Matson, J.M. Mellilo, H.A. Mooney, C.H. Peterson, H.R. Pulliam, L.A. Real, P.J. Regal, and P.G. Risser. 1991. The sustainable biosphere initiative: An ecological research agenda. *Ecology* 72: 371–412.

Magnuson, J.J., T.K. Kratz, and B.J. Benson. 2005. *Long-term dynamics of lakes in the landscape*. Oxford: Oxford University Press.

Millennium Ecosystem Assessment. 2003. *Ecosystems and human well-being – A framework for assessment*. Washington, DC: Island Press.

Mittelbach, G.G., C.F. Steiner, S.M. Scheiner, K.L. Gross, H.L. Reynolds, R.B. Waide, M.R. Willig, S.I. Dodson, and L. Gough. 2001. What is the observed relationship between species richness and productivity? *Ecology* 82: 2381–2396.

National Research Council. 2001. *Grand challenges in environmental sciences*. Washington, DC: National Academies Press.

———. 2003. *NEON: Addressing the nation's environmental challenges*. Washington, DC: National Academies Press.

Pace, M.L., S.R. Carpenter, and J.J. Cole. 2015. With and without warning: Adapting ecosystems to a changing world. *Frontiers in Ecology and the Environment* 13: 460–467.

Palmer, M.A., E. Bernhardt, E. Chornesky, S.L. Collins, A.P. Dobson, C. Duke, B. Gold, R. Jacobson, S. Kingsland, R. Kranz, M. Mappin, M. Martinez, F. Micheli, J. Morse, M. Pace, M. Pascual, S. Palumbi, O.J. Reichman, A. Simons, A. Townsend, and M.G. Turner. 2004. Ecology for a crowded planet. *Science* 304: 1251–1252.

Palmer, M.A., E. Bernhardt, E. Chornesky, S.L. Collins, A. Dobson, C. Duke, B. Gold, R. Jacobson, S. Kingsland, R. Kranz, M. Mappin, F. Micheli, J. Morse, M. Pace, M. Pascual, S. Palumbi, J. Reichman, W.H. Schlesinger, A. Townsend, M. Turner, and M. Vasquez. 2005. Ecological science and sustainability for the 21st century. *Frontiers in Ecology and the Environment* 3: 4–11.

Parton, W., W.L. Silver, I.C. Burke, L. Grassens, M.E. Harmon, W.S. Currie, J.Y. King, E.C. Adair, L.A. Brandt, S.C. Hart, and B. Fasth. 2007. Global-scale similarities in nitrogen release patterns during long-term decomposition. *Science* 315: 361–364.

Peters, D.P.C., P.M. Groffman, K.J. Nadelhoffer, N.B. Grimm, S.L. Collins, W.K. Michener, and M.A. Huston. 2008. Living in an increasingly connected world: A framework for continental-scale environmental science. *Frontiers in Ecology and the Environment* 6: 229–237.

Ratajczak, Z., P. D'Odorico, S.L. Collins, B.T. Bestelmeyer, F. Isbell, and J.B. Nippert. 2017. The interactive effects of press/pulse intensity and duration on regime shifts at multiple scales. *Ecological Monographs* 87: 198–218.

Redman, C., J.M. Grove, and L. Kuby. 2004. Integrating social science into the Long-Term Ecological Research (LTER) Network: Social dimensions of ecological change and ecological dimensions of social change. *Ecosystems* 7: 161–171.

Ripplinger, J., J. Franklin, and S.L. Collins. 2016. When the economic engine stalls–A multi-scale comparison of vegetation dynamics in pre- and post-recession Phoenix, Arizona, USA. *Landscape and Urban Planning* 153: 140–148.

Risser, P.G. 1993. *Long-term ecological research program ten-year review*. Report to the National Science Foundation. https://lternet.edu/wp-content/uploads/2010/12/ten-year-review-of-LTER.pdf. Accessed 20 Apr 2020.

Robertson, G.P., D.C. Coleman, C.S. Bledsoe, and P. Sollins. 1999. *Standard soil methods for long-term ecological research*. Oxford: Oxford University Press.

Robertson, G.P., S.L. Collins, D.R. Foster, N. Brokaw, H.W. Ducklow, T.L. Gragson, C. Gries, S.K. Hamilton, A.D. McGuire, J.C. Moore, E.H. Stanley, R.B. Waide, and M.W. Williams. 2012. Long-term ecological research in a human-dominated world. *Bioscience* 62: 342–353.

Shachak, M., J.R. Gosz, S.T.A. Pickett, and A. Perevolotsky. 2014. *Biodiversity in drylands: Toward a unified framework*. Oxford: Oxford University Press.

Smith, M.D., A.K. Knapp, and S.L. Collins. 2009. A framework for assessing ecosystem dynamics in response to chronic resource alterations induced by global change. *Ecology* 90: 3279–3289.

Smith, M.D., K.J. La Pierre, S.L. Collins, A.K. Knapp, K.L. Gross, J.E. Barrett, S.D. Frey, L. Gough, R.J. Miller, J.T. Morris, L.E. Rustad, and J. Yarie. 2015. Global environmental change and the nature of aboveground net primary productivity responses: Insights from long-term experiments. *Oecologia* 177: 935–947.

Suding, K.N., S.L. Collins, L. Gough, C.M. Clark, E.E. Cleland, K.L. Gross, D.G. Milchunas, and S.C. Pennings. 2005. Functional- and abundance-based mechanisms explain diversity loss due to N fertilization. *Proceedings of the National Academy of Sciences* 102: 4387–4392.

Swank, W.T., and J.R. Webster. 2014. *Long-term response of a forest watershed ecosystem: Clearcutting in the southern Appalachians*. Oxford: Oxford University Press.

Tilman, D., F. Isbell, and J.M. Cowles. 2014. Biodiversity and ecosystem functioning. *Annual Review of Ecology, Evolution and Systematics* 45: 471–493.

Vitousek, P.M., H.A. Mooney, J. Lubchenco, and J.M. Melillo. 1997. Human domination of Earth's ecosystems. *Science* 277: 494–499.

Waide, R.B., M.R. Willig, C.F. Steiner, G. Mittelbach, L. Gough, S.I. Dodson, J.P. Juday, and R. Parmenter. 1999. The relationship between productivity and species richness. *Annual Review of Ecology and Systematics* 30: 257–300.

Willig, M.R., and L. Walker. *Long-term ecological research: Changing the nature of scientists*. Oxford: Oxford University Press.

Chapter 15
Evolving Governance in the U.S. Long Term Ecological Research Network

Ann Zimmerman and Peter M. Groffman

Abstract Governance structures, essential to the function of any network endeavor, can take many forms. The independently funded U.S. National Science Foundation (NSF) Long Term Ecological Research (LTER) sites have posed unique governance challenges since the program's founding, due to evolution in the group of sites involved, changing mandates from the funding agency and the constantly changing science, education and outreach activities of the network. We review the evolution of LTER network governance and examine how different structures have served (or not) the science goals of the network. Governance has clearly influenced the creation, sharing, and dissemination of knowledge in the LTER network and has adapted as the network has grown, while responding to changed priorities at NSF. Although systematic evaluation of the goals, drivers of change, and effectiveness of governance has been limited, governance has been a dynamic contributor to scientific success in the LTER network. Changes in governance were motivated by the desire to do more and better science in response to complex changes, such as the emergence of large-scale questions that required cross-site research, the need for synthesis of complex results, new developments in information management and the emergence of new environmental research networks. The network governance responses have been relatively nimble and effective. The clearest evidence for this success is the continued high productivity of the LTER network and the active engagement of large numbers of scientists in network activities such as Science Council and All Scientists meetings and both standing and ad hoc committees.

Keywords LTER program · Long-term ecological research · Ecological networks · Ecological governance · National Science Foundation · Network science · Governance structures

---

A. Zimmerman (✉)
Ann Zimmerman Consulting, Ann Arbor, MI, USA
e-mail: annzconsulting@gmail.com

P. M. Groffman
City University of New York Advanced Science Research Center, Graduate Center and Cary Institute of Ecosystem Studies, Millbrook, NY, USA
e-mail: Peter.Groffman@asrc.cuny.edu

R. B. Waide, S. E. Kingsland (eds.), *The Challenges of Long Term Ecological Research: A Historical Analysis*, Archimedes 59,
https://doi.org/10.1007/978-3-030-66933-1_15

15.1 Introduction

Governance makes a difference to the way scientific data and knowledge are produced, shared and disseminated. Simply stated, governance is a process through which groups make decisions that direct their collective activity (Graham et al. 2003). Governance determines how power is exercised, how people are given a voice, and how accountability is rendered (Graham et al. 2003). Governance includes both formal and informal elements. The formal components such as constitutions, bylaws, and policies define how the process is supposed to function. In reality, informal practices or unwritten conventions are equally important in determining how governance works. The interplay between formal and informal components is complex and varies markedly in different types of networks.

In this chapter, we trace the evolution of governance in LTER. As LTER's mission has evolved, the governance structure has changed and adapted, especially during the last 15 years of LTER's nearly 40-year history. We focus on this period because it has been a time of comparatively rapid change and has not been covered comprehensively elsewhere.

It is important to state at the outset that this account is written by two people who have different lengths and types of involvement with LTER. One of us (Groffman) has many years at the Hubbard Brook (since 1992) and Baltimore (since 1997) LTER sites, and served as Chair of the LTER Science Council from 2015–2019. The other (Zimmerman), a social scientist, has not been affiliated with an LTER site. Her engagement with LTER has been as an external participant in two efforts related to LTER governance that took place almost 10 years apart, with the latter concluding in 2014. As authors, we blend our personal experiences with internal and external documents, including those authored by individuals in this volume and elsewhere. We describe the history of LTER governance, the factors affecting changes in governance over time, and the ways in which decision-making processes have affected the generation and sharing of knowledge and data.

There remains more to be discovered about the subject of LTER governance. One of our goals in writing this chapter is to point future scholars to the sources, including documents and people, that exist to delve further into LTER governance and increase our understanding about the organization of networks. LTER's experience may be especially timely given some recent focus on the governance of natural resource networks (e.g., Bixler et al. 2016).

The LTER network has some unique aspects that influence the structure and function of its governance. Of primary importance is the fact that the network consists of a group of sites that are independently funded and reviewed by one agency, NSF, which therefore has a major role in determining the nature and extent of network activities and governance over time. It is true that multiple programs within NSF contribute funding to LTER sites, and most LTER sites have multiple sources of funding other than NSF. However, NSF founded the LTER network, conducts reviews of the individual sites and of the network as a whole, and is a major (often primary) source of funding at the sites. It therefore has, does and will continue to have a dominant influence on governance of the LTER network.

Governance has been important to the LTER network in meeting challenges and opportunities that have emerged both from the top (NSF) and from the bottom (new scientific topics, methods, and approaches). Bixler et al. (2016) noted that "network governance emerges when people realize that they (and the organizations they represent) cannot solve a particular problem by acting independently and that their interests may be better served through collaboration, drawing on their diverse capabilities" (p. 115). A major theme that we strike is that the key to reconciling these complex drivers is to focus on science. Indeed, we argue that "the business of science is science" and the governance structure of our scientific networks must reflect this ethos.

The chapter is arranged chronologically according to what we identified as key phases in the history of LTER governance. For each period, we describe the governance structure and the factors that affected the way LTER was organized to make decisions. We then analyze the circumstances that spurred changes in governance and describe the alterations that were made. A major focus for the analysis is if and how network evolution has fostered scientific progress, which is the hallmark of network success. We end the chapter with thoughts about how future needs, conditions, and relationships may drive further evolution in LTER governance.

15.2 1980–1992: LTER's Establishment and Governance in the First Decade

The early history of LTER is well-documented in this volume and elsewhere. In this section, we summarize this work, emphasizing aspects related to LTER governance. When the six initial LTER sites were funded by NSF in 1980, little thought was given by principal investigators (PIs) to network and inter-site research since sites had been selected based on individual, site-specific proposals (Gosz et al. 2010). That view quickly changed in October 1980 when NSF convened a meeting of the sites and "reminded the PIs that the sites constituted a network. The ultimatum received from NSF to cooperate and be a network was a source of tension at later intersite meetings and was intensified when an additional five sites were added to the interactions in 1982" (Magnuson et al. 2006, p. 304).

At the 1980 meeting, the PIs created an initial network governance structure based on NSF's charge. The structure consisted of a steering committee, which was comprised of the PI from each of the six sites, and several inter-site committees (Callahan 1984; Gosz et al. 2010; Magnuson et al. 2006). Richard Marzolf, who was located at Kansas State University (KSU), was chosen as the first Chair of the steering committee. A coordination grant was provided by NSF to KSU. Marzolf served as Chair through 1983, and Jerry Franklin assumed the position in 1984 (Magnuson et al. 2006, p. 304). Franklin (Chap. 2, this volume) said that in 1982, a decision was made that there needed to be a network office. The first office was at Oregon State University (OSU) where Franklin, who worked for OSU and the U.S. Forest Service, was located.

According to James Thomas (Tom) Callahan (1984), who was the first NSF program officer in charge of LTER, NSF believed that networking would be essential to achieve LTER's goals:

> To achieve such goals as comparability of data among projects, representative sampling of national ecosystems, and tests of regional- to national-scale hypotheses, there must be regular communication among researchers working on different projects. One dictionary defines network as "an interconnected or interrelated chain, group, or system." It is in such a context that the funding and operation of LTER projects was planned. Although projects operate virtually autonomously with regard to research at their particular sites, they are prepared to accommodate mutual goals (Callahan 1984, p. 365).

Franklin (Chap. 2, this volume), relates a slightly different version, stating that NSF staff had different opinions at the start of LTER about the degree to which the sites were a "network." Both Franklin and Magnuson et al. (2006) concur that any doubts about this were later erased by Dr. John Brooks, Director of the Division of Environmental Biology at NSF. In late 1986, Brooks attended a meeting of the LTER Network Coordinating Committee where he announced that "...expectations for network science would be part of NSF evaluations of LTER" (Magnuson et al. 2006, p. 305).

As indicated above, and verified by Callahan (1984), at some point between 1980 and 1984, the steering committee became known as the LTER Coordinating Committee. The LTER's first Long-Range Strategic Plan drafted in 1989 noted uncertainty about the exact year in which this name change occurred, writing that the LTER Coordinating Committee was "formed in 198_ (???), to make decisions on behalf of the LTER network of sites".[1] The Strategic Plan briefly described the role of the Coordinating Committee and other elements of LTER governance in place in the 18-site network in 1989, noting that "in 1988 LTER/CC decided to create an executive committee, who would meet more frequently than the LTER/CC, which meets twice a year. The LTER/EXEC would meet 4 times a year and act for the LTER/CC." The Strategic Plan stated that the roles of the Coordinating and Executive Committees and the Network Office had changed over time as the LTER expanded.

Beyond this, no details were given regarding topics such as the composition and powers of the Executive Committee.[2] Considerably more space was given in the Strategic Plan to describing the responsibilities of the Network Office. These included facilitating communication and data sharing among the LTER sites and

---

[1] A Long-Range Strategic Plan for the Long-Term Ecological Research Network, on the LTER History website: https://lternet.edu/network-organization/lter-a-history/2018/. Accessed 24 February 2020.

[2] We know from documents cited later in this chapter that the Coordinating Committee was comprised of representatives from each LTER site, and the Executive Committee was made up of four members plus the Coordinating Committee Chair.

between LTER and other communities; supporting collaborative research efforts; leading some intersite activities; and representing the LTER in its external relationships. During this period, the Network Office was located at the Coordinating Committee Chair's institution and was overseen by him. As noted above, the Network Office's first location was at Oregon State University. On Franklin's move to the University of Washington in 1989, the Network Office and the coordination grant also moved there. The Network Office expanded in size from a single staff person in its early years to a staff of four by 1989.

In 1992 NSF conducted a "ten-year review" of the LTER network. The resulting report included recommendations for LTER governance (Risser and Lubchenco 1993). The review team felt that LTER decision-making structure had not kept pace with the evolution of the research network and that governance of the LTER program needed to evolve as the program developed and matured. There was mention of "a feeling of some disenfranchisement on the part of some of the investigators." The review team recommended that governance processes should be more widely dispersed throughout the program. They also highlighted a need for external advisory groups. The challenge of governing a network of independently funded sites from one source (i.e., NSF) was noted, and there was a call to clarify the roles of NSF, the Executive and Coordinating Committees, advisory committees, and the Network Office in planning new directions and implementing new initiatives and directions.

This first "decade" established the long-standing and still ongoing differences in opinions and expectations about the degree to which LTER is intended to operate as a network. NSF provided supplemental funds to motivate cross-site collaboration and synthesis from the late 1980s until the mid-1990s. Even so, these efforts "developed falteringly" (Gosz et al. 2010, p. 66). Although the authors of the ten-year review report acknowledged the importance of increased attention from NSF to cross-site work, they attributed feedback from reviewers as one reason the goal had been only partially met (Risser and Lubchenco 1993):

> Reviewers of the LTER program, however, have not always placed consistent emphasis or value on cross-site comparisons. As a result, mixed messages have been received by the LTER scientists when the programs are evaluated and when proposals are reviewed.

Differences regarding the degree about which LTER should operate as a network also grew out of NSF's changing expectations over time as well as the extent to which site PIs had an interest in what could be accomplished at a network scale and what they were willing to "spend" – or not – to realize this potential. The use of supplemental funding to spur formal collaboration began during this period and was important for several years.

15.3 1993–2002: LTER's "Second" Decade

The next notable phase in LTER governance was in response to recommendations from the ten-year review and included a new funding mechanism, and ultimately a new location and director, for the Network Office. In 1993, LTER was comprised of 18 sites, so network growth was also a driver in some of the changes in governance.

In his response to the ten-year review to NSF on behalf of the Coordinating Committee, Chair Jerry Franklin said: "The LTER/CC will address the issue of governance during the next year with a view toward developing a detailed proposal on the roles and structure of the full committee, an executive committee, the director of the Network Office and NSF (Franklin 1993)." These changes led to several key decisions that signal the beginning of a new period in governance. Management of the Network Office was separated from the position of Chair, and a distributed structure was proposed for the Network Office. Standing committees were developed, and a fifth member was added to the Executive Committee to increase the involvement of site PIs in operations. In addition, the LTER responded to the review committee's recommendation that it seek outside advice by creating a National Advisory Board (NAB) to provide regular review and advice and guidance on new directions, help to publicize the activities and opportunities in the LTER program, and provide a form of independent oversight for the Network Office (Report of the LTER National Advisory Board 1998). The NAB was chaired by Paul Risser, who had co-led the 10-year review.

At the January 1994 meeting of the LTER Executive Committee, Dr. James Gosz, who was then Director of NSF's Division of Environmental Biology, informed the Committee that beginning in 1995 the funding mechanism for the LTER Network Office would switch from a grant to a cooperative agreement.[3] Unlike a grant, a cooperative agreement presumes substantial interaction between NSF and the recipient, with the details of the relationship negotiated prior to the award. The Executive Committee strongly supported this change. The proposed process was that the LTER Chair, with input from the Executive and Coordinating Committees, would develop objectives and metrics for the Network Office and then negotiate the specifics of the Cooperative Agreement with NSF. After Gosz departed the January 1994 meeting, the Executive Committee met alone to discuss governance issues at length. They developed a set of recommendations to bring to the Coordinating Committee. These included a statement on the desired qualifications, hiring process, and term length and level of effort for the Coordinating Committee Chair. The Executive Committee felt there would be a need for a full-time Executive Director for the Network Office because the Chair was unlikely to be full time. The Chair and Executive Director were also anticipated to be in different places since NSF recommended, and the Executive Committee concurred, that the Network Office should

[3] The information in this section is based on Minutes of the LTER Executive Committee, January 29–30, 1994. Committee minutes are archived on the LTER Document Archive under "Inactive Committees" at https://lternet.edu/intranet/. Accessed 24 February 2020.

not change locations each time the Chair changed. The Executive Committee also went on to discuss the responsibilities of the Network Office, the governance roles of the Coordinating and Executive Committees, and the need to make more use of committees to get work done, including recommending adoption of several standing committees. At the time of these conversations, the Executive Committee consisted of four site members plus the Chair. The Executive Committee's main responsibility was to conduct preliminary work for consideration by the Coordinating Committee. The Coordinating Committee, which met two times per year, made major decisions.

Based on its discussions at the January 1994 meeting, the Executive Committee prepared a governance document that was distributed to each LTER PI prior to the April 1994 Coordinating Committee meeting. Although the notes are unclear in some places, it appears that the following decisions were made:

- LTER Chair: This position would not be limited to an LTER PI, but could include an outside person familiar with the operation of LTER
- LTER Executive Committee: The Committee was increased to five members, and the Chair acts as a non-voting member
- LTER Coordinating Committee: In order to maintain continuity, there should be a regular site representative with a commitment for 3 years
- The Executive Director of the Network Office will reside where the Office is located and will interact directly with the LTER Chair
- LTER committees: An expanded committee structure was adopted

According to Franklin (Chap. 2, this volume), University of Washington submitted a proposal to NSF for the network office; this proposal was not funded. Instead, NSF decided that there should be a competition for the network office. Minutes from subsequent Executive and Coordinating Committee meetings show that the LTER Executive Committee decided to write and submit a proposal for the network office.

Jim Gosz, having left NSF, was elected as Chair at the October 1994 meeting of the Coordinating Committee. He began his tenure in 1995 and would go on to serve in that capacity until 2005 when he resigned to, again, accept a position at NSF (Thomas and Waide 2005). With Gosz's election, Jerry Franklin agreed to act as Executive Director of the Network Office for the next year.[4]

Minutes from the October 1996 Coordinating Committee state that NSF had made an award to the University of New Mexico (UNM) as the site of the LTER Network Office through a cooperative agreement that would last for 6 years.[5] Funds had been provided to bridge the period between the ending of the grant with UW and the transition to the network office's location at UNM in 1997. These minutes also show that the Coordinating Committee was beginning to draft a job description for the LTER Network Office Executive Director.

---

[4] Minutes of the Coordinating Committee Meeting, October 19–21, 1994, Coweeta Hydrologic Laboratory. Coordinating committee minutes are archived on the LTER Document Archive under "Inactive Committees": https://lternet.edu/intranet/. Accessed 24 February 2020.

[5] Archived on the LTER Document Archive website: https://lternet.edu/intranet/

Two candidates were interviewed for LTER Network Office Director in April 1997, and at the end of the two-day meeting Robert Waide was selected for the position.[6] Waide provided an overview of the transition from University of Washington to University of New Mexico in the first issue of *The Network News* that was published from the UNM-based office (Waide 1998).

In July 2001, shortly after LTER's second decade came to an end, NSF's Directorate for Biological Sciences commissioned an international committee of 17 scientists to conduct a 20-year review of the LTER program (Harris and Krishtalka 2002). The overarching recommendation from that review was for the LTER community and NSF to develop a strategic plan. Among its specific suggestions, the review committee stated that "the comprehensive strategic plan must describe an LTER governance and organizational structure that is as adaptive and unifying as the science it will be asked to enable in the next 10 years" (Harris and Krishtalka 2002, p. 4). This comment was based on the review committee's belief that LTER lacked a structure to implement a strategic plan.

Minutes from the September 12, 2002 meeting of the Coordinating Committee show that Chair Jim Gosz identified six topics that needed to be addressed based on the review report.[7] Six committees, including one on Network Governance were established as an outcome of the meeting. In late 2002, Gosz put out a call for volunteers for two working groups to address strategic planning and governance. His description of the charge of the governance group appears below (Waide personal communication):

> The governance group should focus on how we can increase our ability to manage and govern a network of ~1500 people with 24 (or more) sites. This group should have the expertise (or access to it) so that they can think more like a corporation/business. The 20 yr. report suggested that we should not be constrained by our current/past structure. Networking and collaboration are key aspects that can set us apart from the rest of the scientific community.

An additional action at the September 2002 meeting was to expand the Executive Committee to six people because of the increased work projected for strategic planning.

Although there are gaps in the history that we were able to piece together, it appears that by the end of 2002, the main components of the LTER governance structure consisted of:

- A Coordinating Committee composed of representatives from each of the sites, with each representative serving a three-year term. The Coordinating Committee met twice a year, and was, as it had always been, the network's decision-making body

[6] Minutes of the LTER Spring Coordinating Committee Meeting, H.J. Andrews Experimental Forest, Oregon, April 24–25, 1997, archived on LTER Document Archive under "Inactive Committees": https://lternet.edu/intranet/

[7] LTER Coordinating Committee, Niwot Ridge LTER Site, September 12–13, archived on LTER Document Archive under "Inactive Committees": https://lternet.edu/intranet/

- A six-member, elected Executive Committee that met between three to four times per year and was responsible for preparing action items and issues for discussion and making recommendations to the Coordinating Committee. The Chair of the Coordinating Committee was also the Chair of the Executive Committee
- A Network Office located at UNM and led by a full-time Director and staff
- Standing committees
- A National Advisory Board that was intended to meet annually[8] but met only in 1998 and 2001

Collins (Chap. 14, this volume) summarized the ways the Coordinating and Executive Committees worked. The next section describes how LTER governance changed in response to the 20-year review and Gosz's charge for a governance group.

15.4 2003: LTER Bylaws

As the previous sub-sections show, LTER has had an organized approach for making decisions ever since its establishment. It was not until 2003, though, that this process was codified into a set of written bylaws (LTER 2003).[9] By this time, LTER had grown to include 24 sites involving more than 1100 scientists (Hobbie 2003). The bylaws were drafted by the Executive Committee and revised based on comments from the sites. They were submitted to LTER's National Advisory Board for review. In the report it prepared, which was based on findings from its June 2003 review, the National Advisory Board commended the bylaws drafting effort as "a useful way of codifying current practice, and fills a gap that is badly needed now that the Network has evolved to its current size." After further revisions, a review by NSF, a review by the lead PI of each LTER site, and still more revisions, the bylaws were approved by the Coordinating Committee at its September 19, 2003 meeting (Robert Waide, personal communication).[10]

Comparing the bylaws with the structure in place in 2002 (see above), we can see that most of the main elements remained. What changed is that important details were put in writing such as term limits, meeting frequency, the list of Standing Committees and the option for the Coordinating Committee to create ad hoc committees, responsibilities of and review of the Network Office, and responsibilities and supervision of the Executive Director. The bylaws allowed for the Coordinating Committee to decide if any slots on the Executive Committee should be filled by

---

[8] By 2002, it had met twice. The first meeting was in 1998, and the second was in 2001.

[9] Long Term Ecological Research Network Bylaws. September 19, 2003. Archived on LTER Document Archive under "LTER Organization": https://lternet.edu/intranet/. Accessed 24 February 2020.

[10] Minutes from this meeting are not available in the electronic LTER Document archive located at https://lternet.edu/intranet/

specific areas of expertise such as information management. The Network Office Executive Director was an ex-officio member of the Executive Committee. The bylaws also stated that one of the responsibilities of the Network Office was to organize an All Scientists Meeting at approximately three-year intervals and to seek additional funding from NSF to support the meetings. The purpose of the All Scientists Meeting is to facilitate synthesis activities and collaborative research efforts in the LTER Network. The bylaws also specified the process for selecting members of the National Advisory Board and stated that the Board meet at least annually and provide a written report. Finally, in direct response to comments from the 20-year review committee, the bylaws identified the Coordinating Committee as the body with responsibility for developing and updating the strategic plan for the LTER Network and the Network Office.

At the April 2005 Coordinating Committee meeting, Chair Gosz led a discussion of changes to the bylaws proposed by the Executive Committee in response to issues that had arisen since their adoption in 2003.[11] According to the minutes from that meeting, the discussion suggested that the Executive Committee should make better and more frequent use of specialty committees. Beyond this, the minutes do not describe the issues that were discussed. At their fall 2005 meeting, the Coordinating Committee considered and passed revisions to Section 1 of the bylaws that concerned, and were proposed by, the Executive Committee.[12] These were passed by 26–0.[13] Language was added to clarify that the Executive Committee would choose a replacement to complete the term if an elected member of the Executive Committee was unable to fulfill his or her term.[14]

15.5 Major Bylaws Revision: 2006

In 2004, the Network embarked on a planning process. As Collins (Chap. 14, this volume) stated, one of the objectives of the planning process was to "explore alternative governance, planning and evaluation structures for managing LTER Network science." A Governance Working Group was convened as part of this effort. At the April 2005 Coordinating Committee meeting, Chairman Gosz described the composition and charge of the Governance Working Group. The minutes stated that

[11] LTER Coordinating Committee Meeting, Key Largo, FL, April 6–7, 2005. Archived on LTER Document Archive under "Inactive Committees": https://lternet.edu/intranet/. Accessed 24 February 2020.

[12] No mention was made regarding the use of committees.

[13] Minutes of the LTER Coordinating Committee Meeting, Cape Charles, VA, September 19–21, 2005. Archived on the LTER Document Archive under "Inactive Committees": https://lternet.edu/intranet/. Accessed 24 February 2020.

[14] Long Term Ecological Research Network Bylaws: Revision 1.0, September 21, 2005. Archived on the LTER Document Archive under "LTER Organization": https://lternet.edu/intranet/. Accessed 24 February 2020.

"discussion focused around the importance of determining the instrument by which resources for synthesis would reach the scientists conducting synthesis. The Governance Working Group will be asked to discuss various kinds of instruments and present a range of alternatives."[15] The group was comprised of nine individuals with varied backgrounds and perspectives, including five people with many years of experience in LTER and a graduate student associated with one of the sites. The group also included three individuals not affiliated with LTER. One of us (Zimmerman) served as Chair and the other (Groffman) was a member of the working group. Zimmerman was a social scientist who had met one of the Planning Grant leads through her dissertation research and who had invited him to participate in an NSF-funded workshop on collaboration in ecology that she co-organized in 2004. Groffman had participated in a governance effort at Hubbard Brook designed to increase participation and representation by investigators at the site and to facilitate communication between investigators, funding agencies, and the USDA Forest Service, which manages the Hubbard Brook Experimental Forest (Groffman et al. 2004).

The members of the Governance Working Group, along with their organizational affiliations at the time, included:

- Karen Baker, LTER information manager, California Current Ecosystem and Palmer Station
- Daniel Childers, Florida Coastal Everglades LTER and Co-PI of the Planning Grant
- Chelsea Crenshaw, Sevilleta LTER and Co-Chair of the LTER Network Graduate Committee
- Peter Groffman, Hubbard Brook LTER
- Katherine Lawrence, University of Michigan (organizational behaviorist)
- John Magnuson, North Temperate Lakes LTER
- Robert Waide, Executive Director, LTER Network Office
- Lawrence Weider, Organization of Biological Field Stations and Oklahoma University
- Ann Zimmerman, University of Michigan

Nathan Bos, formerly of the University of Michigan, and Sedra Shapiro, formerly Executive Director of the San Diego State University Field Station Programs, contributed to the group's early work.

Our recollection of the group's work is supported by meeting reports and notes and email correspondence. The Governance Working Group met for the first time over a two-day period in April 2005 on the University of Michigan campus. The goals of the first meeting were to 1) familiarize Governance Working Group members with LTER, its governance structure, and the role of the group in the planning effort, 2) analyze the Network's existing structure in light of the planning grant goals, and 3) identify Governance Working Group tasks and an approach to achieve

[15] See footnote 11.

them. A series of presentations formed the basis for discussions that occurred at this and subsequent meetings, and the information presented laid a foundation for a new governance structure. Besides the presentations described below, Zimmerman summarized ideas and literature on governance that seemed relevant to LTER.

John Magnuson, a Principal Investigator (PI) of one of the first LTER sites, and a widely respected scientist and leader, presented an overview of the origin of the LTER Network and its history up to 1991. Bob Waide, through his long association with LTER, first as PI of the Luquillo LTER site, which was established in 1988, and as Network Office Executive Director, combined institutional memory with an up-to-date perspective on LTER governance as of April 2005. He summarized key information from the 2003 bylaws and provided insights on the informal ways that LTER "worked." The latter included the challenge of having Coordinating Committee meetings via phone in between face-to-face meetings due to the size of the group (i.e. 26 sites); the Executive Committee's stronger role in making recommendations to the Coordinating Committee and their role as reviewers for the $50,000 provided to the Network Office for synthesis; and the various ways in which the Network Office encouraged cross-site collaboration, including supplemental funds, non-monetary support in the form of meeting space and training, and organizing the All Scientists Meetings. Waide stated that the use of technology, especially email, to conduct LTER business had grown over time. He also noted that while social scientists had been added to sites over the years, they did not feel as integrated into projects as scientists who were involved in initial project planning.

Dan Childers gave an overview of the LTER Planning Process (see Collins, Chap. 14, this volume). He described the charge to the Governance Working Group as being a) to consider modifications to the LTER organizational structure to promote synthetic research across sites, b) to foster interaction with other research networks, projects, and organizations, and c) to accommodate the Network's evolution and anticipated future growth. As alluded to in the minutes from the April 2005 Coordinating Committee meeting, the Planning Process leads also wanted the Governance Working Group to develop potential models to distribute an additional $10 million to support synthesis activities that might result from a proposal submitted as a product of planning grant activities. This part of our charge was removed over time, and in the end, the Governance Working Group was asked to address the following two questions:

- Is the LTER Network as it is presently constituted well-governed given the scope of present and known future activities?
- Will the present governance structure of the LTER Network accommodate new sites, collaborations with non-LTER sites, and resources that might result from the planning grant?

The Governance Working Group quickly concluded that the answer to these questions was "no" for three main reasons. First, while LTER's growth from six to 26 sites was a sign of success, it presented challenges in terms of governance. As predicted by the organizational science literature and other resources the Governance Working Group reviewed, it was clear that the growth in size and distribution of the network made it challenging for it to function efficiently and effectively. The benefits of a network for an entity such as LTER include flexibility, learning and a decentralized governance structure based on peer relationships rather than authority (Grandori 1997). The disadvantages of a network are that it can be difficult to organize and manage. As the number of members grows and are widely dispersed, making decisions efficiently and effectively becomes more challenging (Jones et al. 1997).

In LTER's case, it was difficult for a group the size of the Coordinating Committee to meet frequently enough to make decisions quickly, when required. Second, the strength of the Network is in its science. Under the existing structure, "the CC's time to engage in discussions about research direction, to set future courses of action, and to implement a higher level of research collaboration, synthesis, and integration was limited" (Zimmerman 2006, p. 5). Third, the current governance framework was not perceived as meeting increased demands for connectivity to facilitate a higher level of research collaboration, synthesis, and integration within and outside the Network. Lastly, cultivating a continuous supply of LTER leaders and new ideas was considered important.

In subsequent phone conferences, face-to-face meetings, and email discussions, the Governance Working Group attempted to address these challenges by exploring governance models that would have a group focused on the science, a more agile decision-making body, a training ground for future leaders, and a tighter relationship between the LTER sites and the Network Office. At the same time, the Governance Working Group recognized that it would be important to retain a democratic form of governance. Ever since it was established, the LTER had operated very much as a network. As Stafford (Chap. 13, this volume) stated, "The LTER Network has always prided itself on having taken a "bottom-up" versus a "top-down" approach (e.g. NEON)." Zimmerman and Nardi (2010) made a similar observation based on their analysis of the fundamentally different ways that LTER and the National Ecological Observatory Network (NEON) took on the objective of making ecology "big." Democracy based on representation was a fundamental component of the LTER approach to this challenge.

Among the literature on governance and networks cited above, the Governance Working Group found two other sources to be especially valuable. These were useful because they combined important concepts and components of governance with practical application for considering the fitness of LTER's decision-making processes given the Governance Working Group's charge. The Institute on Governance is a Canadian non-profit organization whose mission is to share and promote good governance and help organizations of all kinds put it into practice. The Institute on Governance boiled down eight principles of good governance developed by the United Nations Development Program (UNDP) into five categories (Graham et al. 2003, p. 3). The way in which each principle was applied to LTER is described below:

- Legitimacy and voice: Increase representativeness of sites in governance (i.e., greater inclusiveness of sites in LTER governance)

- Direction: Enhance development and visioning of future science agenda (i.e., adequate time for envisioning the future of LTER science)
- Performance: Increase efficiency of network governance (i.e., a smaller decision-making group)
- Accountability: Foster transparency in decision-making and greater accountability of network representatives (i.e., clear expectation of participants in the governance process)
- Fairness: Become more equitable through increased site representation in governance

These principles helped to guide the content of the bylaws that were ultimately proposed to the Coordinating Committee. Lastly, the Governance Working Group drew from the digital Community Tool Box developed and maintained by the Center for Community Health and Development at the University of Kansas. Of particular value was Chap. 9 on developing an organizational structure (Center for Community Health and Development 2017). This chapter included sections that described the important elements of an organization's structure, including governance, and guidance for deciding upon an appropriate framework, particularly the level of formality (Table 15.1).

The major alterations from the 2003 version of the bylaws were described by Zimmerman (2006). The names, responsibilities, and composition of the two key governing bodies were changed. The Coordinating Committee was renamed the "Science Council" and its responsibilities were revised to focus more on network science and synthesis. The Executive Committee was renamed the "Executive Board" and was given a broad mandate to carry out practical governance of the network. The Executive Board was expanded to increase representativeness and

Table 15.1 Conditions favoring more or less formality in organizational structures

Condition	A looser, less formal, less rule bound structure would be favored when...	A tighter, more formal, more rule-bound structure would be favored when...
Stage of organization development	The organization is just starting	The organization is in later stages of development
Prior relationships among members	Many such relationships already exist	Few such relationships already exist
Prior member experience in working together	Many such experiences have occurred	Few such experiences have occurred
Member motivation to be part of the organization	Motivation is high	Motivation is low
Number of organization tasks or issues (broadness of purpose)	There is a single task or issue	There are multiple tasks or issues
Organization size	The organization is small	The organization is large
Organization leadership	The leadership is experienced	The leadership is inexperienced
Urgency for action	There is no particular urgency to take action now	There is strong urgency to take action now

grew to include the Chair of the Science Council, nine members selected by individual sites on a rotating basis, an Information Manager, the Executive Director of the Network Office, and, as needed, a Chair-Elect. The Chair of the Science Council/Executive Board was given greater responsibility. In recognition of the time and effort required of the Chair, the Executive Director of the Network Office, in consultation with NSF, was tasked to negotiate a mechanism for compensation no later than 6 months after the Chair's election. This level of compensation varied from 0.5 to 6 months until this practice was ended by NSF in 2018. Committee mandates were clarified and an annual review process for committees was established. There was a recommendation to expand and diversify the National Advisory Board.

In March 2006, a draft of the bylaws was presented to the Executive Committee for comment, and their suggestions were reflected in a new draft. On May 18, 2006 John Magnuson, who was then Chair of the Coordinating Committee, presented the draft bylaws and led a lively discussion of the document at the Coordinating Committee meeting.[16] After agreeing to incorporate their suggestions, the revised bylaws were adopted by the Coordinating Committee by a vote of 25–0 (with one absence). The new bylaws were summarized and reprinted in full in *The Decadal Plan for LTER* (U.S. Long Term Ecological Research Network 2007).[17]

At the March 2007 meeting of the Executive Board, Chair John Magnuson, stated: "The new bylaws have worked very well, but we have not been able to follow them to the letter. Some items should likely be updated and made more realistic."[18] The minutes state: "The Executive Board decided by consensus that modifications to the bylaws at this time were premature and that another year or two of experience with them should precede submission of recommended changes to the Science Council." At the March 2008 Executive Board meeting, Chair Phil Robertson described five areas of the bylaws that merited discussion.[19] These changes were considered by the Science Council at their May 2008 meeting.[20] One amendment to the proposed changes was adopted by unanimous consent (the addition of "or his or

[16] This discussion is captured in Minutes of the LTER Coordinating Committee Meeting, Cedar Creek, Minnesota, May 17–18 2006. Archived on the LTER Document Archive under "Inactive Committees": https://lternet.edu/intranet/

[17] Robert B. Waide noted that due to an error, the text in *The Decadal Plan* failed to include two approved changes. One was in Article IV, Section 5 (defining a quorum and limiting the vote of the Chair to breaking ties) and the other in Article IX, Section 2 (shifting the responsibility for administering the annual survey of sites from the Network Office to the Executive Board). R. B. Waide, History of revisions to the LTER Bylaws, archived on the LTER Document Archive under "LTER Organization": https://lternet.edu/intranet/. Accessed 24 February 2020.

[18] Minutes of the LTER Executive Board Meeting, Arlington, Virginia, March 7–92,007. Archived on the LTER Document Archive under "Currently Active Committees": https://lternet.edu/intranet/

[19] Minutes of the LTER Executive Board Meeting, March 25, 2008; 2:00–4:00 pm EST Videoconference. Archived on the LTER Document Archive under "Currently Active Committees": https://lternet.edu/intranet/

[20] Minutes of the LTER Science Council Business Meeting, May 8, 2008; 7:30 pm-10:00 pm; Baltimore, Maryland. Archived on the LTER Document Archive under "Currently Active Committees": https://lternet.edu/intranet/

her designate" in Article IV, Section 2).[21] In May 2012, the Science Council was presented with two changes to the LTER Bylaws.[22]

> One set of changes were considered to be "housekeeping" in that they corrected grammar, eliminated contradictions, and added the Past Chair as a non-voting member of the EB [Executive Board] for one year following the end of his/her term. These changes passed unanimously. The second change to the Bylaws was a request for recognition of the Education Committee via a non-voting membership on the LTER EB similar to that of the LTER Data Managers.

The Science Council voted unanimously in favor of this motion and changes to the Bylaws, and this fourth version was adopted on May 17, 2012.[23]

Beyond the Bylaws revisions described above, there has been no formal evaluation of this new governance structure. There is a sense among one of us (Groffman), and those he has discussed this with, that it has been quite effective. Replacing elected members of the Executive Board with a set rotation involving all sites has unquestionably made the governance structure more representative and diverse. The latter is also apparent in the inclusion of an Information Management and an Education representative on the Executive Board. Science Council meetings now focus almost exclusively on synthesis and cross-site research. Network-wide and Targeted Standing Committee members are non-voting members of the Science Council, helping to keep these bodies more closely connected. Monthly meetings of the Executive Board have greatly reduced the amount of time devoted to network business at the Science Council meetings. A more engaged, diverse and representative Executive Board and Chair was able to more effectively interact with both the Network Office and NSF. Institution of term limits (two, two-year terms) for the Chair also served to increase a sense of representativeness within the network.

There was some discussion of the new governance structure during the thirty-year review of the LTER network in 2010. The review committee found that the tension between a less formal and more formal decision making process noted in earlier reviews persisted in spite of changes made to the Network's governance structure. The committee believed a stronger central leadership was necessary in order for LTER to effectively collaborate and cooperate with other initiatives and be a leader in the emerging "network of networks" (Michaels and Powers 2011, p. 2):

> Although the LTER Executive Board has streamlined network-level decision-making, we do see that the LTER network still suffers a perhaps unavoidable tension between bottom-up and top-down control. Decision-making in LTER has long been dominated by a

[21] Long Term Ecological Research Network Bylaws: Revision 3.0, May 8, 2008. Archived on the LTER Document Archive webpage under "LTER Organization": https://lternet.edu/intranet/. Accessed 24 February 2020.

[22] Minutes, LTER Science Council Business Meeting May 17, 2012 (6:30 PM), Andrews Experimental Forest, Oregon. Archived on the LTER Document Archive under "Currently Active Committees": https://lternet.edu/intranet/

[23] Long Term Ecological Research Network Bylaws: Revision 4.0, May 17, 2012. Archived on the LTER Document Archive under "LTER Organization": https://lternet.edu/intranet/. Accessed 24 February 2020.

> bottom-up approach that is very egalitarian and encourages creative within-site research. However, this approach may have compromised the ability of the network to adopt and enforce network-wide policies. We see a need for a greater centralized scientific leadership, empowered by the PIs, but that retains the dedicated involvement of site scientists. Organizational democracy principles may have great relevance here in helping this group find a path towards an empowered leadership team that fosters scientific creativity throughout the network (Michaels and Powers 2011, p. 13).

As noted above, the thirty-year review was commissioned and run by NSF, not by the network, and there was no formal response from the network to the review or these concerns about governance. NSF prepared a response to the 30 year review report that highlighted which of the report recommendations were most important to NSF. Governance was not mentioned in the NSF response.

15.6 Major Revisions to LTER Governance in 2016

Key recommendations from the thirty-year review highlighted by NSF were that "1) Resources are a key limiting factor for the future of the network. The network should: 1) make realistic prioritizations within the existing resource base to create more science per dollar, and 2) engage with NSF and others to pro-actively develop new resources. Prioritizations will include spending more on data and diversity and; 2) the LTER network as a whole must invest in making LTER data comparable across sites and more readily available to those interested in network-wide analyses" (Michaels and Powers 2011, p. 5). These recommendations, perhaps along with other factors, motivated NSF to convene an outside committee, chaired by Diane McKnight and including one of us (Zimmerman) to evaluate LTER governance, with a focus on the role of the network office in facilitating governance, communication and information management. This committee was asked to consider the potential range of network-level services and activities and recommend general approaches and priorities for the future. A major activity of the committee was to solicit input from a broad range of current and potential constituents of the LTER program, including educators and professionals involved in resource management and environmental change adaptation and to examine examples of other scientific network offices to identify successful models and useful innovations.

Based, in part, on the McKnight Committee's report, NSF issued a new solicitation for a LTER National Communications Office and a separate solicitation for an Environmental Data Initiative that would coordinate LTER information management and bring in data from other sources as well. This change effectively split the network office into a LTER National Communications Office (LNCO) based at the National Center for Ecological Analysis and Synthesis (NCEAS) at the University of California Santa Barbara, and an Environmental Data Initiative (EDI) based at the University of Wisconsin, replacing the LTER Network Office (LNO) that had been based at the University of New Mexico from 1997 to 2015. The split had several effects on LTER network governance. The LNCO and EDI now answer directly to NSF, not to the LTER Science Council, which had played a key role in evaluating the function of the LNO and in preparing renewal proposals. The split also reduced resources for network and coordination and governance. NSF was clear that expenditures on governance were not a priority. They prohibited the use of NSF funds to pay the chair of the Science Council and recommended reducing the number of committees. Cross-site research was removed as an evaluation criterion for LTER projects and there was more focus on direct relationships between sites and NSF. The bylaws were revised in 2017 to reflect these changes.[24]

The 2016 changes were quite fundamental and changed the nature and extent of LTER network governance to a more top-down, NSF-driven system. As discussed further below, this change motivates and limits governance to a focus on science. The NCEAS-based LNCO has a strong concentration on synthesis, and the solicitation and support of synthesis working groups has been a major, very well received, focus of the LNCO, with a strong role for the Executive Board and Science Council. The mandate to revisit and reduce the number of committees also motivated a focus on just how these committees were facilitating advances in science, information management, education and outreach, diversity, publications and involvement of graduate students.

15.7 Evaluating Governance

Pinho and Pinho (2016) describe "the production of knowledge as a social process that must be managed" (p. 489). They argue that "there should be an alignment between the knowledge management and the research knowledge governance in order to match common efforts to achieve scientific objectives and desired societal impacts" (p. 490). Authors in a special issue of the journal *Frontiers in Ecology and the Environment* devoted to "network governance and large landscape conservation" (e.g., Bixler et al. 2016) and elsewhere (Mirtl et al. 2018), have emphasized the importance of metrics to assess the strategies and successes of network governance. While such metrics have not been systematically developed for the LTER network, a variety of evaluations have taken place. When the Network Office was located at the University of New Mexico, it carried out annual (or nearly so) surveys of the network primarily to assess satisfaction with the office, but also gather information on the state of the network, including its governance. These surveys were not generally considered to be effective. Response rates were low, the questions varied from year to year and survey professionals were not generally engaged in development of the survey instruments. Since the office moved to Santa Barbara, a very brief (3–5 question) survey instrument has been used to assess satisfaction with

[24] Long Term Ecological Research Network Bylaws: Revision 5.0, May 18, 2017. Archived on the LTER Document Archive under "LTER Organization": https://lternet.edu/intranet/. Accessed 24 February 2020.

the LNCO and EDI, but there has been no effort to assess the specific effects of the governance changes instituted in 2016.

The ten-year reviews of the LTER program have extensively addressed governance and have driven some of the major changes described above, e.g., the development of bylaws, formation of an Executive Committee and then an Executive Board, and conversion of the Coordinating Committee into a Science Council. The approaches and procedures for the ten-year reviews have varied quite a bit however, and there has been no development of metrics to assess the strategies and successes of network governance.

NSF conducts frequent and detailed reviews of individual sites and of the network office, but these have seldom directly addressed network governance. Support for increased governance to facilitate cross-site research or synthesis has been expressed at times, while on other occasions (e.g., 2016), NSF has shown less interest in and has allocated fewer resources for governance. The variation has been driven by changes in the availability of funds, programmatic interests, and personnel changes at NSF.

The lack of systematic evaluation of governance and development of metrics to assess the strategies and successes of network governance leave us with a series of unanswered questions. We do not know how well revised governance structures have served to meet network needs at different times. We have no information on what governance has meant to personnel within the networks. How has it influenced intellectual outcomes, training of the next generation, and the success of individual sites? More generally, what is/has been the interaction between the governance of each site and the governance of the network? How do people who are not part of LTER, but who have studied it or ecology more generally perceive the role of governance?

15.8 Conclusions

Despite the lack of formal and systematic evaluation, we (the authors of this chapter) feel that governance has been a dynamic and important contributor to scientific success in the LTER network. From the very beginning, changes in governance were motivated by the desire to do more and better science. In this chapter, we have been able to link changes in governance to the emergence of large-scale questions that could best be addressed by cross-site research, the need for synthesis of complex results, new developments in information management and the emergence of new environmental research networks. These are fundamental changes, and we argue that the network governance responses have been relatively nimble and effective. Perhaps the clearest evidence for this success is the continued high productivity of the LTER network and the active engagement of large numbers of scientists in network activities such as Science Council and All Scientists Meetings and both standing and ad hoc committees.

Future governance of the LTER network will continue to be driven by a variety of factors including the emergence of a new era of network science, changes in priorities and personnel at NSF, and scientific developments. Much attention has been paid to the emergence of multiple environmental research networks in the U.S. and around the world, heralding a new approach to addressing grand challenges in environmental and ecological science (Richter et al. 2018; Mirtl et al. 2018; Baatz et al. 2018). Developing potential synergies among these networks is exciting but requires extensive time. A major challenge for LTER is who will represent LTER in interactions with NEON, the Critical Zone Observatory and international LTER (ILTER) networks. Will this be done by the LNCO? By the chair of the Executive Board? Ad hoc volunteers?

If the past is any predictor of the future, changes in priorities and personnel at NSF will be a major driver of changes in LTER governance. Certainly, interpreting and implementing these changes will continue to be a major function of network governance. But governance also is needed to further internal and "bottom up" network goals. For LTER, these have been and will most likely continue to be driven by the emergence of new approaches and ideas in science. Recent exciting trends that have emerged in ecological science include a focus on increasingly larger scales, e.g., Macrosystems Biology (Heffernan et al. 2014), incorporation of population, community and evolutionary ecology into the more ecosystem-focused LTER network (Kuebbing et al. 2018), advances in information management and the emergence of integrated sustainability science (Weathers et al. 2016). The LTER governance structure needs to provide a platform for network-wide discussion of these emerging topics and development of ideas of multi-site research. These discussions will always be relevant and independent of changes in focus at NSF.

In addition to facilitating scientific progress, LTER governance will continue to play a role in coordination and development of strategies for managing LTER sites and projects. LTER projects are complex entities, and LTER governance has provided and will continue to provide a platform for sharing ideas and experiences about how to manage sites, experiments, data, personnel, budgets, leadership development, and other components of these entities. Indeed, these discussions become increasingly important as priorities and personnel change at NSF.

Acknowledgments We want to recognize the other members of the Governance Working Group (listed above) who contributed equally to the 2006 bylaws revision described in this chapter. Bob Waide's recollection and insights into LTER's history are remarkable. Although this chapter is written from our perspectives, Bob helped us navigate the information that we drew from to confirm or discover details relevant to LTER history. We hope that Bob will one day write the history from his unique vantage point. Sharon Kingsland and Bob Waide provided comments that greatly improved this chapter. They were patient and encouraging editors. The LTER Document Archive was indispensable in the production of this chapter. This electronic repository is a centralized source of LTER published and unpublished documents. Of particular value for this chapter were minutes from the meetings of LTER's main governing bodies. We acknowledge the foresight and effort required to produce this archive. It is an excellent example of transparency in action. We encourage LTER to continue to add old and new documents to it, so others can use it to understand LTER's history and ongoing evolution.

References

Baatz, Roland, Pamela L. Sullivan, et al. 2018. Steering operational synergies in terrestrial observation networks: Opportunity for advancing Earth system dynamics modelling. *Earth Systems Dynamics* 9 (2): 593–609.

Bixler, R. Patrick, Matthew McKinney, and Lynn Scarlett. 2016. Forging new models of natural resource governance. *Frontiers in Ecology and the Environment* 14 (3): 115. https://doi.org/10.1002/fee.1255.

Callahan, James T. 1984. Long-term ecological research. *BioScience* 34 (6): 363–367.

Center for Community Health and Development. 2017. *Chapter 9, Section 1: Organizational structure: An overview*. Lawrence, Kansas: University of Kansas. https://ctb.ku.edu/en/table-of-contents/structure/organizational-structure/overview/main. Accessed 24 Feb 2020.

Franklin, Jerry.1993. *Long-Term Ecological Research Network 10 year review response*. August 9. https://lternet.edu/wp-content/uploads/2010/12/10year-review-response.pdf. Accessed 24 Feb 2020.

Gosz, James R., Robert B. Waide, and John J. Magnuson. 2010. Twenty-eight years of the US-LTER Program: Experience, results, and research questions. In *Long-term ecological research*, ed. Felix Müller et al., 59–74. New York: Springer.

Graham, John, Bruce Amos, and Tim Plumptre. 2003. *Principles for good governance in the 21st century*. Policy Brief 15. Ottawa: Institute on Governance.

Grandori, Anna. 1997. Governance structures, coordination mechanisms and cognitive models. *Journal of Management and Governance* 1 (1): 29–47.

Groffman, Peter M., Charles T. Driscoll, Christopher Eagar, Melany C. Fisk, Timothy J. Fahey, Richard T. Holmes, Gene E. Likens, and Linda H. Pardo. 2004. A new governance structure for the Hubbard Brook Ecosystem Study. *Bulletin of the Ecological Society of America* 85 (1): 5–6.

Harris, Frank, and Leonard Krishtalka. 2002. *Long-Term Ecological Research Program twenty-year review*. National Science Foundation. Report to NSF of the twenty-year review committee. http://lternet.edu/wp-content/uploads/2014/01/20_yr_review.pdf. Accessed 24 Feb 2020.

Heffernan, James B., Patricia A. Soranno, Michael J. Angilletta, Lauren B. Buckley, Daniel S. Gruner, Tim H. Keitt, James R. Kellner, John S. Kominoski, Adrian V. Rocha, Jingfeng Xiao, Tamara K. Harms, Simon J. Goring, Lauren E. Koenig, William H. McDowell, Heather Powell, Andrew D. Richardson, Craig A. Stow, Rodrigo Vargas, and Kathleen C. Weathers. 2014. Macrosystems ecology: Understanding ecological patterns and processes at continental scales. *Frontiers in Ecology and the Environment* 12 (1): 5–14.

Hobbie, John E. 2003. Scientific accomplishments of the Long Term Ecological Research Program: An introduction. *Bioscience* 53 (1): 17–20.

Jones, Candace, William S. Hesterly, and Stephen P. Borgatti. 1997. A general theory of network governance: Exchange conditions and social mechanisms. *Academy of Management Review* 22 (4): 911–945.

Kuebbing, Sara E., Adam P. Reimer, Seth A. Rosenthal, Geoffrey Feinberg, Anthony Leiserowitz, Jennifer A. Lau, and Mark A. Bradford. 2018. Long-term research in ecology and evolution: A survey of challenges and opportunities. *Ecological Monographs* 88 (2): 245–258. https://doi.org/10.1002/ecm.1289.

LTER National Advisory Board. 1998. Report of the LTER National Advisory Board. https://lternet.edu/wp-content/uploads/2010/12/98nab_report.pdf. Accessed 24 Feb 2020.

———. 2003. Report of the LTER National Advisory Board. https://lternet.edu/wp-content/uploads/2010/12/LTER_NAB.pdf. Accessed 24 Feb 2020.

Magnuson, John, et al. 2006. Origin, operation, evolution, and challenges. In *Long-term dynamics of lakes in the landscape*, ed. John J. Magnuson, Timothy K. Kratz, and Barbara J. Benson, 280–322. New York: Oxford University Press.

Michaels, Anthony, and Alison G. Powers. 2011. *National Science Foundation Long-Term Ecological Research Program: A report of the 30 year review committee*. https://lternet.edu/wp-content/uploads/2011/12/bio12001.pdf. Accessed 24 Feb 2020.

Mirtl, M., E.T. Borer, I. Djukic, M. Forsius, H. Haubold, W. Hugo, J. Jourdan, D. Lindenmayer, W.H. McDowell, H. Muraoka, D.E. Orenstein, J.C. Pauw, J. Peterseil, H. Shibata, C. Wohner, X. Yu, and P. Haase. 2018. Genesis, goals and achievements of Long-Term Ecological Research at the global scale: A critical review of ILTER and future directions. *Science of the Total Environment* 626: 1439–1462.

Pinho, Isabel, and Cláudia Pinho. 2016. Aligning knowledge management with research knowledge governance. In *Handbook of research on innovations in information retrieval, analysis, and management*, ed. Jorge Tiago Martins and Andreea Molnar, 489–504. Hershey: IGI Global.

Richter, D.D., S.A. Billings, P.M. Groffman, E.F. Kelly, K.A. Lohse, W.H. McDowell, T.S. White, S. Anderson, D.D. Baldocchi, S. Banwart, S. Brantley, J.J. Braun, Z.S. Brecheisen, C.W. Cook, H.E. Hartnett, S.E. Hobbie, J. Gaillardet, E. Jobbagy, H.F. Jungkunst, C.E. Kazanski, J. Krishnaswamy, D. Markewitz, K. O'Neill, C.S. Riebe, P. Schroeder, C. Siebe, W.L. Silver, A. Thompson, A. Verhoef, and G. Zhang. 2018. Ideas and perspectives: Strengthening the biogeosciences in environmental research networks. *Biogeosciences* 15 (15): 4815–4832.

Risser, Paul G., and Jane Lubchenco. 1993. Ten-year review of the National Science Foundation Long Term Ecological Research (LTER) Program: Report to NSF of the ten-year review committee. https://lternet.edu/wp-content/uploads/2010/12/ten-year-review-of-LTER.pdf. Accessed 24 Feb 2020.

Thomas, McOwiti O. and Robert Waide. 2005. Jim Gosz resigns as Chair of LTER Coordinating Committee. *Network News* 18(2). http://news.lternet.edu/Article36.html. Accessed 24 Feb 2020.

U.S. Long Term Ecological Research Network (LTER). 2007. *The decadal plan for LTER: Integrative Science for Society and the Environment*. Albuquerque, New Mexico: LTER Network Office Publication Series No. 24. https://lternet.edu/wp-content/uploads/2010/12/TheDecadalPlanReformattedForBook_with_citation.pdf. Accessed 24 Feb 2020.

Waide, Robert. 1998. Network Office survives transition. *The Network News* 11(1). https://lternet.edu/wp-content/uploads/2013/01/volume_11_number_1_Spring_1998.pdf. Accessed 24 Feb 2020.

Weathers, Kathleen C., Peter M. Groffman, E. Van Dolah, E.S. Bernhardt, N.B. Griffm, K. McMahon, Joshua P. Schimel, M. Paolisso, R. Maranger, S. Baer, K. Brauman, and E. Hincklye. 2016. Frontiers in ecosystem ecology from a community perspective: The future is boundless and bright. *Ecosystems* 19: 753–770.

Zimmerman, Ann. 2006. Rapid network evolution prompts changes in LTER governance. *Network News* 19 (2): 5–6.

Zimmerman, Ann, and Bonnie Nardi. 2010. Two approaches to Big Science: An analysis of LTER and NEON. In *Collaboration in the new life sciences*, ed. John N. Parker, Niki Vermeulen, and Bart Penders, 65–84. Surrey: Ashgate.

Chapter 16
Understanding the Fundamental Principles of Ecosystems through a Global Network of Long-Term Ecological Research Sites

Robert B. Waide and Kristin Vanderbilt

Abstract The Long Term Ecological Research (LTER) Network served as a catalyst to promote cooperation among multi-national research programs and networks. The chapter describes the strategic planning process in the 1990s that led to the creation of the International Long Term Ecological Research (ILTER) Network, which expanded rapidly in the late-1990s with support of the National Science Foundation (NSF). Especially under the leadership of James Gosz, cooperative arrangements were made with countries worldwide and 23 LTER networks had formed by 2003. Many U.S. LTER scientists and information management specialists contributed to the expansion of the ILTER Network through site visits and workshops. After NSF scaled back its support, ILTER reorganized its governance structure and grew into a robust, self-sustaining network of networks. The chapter reviews examples of collaborations in research and information management. The ILTER Network today aims to become part of a global infrastructure to address continental and global socio-ecological problems through partnerships with other international networks. Concerns about climate change, biodiversity loss, and the scarcity of research sites producing long-term data led to the creation of the U.S. LTER Network and hence ILTER, and the legacies of data, information, and long-term measurements that have resulted are a critical contribution to resolving current global problems.

Keywords LTER Program · Long-term ecological research · International LTER · International ecology · Ecosystem ecology · Comparative ecology · Networking · ILTER Network · Global infrastructure · Socio-ecological systems

R. B. Waide (✉)
Department of Biology, University of New Mexico, Albuquerque, NM, USA
e-mail: rbwaide@unm.edu

K. Vanderbilt
Florida Coastal Everglades LTER Program, Institute of Environment, Florida International University, Miami, FL, USA
e-mail: krvander@fiu.edu

R. B. Waide, S. E. Kingsland (eds.), *The Challenges of Long Term Ecological Research: A Historical Analysis*, Archimedes 59,
https://doi.org/10.1007/978-3-030-66933-1_16

16.1 Introduction

The rationale that led to the development of the Long Term Ecological Research (LTER) Network is not unique to the United States. The workshops that preceded the establishment of the LTER program called out the general scarcity of long-term ecological data in the literature and the importance of protected research sites where such data could be collected (The Institute of Ecology 1977). Global concern about climate change and loss of biodiversity identified the lack of long-term data as a critical problem in evaluating environmental trends. Long-term data are particularly important in assessing four kinds of ecological phenomena: (1) slow processes, (2) rare or episodic events, (3) processes with high variability, and (4) subtle processes and complex phenomena (Franklin 1989). Existing long-term data sets also emphasized the value of assessing trends at regional (Likens and Bormann 1974), continental (Magnuson et al. 2000), and global scales (Keeling et al. 2005) and implicitly recognized the importance of comparative studies at these scales. These concerns naturally led to a call for cooperation among multi-national research programs and networks (Bledsoe 1993).

The LTER Network was primed to act as a catalyst for the development of such cooperation. The careful planning that had been the precursor to the LTER program identified comparative research and synthesis as important goals of long-term research (National Science Foundation 1979). The persistence of the National Science Foundation (NSF) in emphasizing comparative studies at a network scale stimulated the early development of new approaches to cross-site study (Magnuson and Waide, Chap. 12, this volume). The first decade of the LTER program was an experiment in establishing and sustaining long-term research studies; two of the original 11 sites selected failed after the first funding cycle and a third was discontinued in 1993. As the program stabilized, NSF and LTER leadership collaborated in the formation of a research network that expanded the original focus of the LTER program (Magnuson and Waide, Chap. 12). In 1992, at the request of NSF, the LTER Network developed a strategic plan to expand the scope of long-term ecological research.[1] This plan aimed at capitalizing on the research and organizational experience gained in the development of the U.S. LTER Program to create a global network of research sites of which the U.S. network would be only a part. The plan proposed to complete both broad disciplinary and geographical expansions of the LTER concept by the beginning of the twenty-first century.

The Ten-Year Review of the LTER program (Risser et al. 1993) provided a road map for an expanded and more closely integrated network of sites to address the challenges of designing and operating a sustainable biosphere. Collaboration with other national networks was a key factor in addressing these challenges. The NSF based an expanded trajectory for the U.S LTER Network on the recommendations of the Ten-Year Review. Near-term goals for the LTER Network included: (1)

[1] *LTER 2000: Creating a Global Environmental Research Network*. https://lternet.edu/wp-content/uploads/2010/12/LTER2000.pdf. Accessed 7/3/2020.

Expansion of site activities to a regional scale, (2) Broadening of site research to include the geophysical, social, and economic sciences, and (3) Development of more complete site inventories and histories to better understand current conditions and make future projections (Clutter and Gosz 1994). In addition to these goals, NSF initiated an aggressive program of collaboration with other national research networks, beginning with an invited workshop at the 1993 LTER All Scientists Meeting.

16.2 Origins of the International Long Term Ecological Research (ILTER) Network

The idea to encourage international collaboration in long-term ecological research came from the NSF as a request to the LTER Network Office in Seattle. In 1992, Mary Clutter, Assistant Director for the Biological Sciences Directorate at NSF, proposed the idea of an international LTER program to Jerry Franklin, Chair of the LTER Coordinating Committee.[2] Through communications over the next few months, Jim Gosz, Director of NSF's Division of Environmental Biology, pursued the concept vigorously with Franklin and other members of the Network Office staff. Gosz had come to NSF after a year in Washington working as the Executive Director for the Sustainable Biosphere Initiative for the Ecological Society of America. He was one of the founding scientists of the Sevilleta LTER program and was very familiar with the development of the LTER Network. Between them, Gosz and Franklin settled on the idea of an international summit to be held in conjunction with the LTER All Scientists Meeting in Estes Park, Colorado.

The primary goal of the summit was to examine the potential for developing a global long-term ecological research network. The meeting was by invitation and participants were selected from nominees submitted by NSF (including program managers from the International Program), LTER, and other international agencies like the Program for Man and the Biosphere (MAB). Although the original plan was for a smaller meeting, generous support from NSF allowed participation of 39 scientists and administrators from 16 countries from every continent but Antarctica. Strong representation by NSF emphasized the importance that was attached to the meeting; nine representatives from NSF outnumbered the seven LTER participants (Nottrott et al. 1994). Subsequent to the workshop, NSF provided extensive support to encourage collaboration with the initial group of countries and to expand the number of collaborating countries.

The concept of collaborative long-term research, as embodied by the U.S. LTER Network, was gaining credence in many countries simultaneously. Antecedents of the U.S. LTER program such as the International Biological Program (IBP), research initiatives by private institutions and public agencies, long-term monitoring of

[2] Personal communication from Jerry Franklin 2018.

climate and resources, and singular measurements by dedicated individuals were also operating in other places around the world. The concurrent rise in the importance of long-term ecological research provided an important impetus for the development of multi-national research collaboration. The rationale for collaboration was based on four important points:

- Comparison among different ecosystems is critical to understanding fundamental ecological principles—Generality of ecological principles must be established before they can be applied to issues such as conservation. Comparison of similar systems across national boundaries or continents is one way of establishing such generality.
- National borders are irrelevant to ecological patterns and trends – Issues of pollution, biodiversity loss, invasive species, and climate change require study at scales that transcend national borders.
- Policies that address conservation, resource management, and introduced species must be coordinated internationally—Agreement on common policy objectives requires a common knowledge base.
- The increasing need for solutions to problems such as global environmental change requires international collaboration – Establishing the knowledge base to identify such solutions is a global priority.

16.3 Bridging the Gap between National and International Research Networks

The preface to the proceedings of the international summit (Nottrott et al. 1994), authored by Mary Clutter and Jim Gosz, provides evidence of NSF's commitment to expanding long-term research internationally. They viewed the summit as a springboard to establishing the LTER program as a global leader in comparative ecology. Increased collaboration with well-established and developing research networks worldwide was one of the anticipated outcomes of the meeting, and the attendance of two program officers (William Chang and Christine French) from the Division of International Programs was an indication of broad support within the Foundation. Their goal was to stimulate global scale comparisons and modeling through exchanges of scientists, techniques, data, and information. Expansion of research capabilities through collaboration between established and developing sites and networks was also an important goal.

The summit was a working meeting. After presentations about the status of long-term research in each country, participants identified key issues to be resolved and broke into working groups. The five working groups addressed (1) communication and information access, (2) creating a global directory of research sites, (3) developing LTER programs worldwide, (4) scaling, sampling, and standardization, and (5) education, public relations, and relationships with decision makers. Recommendations in each of these focus areas were presented to the whole group and formed the basis of a working plan. Participants identified six action items and formed two committees (Steering and Connectivity) to follow up on these action items. The Steering Committee, chaired by Jerry Franklin, was charged with identifying approaches to expand the network and to investigate funding opportunities. The Connectivity Committee (chaired by Rudolph Nottrott of the U.S. LTER Network Office) focused on developing a timeline and plan of action to improve electronic communication and international Internet access to information and data.

The U.S. LTER Network Office provided initial support for the ILTER Network by providing resources for meetings of the Steering and Connectivity Committees. The first meeting of the Steering Committee took place in 1994 at Rothamsted in the United Kingdom in association with the International Congress of Ecology (INTECOL). At this meeting, the Steering Committee confirmed the original decision to adopt a regional approach to the growth of ILTER and initiated a program for global networking of ILTER scientists. The Connectivity Committee, drawing on expertise at the Network Office as well as information managers from U.S. LTER sites, maintained e-mail groups for the committees and developed a portable mini-Local Area Network that was used at a global connectivity workshop at INTECOL and demonstrated for the Steering Committee at the Rothamsted meeting. Improved communication among networks quickly became one of the most important tasks of the Connectivity Committee.

The U.S. LTER Network Office assumed much of the responsibility for planning and logistics for the ILTER Network. However, individual LTER sites and scientists were responsible for coordination of international science projects, supported principally by the Division of International Programs at NSF. Funds for exchange of scientists and students or research working groups often came as supplements to site or Network Office awards. The LTER Network Office anticipated a significant expansion in collaboration among scientists, sites, and long-term ecological networks in its renewal proposal prepared in 1994. Six workshops were planned for 1995–96, and the proposal included support for annual ILTER meetings in Hungary (1995), Costa Rica/Panama (1996), Taiwan (1997), and Italy (1998). An inter-network research project between the U.S. LTER and the United Kingdom's Environmental Change Network aimed to identify long-term data that could be compared between the two networks. The initial focus of this study involved the North Temperate Lakes site in the U.S., the Great Lakes, and the U.K.'s Lake District.

Coordination in data management among networks was an important goal of the Connectivity Committee with support from the Network Office and the U.S. LTER's Information Management Committee (Stafford, Chap. 13, this volume). The initial group of sites in the ILTER Network varied significantly in their approaches to communication and their capabilities to manage data. Coordination in these areas required exchanges of ideas and technology that were facilitated by the Network Office. The Network Office also arranged technical assistance for developing sites and networks. Early exchanges took place between U.S. sites and sites in China and Hungary. The U.S. LTER also provided training in techniques of communication and data exchange. The Network Office proposed to establish an ILTER computer server for communication and data exchange, create an ILTER homepage, and carry out workshops at the annual ILTER meeting and in different regions of the world as requested.

16.4 Expansion of the ILTER Network

The U.S. LTER Network also provided leadership in identifying and developing potential new member nations for ILTER and helped to organize regional groupings of networks. The Steering Committee identified existing or potential nodes (institutions, sites, or networks) around which regional groupings of networks could coalesce. This process often involved selecting the most appropriate technically developed nodes to provide a model for less developed national networks. As a result of these efforts, the ILTER Network expanded rapidly. By 1998, over 200 research sites existed in 15 national networks, and another 12 countries were working to establish national programs (Waide et al. 1998). Twenty-one national networks were members of ILTER by 2000 (Gosz et al. 2000).

The early expansion of the ILTER Network was fueled by support from NSF, the U.S. LTER Network, and the Network Office. Jim Gosz took over as chair of the U.S. LTER Coordinating Committee in 1995 after his stint at NSF. He was also elected chair of the multi-national ILTER Network Committee (formerly the Steering Committee). Gosz played a key role in spreading the concept of long-term ecological research internationally, especially in the countries of Eastern Europe. Through his contacts in the Division of International Programs at NSF, he was able to engage program officers whose responsibilities covered most of the globe. These program officers were enthusiastic about expanding the ILTER Network and provided support for many of the exchanges of scientists and students that took place while Gosz was ILTER chair.

Under Gosz's leadership, the ILTER Network Committee developed a mission statement consisting of the following goals (Waide et al. 1998):

1. Promote and enhance the understanding of long-term ecological phenomena across national and regional boundaries;
2. Promote comparative analysis and synthesis across sites;
3. Facilitate interaction among participating scientists across disciplines and sites;
4. Promote comparability of observations and experiments, integration of research and monitoring, and exchange of data;
5. Enhance training and education in comparative long-term ecological research and its relevant technologies;
6. Contribute to the scientific basis for ecosystem management;
7. Facilitate international collaboration among comprehensive, site-based, long-term, ecological research programs; and
8. Facilitate development of such programs in regions where they currently do not exist.

This mission statement provided a framework to identify and recruit new countries to the ILTER Network. The recruitment process was very low-key and depended on identifying those countries that would benefit from long-term ecological studies. Gosz described the approach to potential ILTER networks in the 1998 report on the *International Long Term Ecological Research Network* (Waide et al. 1998):

> Each country must assess its own needs and resources if it wishes to involve itself in an LTER program. Each will have a unique set of opportunities and limitations that are best evaluated by the scientists and policy makers of that country. The typical procedure for a country is for the scientists of that country, along with the funding agencies, to decide whether to endorse the premise that ecology and environmental management are significantly benefited by studies of long-term and broad spatial scales. A plan is then developed that establishes the context and mission for such studies, sites and programs are identified that will contribute to this mission, and support is obtained from within that country or international organizations for implementation and continued maintenance. It is anticipated that each country's program will be part of a global network of scientists and of scientific information that will advance our understanding of not only local and regional, but also global issues and provide solutions to environmental problems at these scales.

In the Preface to the 1998 report, Mary Clutter and Bruce Hayden (Director, Division of Environmental Biology) described NSF's response to the recommendations from the 1993 ILTER Summit:

> The response of the National Science Foundation was to charge the U.S. LTER Network Office with catalyzing the ILTER concept by working with other nations to join in building long-term ecological research programs. The Foundation is pleased with the rapid ILTER progress in 5 short years.

By 1997, the LTER Network Office had relocated to the University of New Mexico with Gosz as the principal investigator on the award (see Chap. 2, this volume). In October of that year, Bob Waide was hired as Executive Director of the Network Office and took over some of the responsibilities for the coordination of the ILTER Network from Gosz. Gosz and Waide shared the task of communicating with countries interested in joining the LTER Network, with Gosz focusing on Eastern Europe and Asia and Waide on Central and South America. Many other scientists from the U.S. LTER Network worked with international colleagues to encourage the formation of new long-term research networks. The Network Office served as the conduit for supplementary awards supporting ILTER activities. Between 1997 and 2002, Waide obtained 14 supplemental awards totaling $747,723 for international exchanges of scientists and students, meetings of the ILTER Network and regional committees, and the development of a public information video about the ILTER Network.

NSF also provided additional support for ILTER by posting Dr. Christine French, from NSF's Office of International Programs, to the LTER Network Office from 1997 to 2000. French played an important role in doubling the size of the ILTER Network by catalyzing numerous exchanges between U.S. and international scientists and students (LTER Network Office 2000). Through support by NSF to the LTER Network Office, she organized groups of U.S. scientists to travel to nascent LTER sites in other countries to promote the LTER model.

The U.S. LTER Network and the Network Office played a critical role in defining and growing the ILTER Network. Paralleling the development of the U.S. LTER network, national networks focusing on long-term ecological research were created in 25 countries between 1993 and 2003. Groups of U.S. LTER scientists acted as consultative bodies regarding interactions with national networks in the East Asia-Pacific, Latin American, Eastern European, and African Regions. U.S. LTER information managers and LTER Network Office staff shared information management expertise with developing networks.

China and Taiwan were among the first countries outside the U.S. to establish LTER Networks in 1988 and 1992, respectively. To encourage collaboration with these Asian LTERs, eight U.S. scientists visited sites in the Chinese Ecological Research Network (CERN) for three weeks in 1991 (Gosz and Leach 1991), followed by a reciprocal visit of scientists from the Chinese Academy of Sciences to several U.S. LTER sites. CERN, the Taiwan Ecological Research Network (TERN), Australia's LTER, and other networks were quick to support the regional LTER model, forming the Eastern Asia-Pacific (EAP) ILTER region in 1995. EAP-ILTER now consists of 10 countries and holds a biennial regional ILTER meeting (Kim et al. 2018). There was strong support from NSF to develop relationships between U.S. and EAP junior scientists. In 1997 12 Japanese and Taiwanese students visited U.S. LTER sites (Vanderbilt 1998) and 14 U.S. students visited TERN and CERN the following year (French 1998a). A group of 10 U.S. LTER students and scientists toured Japanese LTER research sites in 1999 (Sherrod 1999). Both trips were supported by International Programs at NSF (French 1998b).

Jim Gosz worked with National Academies in Eastern and Central European countries to stimulate enthusiasm for LTER and arrange scientist exchanges. U.S. scientists made trips to Hungarian LTER sites (Freckman 1994) and met with scientists from the Czech Republic and Slovakia at nature reserves to discuss forming LTER Networks in their countries (Network Office 1995). Central Europeans made reciprocal visits to U.S. LTER sites in 1995 (Network Office 1996a). Jim Gosz, LTER Network Office personnel, and other researchers attended meetings in Poland in 1998 (French 1998c), and Hungary (Lajtha and Vanderbilt 2000), which solidified the creation of the Central/Eastern Europe regional ILTER network. Some sites that became part of LTER Networks in Poland, Czech Republic, and Hungary had been conducting long-term research since the IBP and MAB programs in the 1970s.

As of 2000, the only two formal LTER Networks in Western Europe were in the United Kingdom and Switzerland (Gosz et al. 2000). Although several U.S. LTER scientists visited possible LTER sites in Spain and Portugal in 1996 (LTER Network Office 1996b), neither country formed an LTER Network until 2006 or later. Many western European countries did not formally establish LTER networks until after LTER-Europe was initiated in 2003 under the auspices of a European Commission's "Networks of Excellence" program entitled "Long-term Biodiversity and Ecosystem Research and Awareness Network (ALTER-Net)" (Mirtl et al. 2013). ALTER-Net emphasized socio-ecological research, and LTER-Europe has many Long-Term Socio-Ecological Research (LTSER) platforms. LTSER platforms usually consist of thousands of hectares with a sizeable human population and may include one or more smaller LTER sites (Mirtl 2010). LTSER research includes humans as an integral part of the study system to better understand societal impacts on the environment and the sustainability of ecosystem services (Haberl et al. 2006). LTER-Europe has subsumed the Central/Eastern Europe regional ILTER network and now has 25 member countries.

U.S. LTER scientists found support for the LTER concept in many other countries in the Americas. Following a meeting in Costa Rica in November 1996, Brazil, Costa Rica, Uruguay, and Venezuela formed LTER networks (French 1998b), and Colombia created a network shortly thereafter. The Latin American regional ILTER Network was established at the ILTER meeting in Brazil in 1997. By 2000, the Latin American regional network comprised five national networks and five other countries (Argentina, Bolivia, Chile, Ecuador, Paraguay; Gosz et al. 2000) were considering forming long-term ecological research programs. A North American regional network, consisting of the U.S., Canada, and Mexico LTERs, was initiated following a workshop at the Ecological Society of America 1998 meeting (Waide et al. 1998). These two regional networks have since merged into a single Americas Region ILTER Network.

Jim Gosz carried the ILTER message to researchers in South Africa in 1997 (Gosz 1997), and in 1999 South Africa hosted the annual ILTER meeting. In 2001, the Environmental Long-term Observatories of Africa (ELTOSA) was born to foster regional synergy among the countries of South Africa, Mozambique, Namibia, Malawi, and Zambia, all of which had established LTER Networks (Henschel et al. 2003). Mozambique LTER hosted the regional ILTER meeting in 2002, which was attended by several scientists from the U.S. Local support for most LTERs in ELTOSA was meager, and the only African country active in the ILTER today is South Africa.

The growth phase of the ILTER Network sparked entities besides NSF to consider development of networks in other regions. In 2002, the U.S. State Department proposed the formation of an LTER network composed of EU-accession states surrounding the Baltic Sea. As a result, Waide was invited by the State Department to provide information on ILTER to relevant scientific organizations in these candidate countries. In 2002, he travelled to Denmark, Sweden, Latvia, Estonia, and Finland in the company of Paul Thorn, a representative of the Nordic/Baltic Regional Environmental Office of the U.S. Embassy in Copenhagen. Thorn contacted scientific organizations in Lithuania and Norway separately. All these countries (except for Estonia and Lithuania) eventually developed long-term ecological research programs and became part of the ILTER Network. Waide also met with Terry Parr of the U.K.'s Environmental Change Network, who was leading an effort to develop a pan-European long-term ecological research network under the Sixth Framework of the European Union. This effort eventually absorbed the Baltic, Central/Eastern Europe, and Western Europe networks into one ILTER region, ALTER-Net, as described above.

As new networks formed, U.S. LTER (through NSF support to the LTER Network Office) engaged international researchers in capacity building workshops to share U.S. LTER's information management expertise. Data sharing is a core principle of LTER research, and U.S. LTER information managers had years of experience to help guide networks just beginning to develop information management systems. The U.S. LTER hosted several information management workshops for scientists throughout the ILTER Network during the 1990s and early 2000s. Four U.S. information managers taught a 2-day information management workshop in China in 1992, which was followed by an intensive month-long training program for Chinese scientists hosted at the Sevilleta LTER's field station in New Mexico (Brunt 1992). Latin American scientists were exposed to U.S. LTER information activities during a workshop at the Sevilleta LTER field station in 1997 and a follow-up workshop in Venezuela (Acevedo 1998). To introduce information management practices to the Central/Eastern European ILTER region, U.S. information managers partnered with Hungarian scientists to organize a week-long workshop in Hungary in 2000 (Porter 2001). A three-day information management workshop was held in Ulaanbaatar, Mongolia in 2001 (Vanderbilt 2001) for scientists from the East Asia-Pacific Region. LTER Network Office personnel and 3 U.S. LTER information managers co-taught a workshop for African ILTER personnel in 2002 (Michener 2002). The Network Office also developed two training workshops in data management for joint Palestinian/Israeli delegations of ecologists. Israel had formed an LTER Network in 1997. The training was organized and led by Network Office staff and two LTER site information managers and funded by Sandia National Laboratory's Cooperative Monitoring Center (Blankman and Vanderbilt 2001). Topics covered in all these workshops included data documentation, quality assurance and quality control, database design, and web page design and implementation.

16.5 Transition in Management of the ILTER Network

The U.S LTER Network was phenomenally successful in meeting the goals set at the 1993 International Summit, and there was considerable enthusiasm in continuing support for international collaboration. LTER scientists and Network Office staff visited 44 countries to transfer the model of long-term ecological research developed in the U.S. to other national scientific communities. More than 200 international LTER sites were designated in seven regions around the globe. Yet, in 2002, the review team evaluating the Network Office renewal proposal recommended a complete de-accession of the ILTER program from the LTER Network and Network Office. The recommendation was based neither on the acknowledged importance of the ILTER Program nor the success of the LTER Network in developing ILTER. The review team felt that management and development of the ILTER program would inevitably compete for Network Office time and resources that should support the U.S. LTER program. They recommended that ILTER management be transitioned to an international organization with a mission more consistent with the growth of international scientific collaboration.

The NSF considered the recommendation for de-accession of the ILTER program to be highly controversial and did not accept it.[3] However, NSF felt that additional clarity was needed in defining the role of the Network Office in the ILTER program and proposed an independent assessment of the effective role the Network Office could play in encouraging and facilitating collaborative field research among U.S. and international sites. Recommendations emerging from this assessment would inform the development of a strategic plan for the international aspects of LTER consistent with the objectives of NSF's research and education program. NSF pursued the idea of an outside assessment for a few months, but by March 2003 they had decided against it. There followed many months of uncertainty as NSF looked for an appropriate mechanism to continue support for ILTER.

One issue that contributed to the decision to limit the involvement of the Network Office in ILTER was the perception that Jim Gosz and Bob Waide were devoting too much time to ILTER business, to the detriment of the U.S. Network. NSF, particularly International Programs, wanted to involve more U.S. scientists in collaborations with ILTER. However, the mechanism to make this transition took a long time to develop, and in the interim Gosz and Waide had to turn down many opportunities to develop and strengthen ILTER networks. In February 2003, Gosz offered to step down as ILTER Chair to try to resolve the situation. His resignation, coupled with NSF's indecision about funding, caused considerable concern among ILTER member networks.

The U.S. LTER Coordinating Committee created the U.S. International LTER Committee[4] in May 2003 to manage U.S. participation in ILTER. The first role of the committee was to represent the U.S. at the annual ILTER meeting, held in association with the LTER All Scientists Meeting in September 2003. Steve Hamburg (Brown University) and Patrick Bougeron (University of Colorado-Boulder) volunteered to serve as co-chairs and were confirmed by the committee in November. Over the next few months, they negotiated support for the activities of the committee with NSF. Eventually, in June 2004, NSF funded U.S. participation in ILTER through an award to Brown University that included a sub-contract to the University of New Mexico for an ILTER coordinator.

NSF continued to provide support to the ILTER Network as it developed its own infrastructure and funding streams. They provided funds for the position of ILTER Coordinator to report directly to the Chair of the ILTER Network. During this period of transition, the LTER Network Office continued to manage a limited number of ILTER activities funded by NSF. However, several other activities were not supported by the Network Office: development of new national networks, support for synthesis workshops, updates for ILTER books and brochures, support for regional ILTER meetings, participation in planning for global information technology infrastructure, and coordination of student or scientific exchanges. Some of these

[3] Letter from H. Gholz to R. Waide, 22 November 2002.

[4] Members: Kristin Vanderbilt, Ted Gragson, Chris Madden, Steve Hamburg, Brian Kloeppel, Bob Waide, Nick Brokaw, Dave Hartnett, Berry Lyons. Dick Lathrop, Patrick Bourgeron.

activities were eventually funded by the award to Brown University or supplements to sites, but the abrupt withdrawal of support from the Network Office still affected many ongoing ILTER plans.

16.6 Strategic Planning

NSF, through the LTER Network Office, was the major source of financial and logistical support for ILTER activities during its first decade. In 2003, however, NSF decided that the ILTER Network needed to develop an organizational structure that did not rely on the LTER Network Office and needed to diversify its funding streams (ILTER 2006). Through the 2004 grant to Brown University, a private company was hired to guide development of strategic, organizational, and funding plans for the ILTER Network. Leaders from all regions of the ILTER Network were engaged in this two-year process, which involved several face-to-face meetings. Hen-biau King (Taiwan) served as ILTER Chair during this planning process, the first non-U.S.-based scientist to hold this position.

The Strategic Plan, published in 2006, identified strengths and weaknesses of the ILTER Network. The greatest strength noted was the ILTER's global network of hundreds of research sites distributed in many ecosystems. The ILTER Network was believed to be uniquely positioned to supply the scientific community with the site-based, long-term data needed to contribute to research on climate change, sustainable development, ecosystem services, and biodiversity loss. Also called out as a strength was the ILTER Network's "bottom-up" governance approach, where the umbrella ILTER Network did not dictate to member-country networks what research to conduct. Weaknesses of the ILTER Network were mostly about structure and public profile. Structural issues included a lack of: staffing; a physical office; a budget; a legal entity to collect and manage funds; global cyberinfrastructure to collect, access, and archive data; clarity about scientific objectives; and branding. The ILTER Network had a very low profile in the international scientific community. Few scientists not involved in the annual ILTER meetings had ever heard of the ILTER Network, and ILTER data were not sought after by international research programs. The strategic plan emphasized that, although unique in an increasingly crowded landscape of international environmental networks, the ILTER Network faced threats to its existence from competition from other networks, a lack of diversified funding sources, poor communication, and lack of name recognition by scientists and policymakers outside the organization itself.

The new ILTER Network organizational structure, designed to address these deficiencies, included four new committees in addition to the existing Executive and Coordinating Committees: Science and Program; Public Policy; Fundraising and Marketing; and Information Management. Establishment of a secretariat with professional staff outside of the U.S. was proposed in the Strategic Plan to provide the ILTER Network with administrative support. A key role of the secretariat would be to help raise and manage funds for ILTER collaborative research, reducing the

ILTER's reliance on NSF. Possible new sources of revenue identified included member dues, the United Nations, private foundations, and scientific funding agencies in non-U.S. ILTER member countries. The ILTER Association, founded in 2007 in Costa Rica, is the legal entity that manages the ILTER Network's funds.

The Strategic Plan was accepted by the ILTER Network at the annual meeting in 2006. The new committee-based governance structure has successfully transformed the ILTER Network from an organization dependent on NSF and the LTER Network Office to a robust, self-supporting and self-directed entity. ILTER settled on member dues as its main source of funds. Countries pay dues based on their ability to do so; U.S. dues were originally $10K per year but are now $5K. Sufficient funding was never available for establishing a secretariat as envisioned in the 2006 Strategic Plan. Instead, the UK's Environmental Change Network (ECN) and later Environment Agency Austria (Umweltbundesamt GmbH) provided operational support, including maintaining the ILTER web page. Money collected from dues is used to support new scientific or information management initiatives and travel to meetings.

16.7 Examples of International Collaboration

Research Collaboration During the ILTER's first decade, support for collaborative research came from a variety of sources. To encourage collaborations between U.S. scientists and their counterparts in new or forming LTER networks, NSF funded a variety of projects. A special competition in 1994 supported a comparison between North Temperate Lakes LTER and lakes in the UK's Lake District, as well as a cross-site study of biodiversity at Coweeta and Luquillo LTERs and La Selva Biological Station in Costa Rica (LTER Network Office 1995; Heneghan et al. 1998). NSF also funded a comparison of grasslands in the U.S. and Mongolia (Ojima et al. 1999). An NSF supplement to the Niwot LTER supported travel of 13 U.S. LTER personnel to a 2001 week-long workshop in France to explore the merits of French (Zones Ateliers)-U.S. LTER collaborations on socio-ecological research (Bourgeron et al. 2002). NSF and the Slovak Academy of Sciences co-funded a nitrogen deposition experiment in the Western Tatra Mountains of Slovakia (Bowman et al. 2008). Students from Hungary and Switzerland received research support from host U.S. LTER sites (Hochstrasser et al. 2002; Kroel-Dulay et al. 2004) while they pursued graduate degrees at U.S. institutions.

Also, in order to help establish and promote local and regional ILTER communities, NSF grants covered the publication of proceedings from regional ILTER meetings (e.g., Lajtha and Vanderbilt 2000). Another source of funding for collaborative ILTER research was the USDA Forest Service, whose scientists studied polluted forested LTER sites in the Carpathian Mountains with Polish colleagues (Bytnerowicz et al. 2003). Support from the International Arid Lands Consortium, an NGO, along with contributions from Ben Gurion University of the Negev, made possible an ILTER workshop in Israel that led to the publication of *Biodiversity in Drylands: Toward a Unified Framework* (Shachak et al. 2004).

Early regional ILTER meetings sometimes catalyzed new collaborative ILTER research projects involving U.S. researchers. One such study, the Detritus Inputs and Removal Treatments (DIRT) network (Lajtha et al. 2018), illustrates the international cross-site research collaborations that NSF envisioned would arise through the ILTER Network. The DIRT Network went international after U.S. scientists met with Hungarian ecologists at the 1999 Central/Eastern European Regional ILTER Network meeting in Hungary. The DIRT project, initiated at the University of Wisconsin in 1956, was designed to test the effects of litter manipulations on soil organic matter. The study was replicated at Harvard Forest LTER in 1990, and at the H.J. Andrews LTER in 1997. Plots were established at Síkfőkút Forest LTER in Hungary in 2000 (Vanderbilt 2002). This collaboration has provided research opportunities in Hungary for U.S. scientists, graduate students, and a post-doc. Many international collaborative papers have been written (e.g., Holub et al. 2005; Tóth et al. 2007; Fekete et al. 2012, 2014; Lajtha et al. 2018). Hungarian graduate students and researchers have also visited the U.S. DIRT sites. Since 2000, the DIRT network has been extended to several non-LTER sites in the U.S., a site in Germany, and four sites in China. The Chinese DIRT sites Qingyuan Forest and ChangBaiShan Forest, which joined the DIRT Network in 2012, are part of CERN, a member of the ILTER Network.

During the period from 2003 to 2013, funds to support U.S. participation in ILTER Network activities were often made available through proposals or supplements to sites. Activities undertaken included supporting ILTER researchers to attend the 2003, 2006 and 2009 LTER All Scientists' meetings. There was a significant ILTER presence at the latter meeting, as many ILTER scientists also self-funded their attendance. Thirty-three posters or presentations were given based on research from all regions of the ILTER Network. Special sessions were organized by U.S., Mexican, and French LTER scientists to: (1) further an ILTER synthesis project on ecosystem services, ecosystem dynamics, and human outcomes; and (2) define international and regional ILTER science initiatives. Information managers from Malaysia, Mexico, Austria, Israel, and Taiwan joined the U.S. LTER information managers for their annual meeting, held in conjunction with the LTER All Scientists meeting, and organized talks and posters about information management in their networks.

Between 2010 and 2014, the U.S. International LTER Committee (co-chaired by Kristin Vanderbilt (University of New Mexico, 2010–2014), Patrick Bourgeron (2003–2011), and Bill McDowell (University of New Hampshire, 2011–present)), implemented new mechanisms for increasing the name recognition of the ILTER Network and stimulating research between ILTER members. To improve visibility of the ILTER Network in the U.S., Vanderbilt organized the annual 2013 NSF-LTER mini-symposium entitled "The Globalization of Long Term Ecological Research", where four U.S. LTER scientists, one graduate student, one information manager, and one educator from LTER Israel spoke about their ILTER collaborative research projects (Thomas and Vanderbilt 2013). To expose the ILTER "brand" to more ecologists worldwide, U.S. ILTER Committee members organized sessions at the 2013 INTECOL meeting to showcase ILTER research projects about phenology and ecosystem carbon budgets. A supplement to the Sevilleta LTER allowed 12 U.S. scientists and information managers to participate in the 20th Anniversary ILTER Science Meeting in South Korea in 2013. Vanderbilt and Gaiser (2017) organized a special issue of *Ecosphere* about ILTER with contributions by authors who participated in the INTECOL and Korean meetings.

Finally, McDowell and four other U.S. International LTER Committee members wrote a successful proposal to the NSF "Catalyzing New International Collaboration" program to conduct a workshop with LTER Mexico to discuss ideas for joint socio-ecohydrology research. LTER Mexico received parallel funding from CONACYT to support their participation in the workshop. The workshop, involving 15 U.S. LTER scientists and graduate students, was held at the Chamela Field Station, Mexico in October 2012 (Koenig and Vanderbilt 2012). A follow-up workshop was held in conjunction with the First All-Scientists LTER Meeting of the Americas, held in Valdivia, Chile in December 2014.

U.S. LTER researchers have co-authored many ILTER publications in recent years. They have contributed to both conceptual and data-driven papers on socio-ecological research and studies of ecosystem services, which have become popular ILTER research topics (Anderson et al. 2008; Rozzi et al. 2014; Maass et al. 2016; Chang et al. 2018; Dick et al. 2018; Haase et al. 2018). Holzer et al. (2018), for example, compared six LTSER platforms, including the Baltimore Ecosystem Study LTER, to explore tensions between local bottom-up research needs versus ILTER Network top-down desires to harmonize research across LTSER platforms. Other notable contributions to ILTER publications from U.S. authors are a review of current phenology science, illustrating how the ILTER Network is poised to contribute at multiple scales of study (Tang et al. 2016), and an analysis of ILTER research outputs (Li et al. 2015).

There is a wealth of research in the ILTER Network that doesn't involve U.S. scientists. Some projects include several national networks, such as a study of climate change impacts on vegetation in 20 montane LTER sites in Italy, Switzerland and Austria (Rogora et al. 2018). A pan-European project included 35 European LTER sites in a study undertaken to define parameters needed to assess ecosystem services (Dick et al. 2014). Individual LTER Networks have also published special issues in *Ecological Indicators* to showcase their own networks (e.g., UK Environmental Change Network (Sier and Monteith 2016) and the German LTER (Haase et al. 2016)). Another way ILTER researchers have collaborated is by co-editing special issues of a journal. Fu and Forsius (2015), from CERN and LTER Finland, respectively, edited a special issue of *Landscape Ecology* that features papers from the European and East Asia-Pacific regional LTER networks. A recent special issue of *Regional Environmental Change* includes papers from LTER networks in Austria, Slovenia, France, Spain, Mexico and South Africa (Dirnböck et al. 2019).

The ILTER Network has initiated research projects that are global in nature. The ILTER Nitrogen Initiative was launched at the 2011 ILTER annual meeting to address the need to understand how ecosystems across the globe respond to anthropogenically derived nitrogen. Shibata et al. (2015) reviewed spatial and temporal patterns and trends in global N emissions to find gaps in knowledge and called for greater integrated and long-term collaborative studies to promote better understanding of the nitrogen cycle. An international workshop on N_2O emissions, organized by U.S., Taiwan, and Japan LTERs, was held in Taiwan in 2017 (Chen et al. 2017). Scientists participating in the International Nitrogen Initiative project have offered an international training program for young researchers to learn state-of-the-art nitrogen analyses and to understand implications of N deposition on ecosystem processes. The class has been held in Japan (2016) and Portugal (2017).

The ILTER Network is beginning to realize its potential as a platform positioned to connect long-term observation data with the results of experiments replicated at many ILTER sites. ILTER scientists are working with existing networks such as the Nutrient Network (Borer et al. 2014) to extend their methodologies to ILTER sites. In collaboration with several other international networks, ILTER is also part of the TeaComposition global initiative which began in 2016 (Djukic et al. 2018). Litters of two qualities have been placed at 450 sites in nine biomes, including several sites in the ILTER Network, to study litter decomposition. The use of standard litter types and common protocols will allow development of a common metric with which to compare decomposition rates across biomes. Studies such as this one demonstrate the power of replicating a simple protocol across many ILTER sites to better understand global processes.

Information Management Collaborations U.S. LTER information managers have had many successful collaborations with information managers from other ILTER member countries, particularly from the EAP-ILTER region. These exchanges began when scientists from Taiwan became interested in developing an information management system for TERN, and, supported by TERN, visited U.S. LTER information managers at their sites. Several scientists from TERN did three-month "internships" with U.S. information managers, most with John Porter (University of Virginia) (Porter 2007). Consequently, TERN was the first ILTER member outside of the U.S. to adopt Ecological Metadata Language (EML), the U.S. LTER's chosen metadata standard, and tools to create, store, and use EML (Lin et al. 2006, 2008a, b). TERN hosted workshops for other members of the EAP-ILTER region, often including U.S. information managers as instructors (Porter 2007), to teach them about EML and the suite of tools used with it. As a result, Taiwan, Malaysia, and Japan now use EML and related software for managing their data (Lin et al. 2016).

Although TERN personnel originally connected with U.S. LTER information managers to learn how information management was accomplished at U.S. LTER sites, they soon became full partners in collaborative informatics research projects (Vanderbilt et al. 2015). One such effort focused on the computational needs of EAP-ILTER researchers who belong to the ForestGeo Network of long-term forest research plots (Anderson-Teixeira et al. 2015). Several plots in the ForestGeo Network are located in EAP-ILTER sites. TERN organized a workshop for U.S. and EAP-ILTER scientists and information managers in which the forest scientists' information management needs were analyzed. Tools were then built to make ForestGeo data discovery, access, analysis and integration easier. In a subsequent workshop hosted by LTER Malaysia, the tools built in the first workshop were "field-tested" during actual data analyses (Vanderbilt and Porter 2010; Lin et al. 2011). Local costs for these workshops were covered by the host, while travel expenses for U.S. participants were covered by international supplements to U.S. LTER grants.

The need for an ILTER network-wide strategy for information management became clear as the ILTER Network matured and each member or regional network began to define how it would make data and other information products accessible for all researchers. LTER-Europe was developing an ontology-based approach for its members, while networks in the EAP-ILTER were using EML-based tools developed in the U.S. In a workshop co-organized by information managers in the U.S. LTER and CERN, experts from throughout the ILTER assembled in China to find common ground. EML was selected as the ILTER's metadata standard, and it was agreed that all ILTER metadata would include discovery level elements in English to facilitate data discovery, regardless of the local language in which the metadata were created (Vanderbilt et al. 2010).

The resolution to provide discovery-level metadata in English for all ILTER datasets was soon deemed impractical, and a better approach to multilingual data discovery was sought. Toward this goal, U.S. LTER and CERN information managers co-organized a workshop to bring together scientists and informaticians from throughout the ILTER to consider how to find datasets documented in different languages (Vanderbilt et al. 2012). Attendees were from the U.S., United Kingdom, China, Japan, South Korea, Taiwan, and Israel LTER Networks. The workshop yielded translations into simple and traditional Chinese, Korean, and Japanese of the roughly 700 terms in the U.S. LTER's Controlled Vocabulary (Porter 2010). This set of terms was chosen because, in addition to being used by the U.S. LTER, they have been included in the Envthes Thesaurus developed by LTER-Europe and other European collaborators (Schentz et al. 2013). During this workshop, a search tool was also prototyped that allowed users to enter a keyword in their local language and have relevant ILTER datasets returned, regardless of the language in which they were documented. (Vanderbilt et al. 2017). Additional translations are needed before a technological approach to multilingual data discovery can be implemented for the ILTER Network. Toward this goal, information specialists from the U.S. and Austrian LTERs led a translation workshop at the 2019 ILTER Open Science Meeting in Leipzig, Germany to assess how well Google Translate and other online tools translate ecological metadata.

An important milestone for the ILTER Network was the launch of the Dynamic Ecological Information Management System-Site and Dataset Registry (DEIMS-SDR), a globally comprehensive registry of long-term ecological research sites with links to associated data. It has been the central site catalog for Europe-LTER and ILTER since 2012 (Wohner et al. 2019). This catalog's infrastructure grew out of a web-based system pioneered by U.S. LTER information managers for managing all information products for an LTER site using the Drupal content management system. U.S. LTER, through NSF supplements to three LTER sites and funds from the American Recovery and Reinvestment Act (ARRA) distributed by the LTER Network Office, paid a Drupal web development firm a total of $340,000 to work with a team of LTER information managers to create the Drupal Ecological Information Management System, or DEIMS (Melendez-Colom 2010; San Gil et al. 2010). DEIMS, an impressive system for managing an LTER site's data, metadata, publications, personnel, images, and research locations, was introduced to LTER-Europe developers and, with input from U.S. information managers, it became the basis of DEIMS-SDR. Since its initial deployment, additional functionality has been added by developers in LTER-Europe. These enhancements have provided for more comprehensive documentation of LTER research sites, including facilities, biome, sensors used, and research themes. The user interface has also been improved to aid discovery of sites and data from the ILTER Network as well as other locations where long-term data are collected.

16.8 The ILTER Today

The ILTER Network today aims to become part of a global infrastructure to address continental and global socio-ecological problems through partnerships with other international networks (Mirtl et al. 2018). NSF continues to support growth in this direction through the International LTER Committee, chaired since 2014 by McDowell and Tiffany Troxler (Florida International University), and the LTER Network Office in Santa Barbara, California. For example, NSF provided funds to the Network Office for travel of U.S. scientists to the 2016 and 2019 ILTER Open Science meetings.

The ILTER Network now comprises 44 member networks, 700 LTER sites and approximately 80 LTSER platforms (Mirtl et al. 2018). Some countries, such as Taiwan and the UK, have been members since the ILTER's inception. Several other countries were members but dropped out for various reasons. Canada was a member through its Ecological Monitoring and Assessment Network (EMAN), but when that network folded and the champion of Canadian involvement in the ILTER Network retired, Canada ceased to be part of the network. Costa Rica also dropped out of ILTER when the leader of that network changed jobs and no one else stepped up to maintain the connection. Some developing countries became ILTER members but, when they were unable to find local support for their own LTER research, they left the ILTER. There is still a wide disparity in resource availability among ILTER members, which influences their level of participation.

The ILTER Network has come a long way from an informal collection of LTER Networks in 1993 to the robust, self-sustaining network of networks it is today (Vanderbilt and Gaiser 2017). It has partially achieved the goals from the 2006 Strategic Plan (Mirtl et al. 2018). The ILTER has been successful in stimulating collaboration among socio-ecological and ecosystem researchers at local to global scales, a major goal in 2006. Other goals, to deliver scientific information to scientists, policymakers and the public, and to support the education of the next generation of LTER scientists, have been met at the level of individual LTER sites or LTSER platforms, but not at the ILTER Network level. A major step for the ILTER Network was holding its first ever ILTER Open Science Meeting in 2016, hosted by South Africa Ecological Observing Network (SAEON). The second ILTER Open Science meeting was held in Leipzig, Germany in 2019. Among the U.S. LTER contingent were representatives from overlapping networks such as CREON (Coral Reef Environmental Observatory Network) who attended the meeting to discuss possible synergies with the ILTER Network.

The ILTER Network has proposed several types of activities for its next 10 years (Mirtl et al. 2018). Until recently, much of the personal interaction among ILTER scientists has been by network leaders who meet annually. ILTER scientists outside that group have had limited or no contact with other ILTER members. The ILTER Network will use new technologies to improve information flow among scientists operating at the site level throughout the ILTER Network. Greater access to data generated by ILTER members is also expected to foster more collaborative projects. ILTER will develop a standards-based global environmental data infrastructure to increase data sharing in keeping with the FAIR (Findable, Accessible, Interoperable, and Reusable) international initiative for Open Data and Open Science (Wilkinson et al. 2016). In a world that now contains many international networks, the global ILTER framework will be re-evaluated to better define the niche for ILTER activities and priorities for cross-site research. The ILTER Network anticipates recruiting additional countries to the network to fill gaps in global coverage of biomes and socio-economic conditions to become ever more relevant as a comprehensive source of information on ecosystem structure, function, and services.

16.9 Conclusions

The ILTER Network owes its existence to NSF and the U.S. LTER Network, but the success of ILTER is the result of nearly three decades of effort by scientists from more than 44 countries. Beginning in 1993, NSF and the U.S. LTER Network invested significant amounts of time and money in promoting long-term ecological research around the globe. As a result of this effort, the U.S. LTER Network is seen as a global leader in comparative and synthetic ecology, and the ILTER Network is cited as a successful example of what NSF terms a "network of networks" (NSF OISE 2018). The number of publications from ILTER member networks has steadily increased since 1993 (Li et al. 2015), reflecting the success of the ILTER as a research enterprise. Moreover, the ILTER Network has evolved into a stable, multinational organization in which the U.S LTER continues to play an important, albeit no longer central, role.

From the perspective of U.S. scientists, participation in ILTER has increased collaboration with well-established and developing research networks worldwide. Through ILTER, many U.S. students and scientists have had the opportunity to interact with international colleagues, visit their research sites, and plan joint research projects. In some cases, long-term partnerships have developed among sites or national networks. Scientists from the U.S. LTER Network have also had the opportunity to compare different models for developing and sustaining networks and funding multi-national research (i.e., ALTER-Net). The interchange of ideas has stimulated new research approaches, generated homogeneous methods for collecting and analyzing data, and enabled the creation of new knowledge. Overall, participation in ILTER has created a richer environment for creativity and synthesis, which is why enthusiasm for ILTER remains strong in the U.S. LTER community.

ILTER has also led to an expansion of research capabilities in parallel to the expansion of research opportunities, particularly with respect to data discovery and accessibility. The U.S. information management community made a significant impact on ILTER during its first decade when it was the major source of information management expertise and training for newly established LTER networks in many countries. U.S. information managers and Network Office staff provided guidance on technologies and methods as many networks established information management systems for the first time. U.S. LTER data sharing experiences and policy were also an important model for scientists throughout the ILTER Network as they sought to synthesize data from multiple sources. Based on this sound information management foundation, ILTER has made great strides by creating DEIMS-SDR, a centralized web portal for disseminating ILTER data and research site information. By providing long-term data from hundreds of research sites, the ILTER has enabled synthesis of global ecological patterns (e.g., Pilotto et al. 2020). The capacity to do such research will accelerate as ILTER adds new semantic, multilingual, and data integration tools to its information management infrastructure.

Although the ILTER Network has not been able to achieve all its strategic goals, it continues to strive to carry out the original mission of ILTER, to stimulate global scale comparisons and modeling through exchanges of scientists, techniques, data, and information. In a world beset by climate change, biodiversity loss, and a global pandemic, the value of knowledge created by an international network of research sites and scientists focusing on socio-ecological problems can only increase. The potential of the ILTER Network to contribute to solutions of global problems has been stymied to date by the absence of funding sources with global scope (Mirtl et al. 2018). However, the ILTER Network stands ready to contribute the human resources, research infrastructure, and long-term data that will be needed to address these problems. Concerns about climate change, biodiversity loss, and the scarcity of research sites producing long-term data led to the creation of the U.S. LTER Network and hence ILTER, and the legacies of data, information, and long-term measurements that have resulted are a critical contribution to resolving current global problems.

References

Acevedo, Miguel F. 1998. Information management: A strong beginning in the Latin American LTER. *LTER Network News* 11 (2): 12. https://lternet.edu/wp-content/uploads/2013/01/volume_11_number_2_Fall_1998.pdf. Accessed 21 July 2020.

Anderson, Christopher B., Gene E. Likens, Ricardo Rozzi, Julio R. Gutiérrez, Juan J. Armesto, and Alexandria Poole. 2008. Integrating science and society through long-term socio-ecological research. *Environmental Ethics* 30 (3): 295–312.

Anderson-Teixeira, Kristina J., Stuart J. Davies, Amy C. Bennett, Erika B. Gonzalez-Akre, Helene C. Muller-Landau, S. Joseph Wright, Kamariah Abu Salim, et al. 2015. CTFS-ForestGEO: A worldwide network monitoring forests in an era of global change. *Global Change Biology* 21 (2): 528–549. https://doi.org/10.1111/gcb.12712. Accessed 31 July 2020.

Blankman, David, and Kristin Vanderbilt. 2001. Information management workshop for scientists from the Middle East. *LTER Network News* 14 (2): 9–11. https://lternet.edu/wp-content/uploads/2010/12/fall01.pdf. Accessed 21 July 2020.

Bledsoe, C. 1993. *Ecological network of networks: Creating a network to study ecological effects of global climate change*. Washington, D.C.: U.S. National Committee on Man and the Biosphere.

Borer, Elizabeth T., W. Stanley Harpole, Peter B. Adler, Eric M. Lind, John L. Orrock, Eric W. Seabloom, and Melinda D. Smith. 2014. Finding generality in ecology: A model for globally distributed experiments. *Methods in Ecology and Evolution* 5 (1): 65–73. https://doi.org/10.1111/2041-210X.12125.

Bourgeron, Patrick S., Hope C. Humphries, and J. Morgan Grove. 2002. *Workshop report: French (Zones Ateliers) and United States (Long Term Ecological Research) network-to-network collaboration in long term integrated environmental research and management*. https://lternet.edu/wp-content/uploads/2010/12/NSF-CNRS-Versailles-Report.pdf. Accessed 21 July 2020.

Bowman, William D., Cory C. Cleveland, Ĺuboš Halada, Juraj Hreško, and Jill S. Baron. 2008. Negative impact of nitrogen deposition on soil buffering capacity. *Nature Geoscience* 1 (11): 767–770. https://doi.org/10.1038/ngeo339.

Brunt, James. 1992. CERN-LTER collaboration advances. *LTER Network News*, Issue 12 (Winter): 5. https://lternet.edu/wp-content/uploads/2013/01/Network_News_Issue_12_Winter_1992-1993.pdf. Accessed 21 July 2020.

Bytnerowicz, Andrzej, Ovidiu Badea, Ion Barbu, Peter Fleischer, Witold Frączek, Vladimir Gancz, Barbara Godzik, et al. 2003. New international long-term ecological research on air pollution effects on the Carpathian Mountain forests, Central Europe. *Environment International* 29 (2): 367–376. https://doi.org/10.1016/S0160-4120(02)00172-1.

Chang, Chung-Te, Matthew A. Vadeboncoeur, and Teng-Chiu Lin. 2018. Resistance and resilience of social-ecological systems to recurrent typhoon disturbance on a subtropical island: Taiwan. *Ecosphere* 9 (1): e02071. https://doi.org/10.1002/ecs2.2071.

Chen, Chiling, Jianwu Tang, Yi-Ching Lin, and Hideaki Shibata. 2017. *International workshop on N_2O emissions in various ecosystems: Workshop summary report*. https://lternet.edu/wp-content/uploads/2018/02/ILTER-workshop-report-N2O-ILTER-INMS-Taiwan-2017-11.pdf. Accessed 3 July 2020.

Clutter, Mary E., and James R. Gosz. 1994. Preface. In *International networking in long-term ecological research*, eds. Rudolph W. Nottrott, Jerry F. Franklin, and John R. Vande Castle, n.p. Seattle/Washington: U. S. LTER Network Office.

Dick, Jan, Amani Al-Assaf, Chris Andrews, Ricardo Díaz-Delgado, Elli Groner, Ľuboš Halada, Zita Izakovičová, et al. 2014. Ecosytem services: A rapid assessment method tested at 35 Sites of the LTER-Europe Network. *Ekológia (Bratislava)* 33 (3): 217–231. https://doi.org/10.2478/eko-2014-0021.

Dick, Jan, Daniel E. Orenstein, Jennifer M. Holzer, et al. 2018. What is socio-ecological research delivering? A literature survey across 25 international LTSER platforms. *Science of The Total Environment* 622 (623): 1225–1240. https://doi.org/10.1016/j.scitotenv.2017.11.324.

Dirnböck, Thomas, Peter Haase, Michael Mirtl, Johan Pauw, and Pamela H. Templer. 2019. Contemporary International Long-Term Ecological Research (ILTER)--from biogeosciences to socio-ecology and biodiversity research. *Regional Environmental Change* 19 (2): 309–311. https://doi.org/10.1007/s10113-018-1445-0.

Djukic, Ika, Sebastian Kepfer-Rojas, Inger Kappel Schmidt, Klaus Steenberg Larsen, Claus Beier, Björn Berg, Kris Verheyen, et al. 2018. Early stage litter decomposition across biomes. *Science of The Total Environment* 628 (629): 1369–1394. https://doi.org/10.1016/j.scitotenv.2018.01.012.

Fekete, István, Zsolt Kotroczó, Csaba Varga, Rita Hargitai, Kimberly Townsend, Gábor Csányi, and Gábor Várbiró. 2012. Variability of organic matter inputs affects soil moisture and soil biological parameters in a European detritus manipulation experiment. *Ecosystems* 15 (5): 792–803.

Fekete, István, Zsolt Kotroczó, Csaba Varga, Péter Tamás Nagy, Gábor Várbíró, Richard D. Bowden, János Attila Tóth, and Kate Lajtha. 2014. Alterations in forest detritus inputs influence soil carbon concentration and soil respiration in a Central-European deciduous forest. *Soil Biology and Biochemistry* 74: 106–114.

Franklin, Jerry F. 1989. Importance and justification of long-term studies in ecology. In *Long-term studies in ecology: Approaches and alternatives*, ed. Gene E. Likens, 3–19. New York: Springer.

Freckman, Diana W. 1994. LTER scientists visit Hungarian nature reserves. *LTER Network News*. Issue 16 (Fall/Winter): 10–11. https://lternet.edu/wp-content/uploads/2013/01/Network_News_Issue_16_Fall_1994.pdf. Accessed 31 July 2020.

French, Christine. 1998a. Science exchange: US students visit Asian LTER Sites. *LTER Network News* 11 (2): 13–15. https://lternet.edu/wp-content/uploads/2013/01/volume_11_number_2_Fall_1998.pdf. Accessed 25 July 2020.

———. 1998b. International Networks Gaining Momentum. *LTER Network News* 11 (1): 20. https://lternet.edu/wp-content/uploads/2013/01/volume_11_number_1_Spring_1998.pdf. Accessed 2 July 2020.

———. 1998c. Central European LTER Network: Development continues. *LTER Network News* 11 (2): 12. https://lternet.edu/wp-content/uploads/2013/01/volume_11_number_2_Fall_1998.pdf. Accessed 21 July 2020.

Fu, Bojie, and Martin Forsius. 2015. Ecosystem services modeling in contrasting landscapes. *Landscape Ecology* 30 (3): 375–379. https://doi.org/10.1007/s10980-015-0176-6.

Gil San, Iñigo, Marshall White, Eda Melendez, and Kristin Vanderbilt. 2010. Case studies of ecological integrative information systems: The Luquillo and Sevilleta information management systems. In *Metadata and semantic research: 4th international conference proceedings*, ed. Salvador Sánchez-Alonso and Ioannis N. Athanasiadis, 18–35. Berlin: Springer. https://doi.org/10.1007/978-3-642-16552-8_3.

Gosz, James R. 1997. International Long-Term Ecological Research Developments. *LTER Network News* 20 (Winter): 15–16. https://lternet.edu/wpcontent/uploads/2013/01/Issue_20_Winter_1997.pdf. Accessed 31 July 2020.

Gosz, James, and Beryl Leach. 1991. U.S.-China exchange, laying the foundations for collaboration. *LTER Network News* 10 (Winter): 1. https://lternet.edu/wp-content/uploads/2013/01/4k.pdf. Accessed 25 October 2018.

Gosz, James R., Christine French, Patricia Sprott, and Marshall White. 2000. The International Long Term Ecological Research Network: Perspectives from participating networks. Albuquerque: U.S. Long Term Ecological Research Network Office. https://lternet.edu/wp-content/themes/ndic/library/pdf/reports/ILTER%202000%20optimized%20ocr.pdf. Accessed 31 July 2020.

Haase, Peter, Mark Frenzel, Stefan Klotz, Martin Musche, and Stefan Stoll. 2016. The long-term ecological research (LTER) network: Relevance, current status, future perspective and examples from marine, freshwater and terrestrial long-term observation. *Ecological Indicators* 65: 1–3. https://doi.org/10.1016/j.ecolind.2016.01.040.

Haase, Peter, Jonathan D. Tonkin, Stefan Stoll, Benjamin Burkhard, Mark Frenzel, Ilse R. Geijzendorffer, Christoph Häuser, et al. 2018. The next generation of site-based long-term ecological monitoring: Linking essential biodiversity variables and ecosystem integrity. *Science of The Total Environment* 613 (614): 1376–1384. https://doi.org/10.1016/j.scitotenv.2017.08.111.

Haberl, Helmut, Verena Winiwarter, Krister Andersson, Robert U. Ayres, Christopher Boone, Alicia Castillo, Geoff Cunfer, et al. 2006. From LTER to LTSER: Conceptualizing the socio-economic dimension of long-term socioecological research. *Ecology and Society* 11 (2) https://www.ecologyandsociety.org/vol11/iss2/art13/.

Heneghan, L., D.C. Coleman, X. Zou, D.A. Crossley, and B.L. Haines. 1998. Soil microarthropod community structure and litter decomposition dynamics: A study of tropical and temperate sites. *Applied Soil Ecology* 9 (1): 33–38. https://doi.org/10.1016/S0929-1393(98)00050-X.

Henschel, Joh, Johan Pauw, Feetham Banyikwa, Rui Brito, Harry Chabwela, Tony Palmer, Sue Ringrose, Luisa Santos, Almeida Sitoe, and Albert van Jaarsveld. 2003. Developing the Environmental Long-Term Observatories Network of Southern Africa (ELTOSA). *South African Journal of Science* 9: 100–108.

Hochstrasser, T., G.Y. Kröel-Dulay, D.P.C. Peters, and J.R. Gosz. 2002. Vegetation and climate characteristics of arid and semi-arid grasslands in North America and their biome transition zone. *Journal of Arid Environments* 51 (1): 55–78. https://doi.org/10.1006/jare.2001.0929.

Holub, Scott M., Kate Lajtha, Julie D.H. Spears, János A. Tóth, Susan E. Crow, Bruce A. Caldwell, Mária Papp, and Péter T. Nagy. 2005. Organic matter manipulations have little effect on gross and net nitrogen transformations in two temperate forest mineral soils in the USA and Central Europe. *Forest Ecology and Management* 214 (1): 320–330. https://doi.org/10.1016/j.foreco.2005.04.016.

Holzer, J.M., M.C. Adamescu, F.J. Bonet-García, R. Díaz-Delgado, J. Dick, J.M. Grove, R. Rozzi, and D.E. Orenstein. 2018. Negotiating local versus global needs in the International Long Term Ecological Research Network's socio-ecological research agenda. *Environmental Research Letters* 13 (10): 105003. https://doi.org/10.1088/1748-9326/aadec8.

ILTER. 2006. *International long-term ecological research network: Strategic plan*. https://www.lter-europe.net/document-archive/central/ECOLEC-D-08-00262.pdf. Accessed 22 July 2020.

Keeling, C.D., S.C. Piper, R.B. Bacastow, M. Wahlen, T.P. Whorf, M. Heimann, and H.A. Meijer. 2005. Atmospheric CO_2 and $^{13}CO_2$ exchange with the terrestrial biosphere and oceans from 1978 to 2000: Observations and carbon cycle implications. In *A history of atmospheric CO_2 and its effects on plants, animals, and ecosystems*, ed. J.R. Ehrlinger, T. Cerling, and M.D. Dearing, 83–113. New York: Springer.

Kim, Eun-Shik, Yongyut Trisurat, Hiroyuki Muraoka, Hideaki Shibata, Victor Amoroso, Bazartseren Boldgiv, Kazuhiko Hoshizaki, et al. 2018. The International Long-Term Ecological Research–East Asia–Pacific Regional Network (ILTER-EAP): History, development, and perspectives. *Ecological Research* 33 (1): 19–34. https://doi.org/10.1007/s11284-017-1523-7.

Koenig, Lauren, and Kristin Vanderbilt. 2012. Mexican and U.S. LTER scientists work to synthesize ideas in socio-ecohydrology. *LTER Network News* 25 (3). http://news.lternet.edu/Article2671.html. Accessed 22 July 2020.

Kröel-Dulay, György, Péter Ódor, Debra P.C. Peters, and Tamara Hochstrasser. 2004. Distribution of plant species at a biome transition zone in New Mexico. *Journal of Vegetation Science* 15 (4): 531–538. https://doi.org/10.1658/1100-9233(2004)015[0531:DOPSAA]2.0.CO;2.

Lajtha, Kate, and Kristin Vanderbilt, eds. 2000. *Cooperation in long term ecological research in Central and Eastern Europe: Proceedings of the ILTER regional workshop, June 22-25, 1999, Budapest, Hungary*. Corvallis, OR: Oregon State University.

Lajtha, Kate, Richard D. Bowden, Susan Crow, István Fekete, Zsolt Kotroczó, Alain Plante, Myrna J. Simpson, and Knute J. Nadelhoffer. 2018. The Detrital Input and Removal Treatment (DIRT) Network: Insights into soil carbon stabilization. *Science of The Total Environment* 640 (641): 1112–1120. https://doi.org/10.1016/j.scitotenv.2018.05.388.

Li, Ben, Terry Parr, and Ricardo Rozzi. 2015. Geographical and thematic distribution of publications generated at the International Long-Term Ecological Research Network (ILTER) sites. In

Earth stewardship, ed. Rozzi Ricardo, F. Stuart Chapin III, J. Baird Callicott, S.T.A. Pickett, Mary E. Power, Juan J. Armesto, and Roy H. May, 195–216. Cham: Springer.

Likens, Gene E., and F. Herbert Bormann. 1974. Acid rain: A serious regional environmental problem. *Science* 14: 1176–1179.

Lin, Chau-Chin, John H. Porter, and Sheng-Shan Lu. 2006. A metadata-based framework for multilingual ecological information management. *Taiwan Journal of Forest Science* 21: 377–382.

Lin, Chau-Chin, John H. Porter, Chi-Wen Hsiao, Sheng-Shan Lu, and Meei-Ru Jeng. 2008a. Establishing an EML-based data management system for automating analysis of field sensor data. *Taiwan Journal of Forest Science* 23 (3): 279–285.

Lin, Chau-Chin, John H. Porter, Sheng-Shan Lu, Meei-Ru Jeng, and Chi-Wen Hsiao. 2008b. Using structured metadata to manage forestry research information: A new approach. *Taiwan Journal of Forest Science* 23: 133–143.

Lin, C.C., A.R. Kassim, K. Vanderbilt, D. Henshaw, E.C. Melendez-Colom, J.H. Porter, K. Niiyama, et al. 2011. An ecoinformatics application for Forest Dynamics Plot data management and sharing. *Taiwan Journal of Forest Science* 26 (4): 357–369.

Lin, Chau-Chin, Guan-Shuo Mai, and Sheng-Shan Lu. 2016. Federation of ecological data repository of EAP-ILTER. *Taiwan Journal of Forest Science* 31 (4): 337–342.

LTER Network Office. 1995. LTER cross-site comparisons and international research awards. *LTER Network News* 17 (Spring/Summer): 12. https://lternet.edu/wp-content/uploads/2013/01/Spring_1995_Issue_17_color.pdf. Accessed 22 July 2020.

———. 1996a. International LTER (ILTER) Interactions: Central Europeans visit NSF and LTER Sites. *LTER Network News* 19 (Spring/Summer): 17. https://lternet.edu/wp-content/uploads/2013/01/Network_News_Issue_19_Spring_1996.pdf. Accessed 22 July 2020.

———. 1996b. International LTER (ILTER) Interactions: U.S. LTER & LMER Scientists Visit Iberian Sites. *LTER Network News* 19 (Spring/Summer): 16. https://lternet.edu/wp-content/uploads/2013/01/Network_News_Issue_19_Spring_1996.pdf. Accessed 22 July 2020.

———. 2000. Changing of the guard at the Network Office. *LTER Network News* 13 (2): 2. https://lternet.edu/wp-content/uploads/2010/12/Network_News_Vol._13_No._2_Fall_2000.pdf. Accessed 18 July 2020.

Maass, Manuel, Patricia Balvanera, Patrick Bourgeron, Miguel Equihua, Jacques Baudry, Jan Dick, Martin Forsius, et al. 2016. Changes in biodiversity and trade-offs among ecosystem services, stakeholders, and components of well-being: The contribution of the International Long-Term Ecological Research Network (ILTER) to Programme on Ecosystem Change and Society (PECS). *Ecology and Society* 21 (3) https://www.ecologyandsociety.org/vol11/iss2/art13/.

Magnuson, J.J., D.M. Robertson, B.J. Benson, R.H. Wynne, D.M. Livingstone, T. Arai, R.A. Assel, R.G. Barry, V. Card, E. Kuusisto, N.G. Grannin, T.D. Prowse, K.M. Stewart, and V.S. Vuglinski. 2000. Historical trends in lake and river ice cover in the Northern Hemisphere. *Science* 289: 1743–1746; Errata 2001, *Science* 291: 254.

Melendez-Colom, Eda. 2010. Developing a Drupal "website-IMS" for Luquillo LTER while learning Drupal. *LTER Databits*. Spring Issue. https://lternet.edu/wp-content/uploads/2017/12/2010-spring-lter-databits.pdf. Accessed 25 July 2020.

Michener, William. 2002. Ecoinformatics training in Maputo, Mozambique. *LTER Network News* 15 (2). http://news.lternet.edu/Article746.html. Accessed 22 July 2020.

Mirtl, M. 2010. Introducing the next generation of ecosystem research in Europe: LTER-Europe's multi-functional and multi-scale approach. In *Long-term ecological research*, ed. F. Müller, C. Baessler, H. Schubert, and S. Klotz, 75–93. Dordrecht: Springer.

Mirtl, Michael, Daniel E. Orenstein, Martin Wildenberg, Johannes Peterseil, and Mark Frenzel. 2013. Development of LTSER platforms in LTER-Europe: Challenges and experiences in implementing place-based long-term socio-ecological research in selected regions. In *Long term socio-ecological research*, ed. Simron Jit Singh, Helmut Haberl, Marian Chertow, Michael Mirtl, and Martin Schmid, 409–442. Dordrecht: Springer.

Mirtl, M., E.T. Borer, I. Djukic, M. Forsius, H. Haubold, W. Hugo, J. Jourdan, et al. 2018. Genesis, goals and achievements of long-term ecological research at the global scale: A critical review of ILTER and future directions. *Science of The Total Environment* 626: 1439–1462. https://doi.org/10.1016/j.scitotenv.2017.12.001.

National Science Foundation 1979. Long-term ecological research: Concept statement and measurement needs. Available from the National Science Foundation LTER Network website, LTER History timeline at https://lternet.edu/network-organization/lter-a-history/, Accessed 29 July 2020.

Nottrott, R.W., J.F. Franklin, and J.R. Vande Castle, eds. 1994. *International networking in long-term ecological research: Proceedings of an international summit*. Seattle, Washington: U.S. LTER Network Office.

NSF OISE. 2018. Input on accelerating research through international network-to-network collaboration. https://www.nsf.gov/od/oise/OISE-AC/Report/InputOnAcceleratingResearchThroughInternationalNetwork-to-NetworkCollaboration.pdf. Accessed 22 July 2020.

Ojima, Dennis, Togtohyn Chuluun, and Wang Yanfen. 1999. Land use in temperate East Asia (LUTEA): Site studies on long-term ecosystem dynamics, 10/01/1995-06/30/1999. In *Proceedings of the 3rd international conference on long-term ecological research (LTER) in East Asia-Pacific Region, 11-16 October, 1999, Sejong Cultural Center, Seoul, Korea*, eds. Eun-Shik Kim and Jeong Soo Oh, 42–43. Seoul: KLTER.

Pilotto, Francesca, Ingolf Kühn, Rita Adrian, et al. 2020. Meta-analysis of multidecadal biodiversity trends in Europe. *Nature Communications* 11: np. https://doi.org/10.1038/s41467-020-17171-y

Porter, John H. 2010. A controlled vocabulary for LTER datasets. *LTER Databits*, Spring Issue. https://lternet.edu/wp-content/uploads/2018/01/2010-spring-lter-databits.pdf. Accessed 22 July 2020.

Porter, John. 2001. International LTER information management workshop in Central Europe. *LTER Network News* 14(1): 9. https://lternet.edu/wp-content/uploads/2010/12/spring01.pdf. Accessed 22 July 2020.

———. 2007. LTER intensifies IM interactions with Taiwan. *LTER Network News* 20 (1): 18–19. https://lternet.edu/wp-content/uploads/2010/12/NN-sPRING07-FINAL2-color.pdf. Accessed 20 July 2020.

Risser, Paul G., Jane Lubchenco, Norman L. Christensen, Philip L. Johnson, Peter J. Dillon, Pamela Matson, Luis Diego Gomez, Nancy A. Moran, Daniel J. Jacob, Thomas Rosswall, and Michael Wright. 1993. Ten-year review of the National Science Foundation's Long-Term Ecological Research program. https://lternet.edu/wp-content/uploads/2010/12/ten-year-review-of-LTER.pdf. Accessed 3 July 2020.

Rogora, M., L. Frate, M.L. Carranza, M. Freppaz, A. Stanisci, I. Bertani, R. Bottarin, et al. 2018. Assessment of climate change effects on mountain ecosystems through a cross-site analysis in the Alps and Apennines. *Science of The Total Environment* 624: 1429–1442. https://doi.org/10.1016/j.scitotenv.2017.12.155.

Rozzi, R., F. Massardo, T. Contador, R.D. Crego, M. Méndez, R. Rijal, L.A. Cavieres, and J.E. Jiménez. 2014. Field environmental philosophy: Ecology and ethics in LTSER-Chile and ILTER networks. *Bosque* 35 (3): 439–447.

Schentz, H., J. Peterseil, and N. Bertrand. 2013. EnvThes-interlinked thesaurus for long term ecological research, monitoring, and experiments. In *Proceedings EnviroInfo 2013: Environmental informatics and renewable energies*, ed. B. Page, A.G. Fleischer, J. Göbel, and V. Volgemuth, 824–832. Aachen: Shaker Verlag.

Shachak, Moshe, James R. Gosz, Stewart T.A. Pickett, and Avi Perevolotsky, eds. 2004. *Biodiversity in drylands: Toward a unified framework*. New York: Oxford University Press.

Sherrod, S. 1999. Students attend US-Japan ILTER workshop 11-19 June, 1999, Otsu and Tomakomai, Japan. *LTER Network News* 12 (2). http://news.lternet.edu/Article993.html. Accessed 21 July 2020.

Shibata, Hideaki, Cristina Branquinho, William H. McDowell, Myron J. Mitchell, Don T. Monteith, Jianwu Tang, Lauri Arvola, et al. 2015. Consequence of altered nitrogen cycles in the coupled human and ecological system under changing climate: The need for long-term and site-based research. *AMBIO* 44 (3): 178–193. https://doi.org/10.1007/s13280-014-0545-4.

Sier, Andrew, and Don Monteith. 2016. The UK Environmental Change Network after twenty years of integrated ecosystem assessment: Key findings and future perspectives. *Ecological Indicators* 68: 1–12. https://doi.org/10.1016/j.ecolind.2016.02.008.

Tang, Jianwu, Christian Körner, Hiroyuki Muraoka, Shilong Piao, Miaogen Shen, Stephen J. Thackeray, and Xi Yang. 2016. Emerging opportunities and challenges in phenology: A review. *Ecosphere* 7 (8): e01436. https://doi.org/10.1002/ecs2.1436.

The Institute of Ecology. 1977. *Experimental ecological reserves: A proposed national network.* Washington, D.C.: U.S. Government Printing Office.

Thomas, McOwiti O., and Kristin Vanderbilt. 2013. 2013 LTER mini-symposium showcases international collaboration as ILTER celebrates 20 years. *LTER Network News* 26 (1). http://news.lternet.edu/Article2714.html. Accessed 21 July 2020.

Tóth, János Attila, Kate Lajtha, Zsolt Kotroczó, Zsolt Krakomperger, Bruce Caldwell, Richard Bowden, and Mária Papp. 2007. The effect of climate change on soil organic matter decomposition. *Acta Silvatica et Lignaria Hungarica* 3: 75–85.

Vanderbilt, Kristin. 1998. ILTER Graduate students visit HJ Andrews Experimental Forest. *LTER Network News* 11 (1): 14. https://lternet.edu/wp-content/uploads/2013/01/volume_11_number_1_Spring_1998.pdf. Accessed 21 July 2020.

———. 2001. Information management outreach in the East Asia-Pacific Region (EAPR-ILTER). *LTER Network News* 14 (2): 11. https://lternet.edu/wp-content/uploads/2010/12/fall01.pdf. Accessed 31 July 2020.

———. 2002. DIRT: Extending an LTER project to ILTER. *LTER Network News* 15 (2): 12. https://lternet.edu/wp-content/uploads/2010/12/fall02.pdf. Accessed 31 July 2020.

Vanderbilt, Kristin, and Evelyn Gaiser. 2017. The International Long Term Ecological Research Network: A platform for collaboration. *Ecosphere* 8 (2): e01697. https://doi.org/10.1002/ecs2.1697.

Vanderbilt, Kristin, and John Porter. 2010. Synthesizing large datasets. *LTER Network News* 23 (2): 21–22. https://lternet.edu/wp-content/uploads/2010/12/Network_News_23_2_Fall_2010_color.pdf. Accessed 3 July 2020.

Vanderbilt, Kristin L., David Blankman, Xuebing Guo, Honglin He, Chau-Chin Lin, Sheng-Shan Lu, Akiko Ogawa, Éamonn Ó. Tuama, Herbert Schentz, and Wen Su. 2010. A multilingual metadata catalog for the ILTER: Issues and approaches. *Ecological Informatics* 5 (3): 187–193. https://doi.org/10.1016/j.ecoinf.2010.02.002.

Vanderbilt, Kristin, Don Henshaw, John Porter, Margaret O'Brien, Sheng-Shan Lu, and Haibo Yang. 2012. Information managers discuss approaches to data discovery during China visit. *LTER Network News* 25 (2). http://news.lternet.edu/Article2540.html. Accessed 7 July 2020.

Vanderbilt, Kristin L., Chau-Chin Lin, Sheng-Shan Lu, Abd Rahman Kassim, Honglin He, Xuebing Guo, Inigo San Gil, David Blankman, and John H. Porter. 2015. Fostering ecological data sharing: Collaborations in the International Long Term Ecological Research Network. *Ecosphere* 6 (10): art204. https://doi.org/10.1890/ES14-00281.1.

Vanderbilt, Kristin, John H. Porter, Sheng-Shan Lu, Nic Bertrand, David Blankman, Xuebing Guo, Honglin He, et al. 2017. A prototype system for multilingual data discovery of International Long-Term Ecological Research (ILTER) Network data. *Ecological Informatics* 40: 93–101. https://doi.org/10.1016/j.ecoinf.2016.11.011.

Waide, Robert B., Christine French, Patricia Sprott, and Louise Williams. 1998. *The International Long Term Ecological Research Network.* Albuquerque: U.S. LTER Network Office.

Wilkinson, Mark D., Michel Dumontier, IJsbrand Jan Aalbersberg, Gabrielle Appleton, Myles Axton, Arie Baak, Niklas Blomberg, et al. 2016. The FAIR Guiding Principles for scientific data management and stewardship. *Scientific Data* 3: 160018. https://doi.org/10.1038/sdata.2016.18.

Wohner, Christoph, Johannes Peterseil, Dimitris Poursanidis, Tomáš Kliment, Mike Wilson, Michael Mirtl, and Nektarios Chrysoulakis. 2019. DEIMS-SDR – A web portal to document research sites and their associated data. *Ecological Informatics* 51: 15–24. https://doi.org/10.1016/j.ecoinf.2019.01.005.

Index

R. B. Waide, S. E. Kingsland (eds.), *The Challenges of Long Term Ecological Research: A Historical Analysis*, Archimedes 59,
https://doi.org/10.1007/978-3-030-66933-1

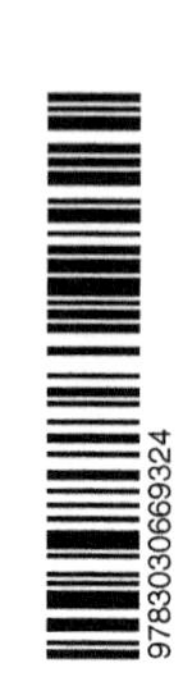
9783030669324